THE OXFORD BOOK OF SCOTTISH SHORT STORIES

R

THE OXFORD BOOK OF
SCOTTISH
SHORT STORIES

Edited by

Douglas Dunn

Oxford New York
OXFORD UNIVERSITY PRESS
1995

Taipei Tokyo Toronto
and associated companies in
Berlin Ibadan

Oxford is a trade mark of Oxford University Press

First published by Oxford University Press 1995

British Library Cataloguing in Publication Data
Data available

Library of Congress Cataloging in Publication Data
Oxford book of Scottish short stories / edited by Douglas Dunn.
p. cm.
Includes bibliographical references and index.
1. Short stories, Scottish. 2. Scotland—Social life and customs—Fiction. I. Dunn, Douglas.
PR8676.O94 1995 823'.01089411—dc20 94–38043
ISBN 0–19–214235–6

10 9 8 7 6 5 4 3 2 1

Typeset by Graphicraft Typesetters Ltd., Hong Kong
Printed in Great Britain
on acid-free paper by
Bookcraft (Bath) Ltd.
Midsomer Norton, Avon

Contents

INTRODUCTION

The Ballads are one of the most distinctive achievements of Scottish culture—and they are stories. In a phrase of John Bayley's—although he was referring to a poem like Thomas Hardy's 'The Frozen Greenhouse'—they can be called 'sung short stories'. Whether sung or spoken, they show supernatural or actual events and persons rendered into what has become permanent fictitiousness. They arose from a whole people; their ancestral roots strike deep. There is no getting away from them.

While the Ballads reinforce the heightened sociability of Scottish literature, the more terrestrial pleasures of fiction are equally ingrained in Scottish psychology. Edwin Muir wrote that,

> For its size, Scottish literature has created a greater number of characters (drawn from life) than any other literature in the world; and it has produced the two greatest biographies in the English language: Boswell's 'Life of Dr Johnson', and Lockhart's 'Life of Scott'. So much, then, for the Scots' boundless, insatiable interest in their fellow men and women; they are tireless gossips and anecdotists.*

Earlier stories, however, were infatuated with wonder rather than the delights and mischief of everyday transactions. 'The Battle of the Birds', for example, contains this sentence:

> The king's son mounted upon the raven, and, before he stopped, he took him over seven Bens, and seven Glens, and seven Mountain Moors.

Magical fiction, with its codes and secrets, sometimes a grisly sensationalism as well, suggests a close-knit community as much as the more modernized society attracted to gossip and anecdotal exchanges. Written down in Gaelic in 1859 from a recitation by a fisherman who lived near Inverary on Loch Fyne, the story harks back to a remarkable Highland tradition of tale-making and telling. 'Death in a Nut', in Duncan Williamson's version, appeals clearly to a similar relish of the unexpected and unlikely. In recent years oral story-telling in Scotland has enjoyed a revival. Stories like these make it easy to see—or, rather, hear—why. Like 'The Wee Bannock', they have long-lasting attractions and an indestructible freshness. They seem to come from the childhood of the world; they stimulate a willingness to believe in astonishments and marvels.

Sir Walter Scott's 'The Two Drovers' (and 'Wandering Willie's Tale' in *Redgauntlet*) have come to be identified as forerunners of the modern or literary short story. (But Pushkin's 'The Negro of Peter the Great' was

* Edwin Muir, *The Scots and their Country* (HMSO, 1946).

also published in 1827, showing, perhaps, how Scott's influence could run ahead of itself.) 'For me, the first modern short story in English', Walter Allen wrote, 'is by Sir Walter Scott, whose fiction marks one of the great watersheds of literature.' Allen continued:

> He was at once a writer of romance in the old style and of novels in the new. He adored the magical, the supernatural, the irrational, all that was sanctified by age and custom. But at the same time he was an extraordinarily acute observer of the behaviour of men in society and of men in specific areas of society.*

As these earlier stories show (and, of course, the ballads) Scott's adorations were thoroughly characteristic and traditional, besides being new to the rest of the world. Eighteenth-century chapbooks offered stories, too, especially those written and published by the remarkable Dougal Graham (1724–79), 'The Skellat Bellman of Glasgow'. Graham fought in the Jacobite army and saw the entire campaign through to its bitter end at Culloden. He wrote a long poem about it. Glasgow, whose contribution to the Jacobite cause was limited to 'ane drunken shoemaker', made Graham the city's town crier, an appointment which was advantageous to the distribution of his chapbooks and stories. Popular, demotic prose story-telling in Scotland began to cross from recitation to private reading more than fifty years before Scott's heralded 'The Two Drovers'. It would have been possible, too, to have included a short story-like episode from Tobias Smollett's *Humphrey Clinker* (1771), which is also considerably prior to Scott. James Hogg and John Galt both published short fiction in advance of 'The Two Drovers' (1827). Hogg's best writing in the form dates from the late 1820s, and Galt's between 1832 and 1836. Journals such as *Blackwood's, Fraser's Magazine*, and *Tait's Edinburgh Magazine*, saw to it that the short story was established very quickly in Scotland, and, it seems probable, in advance of anywhere else.

'The Two Drovers', 'The Cameronian Preacher's Tale' (1828), and 'The Howdie' (1833), illustrate three major sources of the variety and disaccords of Scottish fiction ever since. It is a state of torsion in which several elements contrast to set up the possibility of imbalances, misrepresentations, and the genuine. Scott's tale of Highland pride, passion, and recklessness, up against English stolidity, prejudice, and fair play, expresses a 'British' social vision (as does *Waverley*, and many others of Scott's novels). But its realism is intense, despite the story's involvement with Highland second-sight: in 'The Two Drovers' destiny has a political meaning as well as supernatural glamour. Hogg's story takes for all it's got his own and his readers' willingness to be awed by the uncanny. It should be stressed that he wrote in a more realistic manner, too, as in 'Love Adventures of Mr George Cochrane' (1820), which is too long for a book of this kind.† Galt was another who wrote sustained stories such

* Walter Allen, *The Short Story in English* (Oxford: Clarendon Press, 1981), 9.

† See David Groves (ed.), *James Hogg: Tales of Love and Mystery* (Edinburgh: Canongate, 1985).

as 'Tribulations of the Rev. Cowal Kilmun' and others which support the social and political concerns of novels like *Annals of the Parish* (1821). The demotic, recording realism and humour of 'The Howdie', however, represents fairly enough a third heave and twist in how Scottish fiction struggles with native truths of character, language, and all the implications of place and country.

Much of Scottish literature practically insists on its *demotic* identity. In doing so, it can seem to risk overstating its vernacular impetus and rapport with the people, strengths which can hardly be considered undesirable in themselves. Although strikingly different in what they set out to do, the stories here by Hogg, Scott, and Galt, have in common a conspicuously audible narrator. 'Sit near me, my children, and come nigh, all ye who are not of my kindred, though of my flock . . .' Hogg's tale begins with an ear-catching, biblical invitation to listen to a deathbed recital—'one of those terrible sermons which God preaches to mankind'—delivered by a stern Presbyterian of deepest dye. An immediate audience is the first priority of his artifice, closely followed by a mimetic portrayal of the preacher's voice. There is a strong element of performance to it. Galt's midwife opens her 'autobiography' by saying: 'When my gudeman departed this life, he left me with a heavy handful of seven childer, the youngest but a baby at the breast.' There is a similar spokenness, a proximity of writer to reader which encourages collusion in the tale, and an absence of preliminaries. Scott is more artful in that he delays the identity of the narrator until late in the story. All three writers are in touch with a newer art of story-telling which demanded a written negotiation between the voice and page. A compact between speech and print is an important dimension of short stories in almost every modern tradition; but in the Scottish story it is especially significant as a consequence of the authority of oral story-telling in prose and verse, and of the position of vernacular language stemming from the social and linguistic stresses of Scottish society.

Galt preceded his story 'The Seamstress' with a digression on the Scots language and its vocabulary:

> Besides the beautiful inflexions which help to make the idiomatic differences between the languages of Scotland and England, the former possesses many words which have a particular signification of their own, as well as what may be called the local meaning which they derive from the juxta-position in which they may happen to be placed with respect to others. Owing to this peculiarity, the nation has produced, among the lower classes, several poets, who, in the delicate use of phraseology, equal the most refined students of other countries. Indeed, it is the boast of Scotland, that in the ploughman Burns, she has produced one who, in energy of passion and appropriate expression, has had no superior. No doubt something may be due to the fortunate circumstance of the Scotch possessing the whole range of the English language, as well as their own, by which they enjoy an uncommonly rich vocabulary, and, perhaps, the peculiarity to which we are alluding may have originated in this cause. For example, the English have but the

word 'industry', to denote that constant patience of labour which belongs equally to rough and moderate tasks; but the Scots have also 'eydency', with its derivatives, descriptive of the same constancy and patience, in employments of a feminine and sedentary kind. We never say a ditcher or a drudger is eydent; but the spinster at her wheel, or the seamstress at her sewing, are eydent; . . .

Ironically, though, Galt's polite but proud description disguises the extent to which demotic language can lead a writer into trouble. Galt himself was found by Blackwood to be 'perhaps a little strong'. The 'chaste conventions'* of Maga precluded the publication there of a story like 'The Howdie'. Although not a matter of language alone, but the reflection of the realities of which it is a part, the whole business of truth and idiom has been controversial in Scottish fiction. No rough son of the soil or tenements, Galt's artistic integrity more than class loyalty led him to write as he did. His Tory political views neither inhibited nor distorted his focus on character and place. As the son of a successful shipowner he was very different from Hogg in origins as well as in subsequent experience. Hogg was bitter at what he perceived to be his treatment by the Scottish literati of his day:

I know that I have always been looked on by the learned part of the community as an intruder in the paths of literature, and every opprobrium has been thrown on me from that quarter. The truth is, that I am so. The walks of learning are occupied by a powerful aristocracy, who deem that province their own peculiar right; else, what would avail all their dear-bought collegiate honours and degrees? No wonder that they should view an intruder, from the humble and despised ranks of the community, with a jealous and indignant eye, and impede his progress by every means in their power.†

After acknowledging the possibility of rancorous pique and a dash of self-pity, there is evidence enough that Hogg was not wildly off the mark. Disfavour regurgitated itself in irony long after his death when George Saintsbury (Professor of English at Edinburgh University, 1895–1916) declared that Hogg must have been helped to write *The Private Memoirs and Confessions of a Justified Sinner* by John Gibson Lockhart, probably the worst of Hogg's patronizing detractors. The belief continued until well into the twentieth century. An unfortunate fact of Scottish fiction is the bitterness and controversy which have tended to rack its achievements.

Verisimilitude, of which language is part, as well as local and national identities, and, indeed, the whole question of what Scottish fiction should be like, of what Scotland is or should be, are the issues from which both discontent and achievement have emanated. Terms like 'authenticity' and 'tradition' can be bandied about endlessly and solve nothing. Both notions

* The phrase is Ian A. Gordon's in his *John Galt—The Life of a Writer* (Edinburgh: Oliver & Boyd, 1972), 109–10.

† Quoted in David Groves, *James Hogg: The Growth of a Writer* (Edinburgh: Scottish Academic Press, 1988).

tend to foreclose on a writer's individual freedom and originality. They consist of advocacy brought to bear from outside a writer's imagination as if on the assumption that a writer is incapable of dealing creatively with these issues without strong reminders from non-literary sources. Or else the anxiety arises from the fear that the writer *will* deal with them creatively, but in ways which complicate or challenge the approved hygiene that keeps writing politically and socially on the 'right lines'. 'Authenticity' is prone to assessment by 'political correctness' as much as it is likely to be appreciated by aesthetic criteria. Practically by definition, an insistence on tradition serves only to perpetuate more or less the same kind of thing and obstruct the change and growth that can be introduced by new ways of thinking, showing, and telling.

The rise of prose fiction in Scotland coincided with widespread public interest in the emotional and supernatural thrills of balladry. Nor should it be forgotten that the best known short literary narrative in Scottish literature (although in verse) is Burns's 'Tam o' Shanter'. It could be the case that the literary shift from verse to prose was less complete psychologically than in England or France. Scott's transition from narrative poet to novelist is instructive. So, too, is his erudition in balladry, which he shared with Hogg.

Smollett and Henry Mackenzie look like radical exceptions; but neither had to contend with the full force of a 'vernacular revival' nor with the influence of Scott's genius. In the eighteenth century, expository prose in English was highly developed, whether historiographical, philosophical, scientific, or critical. David Hume (despite his anxiety over 'Scotticisms'), Lord Hailes, Joseph Black, William Robertson, Lord Monboddo, Lord Kames, Adam Smith, and others, all wrote commandingly. Documents such as 'The Solemn League and Covenant' in the century before, as well as sermons and letters by Bishop Leighton and Samuel Rutherford, and the writings of Presbyterian and convenanting divines and pamphleteers, display a confident use of English, invigorated by a passionate certainty that they were communicating with their readers. Sir Thomas Urquhart's translation of Rabelais (1653), his original writings, or the cultivated, conceited elegance of Sir George Mackenzie's *Aretina* (1660), the first Scottish 'novel', or Calderwood's *History of the Kirk of Scotland* (published 1678, but written by 1651) show a similarly communicative sureness in English prose. Even earlier, John Knox's vigorous and expressive Anglicized vernacular suggests that 'Scotticisms' were not an issue either for him or those who re-worked his texts subsequently.

Such a great weight of Scottish prose in English seems to have been easily overturned by the imaginative and realist thrusts of Hogg, Scott, and Galt. All three writers transacted with English narrative and Scots dialogue. But the pull of Scots speech was strong, as was the influence of the ballads, indigenous song-culture, and the poetry of Allan Ramsay, Robert Fergusson, and Robert Burns. For all three prose writers the necessity of reproducing dialectal speech was crucial. That substantial prose tradition from Knox to the Enlightenment luminaries hardly contained the nomadic spark of story-telling even if its remarkable achievements in historiography were to find their way into Scott's warehouse-sized imagination. What provided the spark were poetry and popular, native traditions and pressures playing on individual genius.

Then for the best part of fifty years the Scottish short story endured in a state close to torpor. Pioneered by Scott, Hogg, and Galt, the form flourished elsewhere. Pushkin, Gogol, Poe, Hawthorne, Flaubert, Turgenev, and Bret Harte paved the way for its modernity. R. L. Stevenson's stories played an important part in the form's development, too; but Stevenson's contemporaries, Chekhov and James, were to matter more to how the short story would grow in the twentieth century.

Small nations have their distinctive and sometimes innovative contributions to make to literature—Burns, Scott, Ibsen, and Stevenson are clear examples. It is also of the nature of things that their literary cultures should suffer periodically from decline and provincialism. Often it is due to the influence of large nations on their borders. Where a small country is incorporated into a bigger entity (with or without agreement), and the same or closely related language exists, then the pressures on distinctiveness can be serious. For example, there can grow up kinds of writing peculiar to that country alone and nowhere else. These have been anthologized and described by William Donaldson, drawing from fiction and prose sketches published in Scottish periodicals and newspapers in the Victorian period.* Of the authors involved, William Alexander (1826–94) is by far the most important. His novel *Johnny Gibb of Gushetneuk* (1871) is an acknowledged but neglected landmark. His collection of stories *Sketches of Life Among My Ain Folk* (1875) is couched in what is largely a part-journalistic mode. The impression is one of writing from an urge to expose harsh local conditions. Some years before the Kailyard School got into gear, Alexander recorded rural life with a more pungent realism than writers like J. M. Barrie, 'Ian Maclaren' (Dr John Watson), and S. R. Crockett, were prepared to provide. His story 'Baubie Huie's Bastard Geet' steers skilfully between dialect and English in the narrative voice while opting uncompromisingly for fidelity to local speech in the dialogue. His notation of Buchan talk was appreciated in his time and place. 'Authenticity' enters the picture once again: but the point must be that Alexander's passion drove him to reach for local accuracy out of concern for people and places he knew intimately. Above all, among those he was writing for were readers who could test this work locally and who, if shocked by what they were shown, could not deny that it was true without also denying a part of themselves. It is a long-standing challenge within the fretful, sometimes agitated, context of Scottish writing.

Thomas Hardy wrote that,

* William Donaldson, *The Language of the People: Scots Prose from the Victorian Revival* (Aberdeen: Aberdeen University Press, 1989); and *Popular Literature in Victorian Scotland: Language, Fiction and the Press* (Aberdeen: Aberdeen University Press, 1986).

A certain provincialism of feeling is invaluable. It is of the essence of individuality and is largely made up of that crude enthusiasm without which no great thoughts are thought, no great deeds done.

In Scottish fiction, charges of provincialism stem from language or dialect as much as attitudes to character and locality. By Alexander's time, and before, and since, demotic Scots speech was—as it continues to be—an embarrassment to those for whom it is a marker of low social origins (often their own). Somehow it is more acceptable from the mouths of country or small-town folk than city-dwellers. If it is to be deemed native or ancestral, then let it be rural—that seems to be the psychology. It is one of these complex facts of Scottish life with important consequences for literature. That it issues from Scotland's close association with England within a United Kingdom can be taken for granted.

'I dislike all fine and splendid writing, and admire plain common sense much more,' James Hogg wrote in *Lay Sermons* (1834). Apply that remark to Scottish fiction, with its contrast between demotic, vernacular writing and an equal capacity in some writers (sometimes the same writers) for elegant, forceful English, and the torsion, the trouble, desperation, and complicated course of it, might become clearer. These two possibilities constitute an aboriginal quandary. A writer of fiction in Scotland is obliged to struggle with problems of language which are not easily solved. They are inhibiting, as if the twin pillars of Authenticity and Tradition are too close together for a writer to squeeze through. They can seem to demand the neglect of allegedly eternal verities such as *style* and *imagination* in favour of a brutally recording realism. Failing to press through the narrow gap, a writer can easily opt for the safety of one extreme or the other—a styleless realism, or fraudulent caricatures.

It could have been that Scottish dilemma that led Stevenson into an overt concern for style and the nature of fiction. For him, though, the gap was broad enough to stroll through, perhaps with too much ease. He wrote only one realistic Scottish story. 'The Misadventures of John Nicholson' is too long for this book, but with each re-reading it looks far more impressive (and revealing) than Stevenson's low opinion of it. He thought more highly of 'Thrawn Janet' (1887), which at first sight seems a predictable engagement with the tired old routines of a Satanic visit and ghastly goings-on. To some extent it is. But the narrator is 'one of the older folk', who warms to the task of his story-telling 'over his third tumbler' as he addresses 'strangers who were led by chance or business to that unknown, outlying country'. Narrative set-up, as well as the setting—'the moorland parish of Balweary, in the vale of Dule'—and the Reverend Murdoch Soulis, are all ciphers of a native fictitiousness, like the story itself. It is hard to get away from the suspicion that Stevenson is guying the tale as he writes it. In other words, the narrator is taking the gullible to the cleaners, while leaving plenty of room for doubt. A similar shiftiness of narrative technique—a highly calculated technique—can be seen in an earlier work. 'A Lodging for the Night' (1877) has been disparaged on the grounds that too few readers know who François Villon was. As Hogg's remark suggests, anti-literary prejudice can run riot.

Scottish writers have produced three 'household name' fictions. At one time, though, the plots and characters of Sir Walter Scott's novels were on the lips of the educated of Europe and America. Macpherson's Ossianic figments had their way with multitudes, including tyrants and geniuses. But it is *The Strange Case of Dr Jekyll and Mr Hyde, Treasure Island*, and J. M. Barrie's *Peter Pan* which look as if established with the same permanence as the Homeric stories, Shakespeare's plays, *Don Quixote, Gulliver's Travels, Robinson Crusoe, Les Misérables, The Count of Monte Cristo, Wuthering Heights, Jane Eyre, Crime and Punishment*, and *War and Peace*. The list could be stretched, although it is difficult to think of as many twentieth-century works to add to it. *Dr Jekyll and Mr Hyde* (1885) was, therefore, another temptation; but its form is that of a short novel, which is true also of 'The Beach of Falesá' (1892).

Intense as these two works are, a similar coherence and economy can be felt in 'A Lodging for the Night'. It amounts to much more than atmospherics worked up out of an enthusiasm for Bohemian byways. Sinister in its exposure of an amoral personality—did André Gide know this beguiling tale as well as he knew Hogg's great novel?—it shows the 'dandy in dissent' side of Stevenson's personality. It is 'parable art', in W. H. Auden's definition—'that art which shall teach men to unlearn hatred and learn love'. But the shock of the tale is that Villon is incapable of dismantling cynicism prior to such a learning. The story's success lies in its moral ambiguity, its touch of the perverse, its hints of Poe and Baudelaire. The learning is all down to the reader's perceptiveness and wonder. Apart from 'The Beach of Falesá', it is Stevenson's most 'modern' story. Curiously, though, it's been taken too often as a lump of antiquarian slapstick.

Older than Chekhov by ten years—they died at roughly the same age from the same disease—Stevenson's art is very different from the Russian master's. Where Chekhov depends on psychological insight, mood, gentle but harrowing melancholy, and an unblinkered awareness of the thresholds of human suffering and self-deception, Stevenson is all clarity of outline and episode, structure, and style. Little in his fiction could be described as quotidian or domestic (the exception being 'The Misadventures of John Nicholson'). He once wrote:

> It is stories we want, not the high poetic function which represents the world . . . We want incident, interest, action: to the devil with your philosophy.

These days, that sounds philistine—Jeffrey Archer or Stephen King rather than a 'classic' such as R. L. S. In his 'A Note on Realism' (1883), where he shows himself out of sorts with 'French naturalists' (influential on Chekhov), Stevenson said:

> The question of realism, let it then be clearly understood, regards not in the least degree the fundamental truth, but only technical method, of a work of art. Be as

ideal or as abstract as you please, you will be none the less veracious; but if you be weak, you run the risk of being tedious and inexpressive; and if you be very strong and honest, you may chance upon a masterpiece.

Aiming for 'a more naked, narrative articulation', the succinct, and withdrawing from the 'detail' introduced by Scott and subsequent writers of fiction, and from what he called the 'rancid', Stevenson achieved the glamour and uniqueness of his stories. In terms of literary time, he also pulled off a tragedy. An international view of the short story exposes his conception of narrative and form as one that was rapidly overtaken—already in process when he was writing—by Chekhov, James, Kipling, Joyce, Katherine Mansfield, and D. H. Lawrence. Stevenson's moment was influential; but it was also short-lived. His *Fables*, of which I might have included too few, are a different matter. Certainly, they obey his instinct for 'communicative ardour', boldness of outline and pointedness, as well as his 'quest for the ideal' (that, perhaps, more than anything else). They are very remarkable short, short stories; and they are among the finest things that Stevenson ever put on paper.

It was during Stevenson's lifetime that the so-called Kailyard (cabbage patch) School made its mark with the British, American, and Dominion public. Then, and subsequently, the quarrel concerns the objects and extent of realism. Stevenson was enthusiastic about the earlier work of J. M. Barrie and S. R. Crockett; but as more than one critic has suggested, homesickness in Samoa could have heightened his response to overtly Scottish scenes and characters. It is even possible to understand the Kailyarders as exercising a less talented, less intellectual, and perhaps vulgar modification of Stevenson's principles. In any event, his link with them remains a canonical embarrassment, although it need not be in Barrie's case when you consider his subsequent achievement in the theatre. That, too, however, has been passed over or denigrated by critics.*

'We believe that the surest test of a good novel', 'Ian Maclaren' stated, 'is that it leaves a pleasant flavour behind.' As a Free Church minister (his real name was Revd Dr John Watson, and he served mainly in Liverpool) 'Maclaren' 's pleasant, or moral, view of fiction is hardly surprising. He sketched his convictions at greater length in an essay on 'Ugliness in Fiction' (*Literature*, 1 (1897)):

We are perfectly aware that people swear and do other things which are worse, but without being Pharisees we distinctly object to books which swear on every page between being our companions for the hour when the lamp is lit and the streets are quiet.

Not much room for late nineteenth-century Scottish reality there, clearly. It expresses an overview of fiction founded on what must be left out,

* For detailed accounts of Kailyard fiction, see Ian Campbell, *Kailyard: A New Assessment* (Edinburgh: Ramsay Head Press, 1981); and, esp., Thomas D. Knowles, *Ideology, Art and Commerce: Aspects of Literary Sociology in the Late Victorian Scottish Kailyard* (Gothenburg Studies in English, 54; 1983).

while at the same time—'without being Pharisees'—it lays claim to the pleasantly selective as 'literary'.

J. M. Barrie's *Auld Licht Idylls* (1888), *A Window in Thrums* (1889), and his novel *The Little Minister* (1891) paved the way for S. R. Crockett's *The Stickit Minister and Some Common Men* (1893), to be followed by 'Ian Maclaren' 's *Beside the Bonnie Brier Bush* (1894) and *The Days of Auld Lang Syne* (1895). Their sales were prodigious, especially as export commodities. Extending the phenomenon into the early 1900s, and including George Douglas Brown's anti-Kailyard novel *The House With the Green Shutters* (1900)—'It was antagonism to their method that made me embitter the blackness,' Brown wrote—Kailyard fiction would seem little more than a commercial flash of a few years, complicated by a bleak, hostile rejoinder which amounts to Kailyard-in-reverse, that is, the sentimental in its guise of negative overstatement as opposed to the white-washing distortions or avoidances of writers like Crockett and 'Maclaren'. I've taken my heart in my hands and printed stories by all three of these writers. They are not always available, and as manipulators of emotion they are as skilled as Bret Harte, whose 'The Outcasts of Poker Flats' and 'The Luck of Roaring Camp' strike me as milestones in the history of sentimental literature. Interestingly, Harte was US Consul in Glasgow in the 1880s.

Barrie made his position clear in an essay on Thomas Hardy:

> The professional realists of these times, who wear a giant's robe and stumble in it, see only the seamy side of life, reproducing it with merciless detail, holding the mirror up to the unnatural instead of to nature, and photographing by the light of the policeman's lantern. The difference between them and the man whose name they borrow is that they only see the crack in the cup, while he sees the cup with the crack in it.*

Prissily thoughtful as the passage might seem, it carries with it suggestions of 'bad faith' and a sense of *knowing* that 'the seamy side of life' contains serious causes and circumstances which could have been too difficult for a writer to tackle without having become expert in the Hell's kitchens of Glasgow and Dundee. In retrospect, an ideal of fiction which emphasized propriety looks like one calculated to leave out exactly those details which subsequent criticism demands should have been included. No one else, though, was putting them in with the honesty which would have shocked contemporaries with the truth or satisfied latter-day critics with evidence of creative heroism. For it would have taken literary valour of a very high order for a story-writer or novelist to have tried to penetrate those districts of Scottish urban and industrial experience in the period from around 1850 to 1914.

For all its curious irrelevancy to anywhere else, Barrie's secluded Thrums (Kirriemuir) finds itself carrying a large part of the burden of fictitious Scotland of the 1880s, a role for which it was never designed. His stories

* 'Thomas Hardy: The Historian of Wessex', *Contemporary Review*, 56 (July 1889).

are too intimate and idiosyncratic for a purpose outside their real and imagined locality, and, especially, the narrator's relationship with it. The narrational drama presupposes a subsequent release from a blockaded community. It is its own comment on Scotland and on a town which is the country's momentary cipher.

What irritates his critics most about Barrie is when they think they 'have him', to borrow one of his own expressions, they find he isn't there at all. It is a trick he learned from Peter Pan. When they say superior things about 'Kailyarders', they suddenly realize they are talking about the author of *Dear Brutus*; when they praise *Peter Pan* they suddenly find themselves looking at a dull little public wash-house in the Tenements.*

Each of the three Kailyard stories here is concerned with language or literature. Barrie evokes the mutual trespasses of literature and a community, while Maclaren dwells on local idiom and the psychology it reflects, although paying scant attention to the authorial psychology which exploits it, or the mentality of the readership which found it amusing. Crockett, on the other hand, tries to defend local sincerity in the face of smart journalism. Each story touches on serious, genuine matter which, however, its author erases instinctively or, in Crockett's case, drowns in tears. In the seventy years between Galt's episodic novel *Annals of the Parish* (1821) and the Kailyard writers, the short story had perfected its dynamic. It is their ill-founded ability which exposes their flaws. The literary historian J. H. Millar was to accuse them of a 'misrepresentation of Scottish historical and contemporary reality, and its propagation of a false national image'. To a large extent the accusation is both inevitable and unfair. No one else was able to say at the time how that reality should be shown or described. What Christopher Harvie calls a flight from 'the horror of the real thing' was all too prevalent.

But the Kailyarders were not the only writers of their time. R. B. Cunninghame Graham's local fiction and 'sketches' tend to be sentimental, but not his writing set in foreign parts. At times it is difficult to detach his fiction from travelogue, as in 'Un Infeliz'. What seems remarkable about this socialist and founding member of the National Party of Scotland is that his short stories tend to fail when confronting Scottish topics and characters, a state of affairs quite largely true of John Buchan as well. There is a similar dichotomy in the work of Violet Jacob. Better known as a poet, her verse in the vernacular of the Mearns and Montrose (she was of the family of Erskine of Dun, and, like Graham, high-born) drifts into the balladic, sentimental, and supernatural. She was an inconsistent writer, but at her best in a handful of stories and her novel *Flemington* (1911). 'The Debatable Land' is remarkable for its valiant breaking of barriers. In eloping with a tinker, Jessie-Mary chooses escape from the demeaning circumstances into which she is born. This is exactly the kind of story of which the Kailyarders were incapable—even Barrie,

* J. A. Roy, *James Matthew Barrie, An Appreciation* (Jarrolds, 1937).

the author of that very feminist play *The Twelve-Pound Look* (1910). To write it, he had to become familiar with another society. He could not have written it in his native town. John Davidson's 'Miss Armstrong's Circumstances' (1896; not reprinted until now) is interesting for its command of the form and use of a self-confident female narrator. The first sentence is splendid: 'After all, my friends have been mistaken; my experiences are not nearly so exciting as they appeared when I saw them through their spectacles.' Outlined a little later, these experiences are sensational, including 'the death of my father, a Scotch socialist, on a barricade' during the Paris Commune. Half-hidden in neglect, 'Un Infeliz', 'Miss Armstrong's Circumstances', and 'The Debatable Land' (with other stories by Graham, Davidson, and Jacob) can be seen as a canonical riposte to the Kailyard which was there all along.

James Hogg's impatience with 'fine writing' indicates more than a rattled Scottish attitude to English and some of its stylistic priorities. Oral tradition, poetry, journalism, the essay and sketch, all bear a more formative potency than the novel to the short story's development. In 1874 Stevenson—for all his embroilment with questions of style—vaunted 'the essential interest of a situation' over the literary impulses which could lead a writer to 'cocker up and validify feeble intrigue with incidental fine writing and scenery'. That assumption has been standard throughout the history of modern story-writing. Equally prevalent is Stevenson's belief that the setting should have the status of an inanimate but central character—inseparable from what happens in the story, and bound up with more than atmosphere, but part of its meaning.

Places and their typical inhabitants are the recurring concerns of most modern story-writers; but it is especially true of Scotland. In stories by Lorna Moon, Lewis Grassic Gibbon, Neil Gunn, Robert McLellan, Fred Urquhart, and Jessie Kesson, there is a noticeable heave of the fictitious truth of a place up against unreliable reportage from other sources. Although largely rural or small-town in their focus, their sentient topicality is more hazardous and critical than that of the Kailyarders, for whom time too often seems to have stood still. They do not ignore urban life. Instead, they concentrate on several of the country's many elsewheres. Stories by Moon, Gibbon, Gunn, and others, are rural by virtue of passionate, experienced obligation rather than choice, even if a need to run counter to turn-of-the-century fiction played a part.

A dismal epoch in the literature of a small country can set up repercussions which last for—as in this case—a century. It now *seems* an irreconcilable fact of the Scottish imagination that it is inspired by *either* the rural or the urban. But perhaps it is the short-story form itself which dictates that division. Certainly, it seems rare in the stories of other literatures that writers are familiar equally with city and country. Or when they are it is because place contains both in an experienced proximity. The short story demands a faithful application to what a writer knows best. In a country the size of Scotland, though, it might be expected that more writers aspire to a knowable, national experience, a whole Scotland, rural and urban, Glasgow and Seggat, Edinburgh and Drumorty, Dundee and Mains of Easter Whatnot. (Until recently, Aberdeen was an exception. It really was an entrepôt for a massive agricultural hinterland.) Divisions within Scotland are inhibitingly plural: Highland and Lowland; rural and urban; and between one city and another, as well as identifiable districts. For years, Glasgow, Edinburgh, Dundee, and Aberdeen, were like city states.

Such strong local identities, from diminutive parishes to large cities, make it difficult for writers to cross barriers from local to national. They also tend to ossify class antagonisms; they make it difficult to feel at home with the language of more than one social class. It seems less than surprising, then, that many writers of the period before 1939 should have lived outside Scotland for much of the time—Davidson, Barrie, Moon (she worked and died in Hollywood), Gibbon, Jacob, and Buchan. Stories here by Moon, Gibbon, and Jacob could suggest the form's attachment to what the writer learned in childhood or early in life. Retrospection, especially when dialect is involved, can easily lead to the regressive—a pitfall which Scottish story-writers sometimes have trouble in avoiding.

'The writer operates at a peculiar crossroads where time and place and eternity somehow meet,' Flannery O'Connor wrote. 'His problem is to find that location.'* Her wisdom throws light on the achievements of writers like Jacob (whose place was Angus and Montrose), Gibbon, Moon (whose Drumorty is Strichen in Aberdeenshire), George Mackay Brown (Orkney Islands), or Iain Crichton Smith (Lewis), or Neil Gunn (eastern and northern Highlands). Territorially, they cover a substantial swathe of Scottish geography. It is now the literary fashion in Scotland to denigrate rural and small-town life—and yet many people still live in these places. Like it or lump it, though, the modern Scottish story was born there.

Gibbon's 'Smeddum' is a classic story of spunk, spirit, and fortitude. It has an aboriginal vivacity, tempo, economy, humour, and power; and these are replicated in the cadence and pressure of its idiom. 'Clay' has a similar authority. Like stories here by Jacob, Moon, Urquhart, and Muriel Spark, 'Smeddum' portrays a woman's resistance to circumstances and a determination to go her own way. It is a prominent meaning in Scottish stories by women (and some men). Almost everywhere, the short story has been drawn to the underdog, to characters who are life's exceptions, or up against it, as well as localities which are slightly off the beaten track in the countries or cities where they happen to be (and there's no shortage of these in Scotland). Indignant meanings, though, have imposed an obligatory tact on writers. Humour is one way to achieve that discretion. So, too, is resistance to the overtly didactic, with a preference for showing rather than telling. Writers prefer to leave off a story, as if to finish it with a decisive conclusion were to suggest that he or she is insufficiently in

* Flannery O'Connor, *Mystery and Manners* (Faber & Faber, 1972), 59.

love with the life it depicts, even when the affection is critical. There is a warm, profoundly experienced limit to the form which could be the most wonderful thing about it and the height of its challenge. It thrives on a lived and marvellous intensity, in which the writer's art is directed at creating life's scale and overcoming the miniaturizing biasses of relative brevity.

Although not entirely free of sentimental dips, Edward Gaitens's stories (*Growing Up*, 1942) at long last introduced urban life and idiom at a level higher than that of music hall caricature and jocosity. Insularity, however, was still pervasive: there was too much unfinished native business to attend to for the examples of writing from abroad to alter the practices of Scottish writers by very much. It can seem remarkable when you bear in mind that the first translators of Kant, Nietzsche, Ibsen, Proust, and Kafka, were Scots. Or was there, and does it still exist, a gap between intellectualism and creativity? Although under-articulated, the inner debate in Scottish fiction was still bound up with the question of how to align expressively, coherently, and without compromise, the language of locality with English prose. What would be gained? How much would be lost? These questions had surfaced conspicuously in Gibbon's trilogy and short stories. But the instruction had been there since the 1830s in the form of Galt's 'imaginary autobiography' 'The Howdie'. How could a literary argument have lasted so long without a satisfactory resolution? In 'Smeddum' and 'Clay' Gibbon drew his *narrative* idiom from the speech of the place as well as the dialogue. Robert McLellan's *Linmill Stories* uses a similar technique. The contemporary writer James Kelman has taken the impulse further, perhaps to its limit, in an idiom stylized from Glaswegian speech.

An understanding of Scottish fiction (and poetry for that matter) depends on an appreciation of a tense relationship between the demotic and the cultivated. Galt, Gibbon, and Kelman represent three accomplished and challenging moments in that contest between actual speech and 'standard English' with or without a Scots accent of one kind or another. Stevenson portrayed it in the stylistic clash between the elegant, almost camp mannerisms of *The Suicide Club*, the polished sentence-making and phrase-fondling of his essays, and the idiom of 'Thrawn Janet'. Gibbon wrote that Scottish writers were often unaware of the 'essential foreignness' of their work.

> Seeking an adequate word or phrase he hears an echo in an alien tongue that would adorn his meaning with a richness, a clarity and a conciseness impossible in orthodox English. That echo is from Braid Scots . . .
>
> Further, it is still in most Scots communities (in one or other Anglicized modification) the speech of bed and board and street and plough, the speech of emotional ecstasy and emotional stress. But it is not genteel. It is to the bourgeois of Scotland coarse and low and common and loutish, a matter for laughter, well enough for hinds and the like, but for the genteel to be quoted in vocal inverted commas. It is a thing rigorously elided from their serious intercourse—not only

with the English, but among themselves. It is seriously believed by such stratum of the Scots populace to be an inadequate and pitiful and blunted implement, so that Mr Eric Linklater delivers *ex cathedra* judgement upon it as 'inadequate to deal with the finer shades of emotion'.*

Nationalism, or at least a distrustful attitude towards 'the English', and the Anglicized Scottish middle classes, are as apparent in Gibbon's analysis as his belief in the genuineness and vitality of Braid Scots, that 'alien tongue' to which 'the genteel' is foreign. These elements in what he was saying, however, are all intricately and finely part of the same thought. Also detectable is evidence of a literary psychology which had no option but to oppress itself with anguished arguments and personal criticism as writers observed depressing social and political conditions, for which no remedy seemed conceivable. 'Gentility' (and never mind 'the genteel', which has no place in *any* literature) became a despised luxury, as if it were a lapse of taste to be seen as cultivated, learned, and benign. Curiously, though, in Gibbon's day Scots were making large contributions to the study of literature, notably H. J. C. Grierson, the editor of John Donne and the Metaphysical Poets. You can be a scholar in Scotland, but to be a scholar and gentleman seems to rub against the native grain of its writers.

Gibbon, Gunn, Jacob, and Moon, were all published in London. If a loathed dimension of Kailyarder stories was that they, too, were published in London and looked for a non-Scottish readership, tricking out their work accordingly, with an eye on the main chance, then that side of literary controversy had died down by the 1920s and 1930s. Narrative tone in stories by Gibbon, Gunn, Jacob, and Moon, and the writers who come after them, is measurably more confident, more intimate with both nationality and locale, as well as radical in what they disclose and mean. It could be appropriate to see this heightening of realism in political terms. The subject invites it.

> Literature is necessary to politics above all when it gives a voice to whatever is without a voice, when it gives a name to what as yet has no name, especially to what the language of politics excludes or attempts to exclude. I mean aspects, situations, and languages both of the outer and of the inner world, the tendencies repressed both in individuals and in society. Literature is like an ear that can hear things beyond the understanding of the language of politics; it is like an eye that can see beyond the colour spectrum perceived by politics. Simply because of the solitary individualism of his work, the writer may happen to explore areas that no one has explored before, within himself or outside, and to make discoveries that sooner or later turn out to be vital areas of collective awareness.†

Calvino's heartening wisdom is more generous than the Scottish literary mind is usually prepared to allow; but it is also more benevolent and

* Lewis Grassic Gibbon, *A Scots Hairst: Essays and Short Stories* (Hutchinson, 1967), 144–5.

† Italo Calvino, 'Right and Wrong Political Uses of Literature', *The Literature Machine* (Secker & Warburg, 1987), 98.

encompassing than just about anybody's wisdom on the subject. Nationalism (with which Calvino did not have to grapple) is a contentious business for nationalists and non-nationalists alike, and you would expect to find it expressed in Scottish stories. Here again, though, it could be the form which keeps the explicitly political at arm's length. It manifests itself in individual, grass-roots cases, with very precise locality. It makes itself felt audibly and organically through non-English diction and cadences. In Scottish stories politics is an undoubted presence through the giving of 'a voice to whatever is without a voice', and in the literal sense of 'voice', that is, dialect, and accent, which are indicators of distinctiveness and difference, or, in Gibbon's term, 'foreignness'. But it is that difference which also encourages prejudice. Scots speech of one kind or another is close to English but dissimilar. In 'British' class psychology it is the closeness as much as an ostensible dissimilarity which provokes neglect or ridicule. It is a matter of Class, that very nasty (and sometimes amusing) English invention. There is really no escape from the politics of the matter. Within the British Isles the survival (and resurgence) of deviant or oppositional national and regional cultures can look like proof of the failure of the 'British' political system. Without the opportunism or happenstances of Empire, there was little to bind the various parts of 'Britain' together. 'Political failure', however, is not just a matter of 'England' or 'Britain'. In the social conditions exposed by many of these stories, the failure of politics and governance can be ascribed to Scottish deficiencies as well, whether or not these are seen within a 'British' or Scottish framework. There could be point in suggesting that Scottish writing attempts to redeem that failure by keeping before the public an insistent and consecutive depiction of its consequences. Interestingly, too, some of the stories here depend on colonial settings, those by Linklater—and it could be significant that he writes about an exceptional Anglo-Indian character—Robin Jenkins, and Muriel Spark. Cunninghame Graham's story—and this could be significant too—is about another country's colonialism. But for all the vigorous exploitation of the Empire by Scots, the imperial theme and its taint enter Scottish fiction hardly at all, except through Scott and Stevenson, who took the Empire as a fact of life. For Scott it was an opportunity for younger sons; for Stevenson it was a passport to far-away places, and latterly, in his writing about the South Seas, serious ethical issues.

Demotic, class-bound attitudes to writing, when backed up by little else, can result in unedifying narrow-mindedness. In his *Popular Literature in Victorian Scotland* (1986) William Donaldson suggested that J. M. Barrie's family background was more comfortable than biographical myth-making would have us believe. R. D. S. Jack replied, asking 'why does the self-evident modesty of that tiny house in Kirriemuir matter so much within a literary argument?'* As Jack goes on to outline, a 'literary

* R. D. S. Jack, *The Road to the Never Land* (Aberdeen: Aberdeen University Press, 1991).

argument' in Scotland can find it hard to survive. Instead, there are arguments which might begin as literary but which are swiftly extended into social, political, economic, or nationalist assessments of whatever work or writer is being discussed.

But the sheer prevalence of working-class life in Scottish short stories can hardly be seen as delinquent, or disruptive, despite the bruising candour of work by such recent writers as Kelman or Duncan McLean. Nor is it merely a case of an identification with working-class life being a native aptitude. Actual lives, their circumstances and conditions, *will* tend to dominate the concerns of a country's fiction. Stories reflect the realities of where and when they are written—or, at least, many will, although it is not a prerequisite. On the evidence of the Scottish short story there is a marked pull towards under-expressed, under-described lives; but this is also a commonplace of the form as practised almost everywhere, although rarer in English than in American stories. Gogol's 'The Nose' and 'The Overcoat', and, in a different mode, Flaubert's 'Un cœur simple', or Henry James's middle-class version of the same feeling in 'The Real Thing', are all powerful predecessors. It is a major strand in Scottish writing from Burns and Scott through to the present day. If the emphasis has moved from the country to the city then it was high time that the shift should have been achieved. The remarkable fact is that it took so long. It is as if creative indignation seethed patiently for years before blowing the lid off. Unfortunately, the emphasis on urban working-class stories can appear to be as exaggerated as the agrarian stresses of the past.

Other features of the Scottish story should leave the reader in no doubt that middle-class subjects have not been neglected. Riddled with tell-tale class signposting—Kersland Street, Giffnock, Bearsden, the Malmaison (a restaurant, once considered the best in Britain, in Glasgow's Central Station Hotel), the *Glasgow Herald*, and names of schools—George Friel's 'I'm Leaving You' looks on its surfaces to be a bleak tale of a failed middle-class marriage. It is typical to an extent that makes it a bald fable. Less well recorded than its working-class inhabitants, the Glasgow middle class feels like a recalcitrant subject for Friel. He writes as if he can do so only with the aid of a map—rudimentary social-class cartography—and after a statement of each character's precise circumstances of birth and education. Both intimate and at arm's length, the story finds it impossible to take its characters for granted. What finally shows through that helpless, childless marriage, is the demise of the power which that representative class once possessed in the Second City of the Empire. Economic decline is mirrored by erotic and personal failure.

A comparable disruption can be seen in Ronald Frame's 'Merlewood', where the story shows a transference of ownership from established family to Glaswegian hairdresser. Elspeth Davie's 'A Map of the World' conveys a mood of inventive resistance to what her narrator calls a 'pallor' of anxiety—'looking pale over nothing, or rather over what hasn't happened but might happen tomorrow or the next day, or next month or in ten years' time'. It is a story of how two unmarried sisters have survived in a suffocating household—invalid mother, neglectful father, boorish brother, and visits from a young widower cousin who has nothing else to do except take holidays and travel. Davie's achievement—and she is a very remarkable writer—hints at the family as a microcosm of an entire class poised before change or termination, even if the sisters might manage to snatch a moment of hedonistic release.

Elspeth Davie has been like Fred Urquhart (or James Kelman) in being a relatively prolific and consistent writer of stories. Much of Urquhart's work is set in the rural north-east, including the well-anthologized 'Alicky's Watch', or such stories as 'The Ploughing Match', or his unforgiving portrayal of vindictiveness and attempted escape in 'The Last Sister'. His range is wider. Office life, for example, is touchingly portrayed in 'The Dream Book' (again with the form's affection for a bitter-sweet inconclusiveness). He has also written novels, as well as longer stories like 'The Last G. I. Bride Wore Tartan' and 'Namietnosc—Or the Laundry Girl and the Pole'. In 'All Hens are Neurotic' an attempted escape to England from the confines of Scottish small-town life is seen as an exchange of one kind of unhappiness for another. Or perhaps they are the same.

Neil Paterson, Robin Jenkins, and Muriel Spark (although the same could be said of several others here) show that stories by Scottish writers need not be set in Scotland and do not need to be demotic. Paterson's strange and risky 'The Life and Death of George Wilson' is preferred over his better known 'Scotch Settlement' (which is set in Canada). Like Fred Urquhart, Paterson can be seen to belong to that upsurge in story-writing in mid-century Britain and Ireland which included work by Elizabeth Bowen, Sean O'Faolain, V. S. Pritchett, and the neglected English writer William Sansom. Robin Jenkins writes here of Scottish expats in Borneo—'solid presbyterian worth in this hot lush land of dissipated expatriates, simple-minded pagans, and sinisterly successful Roman Catholic missionaries . . .'. English colonials in Africa are Muriel Spark's subject in such stories as 'Bang-Bang You're Dead' and 'The Go-Away Bird', which she sees through the complicated allure of sex, to which she brings her unique and eccentric insights. Although perfectly natural that Scottish fiction should reflect local subjects and places, and while a number of Spark's stories have a Scottish character or setting (as in 'A Sad Tale's Best for Winter'), there need not be a preliminary national thesis underpinning it. Too strenuous an insistence on the indigenousness of writing obstructs the inclinations of individual gifts. Spark, Urquhart, Paterson, and Jenkins left Scotland either permanently or for long periods of time. Especially in Spark's and Urquhart's stories, where release into intimate, perhaps irregular freedom is part of their meaning, it is easy to see why: local conditions failed to encourage or support a desired liberty, or offer the chances of a literary livelihood. Much that was negative in Scottish life—happily, this is no longer the case—would have sustained the repetitiously negative, of which a writer tires probably well in advance of his or her readers. Full of wit, and full of woe, Spark's characters are like actors and actresses playing the parts of her characters, and it is difficult to see her sportive fiction flourishing without a larger range of experience from which to draw. Cosmopolitan, as well as histrionic, she relies on a personality of 'Scottish formation' (Spark's own phrase) more than a precise national identity. Some writers give the slip to an exact national definition of themselves (and Ronald Frame is another).

Countries remake themselves constantly through their fictions. But it can come about through the sheer use of imagination (as in poetry and drama) rather than its specific purchase on characteristic national issues. In one of Muriel Spark's stories it is said of Lenin that he was 'class-conscious by profession': and it is tempting to think of this as true of many Scottish writers as well. There is really no getting away from the contrast between one kind of 'Scottish formation' and another. It is a matter of language as well as experience and origin, a matter of 'lexical identity', and of class and national affiliations. It is also a matter of personal liberty, of the writer's freedom to write about who, where, and when, whatever, he or she likes.

It is a liberty, then, to be seen in whatever way a writer sees it, which is likely to be plural. There is, or should be, a freedom for Scottish writers to write in the idiom that is natural to them. It can be the self-aware 'middle-class' style of Ronald Frame—a highly polished stylist of English prose—or the 'working-class' idiom of James Kelman. In its own way, each is 'authentic', even if Kelman's pulls from a tradition which in its longevity (from Hogg and Galt) seems irresistible in its definition as native, although the American influences on his work are strong as well as exciting and instructive (Ring Lardner, for example).

'I wanted to write as one of my own people,' Kelman has declared. 'I wanted to write and remain a member of my own community.' Very little in English or Scottish literature supported Kelman's desire for a rooted allegiance. Whenever he encountered someone from his own background, 'they were confined to the margins, kept in their place, stuck in the dialogue'.* In terms of attitude, as well as language, Kelman represents a further development of the indigenousness displayed by Galt, Alexander, and, especially, Gibbon. His objective—like Gibbon's—has been to make the reader aware of his characters being shown from *within* the narrative instead of observed from, as it were, outside, or 'stuck in the dialogue'. It has meant an eschewal of 'fine writing' even more irascible than Hogg's.

Imaginative performance, too, is as typical of the Scottish story as its demotic momentum. Alasdair Gray and Muriel Spark are outstanding creators of one-off voices and idioms: Gray's eccentric, astonishing intelligence is one of the most remarkable in Scottish literature since Hugh MacDiarmid's energizing commands in the 1920s. Muriel Spark's

* James Kelman, *Some Recent Attacks* (A K Press, 1992).

brilliant combination of playful dottiness, skill, and sheer perceptiveness is of the same order of excellence. What links these two otherwise very different writers is the made-upness of much of their work. They delight in personae, and in serious and engaging pretences. They provide an enjoyable and refreshingly lucid belief in the pleasures of fiction, both in the writing and the reading of it. These qualities can seem remote from the politicized demotic challenges of a writer like James Kelman or the feminist purposes of Janice Galloway. However, there is more than one way of telling the truth about any subject you care to name.

The vernacular, demotic thrust of the Scottish story is undeniable. So, too, is a more self-consciously elegant or 'literary' tradition. The picture could have been complicated more thoroughly had George MacDonald's (1824–1905) fiction been represented, especially his well-known 'The Golden Key'. I find it sickly. But fantasy is a strong ingredient in the Scottish imagination. Margaret Elphinstone's 'An Apple from a Tree' represents that side of fiction-making. Margaret Oliphant's 'The Library Window', or 'The Open Door', would have made women writers' mark look stronger. Both, though, are long, available elsewhere, and I needed room for stories that strike me as less antiquarian and 'ghostly'. Mary Findlater's 'Real Estate' is another long story of considerable accomplishment, as is Naomi Mitchison's 'Beyond This Limit'. As it is, stories by Violet Jacob, Lorna Moon, Muriel Spark, Jessie Kesson, and others younger than them ought to assert a feminine presence as noticeable as in the traditions of other literatures. Any student of the short story must be aware that it is a form of writing in which women have excelled, from Katherine Mansfield to Alice Munro.

Jessie Kesson's novels are all relatively short. In their brevity, *The White Bird Passes*, or *Glitter of Mica*, given their internal dynamics, tend to show her as a short-story writer writ large. Like Alexander, Barrie, Jacob, Gibbon, Moon, and Urquhart, she writes about the north-east of Scotland. Her settings should remind us of territorial variousness at a time when demotic vigour has passed to Glasgow and the industrial west of the country, although Duncan McLean reminds us that emotional squalor is an Edinburgh commodity as well. However, it strikes me as pretty well ubiquitous and international; and it is an offensive a mistake to dwell on it as typically 'Scottish'.

In an essay on Janice Galloway's fiction Marjorie Metzstein says that 'Scottish writer' is a masculine identity.* Women writers are excluded by it unless they adhere to codes, myths, and procedures of nationalism, vernacular language justified by ideology, booze, etc. It is an understandable

* G. Wallace and R. Stevenson (eds.), *The Scottish Novel Since the Seventies* (Edinburgh: Edinburgh University Press, 1993), 137. For other stories by women, see M. Burgess (ed.), *The Other Voice: Scottish Women's Writing Since 1808* (Edinburgh: Polygon, 1987). For some criticism see essays by D. P. McMillan, C. Anderson, and M. Elphinstone in C. Gonda (ed.), *Tea & Leg Irons* (Open Letters, 1992). See also D. Dunn, 'The Representation of Women in Scottish Literature', *Scotlands*, 2 (1995).

perspective. That is, there is truth in it. But it is hardly backed up by lack of predecessors. Although neither neglected nor dormant, feminist criticism in Scotland has only just begun to map its terrain and draw its conclusions. Writers like Margaret Elphinstone, Janice Galloway, Dilys Rose, and A. L. Kennedy, are among the most innovative in a literary moment which has seen fiction flourish on a scale unprecedented in modern times in Scotland. It is a moment which also involves the talents of Alasdair Gray, James Kelman, Carl Macdougall, Brian McCabe, Alan Spence, Allan Massie, and Ronald Frame. It is a densely populated period in the history of Scottish fiction. The sheer number of younger writers of interest is exhilarating. So too is the readership which they have attracted. From an editorial point of view, the latter part of this collection involved considerable pain—leaving out admired writers.

I thank warmly Dorothy Black, Helen Kay, and Frances Mullan, the secretaries of the School of English at the University of St Andrews, and the St Andrews Scottish Studies Institute. Their assistance has been invaluable. I'm grateful also to the staff of the St Andrews University Library, especially the Inter-Library Loan Department.

DOUGLAS DUNN

St Andrews
9 May 1994

Scottish

Short Stories

The Battle of the Birds

There was once a time when every creature and bird was gathering to battle. The son of the king of Tethertown said, that he would go to see the battle, and that he would bring sure word home to his father the king, who would be king of the creatures this year. The battle was over before he arrived all but one (fight), between a great black raven and a snake, and it seemed as if the snake would get the victory over the raven. When the King's son saw this, he helped the raven, and with one blow takes the head off the snake. When the raven had taken breath, and saw that the snake was dead, he said, 'For thy kindness to me this day, I will give thee a sight. Come up now on the root of my two wings.' The king's son mounted upon the raven, and, before he stopped, he took him over seven Bens, and seven Glens, and seven Mountain Moors.

'Now,' said the raven, 'seest thou that house yonder? Go now to it. It is a sister of mine that makes her dwelling in it; and I will go bail that thou art welcome. And if she asks thee, Wert thou at the battle of the birds? say thou that thou wert. And if she asks, Didst thou see my likeness? say that thou sawest it. But be sure that thou meetest me to-morrow morning here, in this place.' The king's son got good and right good treatment this night. Meat of each meat, drink of each drink, warm water to his feet, and a soft bed for his limbs.

On the next day the raven gave him the same sight over seven Bens, and seven Glens, and seven Mountain Moors. They saw a bothy far off, but, though far off, they were soon there. He got good treatment this night, as before—plenty of meat and drink, and warm water to his feet, and a soft bed to his limbs—and on the next day it was the same thing.

On the third morning, instead of seeing the raven as at the other times, who should meet him but the handsomest lad he ever saw, with a bundle in his hand. The king's son asked this lad if he had seen a big black raven. Said the lad to him, 'Thou wilt never see the raven again, for I am that raven. I was put under spells; it was meeting thee that loosed me, and for that thou art getting this bundle. Now,' said the lad, 'thou wilt turn back on the self-same steps, and thou wilt lie a night in each house, as thou wert before; but thy lot is not to lose the bundle which I gave thee, till thou art in the place where thou wouldst most wish to dwell.'

The king's son turned his back to the lad, and his face to his father's house; and he got lodging from the raven's sisters, just as he got it when going forward. When he was nearing his father's house he was going through a close wood. It seemed to him that the bundle was growing heavy, and he thought he would look at what was in it.

When he loosed the bundle, it was not without astonishing himself. In a twinkling he sees the very grandest place he ever saw. A great castle, and an orchard about the castle, in which was every kind of fruit and herb. He stood full of wonder and regret for having loosed the bundle—it was not in his power to put it back again—and he would have wished this pretty place to be in the pretty little green hollow that was opposite his father's house; but, at one glance, he sees a great giant coming towards him.

'Bad's the place where thou hast built thy house, king's son,' says the giant. 'Yes, but it is not here I would wish it to be, though it happened to be here by mishap,' says the king's son. 'What's the reward thou wouldst give me for putting it back in the bundle as it was before?' 'What's the reward thou wouldst ask?' says the king's son. 'If thou wilt give me the first son thou hast when he is seven years of age,' says the giant. 'Thou wilt get that if I have a son,' said the king's son.

In a twinkling the giant put each garden, and orchard, and castle in the bundle as they were before. 'Now,' says the giant, 'take thou thine own road, and I will take my road; but mind thy promise, and though thou shouldst forget, I will remember.'

The king's son took to the road, and at the end of a few days he reached the place he was fondest of. He loosed the bundle, and the same place was just as it was before. And when he opened the castle-door he sees the handsomest maiden he ever cast eye upon. 'Advance, king's son,' said the pretty maid; 'everything is in order for thee, if thou wilt marry me this very night.' 'It's I am the man that is willing,' said the king's son. And on the same night they married.

But at the end of a day and seven years, what great man is seen coming to the castle but the giant. The king's son minded his promise to the giant, and till now he had not told his promise to the queen. 'Leave thou (the matter) between me and the giant,' says the queen.

'Turn out thy son,' says the giant; 'mind your promise.' 'Thou wilt get that,' says the king, 'when his mother puts him in order for his journey.' The queen arrayed the cook's son, and she gave him to the giant by the hand. The giant went away with him; but he had not gone far when he put a rod in the hand of the little laddie. The giant asked him—'If thy father had that rod what would he do with it?' 'If my father had that rod he would beat the dogs and the cats, if they would be going near the king's meat,' said the little laddie. 'Thou'rt the cook's son,' said the giant. He catches him by the two small ankles and knocks him—'Sgleog'—against the stone that was beside him. The giant turned back to the castle in rage and madness, and he said that if they did not turn out the king's son to him, the highest stone of the castle would be the lowest. Said the queen to the king, 'We'll try it yet; the butler's son is of the same age as our son.' She arrayed the butler's son, and she gives him to the giant by the hand. The giant had not gone far when he put the rod in his hand. 'If thy father had that rod,' says the giant, 'what would he do with it?' 'He would beat the dogs and the cats when they would be coming near the king's bottles and glasses.' 'Thou art the son of the butler,' says the giant, and dashed his brains out too. The giant returned in very great rage and anger. The earth shook under the sole of his feet, and the castle shook and all that was in it. 'OUT HERE THY SON,' says the giant, 'or in a twinkling the stone that is highest in the dwelling will be the lowest.' So needs must they had to give the king's son to the giant.

The giant took him to his own house, and he reared him as his own son. On a day of days when the giant was from home, the lad heard the sweetest music he ever heard in a room at the top of the giant's house. At a glance he saw the finest face he had ever seen. She beckoned to him to come a bit nearer to her, and she told him to go this time, but to be sure to be at the same place about that dead midnight.

And as he promised he did. The giant's daughter was at his side in a twinkling, and she said, 'To-morrow thou wilt get the choice of my two sisters to marry; but say thou that thou wilt not take either, but me. My father wants me to marry the son of the king of the Green City, but I don't like him.' On the morrow the giant took out his three daughters, and he said, 'Now son of the king of Tethertown, thou hast not lost by living with me so long. Thou wilt get to wife one of the two eldest of my daughters, and with her leave to go home with her the day after the wedding.' 'If thou wilt give me this pretty little one,' says the king's son, 'I will take thee at thy word.'

The giant's wrath kindled, and he said, 'Before thou gett'st her thou must do the three things that I ask thee to do.' 'Say on,' says the king's son. The giant took him to the byre. 'Now,' says the giant, 'the dung of a hundred cattle is here, and it has not been cleansed for seven years. I am going from home to-day, and if this byre is not cleaned before night comes, so clean that a golden apple will run from end to end of it, not only thou shalt not get my daughter, but 'tis a drink of thy blood that will quench my thirst this night.' He begins cleaning the byre, but it was just as well to keep baling the great ocean. After mid-day, when sweat was blinding him, the giant's young daughter came where he was, and she said to him, 'Thou art being punished, king's son.' 'I am that,' says the king's son. 'Come over,' says she, 'and lay down thy weariness.' 'I will do that,' says he, 'there is but death awaiting me, at any rate.' He sat down near her. He was so tired that he fell asleep beside her. When he awoke, the giant's daughter was not to be seen, but the byre was so well cleaned that a golden apple would run from end to end of it. In comes the giant, and he said. 'Thou hast cleaned the byre, king's son?' 'I have cleaned it,' says he. 'Somebody cleaned it,' says the giant. 'Thou didst not clean it, at all events,' said the king's son. 'Yes, yes!' says the giant, 'since thou wert so active to-day, thou wilt get to this time to-morrow to thatch this byre with birds' down—birds with no two feathers of one colour.' The king's son was on foot before the sun; he caught up his bow and his quiver of arrows to kill the birds. He took to the moors, but if he did, the birds were not so easy to take. He was running after them till the sweat was blinding him. About mid-day who should come but the giant's daughter. 'Thou art exhausting thyself, king's son,' says she. 'I am,' said he. 'There fell but these two blackbirds, and both of one colour.' 'Come over and lay down thy weariness on this pretty hillock,' says the giant's daughter. 'It's I am willing,' said he. He thought she would aid him this time, too, and he sat down near her, and he was not long there till he fell asleep.

When he awoke, the giant's daughter was gone. He thought he would go back to the house, and he sees the byre thatched with the feathers. When the giant came home, he said, 'Thou hast thatched the byre, king's son?' 'I thatched it,' says he. 'Somebody thatched it,' says the giant. 'Thou didst not thatch it,' says the king's son. 'Yes, yes!' says the giant. 'Now,' says the giant, 'there is a fir-tree beside that loch down there, and there is a magpie's nest in its top. The eggs thou wilt find in the nest. I must have them for my first meal. Not one must be burst or broken, and there are five in the nest.' Early in the morning the king's son went where the tree was, and that tree was not hard to hit upon. Its match was not in the whole wood. From the foot to the first branch was five hundred feet. The king's son was going all round the tree. She came who was always bringing help to him; 'Thou art losing the skin of thy hands and feet.' 'Ach! I am,' says he. 'I am no sooner up than down.' 'This is no time for stopping,' says the giant's daughter. She thrust finger after finger into the tree, till she made a ladder for the king's son to go up to the magpie's nest. When he was at the nest, she said, 'Make haste now with the eggs, for my father's breath is burning my back.' In his hurry she left her little finger in the top of the tree. 'Now,' says she, 'thou wilt go home with the eggs quickly, and thou wilt get me to marry to-night if thou canst know me. I and my two sisters will be arrayed in the same garments, and made like each other, but look at me when my father says, Go to thy wife, king's son; and thou wilt see a hand without a little finger.' He gave the eggs to the giant. 'Yes, yes!' says the giant, 'be making ready for thy marriage.'

Then indeed there was a wedding, and it *was* a wedding! Giants and gentlemen, and the son of the king of the Green City was in the midst of them. They were married, and the dancing began, and that was a dance? The giant's house was shaking from top to bottom. But bed time came, and the giant said, 'It is time for thee to go to rest, son of the king of Tethertown; take thy bride with thee from amidst those.'

She put out the hand off which the little finger was, and he caught her by the hand.

'Thou hast aimed well this time too; but there is no knowing but we may meet thee another way,' said the giant.

But to rest they went. 'Now,' says she, 'sleep not, or else thou diest. We must fly quick, quick, or for certain my father will kill thee.'

Out they went, and on the blue gray filly in the stable they mounted. 'Stop a while,' says she, 'and I will play a trick to the old hero.' She jumped in, and cut an apple into nine shares, and she put two shares at the head of the bed, and two shares at the foot of the bed, and two shares at the door of the kitchen, and two shares at the big door, and one outside the house.

The giant awoke and called, 'Are you asleep?' 'We are not yet,' said the apple that was at the head of the bed. At the end of a while he called again. 'We are not yet,' said the apple that was at the foot of the bed. A while after this he called again. 'We are not yet,' said the apple at the kitchen door. The giant called again. The apple that was at the big door answered. 'You are now going far from me,' says the giant. 'We are not yet,' says the apple that was outside the house. 'You are flying,' says the giant. The giant jumped on his feet, and to the bed he went, but it was cold—empty.

'My own daughter's tricks are trying me,' said the giant. 'Here's after them,' says he.

In the mouth of day, the giant's daughter said that her father's breath was burning her back. 'Put thy hand, quick,' said she, 'in the ear of the gray filly, and whatever thou findest in it, throw it behind thee.' 'There is a twig of sloe tree,' said he. 'Throw it behind thee,' said she.

No sooner did he that, than there were twenty miles of black thorn wood, so thick that scarce a weasel could go through it. The giant came headlong, and there he is fleecing his head and neck in the thorns.

'My own daughter's tricks are here as before,' said the giant; 'but if I had my own big axe and wood knife here, I would not be long making a way through this.' He went home for the big axe and the wood knife, and sure he was not long on his journey, and he was the boy behind the big axe. He was not long making a way through the black thorn. 'I will leave the axe and the wood knife here till I return,' says he. 'If thou leave them,' said a hoodie that was in a tree, 'we will steal them.'

'You will do that same,' says the giant, 'but I will set them home.' He returned and left them at the house. At the heat of day the giant's daughter felt her father's breath burning her back.

'Put thy finger in the filly's ear, and throw behind thee whatever thou findest in it.' He got a splinter of gray stone, and in a twinkling there were twenty miles, by breadth and height, of great gray rock behind them. The giant came full pelt, but past the rock he could not go.

'The tricks of my own daughter are the hardest things that ever met me,' says the giant; 'but if I had my lever and my mighty mattock, I would not be long making my way through this rock also.' There was no help for it, but to turn the chase for them; and he was the boy to split the stones. He was not long making a road through the rock. 'I will leave the tools here, and I will return no more.' 'If thou leave them,' says the hoodie, 'we will steal them.' 'Do that if thou wilt; there is no time to go back.' At the time of breaking the watch, the giant's daughter said that she was feeling her father's breath burning her back. 'Look in the filly's ear, king's son, or else we are lost.' He did so, and it was a bladder of water that was in her ear this time. He threw it behind him and there was a fresh-water loch, twenty miles in length and breadth, behind them.

The giant came on, but with the speed he had on him, he was in the middle of the loch, and he went under, and he rose no more.

On the next day the young companions were come in sight of his father's house. 'Now,' said she, 'my father is drowned, and he won't trouble us any more; but before we go further,' says she, 'go thou to thy father's house, and tell that thou hast the like of me; but this is thy lot, let neither man nor creature kiss thee, for if thou dost thou wilt not remember that thou hast ever seen me.' Every one he met was giving him welcome and luck, and he charged his father and mother not to kiss him; but as mishap was to be, an old greyhound was in and she knew him, and jumped up to his mouth, and after that he did not remember the giant's daughter.

She was sitting at the well's side as he left her, but the king's son was not coming. In the mouth of night she climbed up into a tree of oak that was beside the well, and she lay in the fork of the tree all that night. A shoemaker had a house near the well, and about mid-day on the morrow, the shoemaker asked his wife to go for a drink for him out of the well. When the shoemaker's wife reached the well, and when she saw the shadow of her that was in the tree, thinking of it that it was her own shadow—and she never thought till now that she was so handsome—she gave a cast to the dish that was in her hand, and it was broken on the ground, and she took herself to the house without vessel or water.

'Where is the water, wife?' said the shoemaker. 'Thou shambling, contemptible old carle, without grace, I have stayed too long thy water and wood thrall.' 'I am thinking, wife, that thou hast turned crazy. Go thou, daughter, quickly, and fetch a drink for thy father.' His daughter went, and in the same way so it happened to her. She never thought till now that she was so loveable, and she took herself home. 'Up with the drink,' said her father. 'Thou hume-spun shoe carle, dost thou think that I am fit to be thy thrall.' The poor shoemaker thought that they had taken a turn in their understandings, and he went himself to the well. He saw the shadow of the maiden in the well, and he looked up to the tree, and he sees the finest woman he ever saw. 'Thy seat is wavering, but thy face is fair,' said the shoemaker. 'Come down, for there is need of thee for a short while at my house.' The shoemaker understood that this was the shadow that had driven his people mad. The shoemaker took her to his house, and he said that he had but a poor bothy, but that she should get a share of all that was in it. At the end of a day or two came a leash of gentlemen lads to the shoemaker's house for shoes to be made them, for the king had come home, and he was going to marry. The glance the lads gave they saw the giant's daughter, and if they saw her, they never saw one so pretty as she. ''Tis thou hast the pretty daughter here,' said the lads to the shoemaker. 'She is pretty, indeed,' says the shoemaker, 'but she is no daughter of mine.' 'St Nail!' said one of them, 'I would give a hundred pounds to marry her.' The two others said the very same. The poor shoemaker said that he had nothing to do with her. 'But,' said they, 'ask her to-night, and send us word to-morrow.' When the gentles went away, she asked the shoemaker—'What's that they were saying about me?' The shoemaker told her. 'Go thou after them,' said she; 'I will marry one of them, and let him bring his purse with him.' The youth returned, and he gave the shoemaker a hundred pounds for tocher. They went to rest, and when she had laid down, she asked the lad for a drink of water from a tumbler that was on the board on the further side of the chamber. He went; but out of that he could not come, as he held the vessel of water the length of the night. 'Thou lad,' said she, 'why wilt thou not lie down?' but out of that he could not drag till the bright morrow's day was. The shoemaker came to the door of the chamber, and she asked him to take away that lubberly boy. This wooer went and betook himself to his home, but he did not tell the other two how it happened to him. Next came the second chap, and in the same way, when she had gone to rest—'Look,' she said, 'if the latch is on the door.' The latch laid hold of his hands, and out of that he could not come the length of the night, and out of that he did not come till the morrow's day was bright. He went, under shame and disgrace. No matter, he did not tell the other chap how it had happened, and on the third night he came. As it happened to the two others, so it happened to him. One foot stuck to the floor; he could neither come nor go, but so he was the length of the night. On the morrow, he took his soles out (of that), and he was not seen looking behind him. 'Now,' said the girl to the shoemaker, 'thine is the sporran of gold; I have no need of it. It will better thee, and I am no worse for thy kindness to me.' The shoemaker had the shoes ready, and on that very day the king was to be married. The shoemaker was going to the castle with the shoes of the young people, and the girl said to the shoemaker, 'I would like to get a sight of the king's son before he marries.' 'Come with me,' says the shoemaker, 'I am well acquainted with the servants at the castle, and thou shalt get a sight of the king's son and all the company.' And when the gentles saw the pretty woman that was here they took her to the wedding-room, and they filled for her a glass of wine. When she was going to drink what is in it, a flame went up out of the glass, and a golden pigeon and a silver pigeon sprung out of it. They were flying about when three grains of barley fell on the floor. The silver pigeon sprang, and he eats that. Said the golden pigeon to him, 'If thou hadst mind when I cleared the byre, thou wouldst not eat that without giving me a share.' Again fell three other grains of barley, and the silver pigeon sprang, and he eats that, as before. 'If thou hadst mind when I thatched the byre, thou wouldst not eat that without giving me my share,' says the golden pigeon. Three other grains fall, and the silver pigeon sprang, and he eats that. 'If thou hadst mind when I harried the magpie's nest, thou wouldst not eat that without giving me my share,' says the golden pigeon; 'I lost my little finger bringing it down, and I want it still.' The king's son minded, and he knew who it was he had got. He sprang where she was, and kissed her from hand to mouth. And when the priest came they married a second time. And there I left them.

The Wee Bannock

[Ayrshire]

Retold by Norah and William Montgomerie

There lived an old man and an old wife at the side of a burn. They had two cows, five hens and a cock, a cat and two kittens. The old man looked after the cows, and the old wife span on the distaff. The kittens often clawed at the old wife's spindle as it danced over the hearth-stone.

'Sho, sho,' she said; 'go away!' And so it danced about.

One day, after porridge time, she thought she would have a bannock. So she baked two oatmeal bannocks, and set them to the fire to toast. After a while, the old man came in, sat down beside the fire, took up one of the bannocks and snapped it through the middle. When the other one saw this, it ran off as fast as it could, and the old wife after it, with the spindle in one hand and the distaff in the other.

But the wee bannock went away, out of sight, and ran till it came to a fine large thatched house, and in it ran till it came to the fireside. There were three tailors sitting on a big table. When they saw the wee bannock come in, they jumped up and went behind the goodwife, who was carding flax beside the fire.

'Don't be frightened,' said she. 'It's only a wee bannock. Catch it, and I'll give you a mouthful of milk with it.'

Up she got with the flax-cards, and the tailor with the smoothing-iron, and the two apprentices, the one with the big shears and the other with the lap-board. But it dodged them and ran about the fire. One of the apprentices, thinking to snap it with the shears, fell into the ash-pit. The tailor threw the smoothing-iron, and the goodwife the flax-cards, but it was no use. The bannock escaped, and ran till it came to a wee house at the roadside. In it ran, and there was a weaver sitting on the loom, and the wife winding a hank of yarn.

'Tibby,' said he, 'what's that?'

'Oh,' said she, 'it's a wee bannock.'

'It's welcome,' said he, 'for our gruel was but thin to-day. Catch it, my woman, catch it!'

'Ay,' said she, 'if I can. That's a clever bannock. Catch it, Willie! Catch, man!'

'Cast the clew at it!' said Willie.

But the bannock ran round about, across the floor and off over the hill, like a new-tarred sheep or a mad cow. On it ran to the next house, and in it ran to the fireside, where the goodwife was churning.

'Come away, wee bannock,' said she. 'I'm having cream and bread to-day.'

But the wee bannock ran round about the churn, the wife after it, and in the hurry she nearly overturned the churn. Before she had it set right again, the wee bannock was off, down the hillside to the mill, and in it ran.

The miller was sifting meal in the trough, but, looking up, he smiled at the wee bannock.

'Ay,' said he, 'it's a sign of plenty when you are running about, and nobody to look after you. I like a bannock and cheese. Come away in, and I'll give you a night's quarters.'

But the wee bannock would not trust itself with the miller and his cheese. So it ran out of the mill, but the miller didn't trouble his head about it.

It ran and it ran till it came to the smiddy. In it went, and up to the anvil. The smith was making horse-nails.

'I like a cog of good ale, and a well-toasted bannock,' said he. 'Come away in here.'

The bannock was frightened when it heard about the ale, and ran off as hard as it could. The smith ran after it, and threw his hammer. But the bannock whirled away, and was out of sight in an instant. It ran till it came to a farmhouse with a large peat-stack at the end of it. In it ran to the fireside. The goodman was separating lint, and the goodwife was dressing flax.

'Janet,' said he, 'there's a wee bannock. I'll have the half of it.'

'Well, John, I'll have the other half. Hit it over the back with a clew.'

The bannock played tig. The old wife threw the heckle at it, but it was too clever for her.

Off and up the stream it ran to the next house, and whirled away in to the fireside. The goodwife was stirring the gruel and the goodman plaiting rush-ropes for the cattle.

'Hey, Jock,' said the goodwife, 'come here! You are always crying about a bannock. Here's one. Come in, hurry now! I'll help you catch it.'

'Ay, wife, where is it?'

'See, there. Run over to that side.'

But the wee bannock ran in behind the goodman's chair. Jock fell among the rushes. The goodwife threw the porridge stick and the goodman a rope, but it was too clever for either of them. It was off and out of sight in an instant, through the whins, and down the road to the next house. In it went to the fireside just as the folk were sitting down to their gruel, and the goodwife was scraping the pot.

'Losh,' said she, 'there's a wee bannock come in to warm itself at our fireside!'

'Fasten the door,' said the goodman, 'and we'll try to get a grip of it.'

When the bannock heard this, it ran into the kitchen, and they after it with their spoons. The goodman threw his bonnet, but it ran, and ran, and faster ran, till it came to another house. When it went in, the folk were just going to their beds. The goodman was casting off his trousers, and the goodwife raking the fire.

'What's that?' said he.

'Oh,' said she, 'it's a wee bannock.'

'I could eat the half of it, for all the porridge I supped,' said he.

'Catch it!' said the wife, 'and I'll have a bit too. Throw your trousers at it! Kep! Kep!'

The goodman threw his trousers at it, and nearly smothered it. But it wrestled out, and ran, the goodman after it without his trousers. There was a rare chase over the croft field, up the yard and in among the whins. There the goodman lost it, and had to go trotting home half naked. But it had grown dark. The wee bannock couldn't see, went through a whin bush, and right into a fox's hole. The fox had had no meat for two days.

'Welcome, welcome,' said the fox, and snapped it in two. And that was the end of the wee bannock.

Death in a Nut

Told by Duncan Williamson

Jack lived with his mother in a little cottage by the shoreside, an his mother kept some ducks an some hens. Jack cuid barely remember his father because his father had died long before he wis born. An they had a small kin o croft, Jack cut a little hay fir his mother's goats. When dher wur no hay tae collect, he spent most of his time along the shoreside as a beach-comber collecting everything that cam in bi the tide, whatever it wad be—any auld drums, any auld cans, pieces of driftwood, something that wis flung off a boat—Jack collectit all these things an brought them in, put them biside his mother's cottage an said, 'Some day they might come in useful.' But the mos thing that Jack ever collected fir his mother was firewood. An Jack wis very happy, he wis jist a young man, his early teens, and he dearly loved his mother. He used tae some days take duck eggs tae the village (his mother wis famed fir er duck eggs) an hen eggs to the village forbyes, they helped them survive, and his mother wad take in a little sewin fir the local people in the village; Jack and his mother lived quite happy. Till one particular day, it wis around about the winter-time, about the month o January, this time o the year now.

Jack used tae always get up early in the mornin an make a cup o tea, he always gev his mother a cup o tea in bed every mornin. An one particular mornin he rose early because he want't tae catch the in-comin tide tae see what it wad bring in fir him. He brought a cup o tea into his mother in her own little bed in a little room, it wis only a two-room little cottage they had.

He says, 'Mother, I've brought you a cup o tea.'

She says, 'Son, I don't want any tea.'

'Mother,' he says, 'why? What's wrong, are you not feelin—'

She says, 'Son, I'm not feelin very well this morning, I'm not feelin very well. I don't think I cuid even drink a cup o tea if ye gev it to me.'

'Oh, Mother,' he says, 'try an take a wee sip,' an he leaned over the bed, held the cup to his mother's mooth an tried to get her . . .

She took two–three sips, 'That's enough, laddie,' she says, 'I don't feel very well.'

He says, 'What's wrong with you, Mother? Are you in pain or somethin?'

'Well, so an no so, Jack, I dinnae ken what's wrong wi me,' she says. 'I'm an ill woman, Jack, an ye're a young man an I cannae go on for ever.'

'But, Mother,' he says, 'you cannae dee an leave me masel! What am I gaunnae dae? I've nae freends, nae naebody in this worl but you, Mother! Ye cannae dee an lea me!'

'Well,' she says, 'Jack, I think I'm no long fir this worl. In fact, I think he'll be comin fir me some o these days . . . soon.'

'Wha, Mother, ye talking about "comin fir me"?'

She says, 'Jack, ye ken wha he is, Jack. Between me an you, we dinna share nae secrets—I'm an auld woman an I'm gaunna dee—Death's gaunna come fir me, Jack, I can see it in ma mind.'

'Oh, Mother, no, Mother,' he says, an he held her hand.

'But,' she says, 'never mind, laddie, ye'll manage to take care o yirsel. Yir mother hes saved a few shillins fir ye an I'm sure some day ye'll meet a nice wee wife when I'm gone, ye'll prob'ly get on in the world.'

'No, Mother,' he says, 'I cuidna get on withoot you.'

She says, 'Laddie, leave me an I'll try an get a wee sleep.'

Bi this time it was daylight as the sun begint tae get up, an Jack walkit up along the shoreway jist in the grey-dark in the mornin, gettin clear. It must hae been about half-past eight–nine o'clock (in the wintertime it took a long while tae get clear in the mornins) when the tide was comin in. Jack walked along the shoreway an lo an behold, the first thing he seen comin a-walkin the shoreway was an auld man with a long grey beard, skinny legs and a ragged coat o'er his back an a scythe on his back. His two eyes were sunk inta his heid, sunk back intae his skull, an he wis the most uglies'-luikin creature that Jack ever seen in his life. But he had on his back *a brand new scythe* an hit was shinin in the light fae the mornin.

Noo, his mother hed always tellt Jack what like Deith luikit an Jack says tae his ainsel, 'That's Deith come fir my auld mother! He's come tae take on'y thing that I love awa fae me, but,' he said, 'he's no gettin awa wi it! He's no gettin away wi hit!' So Jack steps oot aff the shoreside, an up he comes an meets this Auld Man—bare feet, lang ragged coat, lang ragged beard, high cheek bones an his eyes sunk back in his heid, two front teeth sticking out like that—and a shinin scythe on his back, the morning sun wis glitterin on the blade—ready to cut the people's throats an take them away to the Land o Death.

Jack steps up, says, 'Good morning, Auld Man.'

'Oh,' he said, 'good morning, young man! Tell me, is it far tae the next cottage?'

Jack said, 'Ma mother lives i the next cottage just along the shoreway a little bit.'

'Oh,' he says, 'that's her I want to visit.'

'Not this morning,' says Jack, 'ye're not gaunna visit her! I know who you are—you're Death—an you've come tae take my aul mother, kill her an tak her awa an lea me masel.'

'Well,' Death says, 'it's natural. Yir mother, ye know, she's an auld wumman an she's reacht a certain age, I'll no be doin her any harm, I'll be jist doin her a guid turn—she's sufferin in pain.'

'You're no takin my aul mither!' says Jack. And he ran forward, he snappit the scythe aff the Aul Man's back an he walkit tae a big stane, he smashed the scythe against a stane.

An the Auld Man got angrier an angrier an angrier an ugly-luikin, 'My young man,' he says, 'you've done that—but that's not the end!'

'Well,' Jack says, 'it's the end fir you!' An Jack dived o top o him, Jack got a haud o him an Jack pickit a bit stick up the shoreside, he beat him an he weltit him an he weltit him an he beat him an he weltit him. He fought wi Death an Death wis as strong as what Jack was, but finally Jack conquered him! An Jack beat im with a bit stick, and lo an behold the funny thing happened: the more Jack beat him the wee-er he got, an Jack beat him an Jack beat him an Jack beat him—no blood cam fae him or nothing—Jack beat him wi the stick till he got barely the size o that! An Jack catcht im in his hand, 'Now,' he said, 'I got ye! Ye'll no get my aul mither!'

Noo Jack thought in his ainsel, 'What in the worl am I gaunna do wi him? A hev him here, I canna let him go, A beat him, I broke his scythe an I conquered him. But what in the world am I gaunna do wi him? I canna hide him bilow a stane because he'll creep oot an he'll come back tae his normal size again.' An Jack walkit along the shore an he luikit—comin in by the tide was a big hazelnut, that size! But the funny thing about this hazelnut, a squirrel had dug a hole in the nut cause squirrels always dig holes in the nuts—they have sharp teeth—an he eats the kernel oot inside an leas the empty case. An Jack pickit up the hazelnut, he luikit, says, 'The very thing!' An Jack crushed Death in through the wee hole—inta the nut! An squeezed him in heid first, an his wee feet, put him in there, shoved him in. An he walkit aboot, he got a wee plug o stick and he plugged the hole fae the outside. 'Now,' he says, 'Death, you'll never get ma mither.' An he catcht him in his hand, he threw im oot inta the tide! An the heavy waves wis 'whoosh-an-whoosh-an-whoosh-an, whoosh-an-whoosh-in' in an back an forward. An Jack watched the wee nut, hit went a-sailin, floatin an back an forward away wi the tide. 'That's hit!' says Jack, 'that's the end o Death. He'll never bother my mother again, or naebody else forbyes my mither.'

Jack got two–three sticks under his arm an he walkit back. Whan he landed back he seen the reek wis comin fae the chimney, he says, 'My mother must be up, she must be feelin a wee bit better.' Lo an behold he walks in the hoose, there wis his auld mother up, her sleeves rolled up, her face full o flooer, her apron on an she's busy makin scones.

He said, 'How ye feelin, Mother?'

She says, 'Jack, I'm feelin great, I never feelt better in ma life! Laddie, I dinna ken what happened to me, but I wis lyin there fir a minute in pain an torture, and all in a minute I feelt like someone hed come an rumbled all the pains an tuik everything oot o my body, an made me . . . I feel like a lassie again, Jack! I made some scones fir yir breakfast.'

Jack never mentioned to his mother aboot Deith, never said a word. His mother fasselt roon the table, she's pit up her hair . . . Jack never seen his mother in better health in her life! Jack sit doon bi the fire, his mother made some scones. He had a wee bit scone, he says, 'Mother, is that all you've got tae eat?'

'Well,' she says, 'Jack, the're no much, jist a wee puckle flooer an I thocht I'd mak ye a wee scone fir yir breakfast. Go on oot tae the hen house an get a couple eggs, I'll mak ye a couple eggs alang wi yir scone an that'll fill ye up, laddie.'

Jack walks oot to the hen hoose as usual, wee shed beside his mother's hoose. Oh, every nest is full o eggs, hens' eggs, duck eggs, the nests is all full. Jack picks up four o the big beautiful broon eggs oot o the nest, gaes back in an 'Here, Mother, the're fowr,' he said, 'two tae you, two to me.'

De aul wumman says, 'I'll no be a minute, Jack.' It was a open fire they had. The wumman pulled the sway oot, put the fryin pan on, pit a wee bit fat i the pan. She waitit an she waitit an she watcht, but the wee bit fat wadna melt. She poked the fire with the poker but the wee bit fat wadna melt. 'Jack,' she says, 'fire's no kindlin very guid, laddie, it'll no even melt that wee bit fat.'

'Well, pit some mair sticks on, Mother,' he said, 'pit some mair wee bits o sticks on.' Jack pit the best o sticks on, but na! The wee bit o fat sut in the middle o the pan, but it wouldna melt, he says, 'Mother, never mind, pit the egg in an gie it a rummle roon, it'll dae me the way it is. Jis pit it in the pan.'

Aul mother tried—'crack'—na. She hut the egg again—na. An s'pose she cuid hae take a fifty-pun hammer an hut the egg, *that egg would not break!* She says, 'Jack, I cannae break these eggs.'

'An, Mother,' he said, 'I thought ye said ye were feelin weel an feelin guid, an you cannae break an egg! Gie me the egg, I'll break hit!' Jack tuik the egg, went in his big hand, ye ken, Jack big young laddie, catcht the egg one hand—'clank' on the side o the pan—na! Ye're as well tae hit a stane on the side o the pan, *the egg would not break* in no way in this worl! 'Ah, Mother,' he says, 'I dinna ken what's wrong, I dinna ken whit's wrong, Mother, wi these eggs, I don't know. Prob'ly they're no richt eggs, I better go an get another two.'

He walkit oot to the shed again, he brung in two duck eggs. But he tried the same—na, they wadna break, the eggs jist would not break in any way in the worl. 'Mother,' he says, 'pit them in a taste o water an bring them a-boil!'

She says, 'That's right, Jack, I never thocht about that.' The aul wumman got a wee pan an the fire wis goin well bi this time of bonnie shore sticks. She pit the pan on an within seconds the water wis boilan, she poppit the two eggs in. An it bubbled an bubbled an bubbled an bubbled an bubbled, an bubbled, she said, 'They're ready noo, Jack.' She tuik them oot—'crack'—na. As suppose they hed hae tried fir months, they cuidna crack that two eggs.

'Ah, Mother,' he says, 'the're something wrong. Mither, the're something wrong, the're enchantment upon us, that eggs'll no cook. We're gaunna dee wi hunger.'

'Never mind, Jack,' she says, 'eat yir wee bit scone. I'll mak ye a wee drop soup, I'll mak ye a wee pot o soup. Go oot tae the gairden, Jack, an get me a wee taste o vegetables, leeks an a few carrots.'

Noo Jack had a guid garden, he passes all his time makkin a guid garden tae his mother. Ot he goes, he pulls two carrots, a leek, bit parsley an a neep an he brings it tae his mother. Aul wumman washes the pots, pits some water in, pits it on the fire. But she goes tae the table with the knife, but na—every time she touches the carrot, the knife jist skates aff hit. She toucht the leek—it skates aff it an aa. The auld wumman tried her best, an Jack tried his best—there's no way in the world—Jack said, 'That knife's blunt, Mother.'

An Jack had a wee bit o shairpen stane he'd fand in the shoreside, he took the stane an he shairpit the knife, but no way in the world wad hit ever look at the carrots or the neep or the wee bit parsley tae mak a wee pot o soup. She says, 'Jack, the're somethin wrang wi my vegetables, Jack, they must be frozen solid.'

'But,' he said, 'Mother, the're been nae frost tae freeze them! Hoo in the world can this happen?'

'Well,' she says, 'Jack, luik, ye ken I've an awfa cockerels this year, we have an awfa cockerels an we'll no need them aa, Jack. Wad ye gae oot to the shed an pull a cockerel's neck, and A'll pit it in the pot, boil hit for wir supper?'

'Ay,' says Jack. Noo the aul wumman kep a lot o hens. Jack went oot an all i the shed dher wur dozens o them sittin i a raa, cockerels o all description. Jack luikit ti he seen a big fat cockerel sittin on a perch, he put his hand up, catcht hit an he feel'd it, it wis fat. 'Oh,' he says, 'Mother'll be pleased wi this yin.' Jack pullt the neck—na! Pulled again—*no way*. He pullt it, he shakit it, he swung it roond his heid three–five times. He tuik a stick an he battert it i the heid, there's no way—he cuidna touch the cockerel in any way! He pit it bilow his oxter an he walks inta his mother.

She said, 'Ye get a cockerel, Jack?'

'Oh, Mother,' he said, 'I got a cockerel aa right, I got a cockerel. But, Mother, you may care!'*

She says, 'What do you mean, laddie?'

'You may care,' he says, 'I cannae kill hit.'

'Ah, Jack,' she says, 'ye cannae kill a cockerel! I ken, ye killt dozens tae me afore, the hens an ducks an aa.'

'Mother,' he said, 'I cannae kill this one—it'll no dee!'

She says, 'Gie me it ower here, gie me it over here!' An the auld woman had a wee hatchet fir splittin sticks, she kep it by the fire. She says, 'Gie it tae me, Jack, I'll show ye the way tae kill it rictht!' She pit it doon the top o the block an she hut it wi the hatchet, chop its heid aff. She hut it with the hatchet seventeen times, but no—every time the heid jumpit aff—heid jumpit back on! 'Na, Jack,' she says, 'it's nae good. There's something wrang here, the're something terrible gaun a-wrong. Nethin seems tae be richt aboot the place. Here—go out to my purse,

* You may care—there's nothing you can do about it.

laddie, run up tae the village to the butcher! I'm savin this fir a rainy day,' an she tuik a half-croon oot o her purse. 'Jack, gae up tae the butcher an get a wee bit o meat fae the butcher, I'll mak ye a wee bite when ye come back.'

Noo, it wisna far fae the wee hoose to the village, about a quarter o mile Jack hed tae walk. When Jack walkit up the village, all the people were gaithert in the middle o the town square. They're all bletherin an they're chattin and they're bletherin an they're chattin, speakin tae each other. One was sayin, 'A've sprayed ma garden an it's overrun wi caterpillars! An I've tried tae spray hit, it's no good.'

The butcher wis oot wi his apron, he said, 'Three times I tried tae kill a bullock this mornin an three times I killed it, three times it jumpit back on its feet. I don't know what's wrong. The villagers run out o meat! I got a quota o hens in this mornin, ducks, an every time I pull their necks their heads jumps back on. There's somethin terrible is happenin!'

Jack went up to the butcher's, he says, 'Gie me a wee bit o meat fir ma mother.'

He says, 'Laddie, the're no a bit o meat in the shop. Dae ye no ken what I'm tryi' tae tell the people in the village: I've tried ma best this mornin to kill a young bullock tae supply the village an I cannae kill hit!'

'Well,' Jack said, 'the same thing happen to me—I tried tae boil an egg an I cannae boil an egg, I tried tae kill a cockerel—'

'I tried tae kill ten cockerels,' says the butcher, 'but *they'll no dee!*'

'Oh dear-dear,' says Jack, 'we must be in some kin o trouble. Is hit happenin tae other places forbyes this?'

'Well, I jist hed word,' says the butcher, 'the next village up two mile awa an the same thing's happened tae them. Folk cannae even eat an apple—when they sink their teeth inta it, it'll no even bite. They cannae cook a vegetable, they cannae boil water, they cannae dae nothin! The hale worl's gaunna come tae a standstill, the're something gaen terrible wrong—*nothing seems to die anymore.*'

An then Jack thought in his head, he said, 'It's my fault, I'm the cause o't.' He walkit back an he tellt his mother the same story I'm tellin you. He says, 'Mother, there's nae butcher meat fir ye.'

She says, 'Why, laddie, why no?'

He says, 'Luik, the butcher cannae kill nae beef, because hit'll no dee.'

'But Jack,' she says, 'why no—it'll no dee? What's wrang with the country, what's wrang with the world?'

He says, 'Mother, it's all my fault!'

'Your fault,' she says, 'Jack?'

'Ay, Mither, it's my fault,' he says. 'Listen, Mother: this morning when you were no feeling very well, I walkit along the shore tae gather some sticks fir the fire an I met Death comin tae tak ye awa. An I took his scythe fae him an I broke his scythe, I gi'n him a beatin, Mither, an I put him in a nut! An I flung him in the tide an I plugged the nut so's he canna get oot, Mither. An God knows where he is noo. He's floatin in the sea,

Mother, firever an ever an ever, an nothing'll dee—the worl is over-run with caterpillars an worms an everything—Mither, the're nothing can dee! But Mither, I wad rather die with starvation than loss you.'

'Jack, Jack, Jack, laddie,' she says, 'dae ye no ken what ye've done? Ye've destroyed the only thing that keeps the world alive.'

'What do you mean, Mother, "keeps the world alive"? Luik, if I hedna killed him, I hedna hae beat im, Mother, an pit him in that nut—you'd be dead bi this time!'

'I wad be dead, Jack,' she says, 'probably, but the other people would be gettin food, an the worl'd be gaun on—the way it shuid be—only fir you, laddie!'

'But, Mother,' he says, 'what am I gaunna dae?'

She says, 'Jack, there's only thing ye can dae . . . ye're a beach-comber like yir faither afore ye—'

'Aye, Mother,' he says, 'I'm a beach-comber.'

'Well, Jack,' she says, 'there's only thing I can say: ye better gae an get im back an set him free! Because if ye dinnae, ye're gaunna put the whole worl tae a standstill. *Bithout Death there is no life* . . . fir nobody.'

'But, Mother,' he says, 'if I set him free, he's gaunna come fir you.'

'Well, Jack, if he comes fir me,' she said, 'I'll be happy, and go inta another world an be peaceful! But you'll be alive an so will the rest o the world.'

'But Mother,' he says, 'I cuidna live bithoot ye.'

'But,' she says, 'Jack, if ye dinnae set him free, *both* o hus'll suffer, an I cannae stand tae see you suffer fir the want o something to eat: because the're nothing in the world will die unless you set him free, because you cannae eat nothing until it's dead.'

Jack thought in his mind fir a wee while. 'Aa right, Mother,' he says, 'if that's the way it shuid be, that's the way it shuid be. Prob'ly I wis wrong.'

'Of course, Jack,' she says, 'you were wrong.'

'But,' he says, 'Mother, I only done it fir yir sake.'

'Well,' she says, 'Jack, fir *my* sake, wad ye search fir that hazelnut an set him free?'

So the next mornin true tae his word, Jack walks the tide an walks the tide fir miles an miles an miles, day out an day in fir three days an fir three days more. He hedna nothin tae eat, he only hed a drink water. They cuidna cook anything, they cuidna eat any eggs, they couldna fry nothing in the pan if they had it, they cuidna make any soup, they cuidna get nothin. The caterpillars an the worms crawled out o the garden in thousands, an they ett every single vegetable that Jack had. An the're nothing in the world—Jack went out an tried to teem hot water on them but it wis nae good. When he teemed hot water on them it just wis the same as he never poored nothing—no way. At last Jack said, 'I must go an find that nut!' So he walkit an he walkit, an he walkit day an he walkit night mair miles than he ever walked before, but no way cuid Jack fin' this nut! Till

Jack was completely exhaustit an fed up an completely sick, an he cuidna walk another mile. He sat doon bi the shoreside right in front o his mother's hoose to rest, an wonderit, he pit his hand on his jaw an he said tae his ainsel, 'What have I done? I've ruint the world, I've destroyed the world. People disna know,' he said, 'what Death has so good, at Death is such a guid person. I wis wrong tae beat him an put him in a nut.'

An he's luikin all over—an lo and behold he luikit doon—there at his feet he seen a wee nut, an a wee bit o stick stickin oot hit. He liftit hit up in his hand, an Jack wis happy, happier an he'd ever been in his life before! And he pulled the plug an a wee head poppit oot. Jack held im in his two hands and Death spoke tae him, 'Now, Jack,' he said, 'are ye happy?'

'No,' Jack said, 'I'm no happy.'

He said, 'You thought if you beat me an conquered me an killed me—because I'm jist Death—that that wad be the end, everything be all right. Well, Jack, ma laddie, ye've got a lot to learn, Jack. Without me,' he said, 'there's no life.'

An Jack tuik him oot.

'But,' he says, 'Jack, thank you fir settin me free,' an jist like that, after Jack opent the nut, he cam oot an like that, he cam full strength again an stude before Jack—the same Auld Man with the long ragged coat an the sunken eyes an the two teeth in the front an the bare feet. He says, 'Jack, ye broke my scythe.'

Jack said, 'I'll tell ye somethin, while I wis searchin fir you ma mother made me mend it. An I have it in the hoose fir ye, come wi me!' An Jack led him up to the hoose. Lo an behold sure enough, sittin on the front o the porch wis the scythe that Jack broke. Jack had tuik it an he'd mend't it, he sortit it an made it as guid as ever.

Death cam to the door an he ran his hand doon the face o the scythe, he sput on his thumb and he run it up the face o the scythe, an he says tae Jack, 'I see you've sharpened it, Jack, and ye made a good job o it. Well, I hev some people to see in the village, Jack. But remember, I'll come back fir yir mother someday, but seein you been guid to me I'll make it a wee while!' An Death walkit away.

Jack an his mother lived happy till his mother wis about a hundred years of age! An then one day Death cam back tae take his aul mother away, but Jack never saw him. But Jack was happy fir he knew *there is no life bithout Death*. An that is the end o my story.

James Hogg
1770–1835

The Cameronian Preacher's Tale

Sit near me, my children, and come nigh, all ye who are not of my kindred, though of my flock; for my days and hours are numbered; death is with me dealing, and I have a sad and a wonderful story to relate. I have preached and ye have profited; but what I am about to say is far better than man's preaching, it is one of those terrible sermons which God preaches to mankind, of blood unrighteously shed, and most wondrously avenged. The like has not happened in these our latter days. His presence is visible in it; and I reveal it that its burthen may be removed from my soul, so that I may die in peace; and I disclose it, that you may lay it up in your hearts and tell it soberly to your children, that the warning memory of a dispensation so marvellous may live and not perish. Of the deed itself, some of you have heard a whispering; and some of you know the men of whom I am about to speak; but the mystery which covers them up as with a cloud I shall remove; listen, therefore, my children, to a tale of truth, and may you profit by it!

On Dryfe Water, in Annandale, lived Walter Johnstone, a man open hearted and kindly, but proud withal and warm tempered; and on the same water lived John Macmillan, a man of a nature grasping and sordid, and as proud and hot tempered as the other. They were strong men, and vain of their strength; lovers of pleasant company, well to live in the world, extensive dealers in corn and cattle; married too, and both of the same age—five and forty years. They often met, yet they were not friends; nor yet were they companions, for bargain making and money seeking narroweth the heart and shuts up generosity of soul. They were jealous, too, of one another's success in trade, and of the fame they had each acquired for feats of personal strength and agility, and skill with the sword—a weapon which all men carried, in my youth, who were above the condition of a peasant. Their mutual and growing dislike was inflamed by the whisperings of evil friends, and confirmed by the skilful manner in which they negotiated bargains over each other's heads. When they met, a short and surly greeting was exchanged, and those who knew their natures looked for a meeting between them, when the sword or some other dangerous weapon would settle for ever their claims for precedence in cunning and in strength.

They met at the fair of Longtown, and spoke, and no more—with them both it was a busy day, and mutual hatred subsided for a time, in the love of turning the penny and amassing gain. The market rose and fell, and fell and rose; and it was whispered that Macmillan, through the superior skill or good fortune of his rival, had missed some bargains which were very valuable, while some positive losses touched a nature extremely sensible of the importance of wealth. One was elated and the other depressed—but not more depressed than moody and incensed, and in this temper they were seen in the evening in the back room of a public inn, seated apart and silent, calculating losses and gains, drinking deeply, and exchanging dark looks of hatred and distrust. They had been observed, during the whole day, to watch each other's movements, and now when they were met face to face, the labours of the day over, and their natures inflamed by liquor as well as by hatred, their companions looked for personal strife between them, and wondered not a little when they saw Johnstone rise, mount his horse, and ride homewards, leaving his rival in Longtown. Soon afterwards Macmillan started up from a moody fit, drank off a large draught of brandy, threw down a half-guinea, nor waited for change—a thing uncommon with him; and men said, as his horse's feet struck fire from the pavement, that if he overtook Johnstone, there would be a living soul less in the land before sunrise.

Before sunrise next morning the horse of Walter Johnstone came with an empty saddle to his stable door. The bridle was trampled to pieces amongst its feet, and its saddle and sides were splashed over with blood as if a bleeding body had been carried across its back. The cry arose in the country, an instant search was made, and on the side of the public road was found a place where a deadly contest seemed to have happened. It was in a small green field, bordered by a wood, in the farm of Andrew Pattison. The sod was dinted deep with men's feet, and trodden down and trampled and sprinkled over with blood as thickly as it had ever been with dew. Blood drops, too, were traced to some distance, but nothing more was discovered; the body could not be found, though every field was examined and every pool dragged. His money and bills, to the amount of several thousand pounds, were gone, so was his sword—indeed nothing of him could be found on earth save his blood, and for its spilling a strict account was yet to be sought.

Suspicion instantly and naturally fell on John Macmillan, who denied all knowledge of the deed. He had arrived at his own house in due course of time, no marks of weapon or warfare were on him, he performed family worship as was his custom, and he sang the psalm as loudly and prayed as fervently as he was in the habit of doing. He was apprehended and tried, and saved by the contradictory testimony of the witnesses against him, into whose hearts the spirit of falsehood seemed to have entered, in order to perplex and confound the judgment of men—or rather that man might have no hand in the punishment, but that God should bring it about in his own good time and way. 'Revenge is mine, saith the Lord,' which meaneth not because it is too sweet a morsel for man, as the scoffer said, but because it is too dangerous. A glance over this conflicting testimony will show how little was then known of this foul offence, and how that little was rendered doubtful and dark by the imperfections of human nature.

Two men of Longtown were examined. One said that he saw Macmillan insulting and menacing Johnstone, laying his hand on the hilt of his sword with a look dark and ominous; while the other swore that he was present at the time, but that it was Johnstone who insulted and menaced Macmillan, and laid his hand on the hilt of his sword and pointed to the road homewards. A very expert and searching examination could make no more of them; they were both respectable men with characters above suspicion. The next witnesses were of another stamp, and their testimony was circuitous and contradictory. One of them was a shepherd—a reluctant witness. His words were these: 'I was frae hame on the night of the murder, in the thick of the wood, no just at the place which was bloody and trampled, but gaye and near hand it. I canna say I can just mind what I was doing; I had somebody to see I jalouse, but wha it was is naebody's business but my ain. There was maybe ane forbye myself in the wood, and maybe twa; there was ane at ony rate, and I am no sure but it was an auld acquaintance. I see nae use there can be in questioning me. I saw nought, and therefore can say nought. I canna but say that I heard something—the trampling of horses, and a rough voice saying, "Draw and defend yourself." Then followed the clashing of swords and half smothered sort of work, and then the sound of horses' feet was heard again, and that's a' I ken about it; only I thought the voice was Walter Johnstone's, and so thought Kate Pennie, who was with me and kens as meikle as me.' The examination of Katherine Pennie, one of the Pennies of Pennieland, followed, and she declared that she had heard the evidence of Dick Purdie with surprise and anger. On that night she was not over the step of her father's door for more than five minutes, and that was to look at the sheep in the fauld; and she neither heard the clashing of swords nor the word of man or woman. And with respect to Dick Purdie, she scarcely knew him even by sight; and if all tales were true that were told of him, she would not venture into a lonely wood with him, under the cloud of night, for a gown of silk with pearls on each sleeve. The shepherd, when recalled, admitted that Kate Pennie might be right, 'For after a',' said he, 'it happened in the dark, when a man like me, no that gleg of the uptauk, might confound persons. Somebody was with me, I am gaye and sure, frae what took place—if it was nae Kate, I kenna wha it was, and it couldna weel be Kate either, for Kate's a douce quean, and besides is married.' The judge dismissed the witnesses with some indignant words, and, turning to the prisoner, said, 'John Macmillan, the prevarications of these witnesses have saved you; mark my words—saved you from man, but not from God. On the murderer, the Most High will lay his hot right hand, visibly and before men, that we may know that blood unjustly shed will be avenged. You are at liberty to depart.' He left the bar and resumed his station and his pursuits as usual; nor did he appear sensible to the feeling of the country, which was strong against him.

A year passed over his head, other events happened, and the murder of Walter Johnstone began to be dismissed from men's minds. Macmillan went to the fair of Longtown, and when evening came he was seated in the little back room which I mentioned before, and in company with two men of the names of Hunter and Hope. He sat late, drank deeply, but in the midst of the carousal a knock was heard at the door, and a voice called sharply, 'John Macmillan.' He started up, seemed alarmed, and exclaimed, 'What in Heaven's name can *he* want with me?' and opening the door hastily, went into the garden, for he seemed to dread another summons lest his companions should know the voice. As soon as he was gone, one said to the other, 'If that was not the voice of Walter Johnstone, I never heard it in my life; he is either come back in the flesh or in the spirit, and in either way John Macmillan has good cause to dread him.' They listened—they heard Macmillan speaking in great agitation; he was answered only by a low sound, yet he appeared to understand what was said, for his concluding words were, 'Never! never! I shall rather submit to His judgment who cannot err.' When he returned he was pale and shaking, and he sat down and seemed buried in thought. He spread his palms on his knees, shook his head often, then, starting up, said, 'The judge was a fool and no prophet—to mortal man is not given the wisdom of God—so neighbours let us ride.' They mounted their horses and rode homewards into Scotland at a brisk pace.

The night was pleasant, neither light nor dark; there were few travellers out, and the way winded with the hills and with the streams, passing through a pastoral and beautiful country. Macmillan rode close by the side of his companions, closer than was desirable or common; yet he did not speak, nor make answer when he was spoken to; but looked keenly and earnestly before and behind him, as if he expected the coming of some one, and every tree and bush seemed to alarm and startle him. Day at last dawned, and with the growing light his alarm subsided, and he began to converse with his companions, and talk with a levity which surprised them more than his silence had done before. The sun was all but risen when they approached the farm of Andrew Pattison, and here and there the top of a high tree and the summit of a hill had caught light upon them. Hope looked to Hunter silently, when they came nigh the bloody spot where it was believed the murder had been committed. Macmillan sat looking resolutely before him, as if determined not to look upon it; but his horse stopt at once, trembled violently, and then sprung aside, hurling its rider headlong to the ground. All this passed in a moment; his companions sat astonished; the horse rushed forward, leaving him on the ground, from whence he never rose in life, for his neck was broken by the fall, and with a convulsive shiver or two he expired.

Then did the prediction of the judge, the warning voice and summons of the preceding night, and the spot and the time, rush upon their recollection; and they firmly believed that a murderer and robber lay dead beside them. 'His horse saw something,' said Hope to Hunter; 'I never saw such flashing eyes in a horse's head;'—'and *he* saw something too,' replied Hunter, 'for the glance that he gave to the bloody spot, when his horse started, was one of terror. I never saw such a look, and I wish never to see such another again.'

When John Macmillan perished, matters stood thus with his memory. It was not only loaded with the sin of blood and the sin of robbery, with the sin of making a faithful woman a widow and her children fatherless, but with the grievous sin also of having driven a worthy family to ruin and beggary. The sum which was lost was large, the creditors were merciless; they fell upon the remaining substance of Johnstone, sweeping it wholly away; and his widow sought shelter in a miserable cottage among the Dryfesdale hills, where she supported her children by gathering and spinning wool. In a far different state and condition remained the family of John Macmillan. He died rich and unincumbered, leaving an evil name and an only child, a daughter, wedded to one whom many knew and esteemed, Joseph Howatson by name, a man sober and sedate; a member, too, of our own broken remnant of Cameronians.

Now, my dear children, the person who addresses you was then, as he is yet, God's preacher for the scattered kirk of Scotland, and his tent was pitched among the green hills of Annandale. The death of the transgressor appeared unto me the manifest judgment of God, and when my people gathered around me I rejoiced to see so great a multitude, and, standing in the midst of them, I preached in such a wise that they were deeply moved. I took for my text these words, 'Hath there been evil in the land and the Lord hath not known it?' I discoursed on the wisdom of Providence in guiding the affairs of men. How he permitted our evil passions to acquire the mastery over us, and urge us to deeds of darkness; allowing us to flourish for a season, that he might strike us in the midst of our splendour in a way so visible and awful that the wildest would cry out, 'Behold the finger of God.' I argued the matter home to the heart; I named no names, but I saw Joseph Howatson hide his face in his hands, for he felt and saw from the eyes which were turned towards him that I alluded to the judgment of God upon his relative.

Joseph Howatson went home heavy and sad of heart, and somewhat touched with anger at God's servant for having so pointedly and publicly alluded to his family misfortune; for he believed his father-in-law was a wise and a worthy man. His way home lay along the banks of a winding and beautiful stream, and just where it entered his own lands there was a rustic gate, over which he leaned for a little space, ruminating upon earlier days, on his wedded wife, on his children, and finally his thoughts settled on his father-in-law. He thought of his kindness to himself and to many others, on his fulfilment of all domestic duties, on his constant performance of family worship, and on his general reputation for honesty and fair dealing. He then dwelt on the circumstances of Johnstone's disappearance, on the singular summons his father-in-law received in Longtown, and the catastrophe which followed on the spot and on the very day of the year that the murder was supposed to be committed. He was in sore perplexity, and said aloud, 'Would to God that I knew the truth; but the doors of eternity, alas! are shut on the secret for ever.' He looked up and John Macmillan stood before him—stood with all the calmness and serenity and meditative air which a grave man wears when he walks out on a sabbath eve.

'Joseph Howatson,' said the apparition, 'on no secret are the doors of eternity shut—of whom were you speaking?' 'I was speaking', answered he, 'of one who is cold and dead, and to whom you bear a strong resemblance.' 'I am he,' said the shape; 'I am John Macmillan.' 'God of heaven!' replied Joseph Howatson, 'how can that be; did I not lay his head in the grave; see it closed over him; how, therefore, can it be? Heaven permits no such visitations.' 'I entreat you, my son,' said the shape, 'to believe what I say; the end of man is not when his body goes to dust; he exists in another state, and from that state am I permitted to come to you; waste not time, which is brief, with vain doubts, I am John Macmillan.' 'Father, father,' said the young man, deeply agitated, 'answer me, did you kill and rob Walter Johnstone?' 'I did,' said the Spirit, 'and for that have I returned to earth; listen to me.' The young man was so much overpowered by a revelation thus fearfully made, that he fell insensible on the ground; and when he recovered, the moon was shining, the dews of night were upon him, and he was alone.

Joseph Howatson imagined that he had dreamed a fearful dream; and conceiving that Divine Providence had presented the truth to his fancy, he began to consider how he could secretly make reparation to the wife and children of Johnstone for the double crime of his relative. But on more mature reflection he was impressed with the belief that a spirit had appeared to him, the spirit of his father-in-law, and that his own alarm had hindered him from learning fully the secret of his visit to earth; he therefore resolved to go to the same place next sabbath night, seek rather than avoid an interview, acquaint himself with the state of bliss or woe in which the spirit was placed, and learn if by acts of affection and restitution he could soften his sufferings or augment his happiness. He went accordingly to the little rustic gate by the side of the lonely stream; he walked up and down; hour passed after hour, but he heard nothing and saw nothing save the murmuring of the brook and the hares running among the wild clover. He had resolved to return home, when something seemed to rise from the ground, as shapeless as a cloud at first, but moving with life. It assumed a form, and the appearance of John Macmillan was once more before him. The young man was nothing daunted, but looking on the spirit, said, 'I thought you just and upright and devout, and incapable of murder and robbery.' The spirit seemed to dilate as it made answer. 'The death of Walter Johnstone sits lightly upon me. We had crossed each other's purposes, we had lessened each other's gains, we had vowed revenge, we met on fair terms, tied our horses to a gate, and fought fairly and long; and when I slew him, I but did what he sought to do to me. I threw him over his horse, carried him far into the country, sought out a deep quagmire on the north side of the Snipe Knowe, in Crake's Moss, and having secured his bills and other perishable property, with the purpose of returning all to his family, I buried him in the moss, leaving his gold in his purse, and laying his cloak and his sword above him.

'Now listen, Joseph Howatson. In my private desk you will find a little key tied with red twine, take it and go to the house of Janet Mathieson in Dumfries, and underneath the hearthstone in my sleeping room you will get my strong-box, open it, it contains all the bills and bonds belonging to Walter Johnstone. Restore them to his widow. I would have restored them but for my untimely death. Inform her privily and covertly where she will find the body of her husband, so that she may bury him in the churchyard with his ancestors. Do these things, that I may have some assuagement of misery; neglect them, and you will become a world's wonder.' The spirit vanished with these words, and was seen no more.

Joseph Howatson was sorely troubled. He had communed with a spirit, he was impressed with the belief that early death awaited him; he felt a sinking of soul and a misery of body, and he sent for me to help him with counsel, and comfort him in his unexampled sorrow. I loved him and hastened to him; I found him weak and woe-begone, and the hand of God seemed to be sore upon him. He took me out to the banks of the little stream where the shape appeared to him, and having desired me to listen without interrupting him, told me how he had seen his father-in-law's spirit, and related the revelations which it had made and the commands it had laid upon him. 'And now,' he said, 'look upon me. I am young, and ten days ago I had a body strong and a mind buoyant, and gray hairs and the honours of old age seemed to await me. But ere three days pass I shall be as the clod of the valley, for he who converses with a spirit, a spirit shall he soon become. I have written down the strange tale I have told you and I put it into your hands, perform for me and for my wretched parent, the instructions which the grave yielded up its tenant to give; and may your days be long in the land, and may you grow gray-headed among your people.' I listened to his words with wonder and with awe, and I promised to obey him in all his wishes with my best and most anxious judgment. We went home together; we spent the evening in prayer. Then he set his house in order, spoke to all his children cheerfully and with a mild voice, and falling on the neck of his wife, said, 'Sarah Macmillan, you were the choice of my young heart, and you have been a wife to me kind, tender, and gentle.' He looked at his children and he looked at his wife, for his heart was too full for more words, and retired to his chamber. He was found next morning kneeling by his bedside, his hands held out as if repelling some approaching object, horror stamped on every feature, and cold and dead.

Then I felt full assurance of the truth of his communications; and as soon as the amazement which his untimely death occasioned had subsided, and his wife and little ones were somewhat comforted, I proceeded to fulfil his dying request. I found the small key tied with red twine, and I went to the house of Janet Mathieson in Dumfries, and I held up the key and said, 'Woman, knowest thou that?' and when she saw it she said, 'Full well I know it, it belonged to a jolly man and a douce, and mony a merry hour has he whiled away wi' my servant maidens and me.' And when she saw me lift the hearthstone, open the box, and spread out the treasure which it contained, she held up her hands, 'Eh! what o' gowd! what o' gowd! but half's mine, be ye saint or sinner; John Macmillan, douce man, aye said he had something there which he considered as not belonging to him but to a quiet friend; weel I wot he meant me, for I have been a quiet friend to him and his.' I told her I was commissioned by his daughter to remove the property, that I was the minister of that persecuted remnant of the true kirk called Cameronians, and she might therefore deliver it up without fear. 'I ken weel enough wha ye are,' said this worthless woman, 'd'ye think I dinna ken a minister of the kirk; I have seen meikle o' their siller in my day, frae eighteen to fifty and aught have I caroused with divines, Cameronians, I trow, as well as those of a freer kirk. But touching this treasure, give me twenty gowden pieces, else I'se gar three stamps of my foot bring in them that will see me righted, and send you awa to the mountains bleating like a sheep shorn in winter.' I gave the imperious woman twenty pieces of gold, and carried away the fatal box.

Now, when I got free of the ports of Dumfries, I mounted my little horse and rode away into the heart of the country, among the pastoral hills of Dryfesdale. I carried the box on the saddle before me, and its contents awakened a train of melancholy thoughts within me. There were the papers of Walter Johnstone, corresponding to the description which the spirit gave, and marked with his initials in red ink by the hand of the man who slew him. There were two gold watches and two purses of gold, all tied with red twine, and many bills and much money to which no marks were attached. As I rode along pondering on these things, and casting about in my own mind how and by what means I should make restitution, I was aware of a morass, broad and wide, which with all its quagmires glittered in the moonlight before me. I knew I had penetrated into the centre of Dryfesdale, but I was not well acquainted with the country; I therefore drew my bridle, and looked around to see if any house was nigh, where I could find shelter for the night. I saw a small house built of turf and thatched with heather, from the window of which a faint light glimmered. I rode up, alighted, and there I found a woman in widow's weeds, with three sweet children, spinning yarn from the wool which the shepherds shear in spring from the udders of the ewes. She welcomed me, spread bread and placed milk before me. I asked a blessing, and ate and drank, and was refreshed.

Now it happened that, as I sat with the solitary woman and her children, there came a man to the door, and with a loud yell of dismay burst it open and staggered forward crying, 'There's a corse candle in Crake's Moss, and I'll be a dead man before the morning.' 'Preserve me! piper,' said the widow, 'ye're in a piteous taking; here is a holy man who will speak comfort to you, and tell you how all these are but delusions of the eye or exhalations of nature.' 'Delusions and exhalations, Dame Johnstone,' said the piper, 'd'ye think I dinna ken a corse light from an elf candle, an elf candle from a will-o'-wisp, and a will-o'-wisp from all other lights of this wide world.' The name of the morass and the woman's name now flashed upon me, and I was struck with amazement and awe. I looked on the widow, and I looked on the wandering piper, and I said, 'Let me look on those corse lights, for God creates nothing in vain; there is a wise purpose in all things, and a wise aim.' And the piper said, 'Na, na; I have nae wish to see ony mair on't, a dead light bodes the living nae gude; and I am sure if I gang near Crake's Moss it will lair me amang the hags and quags.' And I said, 'Foolish old man, you are equally safe every where; the hand of the Lord reaches round the earth, and strikes and protects according as it was foreordained, for nothing is hid from his eyes—come with me.' And the piper looked strangely upon me and stirred not a foot; and I said, 'I shall go by myself;' and the woman said, 'Let me go with you, for I am sad of heart, and can look on such things without fear; for, alas! since I lost my own Walter Johnstone, pleasure is no longer pleasant: and I love to wander in lonesome places and by old churchyards.' 'Then,' said the piper, 'I darena bide my lane with the bairns; I'll go also; but O! let me strengthen my heart with ae spring on my pipes before I venture.' 'Play,' I said, 'Clavers and his Highlandmen, it is the tune to cheer ye and keep your heart up.' 'Your honour's no cannie,' said the old man; 'that's my favourite tune.' So he played it and said, 'Now I am fit to look on lights of good or evil.' And we walked into the open air.

All Crake's Moss seemed on fire; not illumined with one steady and uninterrupted light, but kindled up by fits like the northern sky with its wandering streamers. On a little bank which rose in the centre of the morass, the supernatural splendour seemed chiefly to settle; and having continued to shine for several minutes, the whole faded and left but one faint gleam behind. I fell on my knees, held up my hands to heaven, and said, 'This is of God; behold in that fearful light the finger of the Most High. Blood has been spilt, and can be no longer concealed; the point of the mariner's needle points less surely to the north than yon living flame points to the place where man's body has found a bloody grave. Follow me,' and I walked down to the edge of the moss and gazed earnestly on the spot. I knew now that I looked on the long hidden resting place of Walter Johnstone, and considered that the hand of God was manifest in the way that I had been thus led blindfold into his widow's house. I reflected for a moment on these things; I wished to right the fatherless, yet spare the feelings of the innocent; the supernatural light partly showed me the way, and the words which I now heard whispered by my companions aided in directing the rest.

'I tell ye, Dame Johnstone,' said the piper, 'the man's no cannie; or what's waur, he may belong to the spiritual world himself, and do us a mischief. Saw ye ever mortal man riding with ae spur and carrying a silver-headed cane for a whip, wi' sic a fleece of hair about his haffets and sic a wild ee in his head; and then he kens a' things in the heavens aboon and the earth beneath. He kenned my favourite tune Clavers; I'se uphaud he's no in the body, but ane of the souls made perfect of the auld Covenanters whom Grahame or Grierson slew; we're daft to follow him.' 'Fool body,' I heard the widow say, 'I'll follow him; there's something about that man, be he in the spirit or in the flesh, which is pleasant and promising. O! could he but, by prayer or other means of lawful knowledge, tell me about my dear Walter Johnstone; thrice has he appeared to me in dream or vision with a sorrowful look, and weel ken I what that means.' We had now reached the edge of the morass, and a dim and uncertain light continued to twinkle about the green knoll which rose in its middle. I turned suddenly round and said, 'For a wise purpose am I come; to reveal murder; to speak consolation to the widow and the fatherless, and to soothe the perturbed spirits of those whose fierce passions ended in untimely death. Come with me; the hour is come, and I must not do my commission negligently.' 'I kenned it, I kenned it,' said the piper, 'he's just one of the auld persecuted worthies risen from his red grave to right the injured, and he'll do't discreetly; follow him, Dame, follow him.' 'I shall follow,' said the widow, 'I have that strength given me this night which will bear me through all trials which mortal flesh can endure.'

When we reached the little green hillock in the centre of the morass, I looked to the north and soon distinguished the place described by my friend Joseph Howatson, where the body of Walter Johnstone was deposited. The moon shone clear, the stars aided us with their light, and some turfcutters having left their spades standing near, I ordered the piper to take a spade and dig where I placed my staff. 'O dig carefully,' said the widow, 'do not be rude with mortal dust.' We dug and came to a sword; the point was broken and the blade hacked. 'It is the sword of my Walter Johnstone,' said his widow, 'I could swear to it among a thousand.' 'It is my father's sword,' said a fine dark haired boy who had followed us unperceived, 'it is my father's sword, and were he living who wrought this, he should na be lang in rueing it.' 'He is dead, my child,' I said, 'and beyond your reach, and vengeance is the Lord's.' 'O, Sir,' cried his widow, in a flood of tears, 'ye ken all things; tell me, is this my husband or no?' 'It is the body of Walter Johnstone,' I answered, 'slain by one who is passed to his account, and buried here by the hand that slew him, with his gold in his purse and his watch in his pocket.' So saying we uncovered the body, lifted it up, laid it on the grass; the embalming nature of the morass had preserved it from decay, and mother and child, with tears and with cries, named his name and lamented over him. His gold watch and his money, his cloak and his dress, were untouched and entire, and we bore him to the cottage of his widow, where with clasped hands she sat at his feet and his children at his head till the day drew nigh the dawn; I then rose and said, 'Woman, thy trials have been severe and manifold; a good wife, a good mother, and a good widow hast thou been, and thy reward will be where the blessed alone are admitted. It was revealed to me by a mysterious revelation that thy husband's body was where we found it; and I was commissioned by a voice, assuredly not of this world, to deliver thee this treasure, which is thy own, that thy children may be educated, and that bread and raiment may be thine.' And I delivered her husband's wealth into her hands, refused gold which she offered, and mounting my horse, rode over the hills and saw her no more. But I soon heard of her, for there rose a strange sound in the land, that a Good Spirit had appeared to the widow of Walter Johnstone, had disclosed where her husband's murdered body lay, had enriched her with all his lost wealth, had prayed by her side till the blessed dawn of day, and then vanished with the morning light. I closed my lips on the secret till now; and I reveal it to you, my children, that you may know there is a God who ruleth this world by wise and invisible means, and punisheth the wicked, and cheereth the humble of heart and the lowly minded.

Such was the last sermon of the good John Farley, a man whom I knew and loved. I think I see him now, with his long white hair and his look mild, eloquent, and sagacious. He was a giver of good counsel, a sayer of wise sayings, with wit at will, learning in abundance, and a gift in sarcasm which the wildest dreaded.

Sir Walter Scott

1771–1832

The Two Drovers

Chapter I

It was the day after Doune Fair when my story commences. It had been a brisk market, several dealers had attended from the northern and midland counties in England, and English money had flown so merrily about as to gladden the hearts of the Highland farmers. Many large droves were about to set off for England, under the protection of their owners, or of the topsmen whom they employed in the tedious, laborious, and responsible office of driving the cattle for many hundred miles, from the market where they had been purchased, to the fields or farm-yards where they were to be fattened for the shambles.

The Highlanders in particular are masters of this difficult trade of driving, which seems to suit them as well as the trade of war. It affords exercise for all their habits of patient endurance and active exertion. They are required to know perfectly the drove-roads, which lie over the wildest tracts of the country, and to avoid as much as possible the highways, which distress the feet of the bullocks, and the turnpikes, which annoy the spirit of the drover; whereas on the broad green or grey track, which leads across the pathless moor, the herd not only move at ease and without taxation, but, if they mind their business, may pick up a mouthful of food by the way. At night, the drovers usually sleep along with their cattle, let the weather be what it will; and many of these hardy men do not once rest under a roof during a journey on foot from Lochaber to Lincolnshire. They are paid very highly, for the trust reposed is of the last importance, as it depends on their prudence, vigilance and honesty, whether the cattle reach the final market in good order, and afford a profit to the grazier. But as they maintain themselves at their own expense, they are especially economical in that particular. At the period we speak of, a Highland drover was victualled for his long and toilsome journey with a few handfuls of oatmeal and two or three onions, renewed from time to time, and a ram's horn filled with whisky, which he used regularly, but sparingly, every night and morning. His dirk, or *skene-dhu* (*i.e.* black-knife), so worn as to be concealed beneath the arm, or by the folds of the plaid, was his only weapon, excepting the cudgel with which he directed the movements of the cattle. A Highlander was never so happy as on these occasions. There was a variety in the whole journey, which exercised the

Celt's natural curiosity and love of motion; there were the constant change of place and scene, the petty adventures incidental to the traffic, and the intercourse with the various farmers, graziers, and traders, intermingled with occasional merry-makings, not the less acceptable to Donald that they were void of expense;—and there was the consciousness of superior skill; for the Highlander, a child amongst flocks, is a prince amongst herds, and his natural habits induce him to disdain the shepherd's slothful life, so that he feels himself nowhere more at home than when following a gallant drove of his country cattle in the character of their guardian.

Of the number who left Doune in the morning, and with the purpose we have described, not a *Glunamie* of them all cocked his bonnet more briskly, or gartered his tartan hose under knee over a pair of more promising *spiogs* (legs), than did Robin Oig M'Combich, called familiarly Robin Oig, that is young, or the Lesser, Robin. Though small of stature, as the epithet Oig implies, and not very strongly limbed, he was as light and alert as one of the deer of his mountains. He had an elasticity of step, which, in the course of a long march, made many a stout fellow envy him; and the manner in which he busked his plaid and adjusted his bonnet, argued a consciousness that so smart a John Highlandman as himself would not pass unnoticed among the Lowland lasses. The ruddy cheek, red lips, and white teeth, set off a countenance, which had gained by exposure to the weather a healthful and hardy rather than a rugged hue. If Robin Oig did not laugh, or even smile frequently, as indeed is not the practice among his countrymen, his bright eyes usually gleamed from under his bonnet with an expression of cheerfulness ready to be turned into mirth.

The departure of Robin Oig was an incident in the little town, in and near which he had many friends, male and female. He was a topping person in his way, transacted considerable business on his own behalf, and was intrusted by the best farmers in the Highlands, in preference to any other drover in that district. He might have increased his business to any extent had he condescended to manage it by deputy; but except a lad or two, sister's sons of his own, Robin rejected the idea of assistance, conscious, perhaps, how much his reputation depended upon his attending in person to the practical discharge of his duty in every instance. He remained, therefore, contented with the highest premium given to persons of his description, and comforted himself with the hopes that a few journeys to England might enable him to conduct business on his own account, in a manner becoming his birth. For Robin Oig's father, Lachlan M'Combich (or *son of my friend*, his actual clan-surname being M'Gregor), had been so called by the celebrated Rob Roy, because of the particular friendship which had subsisted between the grandsire of Robin, and that renowned cateran. Some people even say, that Robin Oig derived his Christian name from one as renowned in the wilds of Lochlomond as ever was his namesake Robin Hood, in the precincts of merry Sherwood. 'Of such ancestry,' as James Boswell says, 'who would not be proud?'

Robin Oig was proud accordingly; but his frequent visits to England and to the Lowlands had given him tact enough to know that pretensions, which still gave him a little right to distinction in his own lonely glen, might be both obnoxious and ridiculous if preferred elsewhere. The pride of birth, therefore, was like the miser's treasure, the secret subject of his contemplation, but never exhibited to strangers as a subject of boasting.

Many were the words of gratulation and good-luck which were bestowed on Robin Oig. The judges commended his drove, especially Robin's own property, which were the best of them. Some thrust out their snuff-mulls for the parting pinch—others tendered the *doch-an-dorrach*, or parting cup. All cried—'Good-luck travel out with you and come home with you.—Give you luck in the Saxon market—brave notes in the *leabhar-dhu*' (black pocketbook), 'and plenty of English gold in the *sporran*' (pouch of goat-skin).

The bonny lasses made their adieus more modestly, and more than one, it was said, would have given her best brooch to be certain that it was upon her that his eye last rested as he turned towards the road.

Robin Oig had just given the preliminary '*Hoo-hoo*!' to urge forward the loiterers of the drove, when there was a cry behind him.

'Stay, Robin—bide a blink. Here is Janet of Tomahourich—auld Janet, your father's sister.'

'Plague on her, for an auld Highland witch and spaewife,' said a farmer from the Carse of Stirling; 'she'll cast some of her cantrips on the cattle.'

'She canna do that,' said another sapient of the same profession—'Robin Oig is no the lad to leave any of them, without tying Saint Mungo's knot on their tails, and that will put to her speed the best witch that ever flew over Dimayet upon a broomstick.'

It may not be indifferent to the reader to know that the Highland cattle are peculiarly liable to be *taken*, or infected, by spells and witchcraft, which judicious people guard against by knitting knots of peculiar complexity on the tuft of hair which terminates the animal's tail.

But the old woman who was the object of the farmer's suspicion seemed only busied about the drover, without paying any attention to the drove. Robin, on the contrary, appeared rather impatient of her presence.

'What auld-world fancy', he said, 'has brought you so early from the ingle-side this morning, Muhme? I am sure I bid you good-even, and had your God-speed last night.'

'And left me more siller than the useless old woman will use till you come back again, bird of my bosom,' said the sibyl. 'But it is little I would care for the food that nourishes me, or the fire that warms me, or for God's blessed sun itself, if aught but weal should happen to the grandson of my father. So let me walk the *deasil* round you, that you may go safe out into the foreign land, and come safe home.'

Robin Oig stopped, half embarrassed, half laughing, and signing to those around that he only complied with the old woman to soothe her humour. In the meantime, she traced around him, with wavering steps, the propitiation which some have thought has been derived from the Druidical mythology. It consists, as is well known, in the person who makes the *deasil* walking three times round the person who is the object of the ceremony, taking care to move according to the course of the sun. At once, however, she stopped short, and exclaimed, in a voice of alarm and horror, 'Grandson of my father, there is blood on your hand.'

'Hush, for God's sake, aunt,' said Robin Oig; 'you will bring more trouble on yourself with this Taishataragh' (second sight) 'than you will be able to get out of for many a day.'

The old woman only repeated, with a ghastly look, 'There is blood on your hand, and it is English blood. The blood of the Gael is richer and redder. Let us see—let us—'

Ere Robin Oig could prevent her, which, indeed, could only have been by positive violence, so hasty and peremptory were her proceedings, she had drawn from his side the dirk which lodged in the folds of his plaid, and held it up, exclaiming, although the weapon gleamed clear and bright in the sun, 'Blood, blood—Saxon blood again. Robin Oig M'Combich, go not this day to England!'

'Prutt, trutt,' answered Robin Oig, 'that will never do neither—it would be next thing to running the country. For shame, Muhme—give me the dirk. You cannot tell by the colour the difference betwixt the blood of a black bullock and a white one, and you speak of knowing Saxon from Gaelic blood. All men have their blood from Adam, Muhme. Give me my skene-dhu, and let me go on my road. I should have been half way to Stirling brig by this time—Give me my dirk, and let me go.'

'Never will I give it to you,' said the old woman—'Never will I quit my hold on your plaid, unless you promise me not to wear that unhappy weapon.'

The women around him urged him also, saying few of his aunt's words fell to the ground; and as the Lowland farmers continued to look moodily on the scene, Robin Oig determined to close it at any sacrifice.

'Well, then,' said the young drover, giving the scabbard of the weapon to Hugh Morrison, 'you Lowlanders care nothing for these freats. Keep my dirk for me. I cannot give it you, because it was my father's; but your drove follows ours, and I am content it should be in your keeping, not in mine.—Will this do, Muhme?'

'It must,' said the old woman—'that is, if the Lowlander is mad enough to carry the knife.'

The strong westlandman laughed aloud.

'Goodwife,' said he, 'I am Hugh Morrison from Glenae, come of the Manly Morrisons of auld langsyne, that never took short weapon against a man in their lives. And neither needed they: They had their broadswords, and I have this bit supple,' showing a formidable cudgel—'for dirking ower the board, I leave that to John Highlandman.—Ye needna snort, none of you Highlanders, and you in especial, Robin. I'll keep the bit knife, if you are feared for the auld spaewife's tale, and give it back to you whenever you want it.'

Robin was not particularly pleased with some part of Hugh Morrison's speech; but he had learned in his travels more patience than belonged to his Highland constitution originally, and he accepted the service of the descendant of the Manly Morrisons, without finding fault with the rather depreciating manner in which it was offered.

'If he had not had his morning in his head, and been but a Dumfries-shire hog into the boot, he would have spoken more like a gentleman. But you cannot have more of a sow than a grumph. It's shame my father's knife should ever slash a haggis for the like of him.'

Thus saying (but saying it in Gaelic), Robin drove on his cattle, and waved farewell to all behind him. He was in the greater haste, because he expected to join at Falkirk a comrade and brother in profession, with whom he proposed to travel in company.

Robin Oig's chosen friend was a young Englishman, Harry Wakefield by name, well known at every northern market, and in his way as much famed and honoured as our Highland driver of bullocks. He was nearly six feet high, gallantly formed to keep the rounds at Smithfield, or maintain the ring at a wrestling match; and although he might have been overmatched, perhaps, among the regular professors of the Fancy, yet, as a yokel or rustic, or a chance customer, he was able to give a bellyful to any amateur of the pugilistic art. Doncaster races saw him in his glory, betting his guinea, and generally successfully; nor was there a main fought in Yorkshire, the feeders being persons of celebrity, at which he was not to be seen, if business permitted. But though a *sprack* lad, and fond of pleasure and its haunts, Harry Wakefield was steady, and not the cautious Robin Oig M'Combich himself was more attentive to the main chance. His holidays were holidays indeed; but his days of work were dedicated to steady and persevering labour. In countenance and temper, Wakefield was the model of Old England's merry yeomen, whose clothyard shafts, in so many hundred battles, asserted her superiority over the nations, and whose good sabres, in our own time, are her cheapest and most assured defence. His mirth was readily excited; for, strong in limb and constitution, and fortunate in circumstances, he was disposed to be pleased with every thing about him; and such difficulties as he might occasionally encounter, were, to a man of his energy, rather matter of amusement than serious annoyance. With all the merits of a sanguine temper, our young English drover was not without his defects. He was irascible, sometimes to the verge of being quarrelsome; and perhaps not the less inclined to bring his disputes to a pugilistic decision, because he found few antagonists able to stand up to him in the boxing ring.

It is difficult to say how Harry Wakefield and Robin Oig first became intimates; but it is certain a close acquaintance had taken place betwixt them, although they had apparently few common subjects of conversation or of interest, so soon as their talk ceased to be of bullocks. Robin Oig, indeed, spoke the English language rather imperfectly upon any other topics but stots and kyloes, and Harry Wakefield could never bring his broad Yorkshire tongue to utter a single word of Gaelic. It was in vain Robin spent a whole morning, during a walk over Minch Moor, in attempting to teach his companion to utter, with true precision, the shibboleth *Llhu*, which is the Gaelic for a calf. From Traquair to Murdercairn, the hill rung with the discordant attempts of the Saxon upon the unmanageable monosyllable, and the heartfelt laugh which followed every failure. They had, however, better modes of awakening the echoes; for Wakefield could sing many a ditty to the praise of Moll, Susan, and Cicely, and Robin Oig had a particular gift at whistling interminable pibrochs through all their involutions, and what was more agreeable to his companion's southern ear, knew many of the northern airs, both lively and pathetic, to which Wakefield learned to pipe a bass. Thus, though Robin could hardly have comprehended his companion's stories about horse-racing, and cock-fighting, or fox-hunting, and although his own legends of clan-fights and *creaghs*, varied with talk of Highland goblins and fairy folk, would have been caviare to his companion, they contrived nevertheless to find a degree of pleasure in each other's company, which had for three years back induced them to join company and travel together, when the direction of their journey permitted. Each, indeed, found his advantage in this companionship; for where could the Englishman have found a guide through the Western Highlands like Robin Oig M'Combich? and when they were on what Harry called the *right* side of the Border, his patronage, which was extensive, and his purse, which was heavy, were at all times at the service of his Highland friend, and on many occasions his liberality did him genuine yeoman's service.

CHAPTER II

Were ever two such loving friends!—
  How could they disagree?
O thus it was, he loved him dear,
  And thought how to requite him,
And having no friend left but he,
  He did resolve to fight him.

Duke upon Duke

The pair of friends had traversed with their usual cordiality the grassy wilds of Liddesdale, and crossed the opposite part of Cumberland, emphatically called The Waste. In these solitary regions, the cattle under the charge of our drovers derived their subsistence chiefly by picking their food as they went along the drove-road, or sometimes by the tempting opportunity of a *start and owerloup*, or invasion of the neighbouring pasture, where an occasion presented itself. But now the scene changed before them; they were descending towards a fertile and enclosed country, where no such liberties could be taken with impunity, or without a previous arrangement and bargain with the possessors of the ground. This was more especially the case, as a great northern fair was upon the eve of taking place, where both the Scotch and English drover expected to dispose of a part of their cattle, which it was desirable to produce in the market, rested and in good order. Fields were therefore difficult to be obtained, and only upon high terms. This necessity occasioned a temporary separation betwixt the two friends, who went to bargain, each as he could, for the separate accommodation of his herd. Unhappily it chanced that both of them, unknown to each other, thought of bargaining for the ground they wanted on the property of a country gentleman of some fortune, whose estate lay in the neighbourhood. The English drover applied to the bailiff on the property, who was known to him. It chanced that the Cumbrian Squire, who had entertained some suspicions of his manager's honesty, was taking occasional measures to ascertain how far they were well founded, and had desired that any enquiries about his enclosures, with a view to occupy them for a temporary purpose, should be referred to himself. As, however, Mr Ireby had gone the day before upon a journey of some miles' distance to the northward, the bailiff chose to consider the check upon his full powers as for the time removed, and concluded that he should best consult his master's interest, and perhaps his own, in making an agreement with Harry Wakefield. Meanwhile, ignorant of what his comrade was doing, Robin Oig, on his side, chanced to be overtaken by a good-looking smart little man upon a pony, most knowingly hogged and cropped, as was then the fashion, the rider wearing tight leather breeches, and long-necked bright spurs. This cavalier asked one or two pertinent questions about markets and the price of stock. So Robin, seeing him a well-judging civil gentleman, took the freedom to ask him whether he could let him know if there was any grassland to be let in that neighbourhood, for the temporary accommodation of his drove. He could not have put the question to more willing ears. The gentleman of the buck-skins was the proprietor, with whose bailiff Harry Wakefield had dealt, or was in the act of dealing.

'Thou art in good luck, my canny Scot,' said Mr Ireby, 'to have spoken to me, for I see thy cattle have done their day's work, and I have at my disposal the only field within three miles that is to be let in these parts.'

'The drove can pe gang two, three, four miles very pratty weel indeed'—said the cautious Highlander; 'put what would his honour pe axing for the peasts pe the head, if she was to tak the park for twa or three days?'

'We won't differ, Sawney, if you let me have six stots for winterers, in the way of reason.'

'And which peasts wad your honour pe for having?'

'Why—let me see—the two black—the dun one—yon doddy—him with the twisted horn—the brockit—How much by the head?'

'Ah,' said Robin, 'your honour is a shudge—a real shudge—I couldna have set off the pest six peasts petter mysell, me that ken them as if they were my pairns, puir things.'

'Well, how much per head, Sawney,' continued Mr Ireby.

'It was high markets at Doune and Falkirk,' answered Robin.

And thus the conversation proceeded, until they had agreed on the *prix juste* for the bullocks, the Squire throwing in the temporary accommodation of the enclosure for the cattle into the boot, and Robin making, as he thought, a very good bargain, provided the grass was but tolerable. The Squire walked his pony alongside of the drove, partly to show him the way, and see him put into possession of the field, and partly to learn the latest news of the northern markets.

They arrived at the field, and the pasture seemed excellent. But what was their surprise when they saw the bailiff quietly inducting the cattle of Harry Wakefield into the grassy Goshen which had just been assigned to those of Robin Oig M'Combich by the proprietor himself! Squire Ireby set spurs to his horse, dashed up to his servant, and learning what had passed between the parties, briefly informed the English drover that his bailiff had let the ground without his authority, and that he might seek grass for his cattle wherever he would, since he was to get none there. At the same time he rebuked his servant severely for having transgressed his commands, and ordered him instantly to assist in ejecting the hungry and weary cattle of Harry Wakefield, which were just beginning to enjoy a meal of unusual plenty, and to introduce those of his comrade, whom the English drover now began to consider as a rival.

The feelings which arose in Wakefield's mind would have induced him to resist Mr Ireby's decision; but every Englishman has a tolerably accurate sense of law and justice, and John Fleecebumpkin, the bailiff, having acknowledged that he had exceeded his commission, Wakefield saw nothing else for it than to collect his hungry and disappointed charge, and drive them on to seek quarters elsewhere. Robin Oig saw what had happened with regret, and hastened to offer to his English friend to share with him the disputed possession. But Wakefield's pride was severely hurt, and he answered disdainfully, 'Take it all, man—take it all—never make two bites of a cherry—thou canst talk over the gentry, and blear a plain man's eye—Out upon you, man—I would not kiss any man's dirty latchets for leave to bake in his oven.'

Robin Oig, sorry but not surprised at his comrade's displeasure, hastened to entreat his friend to wait but an hour till he had gone to the Squire's house to receive payment for the cattle he had sold, and he would come back and help him to drive the cattle into some convenient place of rest, and explain to him the whole mistake they had both of them fallen into. But the Englishman continued indignant: 'Thou hast been selling, hast thou? Ay, ay—thou is a cunning lad for kenning the hours of bargaining. Go to the devil with thyself, for I will ne'er see thy fause loon's visage again—thou should be ashamed to look me in the face.'

'I am ashamed to look no man in the face,' said Robin Oig, something moved; 'and, moreover, I will look you in the face this blessed day, if you will bide at the Clachan down yonder.'

'Mayhap you had as well keep away,' said his comrade; and turning his back on his former friend, he collected his unwilling associates, assisted by the bailiff, who took some real and some affected interest in seeing Wakefield accommodated.

After spending some time in negotiating with more than one of the neighbouring farmers, who could not, or would not, afford the accommodation desired, Henry Wakefield at last, and in his necessity, accomplished his point by means of the landlord of the alehouse at which Robin Oig and he had agreed to pass the night, when they first separated from each other. Mine host was content to let him turn his cattle on a piece of barren moor, at a price little less than the bailiff had asked for the disputed enclosure; and the wretchedness of the pasture, as well as the price paid for it, were set down as exaggerations of the breach of faith and friendship of his Scottish crony. This turn of Wakefield's passions was encouraged by the bailiff (who had his own reasons for being offended against poor Robin, as having been the unwitting cause of his falling into disgrace with his master), as well as by the innkeeper, and two or three chance guests, who stimulated the drover in his resentment against his quondam associate—some from the ancient grudge against the Scots, which, when it exists anywhere, is to be found lurking in the Border counties, and some from the general love of mischief, which characterises mankind in all ranks of life, to the honour of Adam's children be it spoken. Good John Barleycorn also, who always heightens and exaggerates the prevailing passions, be they angry or kindly, was not wanting in his offices on this occasion; and confusion to false friends and hard masters was pledged in more than one tankard.

In the meanwhile Mr Ireby found some amusement in detaining the northern drover at his ancient hall. He caused a cold round of beef to be placed before the Scot in the butler's pantry, together with a foaming tankard of home-brewed, and took pleasure in seeing the hearty appetite with which these unwonted edibles were discussed by Robin Oig M'Combich. The Squire himself lighting his pipe, compounded between his patrician dignity and his love of agricultural gossip, by walking up and down while he conversed with his guest.

'I passed another drove,' said the Squire, 'with one of your countrymen behind them—they were something less beasts than your drove, doddies most of them—a big man was with them—none of your kilts though, but a decent pair of breeches—D'ye know who he may be?'

'Hout aye—that might, could, and would be Hughie Morrison—I didna think he could hae peen sae weel up. He has made a day on us; but his Argyleshires will have wearied shanks. How far was he pehind?'

'I think about six or seven miles,' answered the Squire, 'for I passed them at the Christenbury Crag, and I overtook you at the Hollan Bush. If his beasts be leg-weary, he will be maybe selling bargains.'

'Na, na, Hughie Morrison is no the man for pargains—ye maun come to some Highland body like Robin Oig hersell for the like of these—put I maun be wishing you goot night, and twenty of them let alane ane, and I maun down to the Clachan to see if the lad Harry Waakfelt is out of his humdudgeons yet.'

The party at the alehouse were still in full talk, and the treachery of Robin Oig still the theme of conversation, when the supposed culprit entered the apartment. His arrival, as usually happens in such a case, put an instant stop to the discussion of which he had furnished the subject, and he was received by the company assembled with that chilling silence, which, more than a thousand exclamations, tells an intruder that he is unwelcome. Surprised and offended, but not appalled by the reception which he experienced, Robin entered with an undaunted and even a haughty air, attempted no greeting, as he saw he was received with none, and placed himself by the side of the fire, a little apart from a table, at which Harry Wakefield, the bailiff, and two or three other persons, were seated. The ample Cumbrian kitchen would have afforded plenty of room, even for a larger separation.

Robin, thus seated, proceeded to light his pipe, and call for a pint of twopenny.

'We have no twopence ale,' answered Ralph Heskett the landlord; 'but as thou find'st thy own tobacco, it's like thou mayst find thy own liquor too—it's the wont of thy country, I wot.'

'Shame, goodman,' said the landlady, a blithe bustling housewife, hastening herself to supply the guest with liquor—'Thou knowest well enow what the strange man wants, and it's thy trade to be civil, man. Thou shouldst know, that if the Scot likes a small pot, he pays a sure penny.'

Without taking any notice of this nuptial dialogue, the Highlander took the flagon in his hand, and addressing the company generally, drank the interesting toast of 'Good markets,' to the party assembled.

'The better that the wind blew fewer dealers from the north,' said one of the farmers, 'and fewer Highland runts to eat up the English meadows.'

'Saul of my pody, put you are wrang there, my friend,' answered Robin, with composure; 'it is your fat Englishmen that eat up our Scots cattle, puir things.'

'I wish there was a summat to eat up their drovers,' said another; 'a plain Englishman canna make bread within a kenning of them.'

'Or an honest servant keep his master's favour, but they will come sliding in between him and the sunshine,' said the bailiff.

'If these pe jokes,' said Robin Oig, with the same composure, 'there is ower mony jokes upon one man.'

'It is no joke, but downright earnest,' said the bailiff. 'Harkye, Mr Robin Ogg, or whatever is your name, it's right we should tell you that we are all of one opinion, and that is, that you, Mr Robin Ogg, have behaved to our friend Mr Harry Wakefield here, like a raff and a blackguard.'

'Nae doubt, nae doubt,' answered Robin, with great composure; 'and you are a set of very pretty judges, for whose prains or pehaviour I wad not gie a pinch of sneeshing. If Mr Harry Waakfelt kens where he is wranged, he kens where he may be righted.'

'He speaks truth,' said Wakefield, who had listened to what passed, divided between the offence which he had taken at Robin's late behaviour, and the revival of his habitual feelings of regard.

He now rose, and went towards Robin, who got up from his seat as he approached, and held out his hand.

'That's right, Harry—go it—serve him out,' resounded on all sides—'tip him the nailer—show him the mill.'

'Hold your peace all of you, and be—,' said Wakefield; and then addressing his comrade, he took him by the extended hand, with something alike of respect and defiance. 'Robin,' he said, 'thou hast used me ill enough this day; but if you mean, like a frank fellow, to shake hands, and take a tussle for love on the sod, why I'll forgie thee, man, and we shall be better friends than ever.'

'And would it not pe petter to pe cood friends without more of the matter?' said Robin; 'we will be much petter friendships with our panes hale than proken.'

Harry Wakefield dropped the hand of his friend, or rather threw it from him.

'I did not think I had been keeping company for three years with a coward.'

'Coward pelongs to none of my name,' said Robin, whose eyes began to kindle, but keeping the command of his temper. 'It was no coward's legs or hands, Harry Waakfelt, that drew you out of the fords of Frew, when you was drifting ower the plack rock, and every eel in the river expected his share of you.'

'And that is true enough, too,' said the Englishman, struck by the appeal.

'Adzooks!' exclaimed the bailiff—'sure Harry Wakefield, the nattiest lad at Whitson Tryste, Wooler Fair, Carlisle Sands, or Stagshaw Bank, is not going to show white feather? Ah, this comes of living so long with kilts and bonnets—men forget the use of their daddles.'

'I may teach you, Master Fleecebumpkin, that I have not lost the use of mine,' said Wakefield, and then went on. 'This will never do, Robin. We must have a turn-up, or we shall be the talk of the country side. I'll be d——d if I hurt thee—I'll put on the gloves gin thou like. Come, stand forward like a man.'

'To be peaten like a dog,' said Robin; 'is there any reason in that? If you think I have done you wrong, I'll go before your shudge, though I neither know his law nor his language.'

A general cry of 'No, no—no law, no lawyer! a bellyful and be friends,' was echoed by the bystanders.

'But,' continued Robin, 'if I am to fight, I have no skill to fight like a jackanapes, with hands and nails.'

'How would you fight then?' said his antagonist; 'though I am thinking it would be hard to bring you to the scratch anyhow.'

'I would fight with proadswords, and sink point on the first plood drawn—like a gentlemans.'

A loud shout of laughter followed the proposal, which indeed had rather escaped from poor Robin's swelling heart, than been the dictate of his sober judgment.

'Gentleman, quotha!' was echoed on all sides, with a shout of unextinguishable laughter; 'a very pretty gentleman, God wot—Canst get two swords for the gentlemen to fight with, Ralph Heskett?'

'No, but I can send to the armoury at Carlisle, and lend them two forks, to be making shift with in the meantime.'

'Tush, man,' said another, 'the bonny Scots come into the world with the blue bonnet on their heads, and dirk and pistol at their belt.'

'Best send post,' said Mr Fleecebumpkin, 'to the Squire of Corby Castle, to come and stand second to the *gentleman*.'

In the midst of this torrent of general ridicule, the Highlander instinctively gripped beneath the folds of his plaid.

'But it's better not,' he said in his own language. 'A hundred curses on the swine-eaters, who know neither decency nor civility!'

'Make room, the pack of you,' he said, advancing to the door.

But his former friend interposed his sturdy bulk, and opposed his leaving the house; and when Robin Oig attempted to make his way by force, he hit him down on the floor, with as much ease as a boy bowls down a nine-pin.

'A ring, a ring!' was now shouted, until the dark rafters, and the hams that hung on them, trembled again, and the very platters on the *bink* clattered against each other. 'Well done, Harry'—'Give it him home, Harry'—'Take care of him now—he sees his own blood!'

Such were the exclamations, while the Highlander, starting from the ground, all his coldness and caution lost in frantic rage, sprung at his antagonist with the fury, the activity, and the vindictive purpose, of an incensed tiger-cat. But when could rage encounter science and temper? Robin Oig again went down in the unequal contest; and as the blow was necessarily a severe one, he lay motionless on the floor of the kitchen. The landlady ran to offer some aid, but Mr Fleecebumpkin would not permit her to approach.

'Let him alone,' he said, 'he will come to within time, and come up to the scratch again. He has not got half his broth yet.'

'He has got all I mean to give him, though,' said his antagonist, whose heart began to relent towards his old associate; 'and I would rather by half give the rest to yourself, Mr Fleecebumpkin, for you pretend to know a thing or two, and Robin had not art enough even to peel before setting to, but fought with his plaid dangling about him.—Stand up, Robin, my man! all friends now; and let me hear the man that will speak a word against you, or your country, for your sake.'

Robin Oig was still under the dominion of his passion, and eager to renew the onset; but being withheld on the one side by the peacemaking

Dame Heskett, and on the other, aware that Wakefield no longer meant to renew the combat, his fury sunk into gloomy sullenness.

'Come, come, never grudge so much at it, man,' said the brave-spirited Englishman, with the placability of his country, 'shake hands, and we will be better friends than ever.'

'Friends!' exclaimed Robin Oig with strong emphasis—'friends!—Never. Look to yourself, Harry Waakfelt.'

'Then the curse of Cromwell on your proud Scots stomach, as the man says in the play, and you may do your worst, and be d——; for one man can say nothing more to another after a tussle, than that he is sorry for it.'

On these terms the friends parted; Robin Oig drew out, in silence, a piece of money, threw it on the table, and then left the alehouse. But turning at the door, he shook his hand at Wakefield, pointing with his forefinger upwards, in a manner which might imply either a threat or a caution. He then disappeared in the moonlight.

Some words passed after his departure, between the bailiff, who piqued himself on being a little of a bully, and Harry Wakefield, who, with generous inconsistency, was now not indisposed to begin a new combat in defence of Robin Oig's reputation, 'although he could not use his daddles like an Englishman, as it did not come natural to him'. But Dame Heskett prevented this second quarrel from coming to a head by her peremptory interference. 'There should be no more fighting in her house,' she said; 'there had been too much already.—And you, Mr Wakefield, may live to learn,' she added, 'what it is to make a deadly enemy out of a good friend.'

'Pshaw, dame! Robin Oig is an honest fellow, and will never keep malice.'

'Do not trust to that—you do not know the dour temper of the Scots, though you have dealt with them so often. I have a right to know them, my mother being a Scot.'

'And so is well seen on her daughter,' said Ralph Heskett.

This nuptial sarcasm gave the discourse another turn; fresh customers entered the tap-room or kitchen, and others left it. The conversation turned on the expected markets, and the report of prices from different parts both of Scotland and England—treaties were commenced, and Harry Wakefield was lucky enough to find a chap for a part of his drove, and at a very considerable profit; an event of consequence more than sufficient to blot out all remembrances of the unpleasant scuffle in the earlier part of the day. But there remained one party from whose mind that recollection could not have been wiped away by the possession of every head of cattle betwixt Esk and Eden.

This was Robin Oig M'Combich.—'That I should have had no weapon,' he said, 'and for the first time in my life!—Blighted be the tongue that bids the Highlander part with the dirk—the dirk—ha! the English blood!—My Muhme's word—when did her word fall to the ground?'

The recollection of the fatal prophecy confirmed the deadly intention which instantly sprang up in his mind.

'Ha! Morrison cannot be many miles behind; and if it were an hundred, what then?'

His impetuous spirit had now a fixed purpose and motive of action, and he turned the light foot of his country towards the wilds, through which he knew, by Mr Ireby's report, that Morrison was advancing. His mind was wholly engrossed by the sense of injury—injury sustained from a friend; and by the desire of vengeance on one whom he now accounted his most bitter enemy. The treasured ideas of self-importance and self-opinion—of ideal birth and quality, had become more precious to him (like the hoard to the miser), because he could only enjoy them in secret. But that hoard was pillaged, the idols which he had secretly worshipped had been desecrated and profaned. Insulted, abused, and beaten, he was no longer worthy, in his own opinion, of the name he bore, or the lineage which he belonged to—nothing was left to him—nothing but revenge; and, as the reflection added a galling spur to every step, he determined it should be as sudden and signal as the offence.

When Robin Oig left the door of the alehouse, seven or eight English miles at least lay betwixt Morrison and him. The advance of the former was slow, limited by the sluggish pace of his cattle; the last left behind him stubble-field and hedgerow, crag and dark heath, all glittering with frostrime in the broad November moonlight, at the rate of six miles an hour. And now the distant lowing of Morrison's cattle is heard; and now they are seen creeping like moles in size and slowness of motion on the broad face of the moor; and now he meets them—passes them, and stops their conductor.

'May good betide us,' said the Southlander—'Is this you, Robin M'Combich, or your wraith?'

'It is Robin Oig M'Combich,' answered the Highlander, 'and it is not.—But never mind that, put pe giving me the skene-dhu.'

'What! you are for back to the Highlands—The devil!—Have you selt all off before the fair? This beats all for quick markets!'

'I have not sold—I am not going north—May pe I will never go north again.—Give me pack my dirk, Hugh Morrison, or there will pe words petween us.'

'Indeed, Robin, I'll be better advised before I gie it back to you—it is a wanchancy weapon in a Highlandman's hand, and I am thinking you will be about some barns-breaking.'

'Prutt, trutt! let me have my weapon,' said Robin Oig impatiently.

'Hooly and fairly,' said his well-meaning friend. 'I'll tell you what will do better than these dirking doings—Ye ken Highlander, and Lowlander, and Border-men, are a' ae man's bairns when you are over the Scots dyke. See, the Eskdale callants, and fighting Charlie of Liddesdale, and the Lockerby lads, and the four Dandies of Lustruther, and a wheen mair grey plaids, are coming up behind; and if you are wranged, there is the hand of a Manly Morrison, we'll see you righted, if Carlisle and Stanwix baith took up the feud.'

'To tell you the truth,' said Robin Oig, desirous of eluding the suspicions of his friend, 'I have enlisted with a party of the Black Watch, and must march off to-morrow morning.'

'Enlisted! Were you mad or drunk?—You must buy yourself off—I can lend you twenty notes, and twenty to that, if the drove sell.'

'I thank you—thank ye, Hughie; but I go with good will the gate that I am going—so the dirk—the dirk!'

'There it is for you then, since less wunna serve. But think on what I was saying.—Waes me, it will be sair news in the braes of Balquidder, that Robin Oig M'Combich should have run an ill gate, and ta'en on.'

'Ill news in Balquidder, indeed!' echoed poor Robin: 'but Cot speed you, Hughie, and send you good marcats. Ye winna meet with Robin Oig again, either at tryste or fair.'

So saying, he shook hastily the hand of his acquaintance, and set out in the direction from which he had advanced, with the spirit of his former pace.

'There is something wrang with the lad,' muttered the Morrison to himself; 'but we will maybe see better into it the morn's morning.'

But long ere the morning dawned, the catastrophe of our tale had taken place. It was two hours after the affray had happened, and it was totally forgotten by almost every one, when Robin Oig returned to Heskett's inn. The place was filled at once by various sorts of men, and with noises corresponding to their character. There were the grave low sounds of men engaged in busy traffic, with the laugh, the song, and the riotous jest of those who had nothing to do but to enjoy themselves. Among the last was Harry Wakefield, who, amidst a grinning group of smock-frocks, hobnailed shoes, and jolly English physiognomies, was trolling forth the old ditty,

> 'What though my name be Roger,
> Who drives the plough and cart'—

when he was interrupted by a well-known voice saying in a high and stern voice, marked by the sharp Highland accent, 'Harry Waakfelt—if you be a man stand up!'

'What is the matter?—what is it?' the guests demanded of each other.

'It is only a d——d Scotsman,' said Fleecebumpkin, who was by this time very drunk, 'whom Harry Wakefield helped to his broth to-day, who is now come to have *his cauld kail* het again.'

'Harry Waakfelt,' repeated the same ominous summons, 'stand up, if you be a man!'

There is something in the tone of deep and concentrated passion, which attracts attention and imposes awe, even by the very sound. The guests shrunk back on every side, and gazed at the Highlander as he stood in the middle of them, his brows bent, and his features rigid with resolution.

'I will stand up with all my heart, Robin, my boy, but it shall be to shake hands with you, and drink down all unkindness. It is not the fault of your heart, man, that you don't know how to clench your hands.'

By this time he stood opposite to his antagonist; his open and unsuspecting look strangely contrasted with the stern purpose, which gleamed wild, dark, and vindictive in the eyes of the Highlander.

''Tis not thy fault, man, that not having the luck to be an Englishman, thou canst not fight more than a school-girl.'

'I *can* fight,' answered Robin Oig, sternly but calmly, 'and you shall know it. You, Harry Waakfelt, showed me to-day how the Saxon churls fight—I show you now how the Highland Dunnièwassel fights.'

He seconded the word with the action, and plunged the dagger, which he suddenly displayed, into the broad breast of the English yeoman, with such fatal certainty and force, that the hilt made a hollow sound against the breast-bone, and the double-edged point split the very heart of his victim. Harry Wakefield fell and expired with a single groan. His assassin next seized the bailiff by the collar, and offered the bloody poniard to his throat, whilst dread and surprise rendered the man incapable of defence.

'It were very just to lay you beside him,' he said, 'but the blood of a base pick-thank shall never mix on my father's dirk, with that of a brave man.'

As he spoke, he cast the man from him with so much force that he fell on the floor, while Robin, with his other hand, threw the fatal weapon into the blazing turf-fire.

'There,' he said, 'take me who likes—and let fire cleanse blood if it can.'

The pause of astonishment still continuing, Robin Oig asked for a peace-officer, and a constable having stepped out, he surrendered himself to his custody.

'A bloody night's work you have made of it,' said the constable.

'Your own fault,' said the Highlander. 'Had you kept his hands off me twa hours since, he would have been now as well and merry as he was twa minutes since.'

'It must be sorely answered,' said the peace-officer.

'Never you mind that—death pays all debts; it will pay that too.'

The horror of the bystanders began now to give way to indignation; and the sight of a favourite companion murdered in the midst of them, the provocation being, in their opinion, so utterly inadequate to the excess of vengeance, might have induced them to kill the perpetrator of the deed even upon the very spot. The constable, however, did his duty on this occasion, and with the assistance of some of the more reasonable persons present, procured horses to guard the prisoner to Carlisle, to abide his doom at the next assizes. While the escort was preparing, the prisoner neither expressed the least interest, nor attempted the slightest reply. Only, before he was carried from the fatal apartment, he desired to look at the dead body, which, raised from the floor, had been deposited upon the large table (at the head of which Harry Wakefield had presided but a few minutes before, full of life, vigour, and animation), until the surgeons should examine the mortal wound. The face of the corpse was decently covered with a napkin. To the surprise and horror of the by-standers, which displayed itself in a general *Ah!* drawn through clenched teeth and half-shut lips, Robin Oig removed the cloth, and gazed with a mournful but steady eye on the lifeless visage, which had been so lately animated, that the smile of good-humoured confidence in his own strength, of conciliation at once, and contempt towards his enemy, still curled his lip. While those present expected that the wound, which had so lately flooded the apartment with gore, would send forth fresh streams at the touch of the homicide, Robin Oig replaced the covering, with the brief exclamation—'He was a pretty man!'

My story is nearly ended. The unfortunate Highlander stood his trial at Carlisle. I was myself present, and as a young Scottish lawyer, or barrister at least, and reputed a man of some quality, the politeness of the Sheriff of Cumberland offered me a place on the bench. The facts of the case were proved in the manner I have related them; and whatever might be at first the prejudice of the audience against a crime so un-English as that of assassination from revenge, yet when the rooted national prejudices of the prisoner had been explained, which made him consider himself as stained with indelible dishonour, when subjected to personal violence; when his previous patience, moderation, and endurance, were considered, the generosity of the English audience was inclined to regard his crime as the wayward aberration of a false idea of honour rather than as flowing from a heart naturally savage, or perverted by habitual vice. I shall never forget the charge of the venerable Judge to the jury, although not at that time liable to be much affected either by that which was eloquent or pathetic.

'We have had,' he said, 'in the previous part of our duty,' (alluding to some former trials), 'to discuss crimes which infer disgust and abhorrence, while they call down the well-merited vengeance of the law. It is now our still more melancholy task to apply its salutary though severe enactments to a case of a very singular character, in which the crime (for a crime it is, and a deep one) arose less out of the malevolence of the heart, than the error of the understanding—less from any idea of committing wrong, than from an unhappily perverted notion of that which is right. Here we have two men, highly esteemed, it has been stated, in their rank of life, and attached, it seems, to each other as friends, one of whose lives has been already sacrificed to a punctilio, and the other is about to prove the vengeance of the offended laws; and yet both may claim our commiseration at least, as men acting in ignorance of each other's national prejudices, and unhappily misguided rather than voluntarily erring from the path of right conduct.

'In the original cause of the misunderstanding, we must in justice give the right to the prisoner at the bar. He had acquired possession of the enclosure, which was the object of competition, by a legal contract with the proprietor Mr Ireby; and yet, when accosted with reproaches undeserved in themselves, and galling doubtless to a temper at least sufficiently susceptible of passion, he offered notwithstanding to yield up half his acquisition, for the sake of peace and good neighbourhood, and his amicable proposal was rejected with scorn. Then follows the scene at Mr Heskett the publican's, and you will observe how the stranger was treated by the deceased, and, I am sorry to observe, by those around, who seem to have urged him in a manner which was aggravating in the highest degree. While he asked for peace and for composition, and offered submission to a magistrate, or to a mutual arbiter, the prisoner was insulted by a whole company, who seem on this occasion to have forgotten the national maxim of "fair play"; and while attempting to escape from the place in peace, he was intercepted, struck down, and beaten to the effusion of his blood.

'Gentlemen of the Jury, it was with some impatience that I heard my learned brother, who opened the case for the crown, give an unfavourable turn to the prisoner's conduct on this occasion. He said the prisoner was afraid to encounter his antagonist in fair fight, or to submit to the laws of the ring; and that therefore, like a cowardly Italian, he had recourse to his fatal stiletto, to murder the man whom he dared not meet in manly encounter. I observed the prisoner shrink from this part of the accusation with the abhorrence natural to a brave man; and as I would wish to make my words impressive, when I point his real crime, I must secure his opinion of my impartiality, by rebutting every thing that seems to me a false accusation. There can be no doubt that the prisoner is a man of resolution—too much resolution—I wish to Heaven that he had less, or rather that he had had a better education to regulate it.

'Gentlemen, as to the laws my brother talks of, they may be known in the Bull-ring, or the Bear-garden, or the Cockpit, but they are not known here. Or, if they should be so far admitted as furnishing a species of proof that no malice was intended in this sort of combat, from which fatal accidents do sometimes arise, it can only be so admitted when both parties are *in pari casu*, equally acquainted with, and equally willing to refer themselves to, that species of arbitrement. But will it be contended that a man of superior rank and education is to be subjected, or is obliged to subject himself, to this coarse and brutal strife, perhaps in opposition to a younger, stronger, or more skilful opponent? Certainly even the pugilistic code, if founded upon the fair play of Merry Old England, as my brother alleges it to be, can contain nothing so preposterous. And, gentlemen of the jury, if the laws would support an English gentleman, wearing, we will suppose, his sword, in defending himself by force against a violent personal aggression of the nature offered to this prisoner, they will not less protect a foreigner and a stranger, involved in the same unpleasing circumstances. If, therefore, gentlemen of the jury, when thus pressed by a *vis major*, the object of obloquy to a whole company, and of direct violence from one at least, and, as he might reasonably apprehend, from more, the panel had produced the weapon which his countrymen, as we are informed, generally carry about their persons, and the same unhappy circumstance had ensued which you have heard detailed in evidence, I could not in my conscience have asked from you a verdict of murder. The prisoner's personal defence might indeed, even in that case, have gone more or less beyond the *Moderamen inculpatae tutelae*, spoken of by lawyers, but the punishment incurred would have been that of manslaughter, not of murder. I beg leave to add, that I should have thought this milder species of charge was demanded in the case supposed, notwithstanding the statute of James I. cap. 8, which takes the case of slaughter by stabbing with a short weapon, even without malice prepense, out of the benefit of clergy. For this statute of stabbing, as it is termed, arose out of a temporary cause; and as the real guilt is the same, whether the slaughter be committed by the dagger, or by sword or pistol, the benignity of the modern law places them all on the same, or nearly the same footing.

'But, gentlemen of the jury, the pinch of the case lies in the interval of two hours interposed betwixt the reception of the injury and the fatal retaliation. In the heat of affray and *chaude melée*, law, compassionating the infirmities of humanity, makes allowance for the passions which rule such a stormy moment—for the sense of present pain, for the apprehension of further injury, for the difficulty of ascertaining with due accuracy the precise degree of violence which is necessary to protect the person of the individual, without annoying or injuring the assailant more than is absolutely necessary. But the time necessary to walk twelve miles, however speedily performed, was an interval sufficient for the prisoner to have recollected himself; and the violence with which he carried his purpose into effect, with so many circumstances of deliberate determination, could neither be induced by the passion of anger, nor that of fear. It was the purpose and the act of predetermined revenge, for which law neither can, will, nor ought to have sympathy or allowance.

'It is true, we may repeat to ourselves, in alleviation of this poor man's unhappy action, that his case is a very peculiar one. The country which he inhabits was, in the days of many now alive, inaccessible to the laws, not only of England, which have not even yet penetrated thither, but to those to which our neighbours of Scotland are subjected, and which must be supposed to be, and no doubt actually are, founded upon the general principles of justice and equity which pervade every civilised country. Amongst their mountains, as among the North American Indians, the various tribes were wont to make war upon each other, so that each man was obliged to go armed for his own protection. These men, from the ideas which they entertained of their own descent and of their own consequence, regarded themselves as so many cavaliers or men-at-arms, rather than as the peasantry of a peaceful country. Those laws of the ring, as my brother terms them, were unknown to the race of warlike mountaineers;

that decision of quarrels by no other weapons than those which nature has given every man, must to them have seemed as vulgar and as preposterous as to the Noblesse of France. Revenge, on the other hand, must have been as familiar to their habits of society as to those of the Cherokees or Mohawks. It is indeed, as described by Bacon, at bottom a kind of wild untutored justice; for the fear of retaliation must withhold the hands of the oppressor where there is no regular law to check daring violence. But though all this may be granted, and though we may allow that, such having been the case of the Highlands in the days of the prisoner's fathers, many of the opinions and sentiments must still continue to influence the present generation, it cannot, and ought not, even in this most painful case, to alter the administration of the law, either in your hands, gentlemen of the jury, or in mine. The first object of civilisation is to place the general protection of the law, equally administered, in the room of that wild justice, which every man cut and carved for himself, according to the length of his sword and the strength of his arm. The law says to the subjects, with a voice only inferior to that of the Deity, "Vengeance is mine." The instant that there is time for passion to cool, and reason to interpose, an injured party must become aware, that the law assumes the exclusive cognisance of the right and wrong betwixt the parties, and opposes her inviolable buckler to every attempt of the private party to right himself. I repeat, that this unhappy man ought personally to be the object rather of our pity than our abhorrence, for he failed in his ignorance, and from mistaken notions of honour. But his crime is not the less that of murder, gentlemen, and, in your high and important office, it is your duty so to find. Englishmen have their angry passions as well as Scots; and should this man's action remain unpunished, you may unsheath, under various pretences, a thousand daggers betwixt the Land's-end and the Orkneys.'

The venerable Judge thus ended what, to judge by his apparent emotion, and by the tears which filled his eyes, was really a painful task. The jury, according to his instructions, brought in a verdict of Guilty; and Robin Oig M'Combich, *alias* M'Gregor, was sentenced to death, and left for execution, which took place accordingly. He met his fate with great firmness, and acknowledged the justice of his sentence. But he repelled indignantly the observations of those who accused him of attacking an unarmed man. 'I give a life for the life I took,' he said, 'and what can I do more?'

John Galt

1779–1839

The Howdie: An Autobiography

Part I—Anent Births

When my gudeman departed this life, he left me with a heavy handful of seven childer, the youngest but a baby at the breast, and the elder a lassie scant of eight years old. With such a small family what could a lanely widow woman do? Greatly was I grieved, not only for the loss of our bread-winner, but the quenching of that cheerful light which was my solace and comfort in straitened circumstances, and in the many cold and dark hours which the needs of our necessitous condition obliged us to endure.

James Blithe was my first and only Jo; and but for that armed man, Poverty, who sat ever demanding at our hearth, there never was a brittle minute in the course of our wedded life. It was my pleasure to gladden him at home, when out-of-door vexations ruffled his temper; which seldom came to pass, for he was an honest young man, and pleasant among those with whom his lot was cast. I have often, since his death, thought, in calling him to mind, that it was by his natural sweet nature that the Lord was pleased, when He took him to Himself, to awaken the sympathy of others for me and the bairns, in our utmost distress.

He was the head gairdner to the Laird of Rigs, as his father had been before him; and the family had him in great respect. Besides many a present of useful things which they gave to us, when we were married, they came to our wedding; a compliment that James often said was like the smell of the sweet briar in a lown and dewy evening, a cherishment that seasoned happiness. It was not however till he was taken away that I experienced the extent of their kindness. The ladies of the family were most particular to me; the Laird himself, on the Sabbath after the burial, paid me a very edifying visit; and to the old Leddy Dowager, his mother, I owe the meal that has ever since been in the basin, by which I have been enabled to bring up my childer in the fear of the Lord.

The Leddy was really a managing motherly character; no grass grew beneath her feet when she had a turn to do, as was testified by my case: for when the minister's wife put it into her head that I might do well in the midwife-line, Mrs Forceps being then in her declining years, she lost no time in getting me made, in the language of the church and gospel, her helper and successor. A blessing it was at the time, and the whole parish has, with a constancy of purpose, continued to treat me far above my deserts; for I have ever been sure of a shortcoming in my best endeavours to give satisfaction. But it's not to speak of the difficulties that the hand of a considerate Providence has laid upon me with a sore weight for an earthly nature to bear, that I have sat down to indite this history book. I only intend hereby to show, how many strange things have come to pass in my douce way of life; and sure am I that in every calling, no matter however humble, peradventures will take place that ought to be recorded for the instruction, even of the wisest. Having said this, I will now proceed with my story.

All the har'st before the year of dearth, Mrs Forceps, my predecessor, had been in an ailing condition; insomuch that, on the Halloween, she was laid up, and never after was taken out of her bed a living woman. Thus it came to pass that, before the turn of the year, the midwifery business of our countryside came into my hands in the natural way.

I cannot tell how it happened that there was little to do in the way of trade all that winter; but it began to grow into a fashion that the genteeler order of ladies went into the towns to have there han'lings among the doctors. It was soon seen, however, that they had nothing to boast of by that manœuvre, for their gudemen thought the cost overcame the profit; and thus, although that was to a certainty a niggardly year, and great part of the next no better, it pleased the Lord, by the scanty upshot of the har'st before spoken of, that, whatever the ladies thought of the doctors, their husbands kept the warm side of frugality towards me and other poor women that had nothing to depend upon but the skill of their ten fingers.

Mrs Forceps being out of the way, I was called in; and my first case was with an elderly woman that was long thought by all her friends to be past bearing; but when she herself came to me, and rehearsed the state she was in, with a great sough for fear, instead of a bairn, it might turn out a tympathy, I called to her mind how Sarah the Patriarchess, the wife of Abraham, was more than fourscore before Isaac was born: which was to her great consolation; for she was a pious woman in the main, and could discern in that miracle of Scripture an admonition to her to be of good cheer.

From that night, poor Mrs Houselycat grew an altered woman; and her gudeman, Thomas Houselycat, was as caidgy a man as could be, at the prospect of having an Isaac in his old age; for neither he nor his wife had the least doubt that they were to be blest with a man-child. At last the fulness of time came; and Thomas having provided a jar of cinnamon brandy for the occasion, I was duly called in.

Well do I remember the night that worthy Thomas himself came for me, with a lantern or a bowit in his hand. It was pitch-dark; the winds rampaged among the trees, the sleet was just vicious, and every drop was as salt as pickle. He had his wife's shawl tied over his hat, by a great knot under the chin, and a pair of huggars drawn over his shoes, and above his knees; he was just a curiosity to see coming for me.

I went with him; and to be sure when I got to the house, there was a gathering; young and old were there, all speaking together; widows and grannies giving advice, and new-married wives sitting in the expectation of getting insight. Really it was a ploy; and no wonder that there was such a collection; for Mrs Houselycat was a woman well-stricken in years, and it could not be looked upon as any thing less than an inadvertency that she was ordained to be again a mother. I very well remember that her youngest daughter of the first clecking was there, a married woman, with a wean at her knee, I'se warrant a year-and-a-half old; it could both walk alone, and say many words almost as intelligible as the minister in the poopit, when it was a frosty morning; for the cold made him there shavelin-gabbit, and every word he said was just an oppression to his feckless tongue.

By and by the birth came to pass: but, och on! the long faces that were about me when it took place; for instead of a lad-bairn it proved a lassie; and to increase the universal dismay at this come-to-pass, it turned out that the bairn's cleading had, in a way out of the common, been prepared for a man child; which was the occasion of the innocent being, all the time of its nursing in appearance a very doubtful creature.

The foregoing case is the first that I could properly say was my own; for Mrs Forceps had a regular finger in the pie in all my heretofores. It was, however, good erls; for no sooner had I got Mrs Houselycat on her feet again, than I received a call from the head inns in the town, from a Captain's lady, that was overtaken there as the regiment was going through.

In this affair there was something that did not just please me in the conduct of Mrs Facings, as the gentlewoman was called; and I jaloused, by what I saw with the tail of my eye, that she was no better than a light woman. However, in the way of trade, it does not do to stand on trifles of that sort; for ours is a religious trade, as witness what is said in the Bible of the midwives of the Hebrews; and if it pleased Providence to ordain children to be, it is no less an ordained duty of the midwife to help them into the world. But I had not long been satisfied in my own mind that the mother was no better than she should be, when my kinder feelings were sorely tried, for she had a most extraordinar severe time o't; and I had but a weak hope that she would get through. However, with my help and the grace of God, she did get through: and I never saw, before nor since, so brave a baby as was that night born.

Scarcely was the birth over, when Mrs Facings fell into a weakly dwam that was very terrifying; and if the Captain was not her gudeman, he was as concerned about her, as any true gudeman could be, and much more so than some I could name, who have the best of characters.

It so happened that this Mrs Facings had been, as I have said, overtaken on the road, and had nothing prepared for a sore foot, although she well knew that she had no time to spare. This was very calamitous, and what was to be done required a consideration. I was for wrapping the baby in a blanket till the morning, when I had no misdoubt of gathering among the ladies of the town a sufficient change of needfu' baby clouts; but among other regimental clanjamphry that were around this left-to-hersel' damsel, was a Mrs Gooseskin, the drum-major's wife, a most devising character. When I told her of our straits and jeopardy, she said to give myself no uneasiness, for she had seen a very good substitute for child-linen, and would set about making it without delay.

What she proposed to do was beyond my comprehension; but she soon returned into the room with a box in her hand, filled with soft-teazed wool, which she set down on a chair at the bed-stock, and covering it with an apron, she pressed the wool under the apron into a hollow shape, like a goldfinch's nest, wherein she laid the infant, and covering it up with the apron, she put more wool over it, and made it as snug as a silk-worm in a cocoon, as it has been described to me. The sight of this novelty was, however, an affliction, for if she had intended to smother the bairn, she could not have taken a more effectual manner; and yet the baby lived and thrived, as I shall have occasion to rehearse.

Mrs Facings had a tedious recovery, and was not able to join him that in a sense was her gudeman, and the regiment, which was to me a great cause of affliction; for I thought that it might be said that her case was owing to my being a new hand, and not skilful enough. It thus came to pass that she, when able to stand the shake, was moved to private lodgings, where, for a season, she dwined and dwindled, and at last her life went clean out; but her orphan bairn was spared among us, and was a great means of causing a tenderness of heart to arise among the lasses, chiefly on account of its most thoughtless and ne'er-do-weel father, who never inquired after he left the town, concerning the puir thing; so that if there had not been a seed of charity bred by its orphan condition, nobody can tell what would have come of it. The saving hand of Providence was, however, manifested. Old Miss Peggy Needle, who had all her life been out of the body about cats and dogs, grew just extraordinar to make a pet, in the place of them all, of the laddie Willie Facings; but, as I have said, I will by and by have to tell more about him; so on that account I will make an end of the second head of my discourse, and proceed to the next, which was one of a most piteous kind.

In our parish there lived a young lad, a sticket minister, not very alluring in his looks; indeed, to say the truth, he was by many, on account of them, thought to be no far short of a haverel; for he was lank and most uncomely, being in-kneed; but, for all that, the minister said he was a young man of great parts, and had not only a streak of geni, but a vast deal of inordinate erudition. He went commonly by the name of Dominie Quarto; and it came to pass, that he set his affections on a weel-faured lassie, the daughter of Mrs Stoups, who keepit the Thistle Inn. In this there was nothing wonderful, for she was a sweet maiden, and nobody ever saw her without wishing her well. But she could not abide the Dominie: and, indeed, it was no wonder, for he certainly was not a man to pleasure a woman's eye. Her affections were settled on a young lad called Jock Sym, a horse-couper, a blithe heartsome young man, of a genteel manner, and in great repute, therefore, among the gentlemen.

He won Mally Stoups' heart; they were married, and, in the fulness of time thereafter, her pains came on, and I was sent to ease her. She lay in a back room, that looked into their pleasant garden. Half up the lower casement of the window, there was a white muslin curtain, made out of one of her mother's old-fashioned tamboured aprons, drawn across from side to side, for the window had no shutters. It would be only to distress the reader to tell what she suffered. Long she struggled, and weak she grew; and a sough of her desperate case went up and down the town like the plague that walketh in darkness. Many came to enquire for her, both gentle and semple; and it was thought that the Dominie would have been in the crowd of callers: but he came not.

In the midst of her suffering, when I was going about my business in the room, with the afflicted lying-in woman, I happened to give a glint to the window, and startled I was, to see, like a ghost, looking over the white curtain, the melancholious visage of Dominie Quarto, with watery eyes glistening like two stars in the candle light.

I told one of the women who happened to be in the way, to go out to the sorrowful young man, and tell him not to look in at the window; whereupon she went out, and remonstrated with him for some time. While she was gone, sweet Mally Stoups and her unborn baby were carried away to Abraham's bosom. This was a most unfortunate thing; and I went out before the straighting-board could be gotten, with a heavy heart, on account of my poor family, that might suffer, if I was found guilty of being to blame.

I had not gone beyond the threshold of the back-door that led into the garden, when I discerned a dark figure between me and the westling scad of the setting moon. On going towards it, I was greatly surprised to find the weeping Dominie, who was keeping watch for the event there, and had just heard what had happened, by one of the women telling another.

This symptom of true love and tenderness made me forget my motherly anxieties, and I did all I could to console the poor lad; but he was not to be comforted, saying, 'It was a great trial when it was ordained that she should lie in the arms of Jock Sym, but it's far waur to think that the kirkyard hole is to be her bed, and her bridegroom the worm.'

Poor forlorn creature! I had not a word to say. Indeed, he made my heart swell in my bosom; and I could never forget the way in which he grat over my hand, that he took between both of his, as a dear thing, that he was prone to fondle and mourn over.

But this cutting grief did not end that night; on Sabbath evening following, as the custom is in our parish, Mrs Sym was ordained to be interred; and there was a great gathering of freends and neighbours; for both she and her gudeman were well thought of. Everybody expected the Dominie would be there, for his faithfulness was spoken of by all pitiful tongues; but he stayed away for pure grief; he hid himself from the daylight and the light of every human eye. In the gloaming, however, after, as the betherel went to ring the eight o'clock bell, he saw the Dominie standing with a downcast look, near the new grave, all which made baith a long and a sad story, for many a day among us: I doubt if it's forgotten yet. As for me, I never thought of it without a pang, but all trades have their troubles and the death of a young wife and her unborn baby, in her nineteenth year, is not one of the least that I have had to endure in mine.

But, although I met like many others in my outset both mortifications and difficulties, and what was worse than all, I could not say that I was triumphant in my endeavours; yet, like the Doctors, either good luck or experience made me gradually gather a repute for skill and discernment, insomuch that I became just wonderful for the request I was in. It is therefore needless for me to make a strive for the entertainment of the reader, by rehearsing all the han'lings that I had; but, as some of them were of a notable kind, I will pass over the generality and only make a Nota Bena here of those that were particular, as well as the births of the babies that afterwards came to be something in the world.

Between the death of Mally Stoups and the Whitsunday of that year, there was not much business in my line, not above two cases; but, on the day after, I had a doing, no less than of twins in a farmer's family, that was already overstocked with weans to a degree that was just a hardship; but, in that case, there was a testimony that Providence never sends mouths into the world without at the same time giving the wherewithal to fill them.

James Mashlam was a decent, douce, hard-working, careful man, and his wife was to all wives the very patron of frugality; but, with all their ettling, they could scarcely make the two ends of the year to meet. Owing to this, when it was heard in the parish that she had brought forth a Jacob and Esau, there was a great condolence; and the birth that ought to have caused both mirth and jocundity was not thought to be a gentle dispensation. But short-sighted is the wisdom of man and even of woman likewise; for, from that day, James Mashlam began to bud and prosper, and is now the toppingest man far or near; and his prosperity sprang out of what we all thought would be a narrowing of his straitened circumstances.

All the gentry of the country-side, when they heard the tydings, sent Mrs Mashlam many presents, and stocked her press with cleeding for her and the family. It happened, also, that, at this time, there was a great concourse of Englishers at the castle with my Lord; and one of them, a rattling young gentleman, proposed that they should raise a subscription for a race-purse; promising, that, if his horse won, he would give the purse for the behoof of the twins. Thus it came to pass, that a shower of gold one morning fell on James Mashlam, as he was holding the plough; for that English ramplor's horse, lo and behold! won the race, and he came over with all the company, with the purse in his hand full of golden guineas galloping upon James; and James and his wife sat cloking on this nest-egg, till they have hatched a fortune; for the harvest following, his eldest son was able to join the shearers, and from that day plenty, like a fat carlin, visited him daily. Year after year his childer that were of the male gender grew better and better helps: so that he enlarged his farm, and has since built the sclate house by the water side; that many a one, for its decent appearance, cannot but think it is surely the minister's manse.

From that time I too got a lift in the world; for it happened, that a grand lady, in the family way, came on a visit to the castle, and by some unaccountable accident she was prematurely brought to bed there. No doctor being at hand nearer than the burgh town, I was sent for and, before one could be brought, I had helped into the world the son and heir of an ancient family; for the which, I got ten golden guineas, a new gown that is still my best honesty, and a mutch that many a one came to see for it is made of a French lace. The lady insisted on me to wear it at the christening; which the Doctor was not overly pleased to hear tell of, thinking that I might in time clip the skirts of his practice.

For a long time after the deliverance of that lady I had a good deal to do in the cottars' houses; and lucky it was for me that I had got the guineas aforesaid, for the commonalty have not much to spare on an occasion; and I could not help thinking how wonderful are the ways of Providence, for the lady's gift enabled me to do my duty among the cottars with a lighter heart than I could have afforded to do had the benison been more stinted.

All the remainder of that year, the winter and the next spring, was without a remarkable: but just on the eve of summer, a very comical accident happened.

There was an old woman that come into the parish, nobody could tell how, and was called Lucky Nanse, who made her bread by distilling peppermint. Some said that now and then her house had the smell of whiskey; but how it came, whether from her still or the breath of her nostrils, was never made out to a moral certainty. This carlin had been in her day a by-ordinair woman, and was a soldier's widow forby.

At times she would tell stories of marvels she had seen in America, where she said there was a moose so big that a man could not lift its head. Once, when old Mr Izet, the precentor, to whom she was telling anent this beast, said it was not possible, she waxed very wroth, and knocking her neives together in his face, she told him that he was no gentleman to misdoubt her honour: Mr Izet, who had not much of the sweet milk of human kindness in his nature, was so provoked at this freedom, that he snapped his fingers as he turned to go away, and said she was not better than a ne'er-do-weel camp-randy. If she was oil before she was flame now, and dancing with her arms extended, she looted down, and, grasping a gowpin of earth in each hand, she scattered it with an air to the wind, and cried with a desperate voice, that she did not value his opinion at the worth of that dirt.

By this time the uproar had disturbed the clachan, and at every door the women were looking out to see what was the hobble-show; some with bairns in their arms and others with weans at their feet. Among the rest that happened to look out was Mrs Izet, who, on seeing the jeopardy that her gudeman was in, from that rabiator woman, ran to take him under her protection. But it was a rash action for Lucky Nanse stood with her hands on her henches and daured her to approach, threatening, with some soldier-like words, that if she came near she would close her day-lights.

Mrs Izet was terrified, and stood still.

Home with you, said Nanse, ye mud that ye are, to think yourself on a par with pipeclay, with other hetradox brags, that were just a sport to hear. In the meantime, the precentor was walking homeward, and called to his wife to come away, and leave that tempest and whirlwind with her own wrack and carry.

Lucky Nanse had, by this time, spent her ammunition, and, unable to find another word sufficiently vicious, she ran up to him and spat in his face.

Human nature could stand no more, and the precentor forgetting himself and his dignity in the parish, lifted his foot and gave her a kick, which caused her to fall on her back. There she lay sprawling and speechless, and made herself at last lie as like a corpse, as it was possible. Every body thought that she was surely grievously hurt, though Mr Izet said his foot never touched her; and a hand-barrow was got to carry her home. All present were in great dismay, for they thought Mr Izet had committed a murder and would be hanged in course of law; but I may be spared from describing the dolorosity that was in our town that night.

Lucky Nanse being carried home on the barrow like a carcase, was put to bed; where, when she had lain some time, she opened a comical eye for a short space, and then to all intents and purposes seemed in the dead throes. It was just then that I, drawn into the house by the din of the straemash, looked over a neighbour's shoulder; but no sooner did the artful woman see my face than she gave a skirle of agony, and cried that her time was come, and the pains of a mother were upon her; at which to hear, all the other women gave a shout, as if a miracle was before them, for Nanse was, to all appearance, threescore; but she for a while so enacted her stratagem that we were in a terrification lest it should be true. At last she seemed quite exhausted, and I thought she was in the natural way, when in a jiffy she bounced up with a derisive gaffaw, and drove us all pell-mell out of the house. The like of such a ploy had never been heard of in our countryside. I was, however, very angry to be made such a fool of in my profession before all the people, especially as it turned out that the old woman was only capering in her cups.

Sometime after this exploit another come-to-pass happened that had a different effect on the nerves of us all. This fell out by a sailor's wife, a young woman that came to lie in from Sandy-port with her mother, a most creditable widow, that kept a huckstry shop for the sale of parliament cakes, candles, bone-combs, and prins, and earned a bawbee by the eydency of her spinning wheel.

Mrs Spritsail, as the young woman was called, had a boding in her breast that she could not overcome, and was a pitiable object of despondency, from no cause; but women in her state are often troubled by similar vapours. Hers, however, troubled everybody that came near her, and made her poor mother almost persuaded that she would not recover.

One night when she expected to be confined, I was called in: but such a night as that was! At the usual hour, the post woman, Martha Dauner, brought a letter to the old woman from Sandy-port, sealed with a black wafer; which, when Mrs Spritsail saw, she grew as pale as a clout, and gave a deep sigh. Alas! it was a sigh of prophecy; for the letter was to tell that her husband, John Spritsail, had tumbled overboard the night before, and was drowned.

For some time the young widow sat like an image, making no moan: it was very frightful to see her. By and by, her time came on, and although it could not be said that her suffering was by common, she fell back again into that effigy state, which made her more dreadful to see than if she had been a ghost in its winding sheet; and she never moved from the posture she put herself in till all was over, and the living creature was turned into a clod of church-yard clay.

This for a quiet calamity is the most distressing in my chronicle, for it came about with little ceremony. Nobody was present with us but only her sorrowful mother, on whose lap I laid the naked new-born babe. Soon after, the young widow departed to join her gudeman in paradise; but as it is a mournful tale, it would only be to hurt the reader's tender feelings to make a more particular account.

All my peradventures were not, however, of the same doleful kind; and there is one that I should mention, for it was the cause of meikle jocosity at the time and for no short season after.

There lived in the parish a very old woman, upwards of fourscore: she was as bent in her body as a cow's horn, and she supported herself with a staff in one hand, and for balance held up her gown behind with the other; in short, she was a very antideluvian, something older than seemed the folk at that time of the earth.

This ancient crone was the grandmother to Lizzy Dadily, a light-headed winsome lassie, that went to service in Glasgow; but many months she had not been there when she came back again, all mouth and een; and on the same night her granny, old Maudelin, called on me. It was at the gloaming: I had not trimmed my crusie, but I had just mended the fire, which had not broken out so that we conversed in an obscurity.

Of the history of old Maudelin I had never before heard ony particulars; but her father, as she told me, was out in the rebellion of Mar's year, and if the true king had gotten his rights, she would not have been a needful woman. This I, however, jealouse was vanity; for although it could not be said that she was positively an ill-doer, it was well known in the town that old as she was, the conduct of her house in many points was not the best. Her daughter, the mother of Lizzy, was but a canary-headed creature. What became of her we never heard, for she went off with the soldiers one day, leaving Lizzy, a bastard bairn. How the old woman thereafter fenn't, in her warsle with age and poverty, was to many a mystery, especially as it was now and then seen that she had a bank guinea note to change, and whom it cam frae was a marvel.

Lizzy coming home, her granny came to me, as I was saying, and after awhile conversing in the twilight about this and that, she told me that she was afraid her oe had brought home her wark, and that she didna doubt they would need the sleight of my hand in a short time, for that Lizzy had only got a month's leave to try the benefit of her native air; that of Glasgow, as with most young women, not agreeing with her.

I was greatly taken aback to hear her talk in such a calm and methodical manner concerning Lizzy, whom I soon found was in that condition that would, I'm sure, have drawn tears of the heart's blood from every other grandmother in the clachan. Really I was not well pleased to hear the sinful carlin talk in such a good-e'en and good-morn way about a guilt of that nature; and I said to her, both hooly and fairly, that I was not sure if I could engage myself in the business, for it went against my righteous opinion to make myself a mean of filling the world with natural children.

The old woman was not just pleased to hear me say this, and without any honey on her lips, she replied,

'Widow Blithe, this is an unco strain! and what for will ye no do your duty to Lizzy Dadily; for I must have a reason, because the minister or the magistrates of the borough shall ken of this.'

I was to be sure a little confounded to hear the frail though bardy old woman thus speak to her peremptors, but in my mild and methodical manner I answered and said,

'That no person in a trade with full hands ought to take a new turn; and although conscience, I would allow, had its weight with me, yet there was a stronger reason in my engagements to others.'

'Very well,' said Maudelin, and hastily rising, she gave a rap with her staff, and said, 'that there soon would be news in the land that I would hear of;' and away she went, stotting out at the door, notwithstanding her age, like a birsled pea.

After she was gone, I began to reflect; and I cannot say that I had just an ease of mind, when I thought of what she had been telling anent her oe: but nothing more came to pass that night.

The following evening, however, about the same hour, who should darken my door but the minister himself, a most discreet man, who had always paid me a very sympathysing attention from the death of my gudeman; so I received him with the greatest respect, wondering what could bring him to see me at that doubtful hour. But no sooner had he taken a seat in the elbow chair than he made my hair stand on end at the wickedness and perfidy of the woman sec.

'Mrs Blithe,' said he, 'I have come to have a serious word with you, and to talk with you on a subject that is impossible for me to believe. Last night that old Maudelin, of whom the world speaks no good, came to me with her grand-daughter from Glasgow, both weeping very bitterly; the poor young lass had her apron tail at her face, and was in great distress.'

'What is the matter with you,' said I, quoth the minister; 'and thereupon the piteous grandmother told me that her oe had been beguiled by a false manufacturing gentleman, and was thereby constrained to come back in a state of ignominy that was heartbreaking.'

'Good Maudelin, in what can I help you in your calamity?'

'In nothing, nothing,' said she; 'but we are come to make a confession in time.'

'What confession? quo' I'—that said the minister.

'Oh, sir,' said she, 'it's dreadful, but your counselling may rescue us from a great guilt. I have just been with Widow Blithe, the midwife, to bespeak her helping hand; oh, sir, speir no questions.'

'But,' said the minister, 'this is not a business to be trifled with; what did Mrs Blithe say to you?'

'That Mrs Blithe,' replied Maudelin, 'is a hidden woman; she made sport of my poor Lizzy's misfortune, and said that the best I could do was to let her nip the craig of the bairn in the hour of its birth.'

'Now, Mrs Blithe,' continued the Minister, 'is it possible that you could suggest such a crime?'

I was speechless; blue sterns danced before my sight, my knees trembled, and the steadfast earth grew as it were coggly aneath my chair; at last I replied,

'That old woman, sir, is of a nature, as she is of age enough, to be a witch—she's no canny! to even me to murder! Sir, I commit myself into your hands and judgment.'

'Indeed, I thought,' said the minister, 'that you would never speak as Maudelin said you had done; but she told me to examine you myself, for that she was sure, if I was put to the straights of a question, I would tell the truth.'

'And you have heard the truth, sir,' cried I.

'I believe it,' said he; 'but, in addition to all she rehearsed, she told me that, unless you, Mrs Blithe, would do your duty to her injured oe, and free gratis for no fee at all, she would go before a magistrate, and swear you had egged her on to bathe her hands in innocent infant blood.'

'Mr Stipend,' cried I; 'the wickedness of the human heart is beyond the computations of man: this dreadful old woman is, I'll not say what; but oh, sir, what am I to do; for if she makes a perjury to a magistrate my trade is gone, and my dear bairns driven to iniquity and beggary?'

Then the minister shook his head, and said, 'It was, to be sure, a great trial, for a worthy woman like me, to be so squeezed in the vice of malice and malignity; but a calm sough in all troubles was true wisdom, and that I ought to comply with the deceitful carlin's terms.'

Thus it came to pass, that, after the bastard brat was born, the old wife made a brag of how she had spirited the worthy minister to terrify me. Everybody laughed at her souple trick: but to me it was, for many a day, a heartburning; though, to the laive of the parish, it was a great mean, as I have said, of daffin and merriment.

No doubt, it will be seen, by the foregoing, that, although in a sense I had reason to be thankful that Providence, with the help of the laird's leddy-mother, had enabled me to make a bit of bread for my family, yet, it was not always without a trouble and an anxiety. Indeed, when I think on what I have come through in my profession, though it be one of the learned, and the world not able to do without it, I have often thought that I could not wish waur to my deadliest enemy than a kittle case of midwifery; for surely it is a very obstetrical business, and far above a woman with common talons to practise. But it would be to make a wearisome tale were I to lengthen my story; and so I mean just to tell of another accident that happened to me last year, and then to make an end, with a word or two of improvement on what shall have been said; afterwards I will give some account of what happened to those that, through my instrumentality, were brought to be a credit to themselves and an ornament to the world. Some, it is very true, were not just of that stamp; for, as the impartial sun shines alike on the wicked and the worthy, I have had to deal with those whose use I never could see, more than that of an apple that falleth from the tree, and perisheth with rottenness.

The case that I have to conclude with was in some sort mystical; and long it was before I got an interpretation thereof. It happened thus:—

One morning in the fall of the year and before break of day, when I was lying wakerife in my bed, I heard a knuckling on the pane of the window and got up to inquire the cause. This was by the porter of the Thistle Inns, seeking my help for a leddy at the crying, that had come to their house since midnight and could go no further.

I made no more ado, but dressed myself off-hand, and went to the Inns; where, to be sure, there was a leddy, for any thing that I then knew to the contrary, in great difficulty. Who she was, and where she had come from, I heard not; nor did I speir; nor did I see her face; for over her whole head she had a muslin apron so thrown and tied, that her face was concealed; and no persuasion could get her to remove that veil. It was therefore plain to me that she wished herself, even in my hands, not to be known; but she did not seem to jalouse that the very obstinacy about the veil would be a cause to make me think that she was afraid I would know her. I was not, however, overly-curious; for, among the other good advices that I got when I was about to take up the trade, from the leddy of Rigs, my patron, I was enjoined never to be inquisitive anent family secrets; which I have, with a very scrupulous care, always adhered to; and thus it happened, that, although the leddy made herself so strange as to make me suspicious that all was not right, I said nothing but I opened both my eyes and my ears.

She had with her an elderly woman; and, before she came to the worst, I could gather from their discourse, that the lady's husband was expected every day from some foreign land. By and by, what with putting one thing together with another, and eiking out with the help of my own imagination, I was fain to guess that she would not be ill pleased to be quit of her burden before the Major came home.

Nothing beyond this patch-work of hints then occurred. She had an easy time of it; and, before the sun was up, she was the mother of a bonny bairn. But what surprised me was, that, in less than an hour after the birth, she was so wonderful hale and hearty, that she spoke of travelling another stage in the course of the day, and of leaving Mrs Smith, that was with her, behind to take care of the babby; indeed, this was settled; and, before noon, at twelve o'clock, she was ready to step into the post-chaise that she had ordered to take herself forward—but mark the upshot.

When she was dressed and ready for the road—really she was a stout woman—another chaise drew up at the Inn's door, and, on looking from the window to see who was in it, she gave a shriek and staggered back to a sofa, upon which she fell like one that had been dumbfoundered.

In the chaise I saw only an elderly weather-beaten gentleman, who, as soon as the horses were changed, pursued his journey. The moment he was off, this mysterious mother called the lady-nurse with the babby, and they spoke for a time in whispers. Then her chaise was brought out and in she stepped, causing me to go with her for a stage. I did so and she very liberally gave me a five pound note of the Royal Bank and made me, without allowing me to alight, return back with the retour-chaise; for the which, on my account, she settled with the driver. But there the story did not rest, as I shall have occasion to rehearse by and by.

PART II—ANENT BAIRNS

Although I have not in the foregoing head of my subject mentioned every extraordinary han'ling that came to me, yet I have noted the most remarkable; and made it plain to my readers by that swatch of my professional work, that it is not an easy thing to be a midwife with repute, without the inheritance from nature of good common sense and discretion, over and above skill and experience. I shall now dedicate this second head, to a make-mention of such things as I have heard and known anent the bairns, that in their entrance into this world, came by the grace of God through my hands.

And here, in the first place, and at the very outset, it behoves me to make an observe, that neither omen nor symptom occurs at a birth, by which any reasonable person or gossip present can foretell what the native, as the unchristened baby is then called, may be ordained to come through in the course of the future. No doubt this generality, like every rule, has an exception; but I am no disposed to think the exceptions often kent-speckle; for although I have heard many a well-doing sagacious carlin notice the remarkables she had seen at some births, I am yet bound to say that my experience has never taught me to discern in what way a-come-to-pass in the life of the man was begotten of the uncos at the birth of the child.

But while I say this, let me no be misunderstood as throwing any doubt on the fact, that births sometimes are, and have been, in all ages, attended with signs and wonders manifest. I am only stating the truth it has fallen out in the course of my own experience; for I never misdoubt that it's in the power of Providence to work miracles and cause marvels, when a child is ordained with a superfluity of head-rope. I only maintain, that it is not a constancy in nature to be regular in that way, and that many prodigies happen at the times of births, of which it would not be a facile thing for a very wise prophet to expound the use. Indeed, my observes would go to the clean contrary; for I have noted that, for the most part, the births which have happened in dread and strange circumstances, were not a hair's-breadth better, than those of the commonest clamjamphry. Indeed, I had a very notable instance of this kind in the very first year of my setting up for myself, and that was when James Cuiffy's wife lay in of her eldest born.

James, as all the parish well knew, was not a man to lead the children of Israel through the Red Sea, nor she a Deborah to sing of butter in a lordly dish; but they were decent folk; and when the fulness of her time was come, it behoved her to be put to bed, and my helping hand to be called for. Accordingly I went.

It was the gloaming when James came for me; and as we walked o'er the craft together, the summer lightning ayont the hills began to skimmer in a woolly cloud: but we thought little o't, for the day had been very warm, and that flabbing of the fire was but a natural outcoming of the same cause.

We had not, however, been under the shelter of the roof many minutes, when we heard a-far off, like the ruff of a drum or the hurl of a cart of stones tumbled on the causey, a clap of thunder, and then we heard another and another, just like a sea-fight of Royal Georges in the skies, till the din grew so desperate, that the crying woman could no more be heard than if she had been a stone image of agony.

I'll no say that I was not in a terrification. James Cuiffy took to his Bible, but the poor wife needed all my help. At last the bairn was born; and just as it came into the world, the thunder rampaged, as if the Prince of the Powers of the air had gaen by himself; and in the same minute, a thunder-bolt fell doun the lum, scattered the fire about the house, whiskit out of the window, clove like a wedge the apple-tree at the house-end, and slew nine sucking pigs and the mother grumphy, as if they had been no better than the host of Sennacherib; which every body must allow was most awful: but for all that, nothing afterwards came to pass; and the bairn that was born, instead of turning out a necromancer or a geni, as we had so much reason to expect, was, from the breast, as silly as a windlestraw. Was not this a plain proof that they are but of a weak credulity who have faith in freats of that kind?

I met, likewise, not in the next year, but in the year after, nearer to this time, another delusion of the same uncertainty. Mrs Gallon, the exciseman's wife, was overtaken with her pains, of all places in the world, in the kirk, on a Sabbath afternoon. They came on her suddenly, and she gave a skirle that took the breath with terror from the minister, as he was enlarging with great bir on the ninth clause of the seventh head of his discourse. Every body stood up. The whole congregation rose upon the seats, and in every face was pale consternation. At last the minister said, that on account of the visible working of Providence in the midst of us, yea in the very kirk itself, the congregation should skail: whereupon skail they did; so that in a short time I had completed my work, in which I was assisted by some decent ladies staying to lend me their Christian assistance; which they did, by standing in a circle round the table seat where the ploy was going on, with their backs to the crying mother, holding out their gowns in a minaway fashion, as the maids of honour are said to do, when the queen is bringing forth a prince in public.

The bairn being born, it was not taken out of the kirk till the minister himself was brought back, and baptized it with a scriptural name; for it was every body's opinion that surely in time it would be a brave minister, and become a great and shining light in the Lord's vineyard to us all. But it is often the will and pleasure of Providence to hamper in the fulfilment the carnal wishes of corrupt human nature. Matthew Gallon had not in after life the seed of a godly element in his whole carcase; quite the contrary, for he turned out the most rank ringing enemy that was ever in our country-side; and when he came to years of discretion, which in a sense he never did, he fled the country as a soldier, and for some splore with the Session, though he was born in the kirk;—another plain fact that shows how little reason there is in some cases to believe that births and prognostifications have no natural connexion. Not that I would condumaciously maintain that there is no meaning in signs sometimes, and may be I have had a demonstration; but it was a sober advice that the auld leddy of Rigs gave me, when she put me in a way of business, to be guarded in the use of my worldy wisdom, and never to allow my tongue to describe what my eyes saw or my ears heard at an occasion, except I was well convinced it would pleasure the family.

'No conscientious midwife', said she, 'will ever make causey-talk of what happens at a birth, if it's of a nature to work dule by repetition on the fortunes of the bairn;' and this certainly was most orthodox, for I have never forgotten her counsel.

I have, however, an affair in my mind at this time; and as I shall mention no names, there can be no harm done in speaking of it here; for it is a thing that would perplex a philosopher or a mathematical man, and stagger the self-conceit of an unbeliever.

There was a young Miss that had occasion to come over the moor by herself one day, and in doing so she met with a hurt; what that hurt was, no body ever heard; but it could not be doubted that it was something most extraordinar; for, when she got home, she took to her bed and was very unwell for several days, and her een were blear't with greeting. At last, on the Sabbath-day following, her mother foregathert with me in coming from the kirk; and the day being showery, she proposed to rest in my house as she passed the door, till a shower that she saw coming would blow over. In doing this, and we being by ourselves, I speired in a civil manner for her daughter; and from less to more she told me something that I shall not rehearse, and, with the tear in her eye, she entreated my advice; but I could give her none, for I thought her daughter had been donsie; so no more was said anent it; but the poor lassie from that day fell as it were into a dwining, and never went out; insomuch that before six months were come and gone, she was laid up in her bed, and there was a wally-wallying on her account throughout the parish, none doubting that she was in a sore way, if not past hope.

In this state was her sad condition, when they had an occasion for a gradawa at my Lord's; and as he changed horses at the Cross Keys when he passed through our town, I said to several of the neighbours, to advise the mother that this was a fine opportunity she ought not to neglect, but should consult him anent her dochter. Accordingly, on the doctor returning from the castle, she called him in; and when he had consulted the ailing lassie as to her complaint, every body rejoiced to hear that he made light of it, and said that she would be as well as ever in a month or two; for that all she had to complain of was but a weakness common to womankind, and that a change of air was the best thing that could be done for her.

Maybe I had given an advice to the same effect quietly before, and therefore was none displeased to hear, when it came to pass, that shortly after, the mother and Miss were off one morning, for the benefit of the air of Glasgow, in a retour chaise, by break of day, before anybody was up. To be sure some of the neighbours thought it an odd thing that they should have thought of going to that town for a beneficial air; but as the report soon after came out to the town that the sick lassie was growing brawly, the wonder soon blew over, for it was known that the air of a close town is very good in some cases of the asthma.

By and by, it might be six weeks or two months after, aiblins more, when the mother and the daughter came back, the latter as slimb as a popular tree, and blooming like a rose. Such a recovery after such an illness was little short of a miracle, for the day of their return was just ten months from the day and date of her hurt.

It is needless for me to say what were my secret thoughts on this occasion, especially when I heard the skill of the gradawa extolled, and far less how content I was when, in the year following, the old lady went herself on a jaunt into the East Country to see a sick cousin, a widow woman with only a bairn, and brought the bairn away with her on the death of the parent. It was most charitable of her so to do, and nothing could exceed the love and ecstasy with which Miss received it from the arms of her mother. Had it been her ain bairn she could not have dandalized it more!

Soon after this the young lady fell in with a soldier officer, that was sent to recruit in the borough, and married him on a short acquaintance, and went away with him a regimenting to Ireland; but 'my cousin's wee fatherless and motherless orphan,' as the old pawkie carlin used to call the bairn, stayed with her, and grew in time to be a ranting birkie; and in the end, my lord hearing of his spirit, sent for him one day to the castle, and in the end bought for him a commission, in the most generous manner, such as well befitted a rich young lord to do; and afterwards, in the army, his promotion was as rapid as if he had more than merit to help him.

Now, is not this a thing to cause a marvelling; for I, that maybe had it in my power to have given an explanation, was never called on so to do; for everything came to pass about it in such an ordained-like way, that really I was sometimes at a loss what to think, and said to myself surely I have dreamt a dream; for, although it could not be said to have been a case of prognostications, it was undoubtedly one of a most kittle sort in many particulars. Remembering, however, the prudent admonition I had received from the auld leddy of Rigs, I shall say no more at present, but keep a calm sough.

It is no doubt the even-down fact that I had no hand in bringing 'my cousin's wee fatherless and motherless orphan' into the world, but maybe I might have had, if all the outs and ins of the story were told. As that, however, is not fitting, I have just said enough to let the courteous reader see, though it be as in a glass darkly, that my profession is no without the need of common sense in its handlings, and that I have not earned a long character for prudence in the line without ettle, nor been without jobs that cannot be spoken of, but, like this, in a far-off manner.

But it behoves me, before I go farther, to request the reader to turn back to where I have made mention of the poor deserted bairn, Willy Facings; how he was born in an unprepared hurry, and how his mother departed this life, while his ne'er-do-weel father went away like a knotless thread. I do not know how it happened, but come to pass it did, that I took a kindness for the forsaken creature, insomuch that, if his luck had been no better with Miss Peggy Needle, it was my intent to have brought him up with my own weans; for he was a winsome thing from the hour of his birth, and made every day a warmer nest for his image in my heart. His cordial temper was a mean devised by Providence as a compensation to him for the need that was in its own courses, that he would never enjoy a parent's love.

When Miss Peggy had skailed the byke of her cats, and taken Billy, as he came to be called, home to her house, there was a wonderment both in the borough-town and our clachan how it was possible for her, an inexperienced old maid, to manage the bairn; for by this time he was weaned, and was as rampler a creature as could well be, and she was a most prejinct and mim lady. But, notwithstanding her natural mimness and prejinkity, she was just out of the body with love and tenderness towards him, and kept him all day at her foot, playing in the inside of a stool whamled up-side down.

It was the sagacious opinion of every one, and particularly both of the doctor and Mr Stipend, the minister, that the bairn would soon tire out the patience of Miss Peggy; but we are all short-sighted mortals, for instead of tiring her, she every day grew fonder and fonder of him, and hired a lassie to look after him, as soon as he could tottle. Nay, she bought a green parrot for him from a sailor, when he was able to run about; and no mother could be so taken up with her own get as kind-hearted Miss Peggy was with him, her darling Dagon; for although the parrot was a most outstrapolous beast, and skrighed at times with louder desperation than a pea-hen in a passion, she yet so loved it on his account, that one day when it bit her lip to the bleeding, she only put it in its cage, and said, as she wiped her mouth, that it was 'a sorrow'.

By and by Miss Peggy put Billy to the school; but, by that time, the condumacious laddie had got the upper hand of her, and would not learn his lesson, unless she would give him an apple or sweeties; and yet, for all that, she was out of the body about him, in so much that the minister was obligated to remonstrate with her on such indulgence; telling her she would be the ruin of the boy, fine creature as he was, if she did not bridle him, and intended to leave him a legacy.

In short, Miss Peggy and her pet were just a world's wonder, when, at last, Captain Facings, seven years after Billy's birth, being sent by the king to Glasgow, came out, one Sunday to our town, and sent for me to learn what had become of his bairn. Though I recollected him at the first sight, yet, for a matter of policy, I thought it convenient to pretend doubtful of my memory, till, I trow, I had made him sensible of his sin in deserting his poor baby. At long and length I made him to know the blessing that had been conferred by the fancy of Miss Peggy, on the deserted child, and took him myself to her house. But, judge of my consternation, and his likewise, when, on introducing him to her as the father of Billy, whom I well recollected, she grew very huffy at me, and utterly denied that Billy was any such boy as I had described, and foundled over him, and was really in a comical distress, till, from less to more, she grew, at last, as obstinate as a graven image, and was not sparing in the words she made use of to get us out of her habitation.

But, not to summer and winter on this very unforeseen come-to-pass, the Captain and I went to the minister, and there made a confession of the whole tot of the story. Upon which he advised the Captain to leave Billy with Miss Peggy, who was a single lady, not ill-off in the world; and he would, from time to time, see that justice was done to the bairn. They then made a paction concerning Billy's education; and, after a sore struggle, Miss Peggy, by the minister's exhortation, was brought to consent that her pet should be sent to a boarding-school, on condition that she was to be allowed to pay for him.

This was not difficult to be agreed to; and, some weeks after, Bill was accordingly sent to the academy at Green Knowes, where he turned out a perfect delight; and Miss Peggy sent him every week, by the carrier, a cake, or some other dainty. At last, the year ran round, and the vacance being at hand, Bill sent word by the carrier, that he was coming home to spend the time with Mamma, as he called Miss Peggy. Great was her joy at the tidings; she set her house in order, and had, at least, twenty weans, the best sort in the neighbourhood, for a ploy to meet him. But, och hone! when Billy came, he was grown such a big creature, that he no longer seemed the same laddie; and, at the sight of him, Miss Peggy began to weep and wail, crying, that it was an imposition they were attempting to put upon her, by sending another callan. However, she became, in the course of the night, pretty well convinced that he was indeed her pet; and, from that time, though he was but eight years old, she turned over a new leaf in her treatment.

Nothing less would serve her, seeing him grown so tall, than that he should be transmogrified into a gentleman; and, accordingly, although he was not yet even a stripling—for that's a man-child in his teens—she sent for a taylor next day and had him put into long clothes, with top boots; and she bought him a watch, and just made him into a curiosity, that nowhere else could be seen.

When he was dressed in his new clothes and fine boots, he went out to show himself to all Miss Peggy's neighbours; and, it happened, that, in going along, he fell in with a number of other childer, who were sliding down a heap of mixed lime, and the thoughtless brat joined them; by which he rubbed two holes in the bottom of his breeks, spoiled his new boots, and, when the holes felt cold behind, he made his hat into a seat, and went careering up the heap and down the slope with it, as if he had been a charioteer.

Everybody who saw the result concluded that certainly now Miss Peggy's favour was gone from him for ever. But she, instead of being angry, just exclaimed and demonstrated with gladness over him; saying, that, till this disaster, she had still suspected that he might turn out an imposture. Was there ever such infatuation? But, as I shall have to speak more anent him hereafter, I need not here say how he was sent back to the academy, on the minister's advice, just dress'd like another laddie.

William Alexander

1826–1894

Baubie Huie's Bastard Geet

Jock Huie's Household—Baubie Enters Life

I am not prepared to say how far Baubie Huie's own up-bringing had been a model of judicious parental nurture. There was ground to fear that it had not been at all times regulated by an enlightened regard to the principle laid down by King Solomon, concerning the training up of children. Jock Huie had a muckle sma' faimily, crammed into limited space, in so far as the matter of house accommodation was concerned. It was a little, clay-built, 'rape-thackit' cot in which Jock, with Eppie, his wife, and their family dwelt; and the 'creaturs' came so thickly, and in such multitude, that Jock, who was a 'darger', and did 'days' warks' here and there, as he could find them, experienced rather queer sensations when an unusually 'coorse' day happened to coop him up at home among the 'smatterie' of youngsters.

'Saul o' me, 'oman,' would Jock exclaim, when patience had reached its limit; 'the din o' that bairns o' yours wud rive a heid o' steen—gar them be quaet, aw'm sayin', or I'll hae to tak' a horse fup to them.'

'Haud yer tongue, man; gin ye war amo' them fae screek o' day till gloamin licht 's I am, ye mith speak. Fat can the creaturs dee fan they canna get leuket owre a door?' Eppie would reply.

Notwithstanding his formidable threat, Jock Huie rarely lifted his hand in the way of active correction of his offspring. His wife, who was not indisposed to govern a little more sharply if she could, knew of only one way of enforcing obedience, or some approach thereto, when matters had come to a decided pass of the character indicated, and which may be best described in plain English as indiscriminate chastisement, applied with sufficient heartiness, though it might be quite as much in accordance with the dictates of temper as of calm reason. And so it came to pass that, as most of the youthful Huies were gifted with pretty definite wills of their own, the progress of physical development on their part might be taken, in a general way, as indicative, in inverse proportion, of the measure of moral and mental sway which the parental will was able to exercise over them.

All that by the way, however. Jock Huie got his family brought up as he best could, and off his hands mainly; and he, personally, continued his dargin' with perhaps a little less vir than aforetime. Jock was a man of large bones and strong bodily frame; when thirty he had physical strength that seemed equal to any task, and endurance against which no amount of rough usage appeared to tell with evil effect. But after all, men of Jock Huie's class do not wear long. Jock was now a man only a few years past fifty; yet digging in wet drains and ditches, and eating a bit of oat cake, washed down with 'treacle ale', to his dinner, day by day, had procured for him a very appreciable touch of 'rheumatics', and other indications that he had fairly passed his prime.

And Baubie, his eldest daughter, though not the eldest member of his family, for Jock had various sons older than she—Baubie had grown up—a buxom, ruddy-cheeked 'quine' of nineteen. She was servan' lass to the farmer of Brigfit—Briggies in short.

I remember very distinctly a bonnie summer gloamin at that time. It was gey late owre i' the evenin'. Baubie had milket the kye, seyt the milk, and wash'n up her dishes. Her day's work was at last fairly done, and why should not Baubie go out to the Toon Loan to enjoy the quiet scene as the cool dews of evening began to fall upon the landscape around the cosy, old-fashioned farm 'steading' of Brigfit.

It matters nothing in this narration where I had been that evening, further than to say that, as I pursued my journey homeward, the road took me past the corner of Briggies' stable, where, altogether unexpectedly to me, I encountered Baubie Huie 'in maiden meditation fancy free'. Though Baubie's junior by a twelvemonth or so, I had developed since we two had last met from a mere herd loon into a sort of rawish second or third horseman. We had known each other more or less from infancy, Baubie and I, and our talk during the short parley that now ensued had a tinge of the byegone time in it; though, of course, we could not help giving fulfilment, in our own way, to the saying that out of the abundance of the heart the mouth speaketh; and, naturally enough, at that season of life, that which most occupied our hearts was the present as it bore on our respective positions and prospects.

My own notion (it may be said in confidence) was that I was climbing up the pathway to maturity of life and definiteness of position with creditable alacrity; but in this direction I speedily found that Baubie Huie had fairly out-distanced me. Why, here was the very same 'quine' who, almost the last time I saw her, was lugging along a big, sulky bairn, half her own size, wrapped in an old tartan plaid, and her weather-bleached hair hanging loosely about her shoulders—and that bairn her own younger brother—that very 'quine', giggling and tossing her head knowingly as she spoke, in what seemed a tone of half masculine licence, about the 'chiels' that were more or less familiarly known as sweethearts among young women in the neighbourhood of Brigfit. In matters of love and courtship, I was, it must be confessed, an entire novice; whereas in such affairs, it was obvious, Baubie had become an adept; and if I had been somewhat put out by the ready candour with which she criticised the physical appearance and general bearing of this and the other young man—hangers on after Baubie, I was given to understand—I was nothing short of completely 'flabbergasted' when, just as we were parting, she said—

'Dinna ye never gae fae hame at even, min? Ye mith come owre the gate some nicht an' see's.'

What my confused and stuttering reply amounted to, I cannot really say—something grotesquely stupid, no doubt. What it called forth on Baubie's part, at any rate, was another round of giggling and the exclamation, as she turned off toward the dwelling-house of Brigfit—

'Weel, weel, Robbie, a' nicht wi' you; an' a file o' the morn's mornin'.' —This was simply the slang form of saying 'good night' among persons of Baubie's class. And she added—'I'll need awa' in; for there 's Briggies, the aul' snot, at the ga'le o' the hoose—he'll be barrin' 's oot again, eenoo.'

Now, far be it from me to say that Baubie was a vicious or immodest young woman. I really am not prepared to say that she was anything of the sort. She had simply got the training that hundreds in her station of life in these northern shires do—home training that is. And after she left the parental roof, her experiences had been the common experiences of her class—that is to associate freely with promiscuous assemblages of farm-servants, male and female; mainly older than herself, without any supervision worth mentioning, as she moved from one situation to another. And how could Baubie, as an apt enough scholar, do other than imbibe the spirit and habits of those in whose companionship she lived day by day? Baubie was simply the natural product of the system under which she had been reared. Her moral tone, as indexed by her speech, might not be very high; and yet, after all, it is very possible to have the mere verbal proprieties fully attended to, where the innate morality is no whit better. Coarseness in the outer form, which is thrust on the view of all, is bad enough; depravity in the inner spirit, which is frequently concealed from many, may be a good deal worse.

Brigfit was a decent man; a very decent man, for he was an elder in the parish kirk, and a bachelor of good repute. He was a careful, industrious farmer, the extent of whose haudin enabled him to 'ca' twa pair'. Briggies was none of your stylish gentlemen farmers; he needed neither gig nor 'shalt' to meet his personal convenience, but did his ordinary business journeys regularly on foot. And he stood on reasonably amicable terms with his servants; but he sought little of their confidence, and as little did he give to them of his own. Only Briggies had certain inflexible rules, and one was that his household should be in bed every night by nine o'clock in winter, and an hour later in summer; when he would himself solemnly put the bar on the door, and then walk as solemnly along to the 'horn en'' to seek repose.

Briggies was a very early riser, and as it was his hand that usually put the bar on the door at night, so, honest man, was it his hand that ordinarily took it off in the morning in time to see that the household proper and the occupants of the outside 'chaum'er', consisting of the male servants, were stirring to begin the labours of the day in due season. According to Baubie Huie's account, the bar was sometimes tampered with during the interval by the 'deems'; only if matters were gone about quietly enough, Briggies, whether or not he might suspect aught in that way, usually said nothing.

'Augh, Robbie, man! Fear't for Briggies kennin? Peer bodie! fan onything comes in 's noddle aboot 's nowte beasts he canna get rest, but 'll be up an' paumerin aboot the toon o' the seelence o' the nicht, fan it's as mark 's pick in winter, forbye o' the simmer evenin's. So ae nicht i' the spring time that me an' my neebour hedna been wuntin to gae to oor beds, we pits oot the lamp in gweed time, an' sits still, as quaet 's pussy, till Briggies hed on the bar an' awa' till 's bed. I'm nae sayin' gin onybody was in ahin that or no, but lang aifter the wee oor hed struck'n, me an' Jinse was thereoot. I suppose the chiels hed made mair noise nor they sud 'a deen, caperin' owre the causeway wi' their muckle tacketie beets. At ony rate in a blink there was Briggies oot an' roon to the byres wi' the booet in 's han'. Fan he hed glampit aboot amo' the beasts till he was satisfeet, he gaes awa' to the hoose again; an' we wusna lang o' bein' aifter 'im. But fudder or no he hed leuket ben to the kitchie to see gin we wus there, he hed pitten the bar siccar aneuch on upo' the door this time, I can tell ye; an' nae an in cud we win for near an oor, till we got an aul' ledder an' pat it up to the en' o' the hoose, an' syne I made oot to creep in at the ga'le winnockie—Fat did he say aifterhin? Feint a thing. Briggies never loot on, though he cudna but 'a hed 's ain think, 'cause gin he didna hear huz, he be 't till 'a kent "gyaun oot" that the bar sudna 'a been aff o' the door at that time o' nicht.'

In this wise did Baubie Huie keep up the colloquy, my own side of which, candour compels me to say, was very badly sustained; for had I been ever so willing to take my part, the requisite fluency and *abandon* had not been attained, to say nothing of the utter absence of knowledge germane to the subject in hand, and personally acquired.

As a matter of course, I did not accept Baubie Huie's invitation to visit Brigfit. If the truth were to be told, I was too much of a greenhorn; one who would have been accurately described by Baubie and her associates as utterly destitute of 'spunk'. My Mentor of that date, a vigorous fellow of some eight and twenty years, whose habits might be not incorrectly described by the word 'haiveless', whose speech was at least as free, as refined, and who occupied the responsible position of first horseman, did not indeed hesitate to characterise my behaviour in relation to such matters, generally, in almost those very words. He knew Baubie Huie, moreover, and his estimate of Baubie was expressed in the words—'Sang, she's a richt quine yon, min; there's nae a deem i' the pairt 'll haud 'er nain wi' ye better nor she'll dee; an' she's a fell ticht gweed-leukin hizzie tee,' which, no doubt, was a perfectly accurate description according to the notions entertained by the speaker of the qualities desirable in the female sex.

However these things may be, Baubie Huie continued to perform her covenanted duties to the farmer of Brigfit; and, so far as known, yielding the elder average satisfaction as a servant during the summer 'half-year'.

CHAPTER II

BAUBIE RETURNS HOME

It was nearing the term of Martinmas, and Jock Huie, who had been laid off work for several days by a 'beel't thoom', was discussing his winter prospect with Eppie, his wife. Meal was 'fell chape', and the potato crop untouched by disease; but Jock's opinion was that, as prices were low for the farmer, feein' would be slack. Cattle were down too, and though the price of beef and mutton was a purely abstract question for him personally—he being a strict vegetarian in practice, not by choice but of necessity—Jock was economist enough to know that the fact bore adversely on the farmer's ability to employ labour; so that, altogether, with a superfluity of regular servants unengaged, and a paucity of work for the common 'darger' in the shape of current farming improvements going on, he did not regard the aspect of things as cheering for his class.

'Aw howp neen o' that loons o' oors 'll throw themsel's oot o' a place,' said Jock. 'Wud ye think ony o' them wud be bidin'?'

'That wud be hard to say, man,' replied Eppie.

'That widdifus o' young chiels 's aye sae saucy to speak till,' said Jock; whether he meant that the sauciness would be exhibited in the concrete from his own sons toward himself, or if the remark applied to the bearing of servant chiels generally on the point under consideration, was not clear. 'But better to them tak' a sma' waage nor lippen to orra wark; an' hae to lie aboot idle the half o' the winter.'

'Weel ken we that,' said Eppie, with a tolerably lively recollection of her experiences in having previously had one or two of her sons 'at hame' during the winter season. 'Mere ate-meats till Can'lesmas; I'm seer fowk hae's little need o' that; but creaturs'll tak' their nain gate for a' that.'

'Aw howp Baubie's bidin' wi' Briggies, ony wye,' added Jock.

'I ken naething aboot it,' said Eppie, in a tone that might be described as dry; 'Baubie's gey an' gweed at keepin' 'er coonsel till 'ersel'.'

It was only a fortnight to the term, and Jock would not be kept long in suspense regarding those questions affecting the family arrangements on which he had thus incidentally touched. In point of fact, his mind was set at rest so far when only half the fortnight had run. For the feeing market came in during that period, and as Jock's thumb had not yet allowed him to resume work, he 'took a step doon' to the market, where he had the satisfaction of finding that his sons had all formed engagements as regular farm servants. As for Baubie, though Jock learned on sufficient authority that she was present in the market, he failed to 'meet in' with her.

Concerning Baubie's intended movements, he learnt, too, that she was *not* staying with Briggie's; Briggies himself had indeed told him so; but beyond that Jock's inquiries on the subject did not produce any enlightenment for him.

Subsequently to the feeing market, Jock Huie had once and again reverted to the subject of Baubie's strange behaviour in keeping the family in ignorance of her movements and intentions, but without drawing forth much in the way of response from his wife beyond what she had generally expressed in her previous remark.

The afternoon of the term day had come, and servants who were flittin' were moving here and there. I cannot state the nature of the ruminations that had passed, or were passing, through the mind of either Jock Huie or his wife Eppie concerning their daughter Baubie; but Jock, honest man, had just left his cottage in the grey gloamin to go to the smiddy and get his tramp-pick sharpened with the view of resuming work next day in full vigour, when Baubie dressed in her Sunday garments, and carrying a small bundle, entered. There was a brief pause; and then Baubie's mother, in a distinct and very deliberate tone, said—

'Weel, Baubie, 'oman; an' *ye're* here neist.'

At these words, Baubie, who had just laid aside her bundle, threw herself down beside it, on the top of the family 'deece', with the remark,

'Ay; faur ither wud aw gae?'

And then she proceeded silently to untie the strings of her bonnet. Neither Baubie nor her mother was extremely agitated, but there was a certain measure of restrained feeling operating upon both the one and the other. The mother felt that a faithful discharge of the maternal duty demanded that she should give utterance to a reproof as severe as she could properly frame, accompanied by reproaches, bearing on the special wickedness and ingratitude of the daughter; and, on the part of the daughter along with a vague sense of the fitness of all this, in a general way, there were indications of a volcanic state of temper, which might burst out with considerable, if misplaced fierceness, on comparatively slight provocation. And wherefore create a scene of verbal violence; for deep down, below those irascible feelings, did there not lurk in Eppie Huie's bosom a kind of latent sense that if such crises as that which had now emerged were not to be regarded as absolutely certain, they were assuredly to be looked upon as very much in the nature of events inevitable in the ordinary history of the family? And thus it was that Eppie Huie, virtually accepting the situation as part of the common lot, went no further than a general rasping away at details, and the consequences arising out of the main fact.

'Weel, weel, Baubie, 'oman, ye've begun to gae the aul' gate in braw time—ye'll fin't a hard road to yersel', as weel's to them 't's near conneckit wi' you. Fat gar't ye keep oot o' yer fader's sicht at the market—haudin 'im gyaun like a wull stirk seekin' ye, an' makin' a feel o' 'im?'

'Aw 'm seer ye needna speer that—'s gin ye hedna kent to tell 'im yersel'.'

'That's a bonnie story to set up noo, ye limmer—that I sud say the like,' said Eppie with some heat. 'Didnin ye deny 't i' my face the vera last time that ye was here?'

'H-mph! an' aw daursay ye believ't 's!'

'Weel, Baubie, 'oman, it 's a sair say 't we sud be forc't to tak' for a muckle black lee fat's been threepit, an' yea-threepit i' oor witters b' them that 's sibbest till 's.'

To this observation Baubie made no reply: and after a short silence Eppie Huie continued in a dreary monotone—

'Ay, ay! An' this is fat folk gets for toilin' themsel's to deith feshin up a faimily! There's little aneuch o' peace or rest for's till oor heid be aneth the green sod—jist oot o' ae tribble in till anither. Little did I or yer peer fader think short syne that *ye* was to be hame to be a burden till 's.'

'Aw ha'ena been a burden yet ony wye,' said Baubie with some sharpness, 'ye needna be sae ready speakin' that gate.'

To this retort Eppie Huie made some reply to the effect that others similarly circumstanced had uttered such brave words, and that time would tell in Baubie's case as it had told in theirs. She then rose and put some water in a small pot, which she hung upon the 'crook' over the turf fire, in the light of which Baubie and she had hitherto sat.

'Fa's the fader o' 't than?' said Eppie Huie, as she turned about from completing the operation just mentioned; but though the words were uttered in a very distinct as well as abrupt tone, there was no answer till she repeated her question in the form of a sharp 'Aw 'm sayin'?'

'Ye 'll ken that a-time aneuch,' answered Baubie.

'Ken 't a-time aneuch!—an' you here'—

'Ay an' me here—an' fat aboot it? *It* winna be here the morn, nor yet the morn's morn,' said Baubie in a harder and more reckless tone than she had yet assumed.

Eppie Huie had, no doubt, a sense of being baffled, more or less. She resumed her seat, uttering as she did so, something between a sigh and a groan. There was nothing more said until the water in the little pot having now got to 'the boil', Eppie rose, and lighting the rush wick in the little black lamp that hung on the shoulder of the 'swye' from which the crook depended, proceeded to 'mak' the sowens'. When the lamp had been lighted, Baubie rose from her place on the deece, and lifting her bonnet, which now lay beside her, and her bundle, said,

'Aw 'm gyaun awa' to my bed.'

'Ye better wyte an' get yer sipper—the sowens 'll be ready eenoo.'

'Aw 'm nae wuntin' nae sipper,' said Baubie, turning to go as she spoke. 'There's nae things lyin' i' the mid-hoose bed, is there?'

'Naething; oonless it be the muckle basket, wi' some o' yer breeders' half-dry't claes. Tak' that bit fir i' yer han'—ye'll need it, ony wye, to lat ye see to haud aff o' the tubs an' the backet.'

And Baubie went off to bed forthwith, notwithstanding a sort of second invitation, as she was lighting the fir, to wait for some supper. I rather think that after all she did not relish the comparative light so much as the comparative darkness. And then if she stayed to get even the first practicable mouthful of 'sowens', was there not considerable risk that Jock Huie, her father, might drop in upon her on his return from the smiddy? Not that Baubie had an unreasonably sensitive dread of facing her father. But having now got over what she would have called 'the warst o' 't', with her mother, she felt that her mother, being on the whole so well 'posted up', might be left with advantage to break the ice, at least, to the old man.

When Jock Huie returned from the smiddy that evening, an event that happened in about half an hour after his daughter Baubie had gone to bed, he seemed to be moody, and in a measure out of temper. He put aside his bonnet, and sat down in his usual corner, while Eppie set the small table for his supper, only one or two remarks of a very commonplace sort having been made up to that point.

'Ye 'll better say awa', man; they've been made this file,' said Eppie, as she lifted the dish with the 'sowens' to the table from the hearthstone, where it had been placed in order to retain warmth in the mess.

'Aw 'm sayin', 'oman,' quoth Jock, apparently oblivious to his wife's invitation, 'div ye ken onything about that jaud Baubie—there's something or anither nae richt, ere she wud haud oot o' fowk's road this gate?'

'Baubie's *here*, man,' said Eppie Huie; and the brevity of her speech was more than made up by the significance of the words and the tone in which they were uttered.

'Here?' exclaimed Jock in a tone of inquiry, and looking towards his wife as he spoke.

'She's till 'er bed i' the mid-hoose,' said Eppie in reply; and, perceiving that Jock's look was only half answered, she added, 'Aw daursay she wasna owre fain to see you.'

'Fat!' cried Jock, 'she'll be wi' a geet to some chiel, is she?'

'Ou ye needna speer,' said Eppie in a tone of 'dowie' resignation.

'Weel, that does cowe the gowan—a quine o' little mair nor nineteen! But aw mith 'a been seer o' 't. It wasna for naething that she was playin' hide-an'-seek wi' me yon gate. Brawlie kent I that she was i' the market wi' a set o' them. Deil speed them a', weel-a-wat!'

Jock Huie was not a model man exactly in point of moral sentiment; neither was he a man of keen sensibility. But he did nevertheless possess a certain capability of sincere, if it might be uncultured feeling; and he now placed his rough, weather-beaten face against the horny palms of his two hands, and, resting his two elbows on his knees, gave utterance to a prolonged 'Hoch-hey?' Jock maintained this attitude for sometime, and probably would have maintained it a good deal longer, but for the practical view of matters taken by his wife, and the practical advice urgently pressed upon him by her when her patience had got exhausted:—

'Aw 'm sayin', man, ye needna connach yer sipper; that'll dee nae gweed to naebody.—Tak' your sowens! Ye're lattin them grow stiff wi' caul', for a' the tribble 't aw was at keepin' them het to you.'

Thus admonished, Jock Huie took his supper in silence; and, thereafter, with little more talk beyond one or two questions from Jock of a like nature with those which had been so ineffectually addressed to Baubie by her mother, the husband and wife retired to bed.

CHAPTER III

THE GEET'S ADVENT—INITIAL DIFFICULTIES IN ACQUIRING AN ECCLESIASTICAL STATUS

That Jock Huie's daughter, Baubie, had returned home to her father and mother was a fact about which there could be no manner of doubt or equivocation; as to the cause of Baubie's return, there was a general concurrence of opinion in the neighbourhood; indeed, it had been a point settled long before, among elderly and sagacious females who knew her, that Baubie would speedily appear in her true colours. Yet were there a few of this same class of people in whose sides Baubie was still somewhat of a thorn. For when the first few days were over after her return, so far from shrinking out of their sight, Baubie flung herself across their path at the most unexpected times, and exhibited an unmistakeable readiness to meet their friendly criticisms with a prompt retort. Or was it a staring personal scrutiny—well, Baubie was almost ostentatiously ready to stand that ordeal, and stare with the best of her starers in return. Baubie was perfectly able to take care of herself, and if a young woman of her spirit chose to remain six months out of the 'hire house', whose business was that but her own? Baubie would like to know that.

It is not to be supposed that this bravado went far in the way of deceiving any but very inexperienced people, if it deceived even them, which is more than doubtful. And in the nature of the case, it would at any rate deceive no one very long.

It was just at Candlemas when it was reported that Jock Huie had become a grandfather; a genealogical dignity the attainment of which did not seem to excite in Jock's breast any particular feeling of elation. Such an idea as that of apprehension lest the line of Huies in his branch should become extinct had certainly never troubled Jock to the extent that would have made him anxious to welcome a grandchild, legitimate or illegitimate; and the belief that this particular bairn was born to be a direct and positive burden upon him hardly tended to make its advent either auspicious or cheering. Jock knew full well the 'tyauve' he had had in bringing up his own family proper; and now, ere the obstreperous squalling of the younger of them was well out of his ears, why here was another sample of the race, ready to renew and continue all that turmoil and uproar, by night and by day, from which his small hut had never been free for a good twenty years of his lifetime.

'An' it's a laddie, ye say, that the quine Huie's gotten?'

'A laddie; an' a-wat a richt protty gate-farrin bairnie 's ever ye saw wi' yer twa een.'

'Fan cam' 't hame no?'

'It was jist the streen, nae langer gane. Aifter 't was weel gloam't, I hears a chap at the window, an' fa sud this be but Eppie 'ersel', peer creatur. I pat my tartan shawl aboot my heid immedantly, an' aifter tellin' the littleans to keep weel ootbye fae the fire, an' biddin' their sister pit them to their beds shortly, I crap my wa's roun' as fest 's aw cud. Jock was nae lang come hame fae 's day's wark, an' was sittin' i' the neuk at 's bit sipper. "He's jist makin' ready to gae for Mrs Slorach," says she. Awat I was rael ill-pay't for 'im, peer stock, tir't aneuch nae doot, jist aff o' a sair day's wark. It was a freely immas nicht, wi' byous coorse ploiterie road; an' it's three mile gweed, but I can asseer ye Jock hed gane weel, for it wasna muckle passin' twa oors fan he's back an' Mrs Slorach wi' 'im.'

'Weel, weel, Jock 'll get 's nain o' 't lickly, honest man. It'll be a won'er an' they hinna the tsil' to fesh up.'

'Ou weel-a-wat that's true aneuch; but there's never a hicht but there 's a howe at the boddom o' 't, as I said to Eppie fan she first taul' me o' Baubie's misfortune; an' there 's never a mou' sen' but the maet 's sen' wi' 't.'

'Div they ken yet fa 's the fader o' the creatur?'

'Weel, she hed been unco stubborn aboot it no; but aw 'm thinkin' she hed taul' 'er mither at the lang len'th. At a roch guess, a body mith gae farrer agley, aw daursay, nor licken 't to ane o' yon chiels 't was aboot the toon wi' 'er at Briggies'—yon skyeow-fittet breet.'

The foregoing brief extract from the conversation of a couple of those kindly gossips who had all along taken a special interest in her case will indicate with sufficient distinctness the facts surrounding the birth of Baubie Huie's Geet.

The reputed father of the geet was a sort of nondescript chap, whose habit it was to figure at one time as an indifferent second or third 'horseman', and next time as an 'orra man'; a bullet-headed bumpkin, with big unshapely feet, spreading considerably outward as he walked; a decided taste for smoking tobacco; of somewhat more than average capability in talking bucolic slang of a gross sort; yet possessing withal a comfortable estimate of his own graces of person and manner in the eyes of the fair sex. Such was the—sweetheart, shall we say?—of Baubie Huie.

How one might best define the precise relationship existing between the nondescript chiel and Baubie, it would not be easy to say. It was believed that on the feeing market night he had taken Baubie home to

Briggies', he being not greatly the worse of drink, and that on the term night he had accompanied her part of the way toward her father's house. There was also a sort of vague impression that he had since then come once or twice to visit Baubie, keeping as well out of sight and ken of Jock Huie and his wife as might be. Be that as it may, now that the child was born, Jock, who was very much of a practical man, desired to know articulately from the man himself whether he was to 'tak' wi' 't an' pay for 't'. The idea of asking whether the fellow had any intention of doing the one thing which a man with a shred of honour about him would have felt bound to do in the circumstances—viz., marrying his daughter—had really not occurred to Jock Huie. And so it came to pass, that after a certain amount of rather irritating discussion between himself and the female members of his family, and as the nondescript took very good care not to come to him, Jock 'took road' to hunt up the nondescript, who, as he discovered after some trouble, was now serving on a farm some five or six miles off. He found him as third horseman at the plough in a field of 'neep-reet', along with his two fellow ploughmen. The nondescript had a sufficient aspect of embarrassment when Jock Huie caught him up at the end rig, where he had been waiting till the ploughs should come out, to indicate that he would not have been disappointed had the visit been omitted; and it seemed not improbable that his two companions might thereafter offer one or two interrogatory remarks on the subject, which would not be a great deal more welcome. At any rate, Jock Huie had the satisfaction of finding that the nondescript 'wasna seekin' to deny't'; nay, that he did not refuse to 'pay for 't', any backwardness on his part in that respect up to the date of visit, being readily accounted for by the fact that it was the middle of the half year, when a man was naturally run of cash. Threats about ''reestin waages', therefore, were perfectly uncalled-for; and, indeed, a sort of unjust aspersion on the general character of the nondescript. It was right that Jock Huie should know that.

'Ye sud hae the civeelity to lat fowk ken faur ye are than; an' ye think ony ill o' that. Bonnie story to haud me trailin' here, lossin half a day seekin' ye',' retorted Jock with some roughness of tone.

Between the date of Jock Huie's visit just mentioned and the term of Whitsunday, the father of Baubie Huie's geet visited the abode of the Huies once at any rate; and in course of the conference that ensued, it so happened that the subject of getting the geet christened came up—the needful preliminary to that being, as Jock explained, to appear and give satisfaction to that grave Church Court, the Kirk-Session. This was a point which both the paternal and maternal Huie were a good deal more eager to discuss and settle about than either of the immediate parents of the geet. Indeed, the nondescript seemed penetrated with a sort of feeling that that was a part of the business hardly in his line. Not that he objected on principle to the geet being christened; far from it; for when Eppie Huie had stated the necessity of getting themselves 'clear't', and having that rite performed, and Jock Huie had vigorously backed up her statement, the nondescript assented with a perfectly explicit 'Ou ay'; only he showed a decided tendency always to let the matter drop again. This did not suit Jock Huie's book in the least, however, and he manifested a determination to have the business followed out that was not at all comfortable to the nondescript.

When the nondescript had pondered over the situation for a few days, and all along with the feeling that something must really be done, for he did not in the least relish the idea of further calls from Jock Huie, the happy thought occurred to him of calling on his old master, Briggies, who was one of the elders of the Kirk, and, being after all a humane man, would no doubt be prevailed upon to pave the way for him and Baubie making penitential appearance before the session, and receiving censure and 'absolution'. So he called on Briggies, and was rather drily told that, neither Baubie nor he being 'commeenicants', apart from the censure of the session, which had to be encountered in the first place, he, at any rate, 'as the engaging parent' (and perhaps Baubie too), would have to undergo an examination, at the hands of the minister, as to his knowledge of the cardinal doctrines of the Christian faith, and the significance of the rite of baptism in particular.

'Fat wye cud ye expeck to win throw itherweese, min?' Briggies felt bound to speak as an elder in this case—'Gin fowk winna leern to behave themsel's they maun jist stan' the consequences. The vera Kirk-Session itsel' cudna relieve ye, man, upo' nae ither precunnance.'

The nondescript returned much pondering on this disheartening information, which he got opportunity, by and bye, of communicating to Baubie. In private conference the two agreed that 'a scaulin' fae the session', by itself—a thing they had been both accustomed to hear spoken of with extreme jocularity, not less than they had seen those who had undergone the same, regarded as possessing something of the heroism that is rather to be envied—a scaulin' fae the session might well be borne; but to stand a formal examination before the minister in cold blood was another affair. The dilemma having occurred, the two horns were presented to Jock Huie, who was so relentlessly forcing them on to impalement, in the hope of softening his heart, or at any rate awakening his sympathy; but Jock was just as determined as ever that they must go forward in the performance of their Christian duty, and his one reply was, 'Ou, deil care; ye maun jist haud at the Catechis.'

CHAPTER IV

THE GEET'S STATUS, ECCLESIASTICAL AND SOCIAL, DEFINED

'Aw 'm sayin', 'oman, that geet maun be kirsen't some wye or anither; we canna lat the creatur grow up like a haethen.'

The speaker in this case was Jock Huie, and the person addressed his wife Eppie. It was a fine Saturday evening toward the latter end of June, and Jock who had got home from his work at the close of the week, was now in a deliberative mood.

'Weel, man, ye'll need to see fat wye 't's to be manag't,' was Eppie's reply.

'They'll jist need 'o tak' her 'er leen; that's a' that I can say aboot it,' said Jock.

'Ah-wa, man; aw won'er to hear ye speak.'

'Weel fat else can ye dee? Aw tell ye the littlean 'll be made a moniment o' i' the kwintra side.'

'Ou, weel, ye maun jist gae to the minaister yersel', man, an' tell 'im fat gate her an' huz tee 's been guidet; he 's a rael sympatheesin person, an' there 's nae doot he 'll owreleuk onything as far 's he can.'

'Sorra set 'im, weel-a-wat!' said Jock Huie emphatically, as he knocked the half-burnt 'dottal' of tobacco out of his pipe into the palm of his hand, with a sort of savage thump.

Whether Jock Huie's portentous objurgation on the subject of the Catechism had much or anything to do with the result it would perhaps be difficult to say, but it was a simple matter of fact that after it had been uttered, the father of Baubie's geet exhibited even more than previously a disposition to fight shy of the path of duty on which Jock sought to impel him. The Whitsunday term was drawing on; the Whitsunday term had arrived and the geet still unchristened. Then it was found that the father of the geet had deemed it an expedient thing to seek an appreciable change of air by 'flittin' entirely beyond 'kent bounds'. True it was, that on the very eve of his departure he had by the hands of a third party transmitted to Baubie for the maintenance of her geet a 'paper note' of the value of one pound, and along with it a verbal message to the effect that he was 'gyaun to the pairis' o' Birse'; but as it had been a not infrequent practice among the witty to mention the parish named as a sort of mythical region to which one might be condemned to go, for whom no other sublunary use was apparent, Baubie herself was far from assured that the literal Birse was meant; and we may add was equally at a loss as to whether she had further remittances to look for, or if the note was a once and single payment, in full discharge of the nondescript's obligations in respect to the present maintenance, and prospective up-bringing of his son—the Bastard Geet.

Baubie Huie's Bastard Geet had now reached the age of fully four months; no wonder if the grand-paternal anxieties should be aroused as to the danger of the 'peer innocent' merging into heathenism and becoming a bye-word to the parish. And as Jock Huie had expressed his sense of the importance of kirsenin as a preventative, so after all, it fell to Jock's lot to take the responsible part in getting the rite performed. The name was a matter of difficulty; had there been an available father, it would have been his duty to confer with the mother on the point, and be fully instructed what name to bestow on the infant; and in the case of his own children, the male part of them at any rate, Jock Huie had never been much at a loss about the names. Among his sons, Tam, Sawney, and Jock, came in, in orderly succession; but, ponder as he would, the naming of Baubie's geet puzzled him long. Its reputed father bore the name of Samuel—cut down to Samie—Caie, and Jock rejected promptly and with scorn the suggestion, coming from its mother, to inflict upon the bairn any such name, which he, in strong language, declared to be nauseous enough to serve as an emetic to a dog. Indeed, Jock's honest hatred of the nondescript had now reached a pitch that made him resolutely decline to pronounce his name at all; a practice in which, as a rule, he was tacitly imitated by his wife and daughter. Partly from this cause, and partly by reason of the still further delay that occurred in getting the christening over, it came to pass that the poor youngster began to have attached to it, with a sort of permanency, the title of Baubie Huie's Bastard Geet; and when at last the parson had done the official duty in question, and Jock Huie, with a just sense of his position in the matter, had boldly named the bairn after himself, it only led to the idle youth of the neighbourhood ringing the changes on the geet in this fashion—

> Aul' Jock, an' young Jock, an' Jock comin' tee;
> There 'll never be a gweed Jock till aul' Jock dee.

But notwithstanding of all these things the geet throve and grew as only a sturdy scion of humanity could be expected to do.

To say that Baubie Huie was passionately attached to her child, would perhaps be rather an over-statement; yet was she pleased to nurse the poor geet with a fair amount of kindness; and physically the geet seemed to make no ungrateful return. It was edifying to note the bearing of the different members of the family towards the geet. The practical interest taken in its spiritual welfare by old Jock Huie has been mentioned; and despite the trouble it had caused him, Jock was equally prepared now to let the geet have the first and tenderest 'bite' from his hard-won daily crust to meet its temporal wants; a measure of self-denial such as many a philanthropist of higher station and greater pretensions has never set before himself. The nature of Eppie Huie's feelings towards the geet was sufficiently indicated by the skilled and careful nursing she would expend upon it at those times when Baubie, tired of her charge, with an unceremonious—'Hae, tak' 'im a file, mither,'—would hand over the geet 'body bulk' to the charge of its grannie. When any of Jock Huie's grown-up sons happened to visit home, their cue was simply to ignore the geet altogether. Even when it squalled the loudest they would endeavour to retain the appearance of stolid obliviousness of its presence; just as they did when the hapless geet crowed and 'walloped' its small limbs in the superabundance of its joy at being allowed the novel pleasure of gazing at them. The members of the family who were Baubie's juniors, did not profess indifference; only their feeling toward the geet, when it came under their notice on these temporary visits home, was in the main the reverse of amicable. Her younger sister, indeed, in Baubie's hearing, designated the unoffending geet a 'nasty brat', whereat Baubie flared up hotly and reminded her that it was not so very long since she, the sister, was an equally 'nasty brat', to say the very least of it; as she, Baubie, could very well testify from ample experience of the degrading office of nurse to her. 'Fat ever 't be, ye may haud yer chat ony wye,' said Baubie, and the sister stood rebuked.

When harvest came, the geet being now six months old, was 'spean't', and Baubie 'took a hairst'. Handed over to the exclusive custody of its grannie for the time being, the geet was destined thenceforth to share both bed and board, literally, with Eppie Huie and Jock her husband. The tail of the speaning process when the geet got 'fretty', and especially overnight, brought back to Jock Huie a lively remembrance of by-gone experiences of a like nature; and he once or twice rather strongly protested against the conduct of 'that ablich' in 'brakin' 's nicht's rest' with its outcries. But, on the whole, Jock bore with the geet wonderfully.

When her hairst was finished, it was Baubie's luck to get continuous employment from the same master till Martinmas. When that period had arrived, Baubie, of her own free will and choice, again stood the feeing market, and found what she deemed a suitable engagement at a large farm several miles off, whither she went in due time; and where, as was to be expected, she found the domestic supervision of the male and female servants less stringent on the whole than it had been at the elder's at Brigfit. In so far as her very moderate wages allowed, after meeting her own needs in the matter of dress, Baubie Huie was not altogether disinclined to contribute toward the support of her bastard geet. As a matter of course, nothing further was heard of or from the nondescript father of the geet. He had moved sufficiently far off to be well out of sight at any rate, and Jock Huie had no means of finding him out and pressing the claim against him in respect of the child's maintenance, except by means of the Poor Law Inspector; and Jock, being a man of independent spirit, had not yet thought of calling in the services of the 'Boord'. As time went on, Baubie's maternal care did not manifest itself in an increasing measure in this particular of furnishing the means to support the geet more than it did in any other respect affecting her offspring.

After one or two more flittings from one situation to another, it became known that Baubie Huie was about to be married. At another Martinmas term—there had been an interval of two years—Baubie once more returned home; but this time frankly to announce to Jock and Eppie Huie that she was 'gyaun to be mairriet' to one Peter Ga', who had been a fellow-servant with her during a recent half-year. From considerate regard for the convenience of her parents, and other causes, the happy day would not be delayed beyond a fortnight; and there would be no extensive 'splore' on the occasion, to disturb materially the domestic arrangements of the Huies.

On this latter point certain of the neighbours were keenly disappointed.

Because there were no marriage rejoicings to speak of, they missed an invitation to join in the same, and they spoke in this wise:—

'An' there's to be nae mairriage ava, ye was sayin'?'

'Hoot—fat wye cud there? The bridegreem an aul' widow man't mith be 'er fader, wi' three–four o' a faimily.'

'Na, sirs; a bonny bargaine she'll be to the like o' 'im—three or four o' a faimily, ye say?'

'So aw b'lieve; an' aw doot it winna be lang ere Baubie gi'e 'im ane mair to haud it haill wi'.'

'Weel, weel! Only fat ither cud ye expeck; but the man maun hae been sair misguidet 't loot 's een see the like o' 'er.'

'An' ye may say 't.'

'Fat siclike o' a creatur is he, ken ye?'

'Ou weel, he's a byous quate man it wud appear, an' a gweed aneuch servan', but sair haud'n doon naitrally. Only the peer stock maun be willin' to dee the richt gate in a menner, or he wud a never propos't mairryin Baubie.'

'Gweed pity 'im wi' the like o' 'er, weel-a-wat—senseless cuttie.'

Naturally, and by right, when Baubie Huie had got a home of her own, she ought to have resumed the custody of her Bastard Geet, now a 'gangrel bairn' of fully two years; but on the one hand, it was evident that Mr and Mrs Ga' had the prospect of finding the available accommodation in a hut, whose dimensions afforded scope for only a very limited but and ben, sufficiently occupied by and bye without the geet; and on the other, Eppie Huie, though abundantly forfough'en for a woman of her years in keeping her house, attending to the wants of her husband, Jock, and meeting such demands as her own family made upon her exertions as general washerwoman, would have rather demurred to parting with the geet, to whom she had become, as far as the adverse circumstances of the case allowed, attached. And thus the geet was left in the undisputed possession of Jock and Eppie Huie, to be trained by them as they saw meet.

Unlucky geet, say you? Well, one is not altogether disposed to admit that without some qualification. Sure enough, Jock Huie, senior, would and did permit Baubie's geet to grow up an uncouth, unkempt, and, in the main, untaught bairn; yet was there from him, even, a sort of genuine, if somewhat rugged affection, flowing out toward little Jock Huie (as the geet was alternatively styled); as when he would dab the shaving brush playfully against the geet's unwhiskered cheek, while sternly refusing him a grip of the gleaming razor, as he lifted the instrument upward for service on his own face; or, at another time, would quench the geet's aspiration after the garments of adult life, manifested in its having managed to thrust its puny arms into a huge sleeved moleskin vest belonging to Jock himself, by dropping his big 'wyv'n bonnet' over the toddling creature's head, and down to his shoulders. Bitter memories of Samie

Caie had faded into indistinctness more or less. And when the neighbour wives, as they saw the geet with an old black 'cutty' in his hand, gravely attempting to set the contents of the same alight with a fiery sod in imitation of its grandfather, would exclaim, admiringly, 'Na, but that laddie is a bricht Huie, Jock, man,' Jock would feel a sort of positive pride in the youngster, who bade so fairly to do credit to his upbringing.

No; it might be that meagre fare—meagre even to pinching at times—was what the inmates of Jock Huie's cot had to expect; it might be that in a moral and intellectual point of view the nourishment going was correspondingly scanty and insufficient, to say the least of it; but in being merely left to grow up under these negatively unfavourable conditions, a grotesque miniature copy of the old man at whose heels he had learnt to toddle about with such assiduity, I can by no means admit that, as compared with many and many a geet whose destiny it is to come into the world in the like irregular fashion, the lot of Baubie Huie's Bastard Geet could be justly termed unlucky.

'Ian MacLaren'

1850–1907

The Cunning Speech of Drumtochty

Speech in Drumtochty distilled slowly, drop by drop, and the faces of our men were carved in stone. Visitors without discernment used to pity our dullness, and lay themselves out for missionary work. Before their month was over they spoke bitterly of us, as if we had deceived them, and departed with a grudge in their hearts. When Hillocks scandalised the Glen by letting his house and living in the bothie—through sheer greed of money—it was taken by a fussy little man from the South, whose control over the letter 'h' was uncertain, but whose self-confidence bordered on the miraculous. As a deacon of the Social Religionists—a new denomination, which had made an 'it with Sunday Entertainments—and Chairman of the Amalgamated Sons of Rest—a society of persons with conscientious objections to work between meals—he was horrified at the primeval simplicity of the Glen, where no meeting of protest had been held in the memory of living man, and the ministers preached from the Bible. It was understood that he was to do his best for us, and there was curiosity in the kirkyard.

'Whatna like man is that English veesitor ye've got, Hillocks? a' hear he's fleein' ower the glen, yammerin' and haverin' like a starlin'.'

'He's a gabby (talkative) body, Drumsheugh, there's nae doot o' that, but terrible ignorant.

'Says he tae me nae later than yesterday, "That's a fine field o' barley ye've there, Maister Harris," an' as sure as deith a' didna ken whaur tae luik, for it was a puckle aits.'

'Keep's a',' said Whinnie; 'he's been awfu' negleckit when he was a bairn, or maybe there's a want in the puir cratur.'

Next Sabbath Mr Urijah Hopps appeared in person among the fathers—who looked at each other over his head—and enlightened them on supply and demand, the Game Laws, the production of cabbages for towns, the iniquity of an Established Church, and the bad metre of the Psalms of David.

'You must 'ave henterprise, or it's hall hup with you farmers.'

'Ay, ay,' responded Drumsheugh, after a long pause, and then every man concentrated his attention on the belfry of the kirk.

'Is there onything ava' in the body, think ye, Domsie,' as Mr Hopps bustled into kirk, 'or is't a' wind?'

'Three wechtfu's o' naething, Drumsheugh; a' peety the puir man if Jamie Soutar gets a haud o' him.'

Jamie was the cynic of the Glen—who had pricked many a wind bag—and there was a general feeling that his meeting with Mr Hopps would not be devoid of interest. When he showed himself anxious to learn next Sabbath, any man outside Drumtochty might have been deceived, for Jamie could withdraw every sign of intelligence from his face, as when shutters close upon a shop window. Our visitor fell at once into the trap, and made things plain to the meanest capacity, until Jamie elicited from the guileless Southron that he had never heard of the Act of Union; that Adam Smith was a new book he hoped to buy; that he did not know the difference between an Arminian and a Calvinist, and that he supposed the Confession of Faith was invented in Edinburgh. This in the briefest space of time, and by way of information to Drumtochty, James was making for general literature, and had still agriculture in reserve, when Drumsheugh intervened in the humanity of his heart.

'A' dinna like tae interrupt yir conversation, Maister Hopps, but it's no verra safe for ye tae be stannin' here sae lang. Oor air hes a bit nip in't, and is mair searchin' than doon Sooth. Jamie 'ill be speirin' a' mornin' gin ye 'ill answer him, but a'm thinkin' ye'ill be warmer in the kirk.'

And Drumsheugh escorted Mr Hopps to cover, who began to suspect that he had been turned inside out, and found wanting.

Drumtochty had listened with huge delight, but without a trace of expression, and, on Mr Hopps reaching shelter, three boxes were offered Jamie.

The group was still lost in admiration when Drumsheugh returned from his errand of mercy.

'Sall, ye've dune the job this time, Jamie. Ye're an awfu' creetic. Yon man 'ill keep a quiet cheep till he gets Sooth. It passes me hoo a body wi' sae little in him hes the face tae open his mooth.'

'Ye did it weel, Jamie,' Domsie added, 'a clean furrow frae end tae end.'

'Toots, fouk, yir makin' ower muckle o' it. It wes licht grund, no worth puttin' in a ploo.'

Mr Hopps explained to me, before leaving, that he had been much pleased with the scenery of our Glen, but disappointed in the people.

'They may not be hignorant,' said the little man doubtfully, 'but no man could call them haffable.'

It flashed on me for the first time that perhaps there may have been the faintest want of geniality in the Drumtochty manner, but it was simply the reticence of a subtle and conscientious people. Intellect with us had been brought to so fine an edge by the Shorter Catechism that it could detect endless distinctions, and was ever on the watch against inaccuracy.

Farmers who could state the esoteric doctrine of 'spiritual independence' between the stilts of the plough, and talked familiarly of 'co-ordinate jurisdiction with mutual subordination', were not likely to fall into the vice of generalisation. When James Soutar was in good fettle, he could trace the whole history of Scottish secession from the beginning, winding his way through the maze of Original Seceders and Cameronians, Burghers and Anti-Burghers—there were days when he would include the Glassites—with unfaltering step; but this was considered a feat even in Drumtochty, and it was admitted that Jamie had 'a gift o' discreemination'. We all had the gift in measure, and dared not therefore allow ourselves the expansive language of the South. What right had any human being to fling about superlative adjectives, seeing what a big place the world is, and how little we know? Purple adjectives would have been as much out of place in our conversation as a bird of paradise among our muirfowl.

Mr Hopps was so inspired by one of our sunsets—to his credit let that be told—that he tried to drive Jamie into extravagance.

'No bad! I call it glorious, and if it hisn't, then I'd like to know what his.'

'Man,' replied Soutar austerely, 'ye 'ill surely keep ae word for the twenty-first o' Reevelation.'

Had any native used 'magnificent', there would have been an uneasy feeling in the Glen; the man must be suffering from wind in the head, and might upset the rotation of crops, sowing his young grass after potatoes, or replacing turnip with beetroot. But nothing of that sort happened in my time; we kept ourselves well in hand. It rained in torrents elsewhere, with us it only 'threatened tae be weet'—some provision had to be made for the deluge. Strangers, in the pride of health, described themselves as 'fit for anything', but Hillocks, who died at ninety-two, and never had an hour's illness, did not venture, in his prime, beyond 'Gaein' aboot, a'm thankfu' to say, gaein' aboot.'

When one was seriously ill, he was said to be 'gey an' sober', and no one died in Drumtochty—'he slippit awa'.

Hell and heaven were pulpit words; in private life we spoke of 'the ill place' and 'oor lang hame'.

When the corn sprouted in the stooks one late wet harvest, and Burnbrae lost half his capital, he only said, 'It's no lichtsome,' and no congratulations on a good harvest ever extracted more from Drumsheugh than 'A' daurna complain.'

Drumsheugh might be led beyond bounds in reviewing a certain potato transaction, but, as a rule, he was a master of measured speech. After the privilege of much intercourse with that excellent man, I was able to draw up his table of equivalents for the three degrees of wickedness. When there was just a suspicion of trickiness—neglecting the paling between your cattle and your neighbour's clover field—'He's no juist the man for an elder.' If it deepened into deceit—running a 'greasy' horse for an hour before selling—'He wud be the better o' anither dip.' And in the case of downright fraud—finding out what a man had offered for his farm and taking it over his head—the offender was 'an ill gettit wratch'. The two latter phrases were dark with theology, and even the positive degree of condemnation had an ecclesiastical flavour.

When Drumsheugh approved anyone, he was content to say, 'He micht be waur,' a position beyond argument. On occasion he ventured upon bolder assertions: 'There's nae mischief in Domsie'; and once I heard him in a white heat of enthusiasm pronounce Dr Davidson, our parish minister, 'A graund man ony wy ye tak him.' But he seemed ashamed after this outburst, and 'shooed' the crows off the corn with needless vigour.

No Drumtochty man would commit himself to a positive statement on any subject if he could find a way of escape, not because his mind was confused, but because he was usually in despair for an accurate expression. It was told for years in the Glen, with much relish and almost funereal solemnity, how a Drumtochty witness had held his own in an ecclesiastical court.

'You are beadle in the parish of Pitscourie,' began the advocate with a light heart, not knowing the witness's birthplace.

'It's a fac',' after a long pause and a careful review of the whole situation.

'You remember that Sabbath when the minister of Netheraird preached.'

'Weel, a'll admit that,' making a concession to justice.

'Did ye see him in the vestry?'

'A' canna deny it.'

'Was he intoxicated?'

The crudeness of this question took away Drumtochty's breath, and suggested that something must have been left out in the creation of that advocate. Our men were not bigoted abstainers, but I never heard any word so coarse and elementary as intoxicated used in Drumtochty. Conversation touched this kind of circumstance with delicacy and caution, for we keenly realised the limitations of human knowledge.

'He hed his mornin',' served all ordinary purposes, and in cases of emergency, such as Muirtown market:

'Ye cud see he hed been tastin'.'

When an advocate forgot himself so far as to say intoxicated, a Drumtochty man might be excused for being upset.

'Losh, man,' when he had recovered, 'hoo cud ony richt-thinkin' man sweer tae sic an awfu' word? Na, na, a' daurna use that kin' o' langidge; it's no cannie.'

The advocate tried again, a humbler, wiser man.

'Was there a smell of drink on him?'

'Noo, since ye press me, a'll juist tell ye the hale truth; it wes doonricht stupid o' me, but, as sure as a'm livin', a' clean forgot tae try him.'

Then the chastened counsel gathered himself up for his last effort.

'Will you answer one question, sir? you are on your oath. Did you see anything unusual in Mr MacOmish's walk? Did he stagger?'

'Na,' when he had spent two minutes in recalling the scene. 'Na, I cudna say stagger, but he micht gie a bit trimmil.'

'We are coming to the truth now; what did you consider the cause of the trimmiling, as you call it?' and the innocent young advocate looked round in triumph.

'Weel,' replied Drumtochty, making a clean breast of it, 'since ye maun hae it, a' heard that he wes a very learned man, and it cam intae ma mind that the Hebrew, which, a'm telt, is a very contrairy langidge, hed gaen doon and settled in his legs.'

The parish of Netheraird was declared vacant, but it was understood that the beadle of Pitscourie had not contributed to this decision.

His own parish followed the trial with intense interest, and were much pleased with Andra's appearance.

'Sall,' said Hillocks, 'Andra has mair gumption than ye wud think, and yon advocat didna mak muckle o' him. Na, na, Andra wesna brocht up in the Glen for naethin'. Maister MacOmish may hae taen his gless atween the Hebrew and the Greek, and its no verra suitable for a minister, but that's anither thing frae bein' intoxicat.

'Keep's a', if ye were tae pit me in the box this meenut, a' cudna sweer a' hed ever seen a man intoxicat in ma life, except a puir body o' an English bag-man at Muirtown Station. A' doot he hed bin meddlin' wi' speerits, and they were wheelin' him tae his kerridge in a luggage barrow. It wes a fearsome sicht, and eneugh tae keep ony man frae speaking aboot intoxicat in yon louse wy.'

Archie Moncur fought the drinking customs of the Glen night and day with moderate success, and one winter's night he gave me a study in his subject which, after the lapse of years, I still think admirable for its reserve power and Dantesque conclusion.

'They a' begin in a sma' wy,' explained Archie, almost hidden in the depths of my reading chair, and emphasising his points with a gentle motion of his right hand; 'naethin' tae mention at first, juist a gless at an orra time—a beerial or a merridge—and maybe New Year. That's the first stage; they ca' that moderation. Aifter a while they tak a mornin' wi' a freend, and syne a gless at the public-hoose in the evenin', and they treat ane anither on market days. That's the second stage; that's tastin'. Then they need it reg'lar every day, nicht an' mornin', and they'll sit on at nicht till they're turned oot. They 'ill fecht ower the Confession noo, and laist Sabbath's sermon, in the Kildrummie train, till it's clean reediklus. That's drammin', and when they've hed a year or twa at that they hae their first spatie (spate is a river flood), and that gies them a bit fricht. But aff they set again, and then comes anither spatie, and the doctor hes tae bring them roond. They ca' (drive) cannie for a year or sae, but the feein' market puts the feenishin' titch. They slip aff sudden in the end, and then they juist gang plunk—aye,' said Archie in a tone of gentle meditation, looking, as it were, over the edge, 'juist plunk'.

Nothing ever affected my imagination more powerfully than the swift surprise and gruesome suggestion of that 'plunk'.

But the literary credit of Drumtochty rested on a broad basis, and no one could live with us without having his speech braced for life. You felt equal to any emergency, and were always able to express your mind with some degree of accuracy, which is one of the luxuries of life. There is, for instance, a type of idler who exasperates one to the point of assault, and whom one hungers to describe after a becoming manner. He was rare in the cold air of the North, but we had produced one specimen, and it was my luck to be present when he came back from a distant colony, and Jamie Soutar welcomed him in the kirkyard.

'Weel, Chairlie,' and Jamie examined the well-dressed prodigal from top to toe, 'this is a prood moment for Drumtochty, and an awfu' relief tae ken yir safe. Man, ye hevna wanted meat nor claithes; a' tak it rael neeburly o' ye tae speak ava wi' us auld-fashioned fouk.

'Ye needna look soor nor cock yir nose in the air, for you an' me are auld freends, and yir puir granny wes na mair anxious aboot ye than a' wes.

' "A'm feared that laddie o' Bell's 'ill kill himsel' oot in Ameriky," were ma verra words tae Hillocks here; "he 'ill be slavin' his flesh aff his banes tae mak a fortune and keep her comfortable."

'It was a rael satisfaction tae read yir letter frae the backwoods—or was't a public-hoose in New York? ma memory's no what it used to be—telling hoo ye were aye thinking o' your auld granny, and wantin' tae come hame and be a comfort tae her if she wud send ye out twenty pund.

'The bit that affeckit me maist wes the text frae the Prodigal Son—it cam in sae natural. Mony a broken hert hes that story bund up, as we ken weel in this Glen; but it's dune a feck o' mischief tae—that gude word o' the Maister. Half the wastrels in the warld pay their passage hame wi' that Parable, and get a bran new outfit for anither start in the far country.

'Noo dinna turn red, Chairlie, for the neeburs ken ye were tae work yir wy hame had it no been for yir health. But there's a pack of rascals 'ill sorn on their father as lang as he's livin', and they 'ill stairve a weedowed mither, and they 'ill tak a sister's wages, and if they canna get ony better a dune body o' eighty 'ill serve them.

'Man, Chairlie, if a' hed ma wull w' thae wawfies, I wud ship them aff tae a desert island, wi' ae sack o' seed potatoes and anither o' seed corn, and let them work or dee. A' ken yir wi' me there, for ye aye hed an independent spirit, and wesna feared tae bend yir back.

'Noo, if a' cam across ane o' thae meeserable objects in Drumtochty, div ye ken the advice I wud gie him?

'A' wud tell the daidlin', thowless, feckless, fushionless wratch o' a cratur tae watch for the first spate and droon himsel' in the Tochty.'

'What's he aff through the graves for in sic a hurry?' and Jamie followed Charlie's retreating figure with a glance of admirable amazement; 'thae's no very gude mainners he's learned in Americky.'

'Thank ye, Jamie, thank ye; we're a' obleeged tae ye,' said Drumsheugh. 'A' wes ettlin' tae lay ma hands on the whup-ma-denty (fop) masel, but ma certes, he's hed his kail het this mornin'. Div ye think he 'ill tak yir advice?'

'Nae fear o' him; thae neer-dae-weels haena the spunk; but a'm expeckin' he 'ill flee the pairish.'

Which he did. Had you called him indolent or useless he had smiled, but 'daidlin', thowless, feckless, fushionless wratch,' drew blood at every stroke, like a Russian knout.

We had tender words also, that still bring the tears to my eyes, and chief among them was 'couthy'. What did it mean? It meant a letter to some tired townsman, written in homely Scotch, and bidding him come to get new life from the Drumtochty air; and the grip of an honest hand on the Kildrummie platform whose warmth lasted till you reached the Glen; and another welcome at the garden-gate that mingled with the scent of honeysuckle, and moss-roses, and thyme, and carnations; and the best of everything that could be given you; and motherly nursing in illness, with skilly remedies of the olden time; and wise, cheery talk that spake no ill of man or God; and loud reproaches if you proposed to leave under a month or two; and absolute conditions that you must return; and a load of country dainties for a bachelor's bare commons; and far more, that cannot be put into words, of hospitality, and kindness, and quietness, and restfulness, and loyal friendship of hearts now turned to dust in the old kirkyard.

But the best of all our words were kept for spiritual things, and the description of a godly man. We did not speak of the 'higher life', nor of a 'beautiful Christian', for this way of putting it would not have been in keeping with the genius of Drumtochty. Religion there was very lowly and modest—an inward walk with God. No man boasted of himself, none told the secrets of the soul. But the Glen took notice of its saints, and did them silent reverence, which they themselves never knew. Jamie Soutar had a wicked tongue, and, at a time, it played round Archie's temperance schemes, but when that good man's back was turned Jamie was the first to do him justice.

'It wud set us better if we did as muckle gude as Archie; he's a richt livin' man and weel prepared.'

Our choicest tribute was paid by general consent to Burnbrae, and it may be partiality, but it sounds to me the deepest in religious speech. Every cottage, strangers must understand, had at least two rooms—the kitchen where the work was done, that we called the 'But', and there all kinds of people came; and the inner chamber which held the household treasures, that we called the 'Ben', and there none but a few honoured visitors had entrance. So we imagined an outer court of the religious life where most of us made our home, and a secret place where only God's nearest friends could enter, and it was said of Burnbrae, 'He's far ben.' His neighbours had watched him, for a generation and more, buying and selling, ploughing and reaping, going out and in the common ways of a farmer's life, and had not missed the glory of the soul. The cynic of Drumtochty summed up his character: 'There's a puckle gude fouk in the pairish, and ane or twa o' the ither kind, and the maist o' us are half and between,' said Jamie Soutar, 'but there's ae thing ye may be sure o', Burnbrae is "far ben".'

ROBERT LOUIS STEVENSON

1850–1894

A LODGING FOR THE NIGHT

A STORY OF FRANCIS VILLON

It was late in November 1456. The snow fell over Paris with rigorous, relentless persistence; sometimes the wind made a sally and scattered it in flying vortices; sometimes there was a lull, and flake after flake descended out of the black night air, silent, circuitous, interminable. To poor people, looking up under moist eyebrows, it seemed a wonder where it all came from. Master Francis Villon had propounded an alternative that afternoon at a tavern window: was it only Pagan Jupiter plucking geese upon Olympus? or were the holy angels moulting? He was only a poor Master of Arts, he went on; and as the question somewhat touched upon divinity, he durst not venture to conclude. A silly old priest from Montargis, who was among the company, treated the young rascal to a bottle of wine in honour of the jest and the grimaces with which it was accompanied, and swore on his own white beard that he had been just such another irreverent dog when he was Villon's age.

The air was raw and pointed, but not far below freezing; and the flakes were large, damp, and adhesive. The whole city was sheeted up. An army might have marched from end to end and not a footfall given the alarm. If there were any belated birds in heaven, they saw the island like a large white patch, and the bridges like slim white spars, on the black ground of the river. High up overhead the snow settled among the tracery of the cathedral towers. Many a niche was drifted full; many a statue wore a long white bonnet on its grotesque or sainted head. The gargoyles had been transformed into great false noses, drooping towards the point. The crockets were like upright pillows swollen on one side. In the intervals of the wind there was a dull sound of dripping about the precincts of the church.

The cemetery of St John had taken its own share of the snow. All the graves were decently covered; tall white housetops stood around in grave array; worthy burghers were long ago in bed, be-nightcapped like their domiciles; there was no light in all the neighbourhood but a little peep from a lamp that hung swinging in the church choir, and tossed the shadows to and fro in time to its oscillations. The clock was hard on ten when the patrol went by with halberds and a lantern, beating their hands; and they saw nothing suspicious about the cemetery of St John.

Yet there was a small house, backed up against the cemetery wall, which was still awake, and awake to evil purpose, in that snoring district. There was not much to betray it from without; only a stream of warm vapour from the chimney-top, a patch where the snow melted on the roof, and a few half-obliterated footprints at the door. But within, behind the shuttered windows, Master Francis Villon the poet, and some of the thievish crew with whom he consorted, were keeping the night alive and passing round the bottle.

A great pile of living embers diffused a strong and ruddy glow from the arched chimney. Before this straddled Dom Nicolas, the Picardy monk, with his skirts picked up and his fat legs bared to the comfortable warmth. His dilated shadow cut the room in half; and the firelight only escaped on either side of his broad person, and in a little pool between his outspread feet. His face had the beery, bruised appearance of the continual drinker's; it was covered with a network of congested veins, purple in ordinary circumstances, but now pale violet, for even with his back to the fire the cold pinched him on the other side. His cowl had half-fallen back, and made a strange excrescence on either side of his bull-neck. So he straddled, grumbling, and cut the room in half with the shadow of his portly frame.

On the right, Villon and Guy Tabary were huddled together over a scrap of parchment; Villon making a ballade which he was to call the 'Ballade of Roast Fish', and Tabary spluttering admiration at his shoulder. The poet was a rag of a man, dark, little, and lean, with hollow cheeks and thin black locks. He carried his four-and-twenty years with feverish animation. Greed had made folds about his eyes, evil smiles had puckered his mouth. The wolf and pig struggled together in his face. It was an eloquent, sharp, ugly, earthly countenance. His hands were small and prehensile, with fingers knotted like a cord; and they were continually flickering in front of him in violent and expressive pantomime. As for Tabary, a broad, complacent, admiring imbecility breathed from his squash nose and slobbering lips: he had become a thief, just as he might have become the most decent of burgesses, by the imperious chance that rules the lives of human geese and human donkeys.

At the monk's other hand, Montigny and Thevenin Pensete played a game of chance. About the first there clung some flavour of good birth and training, as about a fallen angel; something long, lithe, and courtly in the person; something aquiline and darkling in the face. Thevenin, poor soul, was in great feather: he had done a good stroke of knavery that afternoon in the Faubourg St Jacques, and all night he had been gaining from Montigny. A flat smile illuminated his face; his bald head shone rosily in a garland of red curls; his little protuberant stomach shook with silent chucklings as he swept in his gains.

'Doubles or quits?' said Thevenin.

Montigny nodded grimly.

'*Some may prefer to dine in state*,' wrote Villon, '*On bread and cheese on silver plate*. Or—or—help me out, Guido!'

Tabary giggled.

'*Or parsley on a golden dish*,' scribbled the poet.

The wind was freshening without; it drove the snow before it, and sometimes raised its voice in a victorious whoop, and made sepulchral grumblings in the chimney. The cold was growing sharper as the night went on. Villon, protruding his lips, imitated the gust with something between a whistle and a groan. It was an eerie, uncomfortable talent of the poet's, much detested by the Picardy monk.

'Can't you hear it rattle in the gibbet?' said Villon. 'They are all dancing the devil's jig on nothing, up there. You may dance, my gallants, you'll be none the warmer! Whew! what a gust! Down went somebody just now! A medlar the fewer on the three-legged medlar-tree!—I say, Dom Nicolas, it'll be cold to-night on the St Denis Road?' he asked.

Dom Nicolas winked both his big eyes, and seemed to choke upon his Adam's apple. Montfaucon, the great grisly Paris gibbet, stood hard by the St Denis Road, and the pleasantry touched him on the raw. As for Tabary, he laughed immoderately over the medlars; he had never heard anything more light-hearted; and he held his sides and crowed. Villon fetched him a fillip on the nose, which turned his mirth into an attack of coughing.

'Oh, stop that row,' said Villon, 'and think of rhymes to "fish."'

'Doubles or quits?' said Montigny doggedly.

'With all my heart,' quoth Thevenin.

'Is there any more in that bottle?' asked the monk.

'Open another,' said Villon. 'How do you ever hope to fill that big hogshead, your body, with little things like bottles? And how do you expect to get to heaven? How many angels, do you fancy, can be spared to carry up a single monk from Picardy? Or do you think yourself another Elias—and they'll send the coach for you?'

'*Hominibus impossibile*,' replied the monk, as he filled his glass.

Tabary was in ecstasies.

Villon filliped his nose again.

'Laugh at my jokes, if you like,' he said.

'It was very good,' objected Tabary.

Villon made a face at him. 'Think of rhymes to "fish",' he said. 'What have you to do with Latin? You'll wish you knew none of it at the great assizes, when the devil calls for Guido Tabary, clericus—the devil with the hump-back and red-hot finger-nails. Talking of the devil,' he added in a whisper, 'look at Montigny!'

All three peered covertly at the gamester. He did not seem to be enjoying his luck. His mouth was a little to a side; one nostril nearly shut, and the other much inflated. The black dog was on his back, as people say, in terrifying nursery metaphor; and he breathed hard under the gruesome burden.

'He looks as if he could knife him,' whispered Tabary, with round eyes.

The monk shuddered, and turned his face and spread his open hands to the red embers. It was the cold that thus affected Dom Nicolas, and not any excess of moral sensibility.

'Come now,' said Villon—'about this ballade. How does it run so far?' And beating time with his hand, he read it aloud to Tabary.

They were interrupted at the fourth rhyme by a brief and fatal movement among the gamesters. The round was completed, and Thevenin was just opening his mouth to claim another victory, when Montigny leaped up, swift as an adder, and stabbed him to the heart. The blow took effect before he had time to utter a cry, before he had time to move. A tremor or two convulsed his frame; his hands opened and shut, his heels rattled on the floor; then his head rolled backwards over one shoulder with the eyes wide open; and Thevenin Pensete's spirit had returned to Him who made it.

Every one sprang to his feet; but the business was over in two twos. The four living fellows looked at each other in rather a ghastly fashion; the dead man contemplating a corner of the roof with a singular and ugly leer.

'My God!' said Tabary; and he began to pray in Latin.

Villon broke out into hysterical laughter. He came a step forward and ducked a ridiculous bow at Thevenin, and laughed still louder. Then he sat down suddenly, all of a heap, upon a stool, and continued laughing bitterly as though he would shake himself to pieces.

Montigny recovered his composure first.

'Let's see what he has about him,' he remarked; and he picked the dead man's pockets with a practised hand, and divided the money into four equal portions on the table. 'There's for you,' he said.

The monk received his share with a deep sigh, and a single stealthy glance at the dead Thevenin, who was beginning to sink into himself and topple sideways off the chair.

'We're all in for it,' cried Villon, swallowing his mirth. 'It's a hanging job for every man jack of us that's here—not to speak of those who aren't.' He made a shocking gesture in the air with his raised right hand, and put out his tongue and threw his head on one side, so as to counterfeit the appearance of one who has been hanged. Then he pocketed his share of the spoil, and executed a shuffle with his feet as if to restore the circulation.

Tabary was the last to help himself; he made a dash at the money, and retired to the other end of the apartment.

Montigny stuck Thevenin upright in the chair, and drew out the dagger, which was followed by a jet of blood.

'You fellows had better be moving,' he said, as he wiped the blade on his victim's doublet.

'I think we had,' returned Villon, with a gulp. 'Damn his fat head!' he broke out. 'It sticks in my throat like phlegm. What right has a man to have red hair when he is dead?' And he fell all of a heap again upon the stool, and fairly covered his face with his hands.

Montigny and Dom Nicolas laughed aloud, even Tabary feebly chiming in.

'Cry baby,' said the monk.

'I always said he was a woman,' added Montigny with a sneer. 'Sit up, can't you?' he went on, giving another shake to the murdered body. 'Tread out that fire, Nick!'

But Nick was better employed; he was quietly taking Villon's purse, as the poet sat, limp and trembling, on the stool where he had been making a ballade not three minutes before. Montigny and Tabary dumbly demanded a share of the booty, which the monk silently promised as he passed the little bag into the bosom of his gown. In many ways an artistic nature unfits a man for practical existence.

No sooner had the theft been accomplished than Villon shook himself, jumped to his feet, and began helping to scatter and extinguish the embers. Meanwhile Montigny opened the door and cautiously peered into the street. The coast was clear; there was no meddlesome patrol in sight. Still it was judged wiser to slip out severally; and as Villon was himself in a hurry to escape from the neighbourhood of the dead Thevenin, and the rest were in a still greater hurry to get rid of him before he should discover the loss of his money, he was the first by general consent to issue forth into the street.

The wind had triumphed and swept all the clouds from heaven. Only a few vapours, as thin as moonlight, fleeted rapidly across the stars. It was bitter cold; and by a common optical effect, things seemed almost more definite than in the broadest daylight. The sleeping city was absolutely still: a company of white hoods, a field full of little Alps, below the twinkling stars. Villon cursed his fortune. Would it were still snowing! Now, wherever he went, he left an indelible trail behind him on the glittering streets; wherever he went he was still tethered to the house by the cemetery of St John; wherever he went he must weave, with his own plodding feet, the rope that bound him to the crime and would bind him to the gallows. The leer of the dead man came back to him with a new significance. He snapped his fingers as if to pluck up his own spirits, and choosing a street at random, stepped boldly forward in the snow.

Two things preoccupied him as he went: the aspect of the gallows at Montfaucon in this bright windy phase of the night's existence, for one; and for another, the look of the dead man with his bald head and garland of red curls. Both struck cold upon his heart, and he kept quickening his pace as if he could escape from unpleasant thoughts by mere fleetness of foot. Sometimes he looked back over his shoulder with a sudden nervous jerk; but he was the only moving thing in the white streets, except when the wind swooped round a corner and threw up the snow, which was beginning to freeze, in spouts of glittering dust.

Suddenly he saw, a long way before him, a black clump and a couple of lanterns. The clump was in motion, and the lanterns swung as though carried by men walking. It was a patrol. And though it was merely crossing his line of march, he judged it wiser to get out of eyeshot as speedily as he could. He was not in the humour to be challenged, and he was conscious of making a very conspicuous mark upon the snow. Just on his left hand there stood a great hotel, with some turrets and a large porch before the door; it was half-ruinous, he remembered, and had long stood empty; and so he made three steps of it and jumped into the shelter of the porch. It was pretty dark inside, after the glimmer of the snowy streets, and he was groping forward with outspread hands, when he stumbled over some substance which offered an indescribable mixture of resistances, hard and soft, firm and loose. His heart gave a leap, and he sprang two steps back and stared dreadfully at the obstacle. Then he gave a little laugh of relief. It was only a woman, and she dead. He knelt beside her to make sure upon this latter point. She was freezing cold, and rigid like a stick. A little ragged finery fluttered in the wind about her hair, and her cheeks had been heavily rouged that same afternoon. Her pockets were quite empty; but in her stocking, underneath the garter, Villon found two of the small coins that went by the name of whites. It was little enough; but it was always something; and the poet was moved with a deep sense of pathos that she should have died before she had spent her money. That seemed to him a dark and pitiable mystery; and he looked from the coins in his hand to the dead woman, and back again to the coins, shaking his head over the riddle of man's life. Henry V of England, dying at Vincennes just after he had conquered France, and this poor jade cut off by a cold draught in a great man's doorway, before she had time to spend her couple of whites—it seemed a cruel way to carry on the world. Two whites would have taken such a little while to squander; and yet it would have been one more good taste in the mouth, one more smack of the lips, before the devil got the soul, and the body was left to birds and vermin. He would like to use all his tallow before the light was blown out and the lantern broken.

While these thoughts were passing through his mind, he was feeling, half mechanically, for his purse. Suddenly his heart stopped beating; a feeling of cold scales passed up the back of his legs, and a cold blow seemed to fall upon his scalp. He stood petrified for a moment; then he felt again with one feverish movement; and then his loss burst upon him, and he was covered at once with perspiration. To spendthrifts money is so living and actual—it is such a thin veil between them and their pleasures! There is only one limit to their fortune—that of time; and a spendthrift with only a few crowns is the Emperor of Rome until they are spent. For such a person to lose his money is to suffer the most shocking reverse, and fall from heaven to hell, from all to nothing, in a breath. And all the more if he has put his head in the halter for it; if he may be hanged to-morrow for that same purse so dearly earned, so foolishly departed! Villon stood and cursed; he threw the two whites into the street; he shook his fist at heaven; he stamped, and was not horrified to find himself trampling the poor corpse. Then he began rapidly to retrace his steps towards the house beside the cemetery. He had forgotten all fear of the patrol, which was long gone by at any rate, and had no idea but that of his lost purse. It was in vain that he looked right and left upon the snow: nothing was to be seen. He had not dropped it in the streets. Had it fallen in the house? He would have liked dearly to go in and see; but the idea of the grisly occupant unmanned him. And he saw besides, as he drew near, that their efforts to put out the fire had been unsuccessful; on the contrary, it had broken into a blaze, and a changeful light played in the chinks of door and window, and revived his terror for the authorities and Paris gibbet.

He returned to the hotel with the porch, and groped about upon the snow for the money he had thrown away in his childish passion. But he could only find one white; the other had probably struck sideways and sunk deeply in. With a single white in his pocket, all his projects for a rousing night in some wild tavern vanished utterly away. And it was not only pleasure that fled laughing from his grasp; positive discomfort, positive pain, attacked him as he stood ruefully before the porch. His perspiration had dried upon him; and though the wind had now fallen, a binding frost was setting in stronger with every hour, and he felt benumbed and sick at heart. What was to be done? Late as was the hour, improbable as was success, he would try the house of his adopted father, the chaplain of St Benoît.

He ran there all the way, and knocked timidly. There was no answer. He knocked again and again, taking heart with every stroke; and at last steps were heard approaching from within. A barred wicket fell open in the iron-studded door, and emitted a gush of yellow light.

'Hold up your face to the wicket,' said the chaplain from within.

'It's only me,' whimpered Villon.

'Oh, it's only you, is it?' returned the chaplain; and he cursed him with foul unpriestly oaths for disturbing him at such an hour, and bade him be off to hell, where he came from.

'My hands are blue to the wrist,' pleaded Villon; 'my feet are dead and full of twinges: my nose aches with the sharp air; the cold lies at my heart. I may be dead before morning. Only this once, father, and before God I will never ask again!'

'You should have come earlier,' said the ecclesiastic coolly. 'Young men require a lesson now and then.' He shut the wicket and retired deliberately into the interior of the house.

Villon was beside himself; he beat upon the door with his hands and feet, and shouted hoarsely after the chaplain.

'Wormy old fox!' he cried. 'If I had my hand under your twist, I would send you flying headlong into the bottomless pit.'

A door shut in the interior, faintly audible to the poet down long passages. He passed his hand over his mouth with an oath. And then the humour of the situation struck him, and he laughed and looked lightly up to heaven, where the stars seemed to be winking over his discomfiture.

What was to be done? It looked very like a night in the frosty streets. The idea of the dead woman popped into his imagination, and gave him a hearty fright; what had happened to her in the early night might very well happen to him before morning. And he so young! and with such immense possibilities of disorderly amusement before him! He felt quite pathetic over the notion of his own fate, as if it had been some one else's, and made a little imaginative vignette of the scene in the morning, when they should find his body.

He passed all his chances under review, turning the white between his thumb and forefinger. Unfortunately he was on bad terms with some old friends who would once have taken pity on him in such a plight. He had lampooned them in verses, he had beaten and cheated them; and yet now, when he was in so close a pinch, he thought there was at least one who might perhaps relent. It was a chance. It was worth trying at least, and he would go and see.

On the way, two little accidents happened to him which coloured his musings in a very different manner. For, first, he fell in with the track of a patrol, and walked in it for some hundred yards, although it lay out of his direction. And this spirited him up; at least he had confused his trail; for he was still possessed with the idea of people tracking him all about Paris over the snow, and collaring him next morning before he was awake. The other matter affected him very differently. He passed a street corner, where, not so long before, a woman and her child had been devoured by wolves. This was just the kind of weather, he reflected, when wolves might take it into their heads to enter Paris again; and a lone man in these deserted streets would run the chance of something worse than a mere scare. He stopped and looked upon the place with an unpleasant interest—it was a centre where several lanes intersected each other; and he looked down them all one after another, and held his breath to listen, lest he should detect some galloping black things on the snow, or hear the sound of howling between him and the river. He remembered his mother telling him the story and pointing out the spot, while he was yet a child. His mother! If he only knew where she lived, he might make sure at least of shelter. He determined he would inquire upon the morrow; nay, he would go and see her too, poor old girl! So thinking, he arrived at his destination—his last hope for the night.

The house was quite dark, like its neighbours, and yet after a few taps, he heard a movement overhead, a door opening, and a cautious voice asking who was there. The poet named himself in a loud whisper, and waited, not without some trepidation, the result. Nor had he to wait long. A window was suddenly opened, and a pailful of slops splashed down upon the doorstep. Villon had not been unprepared for something of the sort, and had put himself as much in shelter as the nature of the porch admitted; but for all that, he was deplorably drenched below the waist.

His hose began to freeze almost at once. Death from cold and exposure stared him in the face; he remembered he was of phthisical tendency, and began coughing tentatively. But the gravity of the danger steadied his nerves. He stopped a few hundred yards from the door where he had been so rudely used, and reflected with his finger to his nose. He could only see one way of getting a lodging, and that was to take it. He had noticed a house not far away, which looked as if it might be easily broken into, and thither he betook himself promptly, entertaining himself on the way with the idea of a room still hot, with a table still loaded with the remains of supper, where he might pass the rest of the black hours, and whence he should issue, on the morrow, with an armful of valuable plate. He even considered on what viands and what wines he should prefer; and as he was calling the roll of his favourite dainties, roast fish presented itself to his mind with an odd mixture of amusement and horror.

'I shall never finish that ballade,' he thought to himself; and then, with another shudder at the recollection, 'Oh, damn his fat head!' he repeated fervently, and spat upon the snow.

The house in question looked dark at first sight; but as Villon made a preliminary inspection in search of the handiest point of attack, a little twinkle of light caught his eye from behind a curtained window.

'The devil!' he thought. 'People awake! Some student or some saint, confound the crew! Can't they get drunk and lie in bed snoring like their neighbours! What's the good of curfew, and poor devils of bell-ringers jumping at a rope's-end in bell-towers? What's the use of day, if people sit up all night? The gripes to them!' He grinned as he saw where his logic was leading him. 'Every man to his business, after all,' added he, 'and if they're awake, by the lord, I may come by a supper honestly for this once, and cheat the devil.'

He went boldly to the door and knocked with an assured hand. On both previous occasions, he had knocked timidly and with some dread of attracting notice; but now, when he had just discarded the thought of a burglarious entry, knocking at a door seemed a mighty simple and innocent proceeding. The sound of his blows echoed through the house with thin, phantasmal reverberations, as though it were quite empty; but these had scarcely died away before a measured tread drew near, a couple of bolts were withdrawn, and one wing was opened broadly, as though no guile or fear of guile were known to those within. A tall figure of a man, muscular and spare, but a little bent, confronted Villon. The head was in massive bulk, but finely sculptured; the nose blunt at the bottom, but refining upward to where it joined a pair of strong and honest eyebrows; the mouth and eyes surrounded with delicate markings, and the whole face based upon a thick white beard, boldly and squarely trimmed. Seen as it was by the light of a flickering hand-lamp, it looked perhaps nobler than it had a right to do; but it was a fine face, honourable rather than intelligent, strong, simple, and righteous.

'You knock late, sir,' said the old man in resonant, courteous tones.

Villon cringed, and brought up many servile words of apology; at a crisis of this sort the beggar was uppermost in him, and the man of genius hid his head with confusion.

'You are cold,' repeated the old man, 'and hungry? Well, step in.' And he ordered him into the house with a noble enough gesture.

'Some great seigneur,' thought Villon, as his host setting down the lamp on the flagged pavement of the entry, shot the bolts once more into their places.

'You will pardon me if I go in front,' he said, when this was done; and he preceded the poet upstairs into a large apartment, warmed with a pan of charcoal and lit by a great lamp hanging from the roof. It was very bare of furniture: only some gold plate on a sideboard; some folios; and a stand of armour between the windows. Some smart tapestry hung upon the walls, representing the crucifixion of our Lord in one piece, and in another a scene of shepherds and shepherdesses by a running stream. Over the chimney was a shield of arms.

'Will you seat yourself,' said the old man, 'and forgive me if I leave you? I am alone in my house to-night, and if you are to eat I must forage for you myself.'

No sooner was his host gone than Villon leaped from the chair on which he had just seated himself, and began examining the room, with the stealth and passion of a cat. He weighed the gold flagons in his hand, opened all the folios, and investigated the arms upon the shield, and the stuff with which the seats were lined. He raised the window curtains, and saw that the windows were set with rich stained glass in figures, so far as he could see, of martial import. Then he stood in the middle of the room, drew a long breath, and retaining it with puffed cheeks, looked round and round him, turning on his heels, as if to impress every feature of the apartment on his memory.

'Seven pieces of plate,' he said. 'If there had been ten, I would have risked it. A fine house, and a fine old master, so help me all the saints!'

And just then, hearing the old man's tread returning along the corridor, he stole back to his chair, and began humbly toasting his wet legs before the charcoal pan.

His entertainer had a plate of meat in one hand and a jug of wine in the other. He set down the plate upon the table, motioning Villon to draw in his chair, and going to the sideboard, brought back two goblets, which he filled.

'I drink to your better fortune,' he said, gravely touching Villon's cup with his own.

'To our better acquaintance,' said the poet, growing bold. A mere man of the people would have been awed by the courtesy of the old seigneur, but Villon was hardened in that matter; he had made mirth for great lords before now, and found them as black rascals as himself. And so he devoted himself to the viands with a ravenous gusto, while the old man, leaning backward, watched him with steady, curious eyes.

'You have blood on your shoulder, my man,' he said.

Montigny must have laid his wet right hand upon him as he left the house. He cursed Montigny in his heart.

'It was none of my shedding,' he stammered.

'I had not supposed so,' returned his host quietly. 'A brawl?'

'Well, something of that sort,' Villon admitted with a quaver.

'Perhaps a fellow murdered?'

'Oh, no—not murdered,' said the poet, more and more confused. 'It was all fair play—murdered by accident. I had no hand in it, God strike me dead!' he added fervently.

'One rogue the fewer, I daresay,' observed the master of the house.

'You may dare to say that,' agreed Villon, infinitely relieved. 'As big a rogue as there is between here and Jerusalem. He turned up his toes like a lamb. But it was a nasty thing to look at. I daresay you've seen dead men in your time, my lord?' he added, glancing at the armour.

'Many,' said the old man. 'I have followed the wars, as you imagine.'

Villon laid down his knife and fork, which he had just taken up again.

'Were any of them bald?' he asked.

'Oh yes, and with hair as white as mine.'

'I don't think I should mind the white so much,' said Villon. 'His was red.' And he had a return of his shuddering and tendency to laughter, which he drowned with a great draught of wine. 'I'm a little put out when I think of it,' he went on. 'I knew him—damn him! And then the cold gives a man fancies—or the fancies give a man cold, I don't know which.'

'Have you any money?' asked the old man.

'I have one white,' returned the poet, laughing. 'I got it out of a dead jade's stocking in a porch. She was as dead as Cæsar, poor wench, and as cold as a church, with bits of ribbon sticking in her hair. This is a hard world in winter for wolves and wenches and poor rogues like me.'

'I,' said the old man, 'am Enguerrand de la Feuillée, seigneur de Brisetout, bailly du Patatrac. Who and what may you be?'

Villon rose and made a suitable reverence. 'I am called Francis Villon,' he said, 'a poor Master of Arts of this university. I know some Latin, and a deal of vice. I can make chansons, ballades, lais, virelais, and roundels, and I am very fond of wine. I was born in a garret, and I shall not improbably die upon the gallows. I may add, my lord, that from this night forward I am your lordship's very obsequious servant to command.'

'No servant of mine,' said the knight; 'my guest for this evening, and no more.'

'A very grateful guest,' said Villon politely; and he drank in dumb show to his entertainer.

'You are shrewd,' began the old man, tapping his forehead, 'very shrewd; you have learning; you are a clerk; and yet you take a small piece of money off a dead woman in the street. Is it not a kind of theft?'

'It is a kind of theft much practised in the wars, my lord.'

'The wars are the field of honour,' returned the old man proudly.

'There a man plays his life upon the cast; he fights in the name of his lord the king, his Lord God, and all their lordships the holy saints and angels.'

'Put it,' said Villon, 'that I were really a thief, should I not play my life also, and against heavier odds?'

'For gain, but not for honour.'

'Gain?' repeated Villon, with a shrug. 'Gain! The poor fellow wants supper, and takes it. So does the soldier in a campaign. Why, what are all these requisitions we hear so much about? If they are not gain to those who take them, they are loss enough to the others. The men-at-arms drink by a good fire, while the burgher bites his nails to buy them wine and wood. I have seen a good many ploughmen swinging on trees about the country; ay, I have seen thirty on one elm, and a very poor figure they made; and when I asked some one how all these came to be hanged, I was told it was because they could not scrape together enough crowns to satisfy the men-at-arms.'

'These things are a necessity of war, which the low-born must endure with constancy. It is true that some captains drive overhard; there are spirits in every rank not easily moved by pity; and indeed many follow arms who are no better than brigands.'

'You see,' said the poet, 'you cannot separate the soldier from the brigand; and what is a thief but an isolated brigand with circumspect manners? I steal a couple of mutton chops, without so much as disturbing people's sleep; the farmer grumbles a bit, but sups none the less wholesomely on what remains. You come up blowing gloriously on a trumpet, take away the whole sheep, and beat the farmer pitifully into the bargain. I have no trumpet; I am only Tom, Dick, or Harry; I am a rogue and a dog, and hanging's too good for me—with all my heart; but just you ask the farmer which of us he prefers, just find out which of us he lies awake to curse on cold nights.'

'Look at us two,' said his lordship. 'I am old, strong, and honoured. If I were turned from my house to-morrow, hundreds would be proud to shelter me. Poor people would go out and pass the night in the streets with their children if I merely hinted that I wished to be alone. And I find you up, wandering homeless, and picking farthings off dead women by the wayside! I fear no man and nothing; I have seen you tremble and lose countenance at a word. I wait God's summons contentedly in my own house, or, if it please the king to call me out again, upon the field of battle. You look for the gallows; a rough, swift death, without hope or honour. Is there no difference between these two?'

'As far as to the moon,' Villon acquiesced. 'But if I had been born lord of Brisetout, and you had been the poor scholar Francis, would the difference have been any the less? Should not I have been warming my knees at this charcoal pan, and would not you have been groping for farthings in the snow? Should not I have been the soldier, and you the thief?'

'A thief!' cried the old man. 'I a thief! If you understood your words, you would repent them.'

Villon turned out his hands with a gesture of inimitable impudence. 'If your lordship had done me the honour to follow my argument!' he said.

'I do you too much honour in submitting to your presence,' said the knight. 'Learn to curb your tongue when you speak with old and honourable men, or some one hastier than I may reprove you in a sharper fashion.' And he rose and paced the lower end of the apartment, struggling with anger and antipathy. Villon surreptitiously refilled his cup, and settled himself more comfortably in the chair, crossing his knees and leaning his head upon one hand and the elbow against the back of the chair. He was now replete and warm; and he was in nowise frightened for his host, having gauged him as justly as was possible between two such different characters. The night was far spent, and in a very comfortable fashion after all; and he felt morally certain of a safe departure on the morrow.

'Tell me one thing,' said the old man, pausing in his walk. 'Are you really a thief?'

'I claim the sacred rights of hospitality,' returned the poet. 'My lord, I am.'

'You are very young,' the knight continued.

'I should never have been so old,' replied Villon, showing his fingers, 'if I had not helped myself with these ten talents. They have been my nursing-mothers and my nursing-fathers.'

'You may still repent and change.'

'I repent daily,' said the poet. 'There are few people more given to repentance than poor Francis. As for change, let somebody change my circumstances. A man must continue to eat, if it were only that he may continue to repent.'

'The change must begin in the heart,' returned the old man solemnly.

'My dear lord,' answered Villon, 'do you really fancy that I steal for pleasure? I hate stealing, like any other piece of work or of danger. My teeth chatter when I see a gallows. But I must eat, I must drink, I must mix in society of some sort. What the devil! Man is not a solitary animal—*Cui Deus fœminam tradit*. Make me king's pantler—make me abbot of St Denis; make me bailly of the Patatrac; and then I shall be changed indeed. But as long as you leave me the poor scholar Francis Villon, without a farthing, why, of course, I remain the same.'

'The grace of God is all-powerful.'

'I should be a heretic to question it,' said Francis. 'It has made you lord of Brisetout and bailly of the Patatrac; it has given me nothing but the quick wits under my hat and these ten toes upon my hands. May I help myself to wine? I thank you respectfully. By God's grace, you have a very superior vintage.'

The lord of Brisetout walked to and fro with his hands behind his back. Perhaps he was not yet quite settled in his mind about the parallel between thieves and soldiers; perhaps Villon had interested him by some cross-thread of sympathy; perhaps his wits were simply muddled by so much unfamiliar reasoning; but whatever the cause, he somehow yearned to convert the young man to a better way of thinking, and could not make up his mind to drive him forth again into the street.

'There is something more than I can understand in this,' he said at length. 'Your mouth is full of subtleties, and the devil has led you very far astray; but the devil is only a very weak spirit before God's truth, and all his subtleties vanish at a word of true honour, like darkness at morning. Listen to me once more. I learned long ago that a gentleman should live chivalrously and lovingly to God, and the king, and his lady; and though I have seen many strange things done, I have still striven to command my ways upon that rule. It is not only written in all noble histories, but in every man's heart, if he will take care to read. You speak of food and wine, and I know very well that hunger is a difficult trial to endure; but you do not speak of other wants; you say nothing of honour, of faith to God and other men, of courtesy, of love without reproach. It may be that I am not very wise—and yet I think I am—but you seem to me like one who has lost his way and made a great error in life. You are attending to the little wants, and you have totally forgotten the great and only real ones, like a man who should be doctoring a toothache on the Judgment Day. For such things as honour and love and faith are not only nobler than food and drink, but indeed I think that we desire them more, and suffer more sharply for their absence. I speak to you as I think you will most easily understand me. Are you not, while careful to fill your belly, disregarding another appetite in your heart, which spoils the pleasure of your life and keeps you continually wretched?'

Villon was sensibly nettled under all this sermonising. 'You think I have no sense of honour!' he cried. 'I'm poor enough, God knows! It's hard to see rich people with their gloves, and you blowing in your hands. An empty belly is a bitter thing, although you speak so lightly of it. If you had had as many as I, perhaps you would change your tune. Any way I'm a thief—make the most of that—but I'm not a devil from hell, God strike me dead! I would have you to know I've an honour of my own, as good as yours, though I don't prate about it all day long, as if it was a God's miracle to have any. It seems quite natural to me; I keep it in its box till it's wanted. Why now, look you here, how long have I been in this room with you? Did you not tell me you were alone in the house? Look at your gold plate! You're strong, if you like, but you're old and unarmed, and I have my knife. What did I want but a jerk of the elbow and here would have been you with the cold steel in your bowels, and there would have been me, linking in the streets, with an armful of gold cups! Did you suppose I hadn't wit enough to see that? And I scorned the action. There are your damned goblets, as safe as in a church; there are you, with your heart ticking as good as new; and here am I, ready to go out again as poor as I came in, with my one white that you threw in my teeth! And you think I have no sense of honour—God strike me dead!'

The old man stretched out his right arm. 'I will tell you what you are,' he said. 'You are a rogue, my man, an impudent and a black-hearted rogue and vagabond. I have passed an hour with you. Oh! believe me, I feel myself disgraced! And you have eaten and drunk at my table. But now I am sick at your presence; the day has come, and the night-bird should be off to his roost. Will you go before, or after?'

'Which you please,' returned the poet, rising. 'I believe you to be strictly honourable.' He thoughtfully emptied his cup. 'I wish I could add you were intelligent,' he went on, knocking on his head with his knuckles. 'Age, age! the brains stiff and rheumatic.'

The old man preceded him from a point of self-respect; Villon followed, whistling, with his thumbs in his girdle.

'God pity you,' said the lord of Brisetout at the door.

'Good-bye, papa,' returned Villon, with a yawn. 'Many thanks for the cold mutton.'

The door closed behind him. The dawn was breaking over the white roofs. A chill, uncomfortable morning ushered in the day. Villon stood and heartily stretched himself in the middle of the road.

'A very dull old gentleman,' he thought. 'I wonder what his goblets may be worth.'

THRAWN JANET

The Reverend Murdoch Soulis was long minister of the moorland parish of Balweary, in the vale of Dule. A severe, bleak-faced old man, dreadful to his hearers, he dwelt in the last years of his life, without relative or servant or any human company, in the small and lonely manse under the Hanging Shaw. In spite of the iron composure of his features, his eye was wild, scared, and uncertain; and when he dwelt, in private admonitions, on the future of the impenitent, it seemed as if his eye pierced through the storms of time to the terrors of eternity. Many young persons, coming to prepare themselves against the season of the Holy Communion, were dreadfully affected by his talk. He had a sermon on 1st Peter v. and 8th, 'The devil as a roaring lion,' on the Sunday after every seventeenth of August, and he was accustomed to surpass himself upon that text both by the appalling nature of the matter and the terror of his bearing in the pulpit. The children were frightened into fits, and the old looked more than usually oracular, and were, all that day, full of those hints that Hamlet deprecated. The manse itself, where it stood by the water of Dule among some thick trees, with the Shaw overhanging it on the one side, and on the other many cold, moorish hill-tops rising towards the sky, had begun, at a very early period of Mr Soulis's ministry, to be avoided in the dusk hours by all who valued themselves upon their prudence; and guidmen sitting at the clachan alehouse shook their heads together at the thought of passing late by that uncanny neighbourhood. There was one spot, to be more particular, which was regarded with especial awe. The manse stood between the high-road and the water of Dule, with a gable to each; its back was towards the kirk-town of Balweary, nearly half a mile away; in front of it, a bare garden, hedged with thorn, occupied the land between the river and the road. The house was two stories high, with two large rooms on each. It opened not directly on the garden, but on a causewayed path, or passage, giving on the road on the one hand, and closed on the other by the tall willows and elders that bordered on the stream. And it was this strip of causeway that enjoyed among the young parishioners of Balweary so infamous a reputation. The minister walked there often after dark, sometimes groaning aloud in the instancy of his unspoken prayers; and when he was from home, and the manse door was locked, the more daring schoolboys ventured, with beating hearts, to 'follow my leader' across that legendary spot.

This atmosphere of terror, surrounding, as it did, a man of God of spotless character and orthodoxy, was a common cause of wonder and subject of inquiry among the few strangers who were led by chance or business into that unknown, outlying country. But many even of the people of the parish were ignorant of the strange events which had marked the first year of Mr Soulis's ministrations; and among those who were better informed, some were naturally reticent, and others shy of that particular topic. Now and again, only, one of the older folk would warm into courage over his third tumbler, and recount the cause of the minister's strange looks and solitary life.

Fifty years syne, when Mr Soulis cam' first into Ba'weary, he was still a young man—a callant, the folk said—fu' o' book-learnin' an' grand at the exposition, but, as was natural in sae young a man, wi' nae leevin' experience in religion. The younger sort were greatly taken wi' his gifts an' his gab; but auld, concerned, serious men and women were moved even to prayer for the young man, whom they took to be a self-deceiver, an' the parish that was like to be sae ill-supplied. It was before the days o' the Moderates—weary fa' them; but ill things are like guid—they baith come bit by bit, a pickle at a time; an' there were folk even then that said the Lord had left the college professors to their ain devices, an' the lads that went to study wi' them wad hae done mair an' better sittin' in a peat-bog, like their forbears o' the persecution, wi' a Bible under their oxter an' a speerit o' prayer in their heart. There was nae doubt, onyway, but that Mr Soulis had been ower lang at the college. He was careful an' troubled for mony things besides the ae thing needful. He had a feck o' books wi' him—mair than had ever been seen before in a' that presbytery; and a sair wark the carrier had wi' them, for they were a' like to have smoored in the De'il's Hag between this an' Kilmackerlie. They were books o' divinity, to be sure, or so they ca'd them; but the serious were of opinion there was little service for sae mony, when the hale o' God's Word would gang in the neuk o' a plaid. Then he wad sit half the day, an' half the nicht forbye, which was scant decent—writin', nae less; an' first, they were feared he wad read his sermons; an' syne it proved he was writin' a book himsel', which was surely no' fittin' for ane o' his years an' sma' experience.

Onyway it behoved him to get an auld, decent wife to keep the manse for him an' see to his bit denners; an' he was recommended to an auld limmer—Janet M'Clour, they ca'd her—an' sae far left to himself as to be ower persuaded. There was mony advised him to the contrar, for Janet was mair than suspeckit by the best folk in Ba'weary. Lang or that, she had had a wean to a dragoon; she hadna come forrit* for maybe thretty year; an' bairns had seen her mumblin' to hersel' up on Key's Loan in the gloamin', whilk was an unco time an' place for a God-fearin' woman. Howsoever, it was the laird himsel' that had first tauld the minister o' Janet; an' in thae days he wad hae gane a far gate to pleesure the laird.

* 'To come forrit'—to offer oneself as a communicant.

When folk tauld him that Janet was sib to the de'il, it was a' superstition by his way o' it; an' when they cast up the Bible to him an' the witch o' Endor, he wad threep it doun their thrapples that thir days were a' gane by, an' the de'il was mercifully restrained.

Weel, when it got about the clachan that Janet M'Clour was to be servant at the manse, the folk were fair mad wi' her an' him thegither; an' some o' the guidwives had nae better to dae than get round her door-cheeks and chairge her wi' a' that was ken't again' her, frae the sodger's bairn to John Tamson's twa kye. She was nae great speaker; folk usually let her gang her ain gate, an' she let them gang theirs, wi' neither Fair-guid-een nor Fair-guid-day: but when she buckled to, she had a tongue to deave the miller. Up she got, an' there wasna an auld story in Ba'-weary but she gart somebody lowp for it that day; they couldna say ae thing but she could say twa to it; till, at the hinder end, the guidwives up and claught haud o' her, an' clawed the coats aff her back, an' pu'd her doun the clachan to the water o' Dule, to see if she were a witch or no, soom or droun. The carline skirled till ye could hear her at the Hangin' Shaw, an' she focht like ten; there was mony a guidwife bure the mark o' her neist day an' mony a lang day after; an' just in the hettest o' the collieshangie, wha suld come up (for his sins) but the new minister.

'Women,' said he (and he had a grand voice), 'I charge you in the Lord's name to let her go.'

Janet ran to him—she was fair wud wi' terror—an' clang to him, an' prayed him, for Christ's sake, save her frae the cummers; an' they, for their pairt, tauld him a' that was ken't, an' maybe mair.

'Woman,' says he to Janet, 'is this true?'

'As the Lord sees me,' says she, 'as the Lord made me, no a word o't. Forbye the bairn,' says she, 'I've been a decent woman a' my days.'

'Will you,' says Mr Soulis, 'in the name of God, and before me, His unworthy minister, renounce the devil and his works?'

Weel, it wad appear that when he askit that, she gave a girn that fairly frichtit them that saw her, an' they could hear her teeth play dirl thegither in her chafts; but there was naething for't but the ae way or the ither; an' Janet lifted up her hand an' renounced the de'il before them a'.

'And now,' says Mr Soulis to the guidwives, 'home with ye, one and all, and pray to God for His forgiveness.'

An' he gied Janet his arm, though she had little on her but a sark, an' took her up the clachan to her ain door like a leddy o' the land; an' her screighin' and laughin' as was a scandal to be heard.

There were mony grave folk lang ower their prayers that nicht; but when the morn cam' there was sic a fear fell upon a' Ba'weary that the bairns hid theirsels, an' even the men-folk stood an' keekit frae their doors. For there was Janet comin' doun the clachan—her or her likeness, nane could tell—wi' her neck thrawn, an' her heid on ae side, like a body that has been hangit, an' a girn on her face like an unstreakit corp. By an' by they got used wi' it, an' even speered at her to ken what was wrang;

but frae that day forth she couldna speak like a Christian woman, but slavered an' played click wi' her teeth like a pair o' shears; an' frae that day forth the name o' God cam' never on her lips. Whiles she wad try to say it, but it michtna be. Them that kenned best said least; but they never gied that Thing the name o' Janet M'Clour; for the auld Janet, by their way o't, was in muckle hell that day. But the minister was neither to haud nor to bind; he preached about naething but the folk's cruelty that had gi'en her a stroke of the palsy; he skelpit the bairns that meddled her; an' he had her up to the manse that same nicht, an' dwalled there a' his lane wi' her under the Hangin' Shaw.

Weel, time gaed by: an' the idler sort commenced to think mair lichtly o' that black business. The minister was weel thocht o'; he was aye late at the writing, folk wad see his can'le doon by the Dule water after twal' at e'en; an' he seemed pleased wi' himsel' an' upsitten as at first, though a' body could see that he was dwining. As for Janet she cam' an' she gaed; if she didna speak muckle afore, it was reason she should speak less then; she meddled naebody; but she was an eldritch thing to see, an' nane wad hae mistrysted wi' her for Ba'weary glebe.

About the end o' July there cam' a spell o' weather, the like o't never was in that countryside; it was lown an' het an' heartless; the herds couldna win up the Black Hill, the bairns were ower weariet to play; an' yet it was gousty too, wi' claps o' het wund that rumm'led in the glens, and bits o' shouers that slockened naething. We aye thocht it but to thun'er on the morn; but the morn cam', an' the morn's morning, an' it was aye the same uncanny weather, sair on folks and bestial. O' a' that were the waur, nane suffered like Mr Soulis; he could neither sleep nor eat, he tauld his elders; an' when he wasna writin' at his weary book, he wad be stravaguin' ower a' the countryside like a man possessed, when a' body else was blithe to keep caller ben the house.

Abune Hangin' Shaw, in the bield o' the Black Hill, there's a bit enclosed grund wi' an iron yett; an' it seems, in the auld days, that was the kirkyaird o' Ba'weary, and consecrated by the Papists before the blessed licht shone upon the kingdom. It was a great howff o' Mr Soulis's onyway; there he wad sit an' consider his sermons; an' indeed it's a bieldy bit. Weel, as he cam' ower the wast end o' the Black Hill ae day, he saw first twa, an' syne fower, an' syne seeven corbie craws fleein' round an' round abune the auld kirkyaird. They flew laigh an' heavy, an' squawked to ither as they gaed; an' it was clear to Mr Soulis that something had put them frae their ordinar'. He wasna easy fleyed, an' gaed straucht up to the wa's; an' what suld he find there but a man, or the appearance o' a man, sittin' in the inside upon a grave. He was of a great stature, an' black as hell, an' his e'en were singular to see.* Mr Soulis had heard tell o' black men, mony's the time; but there was something unco about this black

* It was a common belief in Scotland that the devil appeared as a black man. This appears in several witch trials, and I think in Law's 'Memorials', that delightful storehouse of the quaint and grisly.

man that daunted him. Het as he was, he took a kind o' cauld grue in the marrow o' his banes; but up he spak for a' that; an' says he: 'My friend, are you a stranger in this place?' The black man answered never a word; he got upon his feet, an' begoud to hirsle to the wa' on the far side; but he aye lookit at the minister; an' the minister stood an' lookit back; till a' in a meenit the black man was ower the wa' an' rinnin' for the bield o' the trees. Mr Soulis, he hardly kenned why, ran after him; but he was fair forjeskit wi' his walk an' the het, unhalesome weather; an' rin as he likit, he got nae mair than a glisk o' the black man amang the birks, till he won doun to the foot o' the hillside, an' there he saw him ance mair, gaun hap-step-an'-lowp ower Dule water to the manse.

Mr Soulis wasna weel pleased that this fearsome gangrel suld mak' sae free wi' Ba'weary manse; an' he ran the harder, an', wet shoon, ower the burn, an' up the walk; but the deil a black man was there to see. He stepped out upon the road, but there was naebody there; he gaed a' ower the gairden, but na, nae black man. At the hinder end, an' a bit feared, as was but natural, he lifted the hasp an' into the manse; an' there was Janet M'Clour before his een, wi' her thrawn craig, an' nane sae pleased to see him. An' he aye minded sinsyne, when first he set his een upon her, he had the same cauld and deidly grue.

'Janet,' says he, 'have you seen a black man?'

'A black man?' quo' she. 'Save us a'! Ye're no wise, minister. There's nae black man in a' Ba'weary.'

But she didna speak plain, ye maun understand; but yam-yammered, like a powney wi' the bit in its moo.

'Weel,' says he, 'Janet, if there was nae black man, I have spoken with the Accuser of the Brethren.'

An' he sat down like ane wi' a fever, an' his teeth chittered in his heid.

'Hoots,' says she, 'think shame to yoursel', minister'; an' gied him a drap brandy that she keept aye by her.

Syne Mr Soulis gaed into his study amang a' his books. It's a lang, laigh, mirk chalmer, perishin' cauld in winter, an' no' very dry even in the tap o' the simmer, for the manse stands near the burn. Sae doun he sat, an' thocht o' a' that had come an' gane since he was in Ba'weary, an' his hame, an' the days when he was a bairn an' ran daffin' on the braes; an' that black man aye ran in his heid like the owercome o' a sang. Aye the mair he thocht, the mair he thocht o' the black man. He tried the prayer, an' the words wadna come to him; an' he tried, they say, to write at his book, but he couldna mak' nae mair o' that. There was whiles he thocht the black man was at his oxter, an' the swat stood upon him cauld as well-water; an' there was ither whiles when he cam' to himsel' like a christened bairn an' minded naething.

The upshot was that he gaed to the window an' stood glowrin' at Dule water. The trees are unco thick, an' the water lies deep an' black under the manse; an' there was Janet washin' the cla'es wi' her coats kilted. She had her back to the minister, an' he, for his pairt, hardly kenned what he was lookin' at. Syne she turned round, an' shawed her face; Mr Soulis had the same cauld grue as twice that day afore, an' it was borne in upon him what folk said, that Janet was deid lang syne, an' this was a bogle in her clay-cauld flesh. He drew back a pickle and he scanned her narrowly. She was tramp-trampin' in the cla'es, croonin' to hersel'; and eh! Gude guide us, but it was a fearsome face. Whiles she sang louder, but there was nae man born o' woman that could tell the words o' her sang; an' whiles she lookit side-lang doun, but there was naething there for her to look at. There gaed a scunner through the flesh upon his banes; an' that was Heeven's advertisement. But Mr Soulis just blamed himsel', he said, to think sae ill o' a puir, auld afflicted wife that hadna a freend forbye himsel'; an' he put up a bit prayer for him an' her, an' drank a little caller water—for his heart rose again' the meat—an' gaed up to his naked bed in the gloamin'.

That was a nicht that has never been forgotten in Ba'weary, the nicht o' the seeventeenth o' August, seeventeen hun'er' an' twal'. It had been het afore, as I hae said, but that nicht it was hetter than ever. The sun gaed doun amang unco-lookin' clouds; it fell as mirk as the pit; no' a star, no' a breath o' wund; ye couldna see your han' afore your face, an' even the auld folk cuist the covers frae their beds an' lay pechin' for their breath. Wi' a' that he had upon his mind, it was geyan unlikely Mr Soulis wad get muckle sleep. He lay an' he tummled; the gude, caller bed that he got into brunt his very banes; whiles he slept, an' whiles he waukened; whiles he heard the time o' nicht, an' whiles a tyke yowlin' up the muir, as if somebody was deid; whiles he thocht he heard bogles claverin' in his lug, an' whiles he saw spunkies in the room. He behoved, he judged, to be sick; an' sick he was—little he jaloosed the sickness.

At the hinder end he got a clearness in his mind, sat up in his sark on the bed-side, an' fell thinkin' ance mair o' the black man an' Janet. He couldna weel tell how—maybe it was the cauld to his feet—but it cam' in upon him wi' a spate that there was some connection between thir twa, an' that either or baith o' them were bogles. An' just at that moment, in Janet's room, which was neist to his, there cam' a stramp o' feet as if men were wars'lin', an' then a loud bang; an' then a wund gaed reishling round the fower quarters o' the house; an' then a' was aince mair as seelent as the grave.

Mr Soulis was feared for neither man nor deevil. He got his tinder-box, an' lit a can'le, an' made three steps o't ower to Janet's door. It was on the hasp, an' he pushed it open, an' keekit bauldly in. It was a big room, as big as the minister's ain, an' plenished wi' grand, auld, solid gear, for he had naething else. There was a fower-posted bed wi' auld tapestry; an' a braw cabinet o' aik, that was fu' o' the minister's divinity books, an' put there to be out o' the gate; an' a wheen duds o' Janet's lying here an' there about the floor. But nae Janet could Mr Soulis see; nor ony sign o' a contention. In he gaed (an' there's few that wad hae followed him) an' lookit a' round, an' listened. But there was naething to be heard, neither inside the manse nor in a' Ba'weary parish, an' naething to be seen but the muckle shadows turnin' round the can'le. An' then a' at aince, the minister's heart played dunt an' stood stock-still; an' a cauld wund blew amang the hairs o' his heid. Whaten a weary sicht was that for the puir man's een! For there was Janet hangin' frae a nail beside the auld aik cabinet: her heid aye lay on her shouther, her een were steekit, the tongue projected frae her mouth, an' her heels were twa feet clear abune the floor.

'God forgive us all!' thocht Mr Soulis; 'poor Janet's dead.'

He cam' a step nearer to the corp; an' then his heart fair whammled in his inside. For, by what cantrip it wad ill beseem a man to judge, she was hingin' frae a single nail an' by a single wursted thread for darnin' hose.

It's an awfu' thing to be your lane at nicht wi' siccan prodigies o' darkness; but Mr Soulis was strong in the Lord. He turned an' gaed his ways oot o' that room, an' lockit the door ahint him; an' step by step, doon the stairs, as heavy as leed; an' set doon the can'le on the table at the stairfoot. He couldna pray, he couldna think, he was dreepin' wi' caul' swat, an' naething could he hear but the dunt-dunt-duntin' o' his ain heart. He micht maybe hae stood there an hour, or maybe twa, he minded sae little; when a' o' a sudden, he heard a laigh, uncanny steer upstairs; a foot gaed to an' fro in the chalmer whaur the corp was hingin'; syne the door was opened, though he minded weel that he had lockit it; an' syne there was a step upon the landin', an' it seemed to him as if the corp was lookin' ower the rail an' doun upon him whaur he stood.

He took up the can'le again (for he couldna want the licht), an' as saftly as ever he could, gaed straucht out o' the manse an' to the far end o' the causeway. It was aye pit-mirk; the flame o' the can'le, when he set it on the grund, brunt steedy and clear as in a room; naething moved, but the Dule water seepin' an' sabbin' doun the glen, an' yon unhaly footstep that cam' ploddin' doun the stairs inside the manse. He kenned the foot ower weel, for it was Janet's; an' at ilka step that cam' a wee thing nearer, the cauld got deeper in his vitals. He commended his soul to Him that made an' keepit him; 'and, O Lord,' said he, 'give me strength this night to war against the powers of evil.'

By this time the foot was comin' through the passage for the door; he could hear a hand skirt alang the wa', as if the fearsome thing was feelin' for its way. The saughs tossed an' maned thegither, a lang sigh cam' ower the hills, the flame o' the can'le was blawn aboot; an' there stood the corp o' Thrawn Janet, wi' her grogram goun an' her black mutch, wi' the heid aye upon the shouther, an' the girn still upon the face o't—leevin', ye wad hae said—deid, as Mr Soulis weel kenned—upon the threshold o' the manse.

It's a strange thing that the saul o' man should be that thirled into his perishable body; but the minister saw that, an' his heart didna break.

She didna stand there lang; she began to move again an' cam' slowly towards Mr Soulis whaur he stood under the saughs. A' the life o' his body, a' the strength o' his speerit, were glowerin' frae his een. It seemed she was gaun to speak, but wanted words, an' made a sign wi' the left hand. There cam' a slap o' wund, like a cat's fuff; oot gaed the can'le, the slaughs skreighed like folk; and Mr Soulis kenned that, live or die, this was the end o't.

'Witch, beldame, devil!' he cried, 'I charge you, by the power of God, begone—if you be dead, to the grave—if you be damned, to hell.'

An' at that moment the Lord's ain hand out o' the Heevens struck the Horror whaur it stood; the auld, deid, desecrated corp o' the witch-wife, sae lang keepit frae the grave an' hirsled round by de'ils, lowed up like a brunstane spunk an' fell in ashes to the grund; the thunder followed, peal on dirlin' peal, the rairin' rain upon the back o' that; an' Mr Soulis lowped through the garden hedge, an' ran, wi' skelloch upon skelloch, for the clachan.

That same mornin', John Christie saw the Black Man pass the Muckle Cairn as it was chappin' six; before eicht, he gaed by the change-house at Knockdow; an' no' lang after, Sandy M'Lellan saw him gaun linkin' doun the braes frae Kilmackerlie. There's little doubt but it was him that dwalled sae lang in Janet's body; but he was awa' at last; an' sinsyne the de'il has never fashed us in Ba'weary.

But it was a sair dispensation for the minister; lang, lang he lay ravin' in his bed; an' frae that hour to this he was the man ye ken the day.

from FABLES

II
The Sinking Ship

'Sir,' said the first lieutenant, bursting into the Captain's cabin, 'the ship is going down.'

'Very well, Mr Spoker,' said the Captain; 'but that is no reason for going about half-shaved. Exercise your mind a moment, Mr Spoker, and you will see that to the philosophic eye there is nothing new in our position: the ship (if she is to go down at all) may be said to have been going down since she was launched.'

'She is settling fast,' said the first lieutenant, as he returned from shaving.

'Fast, Mr Spoker?' asked the Captain. 'The expression is a strange one, for time (if you will think of it) is only relative.'

'Sir,' said the lieutenant, 'I think it is scarcely worth while to embark in such a discussion when we shall all be in Davy Jones's Locker in ten minutes.'

'By parity of reasoning,' returned the Captain gently, 'it would never be worth while to begin any inquiry of importance; the odds are always overwhelming that we must die before we shall have brought it to an end. You have not considered, Mr Spoker, the situation of man,' said the Captain, smiling, and shaking his head.

'I am much more engaged in considering the position of the ship,' said Mr Spoker.

'Spoken like a good officer,' replied the Captain, laying his hand on the lieutenant's shoulder.

On deck they found the men had broken into the spirit-room, and were fast getting drunk.

'My men,' said the Captain, 'there is no sense in this. The ship is going down, you will tell me, in ten minutes: well, and what then? To the philosophic eye, there is nothing new in our position. All our lives long, we may have been about to break a blood-vessel or to be struck by lightning, not merely in ten minutes, but in ten seconds; and that has not prevented us from eating dinner, no, nor from putting money in the Savings Bank. I assure you, with my hand on my heart, I fail to comprehend your attitude.'

The men were already too far gone to pay much heed.

'This is a very painful sight, Mr Spoker,' said the Captain.

'And yet to the philosophic eye, or whatever it is,' replied the first lieutenant, 'they may be said to have been getting drunk since they came aboard.'

'I do not know if you always follow my thought, Mr Spoker,' returned the Captain gently. 'But let us proceed.'

In the powder magazine they found an old salt smoking his pipe.

'Good God!' cried the Captain, 'what are you about?'

'Well, sir,' said the old salt apologetically, 'they told me as she were going down.'

'And suppose she were?' said the Captain. 'To the philosophic eye, there would be nothing new in our position. Life, my old shipmate, life, at any moment and in any view, is as dangerous as a sinking ship; and yet it is man's handsome fashion to carry umbrellas, to wear indiarubber overshoes, to begin vast works, and to conduct himself in every way as if he might hope to be eternal. And for my own poor part I should despise the man who, even on board a sinking ship, should omit to take a pill or to wind up his watch. That, my friend, would not be the human attitude.'

'I beg pardon, sir,' said Mr Spoker. 'But what is precisely the difference between shaving in a sinking ship and smoking in a powder magazine?'

'Or doing anything at all in any conceivable circumstances?' cried the Captain. 'Perfectly conclusive; give me a cigar!'

Two minutes afterwards the ship blew up with a glorious detonation.

IV

THE SICK MAN AND THE FIREMAN

There was once a sick man in a burning house, to whom there entered a fireman.

'Do not save me,' said the sick man. 'Save those who are strong.'

'Will you kindly tell me why?' inquired the fireman, for he was a civil fellow.

'Nothing could possibly be fairer,' said the sick man. 'The strong should be preferred in all cases, because they are of more service in the world.'

The fireman pondered a while, for he was a man of some philosophy. 'Granted,' said he at last, as a part of the roof fell in; 'but for the sake of conversation, what would you lay down as the proper service of the strong?'

'Nothing can possibly be easier,' returned the sick man; 'the proper service of the strong is to help the weak.'

Again the fireman reflected, for there was nothing hasty about this excellent creature. 'I could forgive you being sick,' he said at last, as a portion of the wall fell out, 'but I cannot bear your being such a fool.'

And with that he heaved up his fireman's axe, for he was eminently just, and clove the sick man to the bed.

VI

THE PENITENT

A man met a lad weeping. 'What do you weep for?' he asked.

'I am weeping for my sins,' said the lad.

'You must have little to do,' said the man.

The next day they met again. Once more the lad was weeping. 'Why do you weep now?' asked the man.

'I am weeping because I have nothing to eat,' said the lad.

'I thought it would come to that,' said the man.

XII

THE CITIZEN AND THE TRAVELLER

'Look round you,' said the citizen. 'This is the largest market in the world.'

'O, surely not,' said the traveller.

'Well, perhaps not the largest,' said the citizen, 'but much the best.'

'You are certainly wrong there,' said the traveller. 'I can tell you . . . '

They buried the stranger at the dusk.

R. B. CUNNINGHAME GRAHAM
1852–1936

UN INFELIZ

During the somewhat fragmentary meal, I had watched him, seeing a difference between him and the usual French Algerian types. Dressed all in grey, his clothes of that peculiar substance which seems specially constructed for Algeria, Morocco, and the Levant, and which, intended to look like English tweed, yet is as different from its prototype as is 'kincob', his shirt of greenish flannel, his boots apparently made by a portmanteau maker, his scanty hair a yellowish grey, and his thin beard a greyish yellow, he gave you the idea of some pathetic seabeaten boulder, worn hollow by the beating of the waves of life.

As the smart Spanish-looking, but French-speaking, daughter of the landlady brought round the dishes, in which sea-slieve, stewed in high-smelling oil, made the air redolent, and over which myriads of flies kept up a pandemoniac concert, or yielded up their lives in the thick oleaginous black sauce, he paid her all those futile, yet kindly compliments, which only men, who in their youth have never known that ginger may be hot in the mouth, pay womankind. She easily accepted them, whilst smiling at the commercial travellers, who, with napkins tucked into their waist-coats, performed miraculous feats of sleight of hand, taking up pease as dextrously with the broad-pointed, iron-handled knives, as does an elephant transfer the buns which children give him at a travelling circus, from his proboscis to his mouth. Loose-trousered officers of the Chasseurs d'Afrique sat over the high-smelling foods talking regretfully of Paris, and of 'les petites' who there and elsewhere had fallen victims to their all-compelling charms. Detailing all the points both physical and moral of the victims, they pitied them, and spoke regretfully of what they had been, so to speak, impelled to do by the force of circumstances, but still with that well-founded yet chastened pride with which a horseman, once the struggle over, depreciates the efforts of a vicious horse.

Outside, the sandy street, shaded by bella sombra and by China trees, was full of Arabs straying aimlessly about, existing upon sufferance in their own country, each with his hand ready to raise at once to a military salute and his lips twitching with the salutation of 'Bonne chour, Mossi,' if the most abject member of the ruling race should deign to greet him as he passed. Dogs, thin and looking like cross-breeds between a jackal and a fox, slunk furtively about, their ears raw with mange, the sores upon their bodies all alive with flies, squirmed in and out between the people's legs, receiving patiently or with a half-choked yelp, blows with the cudgels which all country Arabs use, or kicks administered between their ribs from seedy, unvarnished patent-leather boots with drab cloth tops. At the corners of the streets, horses blinked sleepily, their high and chair-like saddles sharply outlined against the white-washed walls in the fierce glare of the Algerian sun. The hum compounded of the cries of animals and men, not disagreeable and acute as is the noise which rises from a northern crowd, but which throughout the East blends itself into a sort of chant, rose in the air, and when it ceased, the grating of the pebbles on the beach, tossed in the ceaseless surf, fell on the ear in rhythmic cadences. In all the spaces and streets of the incongruous North-European-looking town, the heterogeneous population lounged about lazily, knowing full well that time was the commodity of which they had the most. Riffians in long white haiks, carrying the sword-shaped sticks with which their ancestors attacked the Roman legionaries, strode to and fro, their heads erect, their faces set like cameos, impassible except their eyes, which lighted for a second in a blaze when a French soldier pushed them roughly, and then became deliberately opaque. Their women with their chins tattooed like Indians, dressed in sprigged muslins, their jet black hair hanging in plaited tails upon their shoulders, walked about staring like half-wild horses at the unfamiliar shops. Wearing no veils, their appearance drew from the wealthier Mohammedans pious ejaculations as to their shamelessness, and aphorisms such as 'the married woman is best with a broken leg at home,' and others more direct and quite unfitted for our European taste, as we have put a veil of cotton wool before our ears, and count all decent, so that we do not hear.

Over the insubstantial French provincial houses hung that absorbing eastern thin white dust, which in Algeria seems to mock the efforts of the conquering race to Europeanise the land, no matter howsoever mathematically correct they build the spire of Congregational Gothic church or façade of the gingerbread town hall. The streets all duly planted with the most shady-foliaged trees, the arms of the Republic, looking as dignified as the tin plates of fire insurance offices upon 'les monuments', even the pomp and circumstance of the military band, crashing out patriotic airs upon the square, were unavailing to remove the feeling that the East was stronger than the West here in its kingdom, and that did some convulsion but remove the interlopers, all would fall back again into its time-worn rut.

Musing upon the instability of accepted facts, and wondering whether after all, if both the English and the French were expelled from India and Algeria, they would leave as much remembered of themselves as have the makers of the tanks in Kandy, or the builders of the walls of Constantine: in fact, having fallen into that state, which we in common with the animals fall into after eating, but which we usually put down to the workings of the spirit when it is nothing but the efforts of digestion, a voice fell on my ear.

'Would I be good enough to share my carriage with a gentleman, an engineer who wanted to regain his mine some thirty miles away, upon the road.'

The stranger was my dissonance in grey, a blot upon the landscape, an outrage in his baggy trousers amongst the white-robed people of the place. He bore upon his face the not to be mistaken mark of failure: that failure which alone makes a man interesting and redeems him from the vulgarity of mere crass success. Gently but with prolixity, he proffered his request. All the timidity which marks the vanquished of the world exhaled from his address as he politely—first tendering his card—apologised both for existence and for troubling me to recognise the fact. I had the only carriage in the town, the diligence did not run more than once a week, and he was old to make the journey on a mule, besides which, though he had been for five and thirty years a dweller in the province of Oran, he spoke but little Arabic, and it was dull to be obliged for a long day to talk in nothing but 'le petit nègre'.

Most willingly I gave consent, and shortly the miserable conveyance drawn by a starveling mule and an apocalyptic horse, and driven by a Jew, dressed in a shoddy suit of European clothes, surmounted by a fez, holding in either hand a rein and carrying for conveniency his whip between his teeth, jangled and rattled to the door. We both stood bowing after the fashion of Don Basilio and Don Bartolo, waving each other in, and making false preparatory steps, only to fall back again, until I fairly shoved my self-invited guest into the carriage, shut to the door, and called upon the Jew to start. He did so, dextrously enveloping his miserable beasts with a well-executed slash of his whip and a few curses, without which no animal will start in any colony, ill-use him as you may.

In a melancholy, low-pitched, cultivated voice my fellow-traveller pointed out the objects of chief interest on the road. Here such and such an officer had been led into an ambush and his men 'massacred' by Arabs posted on a hill. Their tombs, with little cast-iron crosses sticking in the sparse sandy grass, were hung with immortelles, and the shaky cemetery gate, guarded by a plaster lion modelled apparently from a St Bernard dog, was there to supplement his history. A palm tree grew luxuriantly outside, 'its roots in water and its head in fire', as if to typify the resistance of the land to all that comes from Europe, whilst within the walls exotic trees from France withered and drooped their heads, and seemed to pine for their lost rain and mist. The road, well made and bridged, and casting as it were a shadow of the cross upon the land, wound in and out between a range of hills. At intervals it passed through villages, built on the French provincial type, with a wide street and pointed-steepled church, a 'mairie', telegraph station, and a barracks for the troops. An air of discontent, begot of 'maladie du pays' and absinthe, seemed endemic in them all: no one looked prosperous but the two Arab soldiers, who on their horses, sitting erect and motionless, turned out to see the passage of the coach.

Long trains of donkeys and of mules passed on the road, driven by men dressed in mere bundles of white rags, or by Mallorcans or Valencians, who, with their sticks shoved down between their shirts and backs, urged on their beasts with the loud raucous cries which throughout Spain the Moors have left to their descendant muleteers, together with their pack saddles, their baskets of esparto, and the rest of the equipment of the road. Occasionally camels passed by, looking quite out of place on the high road, but still maintaining the same swaying pace with which their ancestors from immemorial time have paced the desert sands.

And as we jangled noisily upon our path, my guest detailed his life, with circumstance, quoting his 'acte de naissance', telling the number of his family, his adventures in the colony, on which he looked half with affection, half with dislike, after the fashion of one mated to a loud-tongued wife, who in recounting all his sufferings never forgets to add, 'But still she was a splendid housekeeper,' thus hoping to deceive his audience and himself.

'The country it is good, you see (he said), but still unsuited for most kinds of crops. Either it rains in torrents and the corn is washed away, or else the drought lasts years, so that the colonist is always grumbling; not that our countrymen as a general rule are agriculturists, no, that they leave to the Mallorcans and Valencians, but still they grumble at their relatively prosperous life.' A comfortable doctrine and a true; for grumbling is as sauce to the hard bread of poverty; without it riches would be bereft of half their charm, and life be rendered tasteless and a mere dream of stertorous content.

As we drove on, the road emerged from woods of greenish-grey Aleppo pine into rough hills clothed with lentiscus and wild olives, and thicketed with cistus and dwarf rhododendrons. Partridges flew across the path continually, occasionally wild boars peeped out, grunted and wheeling back, dived into the recesses of the scrub. Parties of mounted Arabs, their haiks and selhams floating in the wind, carrying hooded hawks on their gloved hands or balancing upon their horses' croups, passed us impassible, making their stallions rear and passage; their reins held high and loosely as they raised themselves almost upright upon their horses' backs. We passed outlying farms, sun-swept and desolate, without the charm of mystery of a ranche in Texas or in Mexico, but looking rather more like bits of railway stations, cut off in lengths, and dropped upon the hills. I learned that most of them were held by officers and soldiers who had served in times gone by against 'les indigènes', and that some of them had grown quite rich by waiting till civilisation had spread up to them; a kind of unearned increment which even dogmatists in points of economics could not be hard on, taking into consideration the time and dulness that the owners had endured. Gourbis of Arabs, mud-built ksour with now and then black goats'-hair tents, each with its horse feeding in front of it, were dotted on the hill sides or on the plains green with palmettos and with camel-thorn. Occasionally white little towns glittered upon the mountain sides or nestled in the corries of the hills. The untiring sun beat down and blended all in one harmonious whole of brown: brown dusty roads, brown shaggy hills and rocks; the animals were all coated with the bright brown dust, and men, scorched copper-coloured, stood leaning on their sticks playing reed pipes and watching goats and sheep, so motionless that they seemed tree-trunks from which floated sound.

Little by little I learned all my companion's life. His college days, his triumphs, medals, and his entry to the world, wise as he said in scientific knowledge, but a child in the mean necessary arts without which no one can achieve success.

'I was,' he said, 'bête comme tous les chastes, and therefore fell a victim to the first pretty face . . . I married and adored her, working day and night to make a home, a stupid story of a stupid man, eh? . . . well, well, the usual thing, the husband all day out, planning and striving, and the devil, no not the devil, but the idle fool, who flattered . . . and the nest empty when the working bird came home. So I forswore all women, and lived miserably, came to this colony, and thought I saw an opening, and then married again, this time an honest woman almost a peasant, and have passed my life, the wolf ever just howling close to the door, but not quite entering the house.

'A happy life, yes happy, for you see I knew that I was born a simple, and holy writ says that we simples are to inherit all the earth . . . well so we do, for we maintain all our illusions green, and after all illusions are the best riches, so I have been rich, that is until a month ago. Not rich, you know, in money, though I have had my chances, but never took them, as when the Germany company offered me fifty thousand francs to discover copper in a mine, where since the beginning of the world no copper ever was. I have seen friends grow rich and have not envied them, for till a month ago I had a treasure in my wife. Yes, a good woman, always equal, ever the same, good year, bad year, smiling but sensible, hard-working, and with just that worldly sense I ever wanted . . . yet looked up to me for my scant book learning . . . No, no . . . I have not wept much, for I have work to do; not that work deadens grief, as you in England say, but that you cannot work and weep.

'The mine is not a rich mine, ten or twelve Spaniards, the foreman and myself, the sole inhabitants. Dull life, you say. . . . Yes, but no duller than in Paris: life is life, no matter where you have to live. No I do not shoot; why should one shoot? rabbits and hares are under every tuft of grass; the Spanish workmen kill them now and then with stones. Ah, there is the mine, that yellowish mark upon the hill, those tunnels, and the huts.'

We rattled down the hill, the miserable jades both galloped for their lives, the carriage bounding after them, checked but by a rusty Arab stirrup fastened to a chain, which acted as a drag. We pulled up sharply, and the drag chain breaking left the stirrup stranded on the road. As the driver went to retrieve it, and to repair the damage, I had full time to contemplate the mine. Twelve or thirteen kilometres from the nearest house just perched above the road, it seemed as if some giant rabbit had burrowed in the hill. Two or three tunnels, one of which vomited yellowish water underneath the road, two or three workings, open-cast and left deserted, two or three heaps of cinders, and a pumping engine broken and left to rust, together with the ten or a dozen cottages flanked by the dreary unsuccessful gardens which in all countries miners seem to own, were its chief features. An iron water tank upon a pile of masonry, and several heaps of coal dumped in the bushes which grew between the dark grey boulders with which the hill was strewn, served as embellishments toward the melancholy scene. Slatternly women washed their husbands' clothes, or stood and looked out listlessly into the driving mist; a mangy goat or two grazed on the prickly shrubs, and a keen wind, whistling and screeching through the gullies of the hills, made the coarse skirts and flannel petticoats crack in the air like whips. The sort of place which might have had a kind of grandeur of its own had not the mine been there, but which disfigured and made vulgar as it was became more desolating than a slum outside a town. The engineer collected his few traps, his carpet bag and shoddy plaid, his bulgy umbrella and his new hat carefully carried in a handbox all the journey on his knee: he tendered me his card, large, limp, and shiny, and with his 'noms', his 'prénoms', and his 'titres', duly set forth upon it.

Then, having thanked me with prolixity, he took his leave of me, and slinging all his things upon his back, struck into a small footpath up the hill, winding his way amongst the boulders, looking so like them in his worn grey clothes that it appeared all were identical, only that one was moving on the ground. I called and waved my hand, but he went upwards towards the huts without once turning, and when I looked again, the bent grey moving figure had disappeared amongst the stones.

JOHN DAVIDSON

1857–1909

MISS ARMSTRONG'S CIRCUMSTANCES

After all, my friends have been mistaken; my experiences are not nearly so exciting as they appeared to be when I saw them through their spectacles. They insisted that I had only to write down an exact chronicle of the days of the years of my life to be the author of a record as interesting as any novel. I was pretty well persuaded of the truth of their judgment when I began to write my history; but I had not proceeded far when doubts began to spring up, and by the time I had arrived at my seventh chapter, and the end of my seventeenth year, I was so tired of writing, and of my subject, that I threw my pen in the fire, and stowed away my papers in an old bandbox, out of sight and out of mind.

I have read somewhere that if a woman once falls in love, and then falls out of it, she has no peace until she is again swimming for life in a high-sea of passion. (I had better state here that I am just nineteen. The English master used to object to my figures of speech; but I am writing this entirely for my own satisfaction, and mean to give my imagination free scope.) It seems to me that literary composition is like love. When one has begun to write something of one's own, it doesn't matter how disgusted one may become, one returns to the ink-pot like a drunkard to his cups. So, after three months, I unearthed the bandbox, and read over my seven chapters. There were only two interesting pages in the whole manuscript, and those were the two last. All the early incidents in my life which my friends thought so wonderful were of no moment to me. My birth in Paris during the siege; the death of my father, a Scotch Socialist, on a barricade; my French mother's penniless journey to London; our life as beggars; my mother's second marriage to a philanthropic City man; my running away when I was seven, and my wanderings for a fortnight; my attempt to poison my baby-brother with matches; my attack on my philanthropic step-papa with a poker; my exile to a suburban boarding-school; my step-papa's fraudulent bankruptcy and disappearance, and the deaths of my poor mother and her little boy—all this was narrated in a dull, frigid manner, quite up to the degree of stupidity that would have registered 'Excellent' on Mr Standard the English master's meter. (I wonder what he would think of that metaphor!) A great deal, doubtless, might be made out of my early life, and when I am older I may be able to embody it in some readable way; but in the meantime it is impossible for me to put myself in the place of the little girl I was. This is simply because I did not begin to be self-conscious until I was seventeen. When my life ceases to be as full as it has been of late, I shall doubtless be able to study myself from the beginning. At present I am driven as if by some power outside me to write an account of a certain day in my life. I don't like writing, so I am going to make it as short as I can.

First of all, I shall quote the last two pages of my manuscript:

'It was at the age of seventeen, when I found myself in a position of dependence in the house of a relative of my stepfather's, that I first began to look upon myself as a *circumstance*. Doubtless this notion arose from something I had read, but I have never been able to trace its origin. One night while I was sitting alone in my room, the thought came to me that the whole world was an experiment. Here was I, a tall, handsome girl, already a woman in appearance, thrust by circumstances into a family that would have preferred to do without me. Were circumstances playing off a serio-comic practical joke on this family and me? But my fancy took a higher flight. I saw circumstances in the shape of the professor of chemistry and his lean assistant shaking up folk and families, and towns and countries, in bottles and beakers; braying stubborn folks like me in mortars; precipitating, calcining, sifting, subliming, filtering powers and principalities, companies and corporations; conducting a stupendous qualitative analysis of the world. I thought, "Since it's all an experiment, how can we help it if we're miserable?" "By joining the experimenters," came the answer pat. This warmed me, and I began to pace my room. "I will be an experimenter," I said to myself. "I will be a circumstance, and cause things." I marched up and down for awhile, thinking how much greater I was than the Prime Minister, who had simply been tossed up there by circumstances; he was only a bit of the experiment, but I was going to be a circumstance. Suddenly I saw that my metaphor had misled me. Circumstances, I perceived, *are* the experiment; everybody and everything is a circumstance. "You donkey!" I said to myself. "You don't need to become a circumstance; you are one." Then I marched up and down the room again, feeling very miserable indeed, till I hit upon an epigram. "People are divided into two classes," I said triumphantly, as I prepared for bed: "those who are circumstances without knowing it, and those who are conscious of the fact." I lay awake for long, overpowered by the tremendous responsibility which this discovery had laid on me. The load was lifted, and I fell asleep the moment I resolved not to submit tamely like a solution or a salt, which is boiled with this, and burned with that, but to have a hand in my own experiment.'

Two remarks I must make with regard to this paragraph. The first is about myself. I say that I was 'a tall, handsome girl, already a woman in appearance'. A romantic statement: the simple truth is that I was big, and rather stout, with a lot of brown, curly hair, pink cheeks, gray eyes, and generally pleasant to look at—at least, I know I liked to look at myself.

The second remark is about the chemists who taught in the school where I was—done something to, not educated. I had, and have, no ill will to these men; it was simply impossible that I could help thinking of them in the connection. The only one of my teachers whom I disliked, and of whom I still cherish hard thoughts, is Mr Standard, who condemned my compositions, and objected strongly to my metaphors.

Well, on the morning after my great discovery, while I was engaged in a large half-furnished room teaching the three little boys of my step-father's relative, a loud knock came to the door, and was followed immediately by the entrance of William Somers, the eldest son. There had come between him and the oldest of my charges three children, but they were dead. I was much astonished to see him, because, although we were on the frankest terms, we seldom met. My astonishment increased, I even felt indignant at his masterful manner, as he gave his little brothers sixpence each, and said:

'Be off with you! They deserve a holiday, don't they, Miss Armstrong?'

The three little scapegraces needed no second bidding; they were half-way downstairs before I had recovered my presence of mind. William Somers closed the door, and came up straight to me as if he had been sent for on important business. I stared at him blankly, and he stood dumb and blushing within a yard of me. At last he said:

'I have a holiday. Will you come with me?'

It was evidently not the thing he had intended to say.

'What have you a holiday for?' I asked.

'It's a Bank Holiday,' he said.

'A Bank Holiday!' I exclaimed with scorn, determined to pay him off for his intrusion. 'What slaves you are, you and the whole of this toiling London! Your very holidays you must take when they come. You can't do anything else.'

'What do you mean?' he said, crestfallen.

'Are you aware that you are a circumstance?' I asked severely.

I deeply resented the laugh, quickly smothered as it was, with which he greeted this question. I see now that it must have sounded funny to him, although after my meditation of the previous night it was a natural thing for me to say in all sincerity.

'I see that you have never realized that you are a circumstance,' I continued coldly. 'The best thing you can do with your holiday is to spend it, the whole of it, hour by hour, minute by minute, in the intensest contemplation possible to you of the fact that you are a circumstance.'

He looked at my eyes for fully half a minute, until I was forced to wink.

'You are not mad,' he said; 'and you don't seem to be joking. Still, I mean to say what I have come to say. Will you sit down?'

His coolness—which was, however, assumed—and his determined tone aggravated me.

'No,' I said; 'I will not sit down. I wish you to understand that *I* have fully realized that *I* am a circumstance, and I am not going to submit except to such other circumstances as please me. You are a circumstance, and don't know it. And what a circumstance! Something in the City—a broker's clerk, I suppose. You needn't tell me; I don't want to know. The prop and stay of your widowed mother and your three little brothers! Did it never strike you what a disagreeable circumstance you are? A good, respectable young man, who never misspends a penny. The very thought of you is like the taste of yarn.'

Now, I didn't mean all I said; I was simply angry without a sufficient reason, as girls and older people will sometimes be. He changed colour at my tirade, and held up his hand deprecatingly; but I went on.

'Don't interrupt me!' I cried. 'And what is it all for, all your toiling and moiling? To feed the mouths of four other circumstances, as unconscious of what they are as if they didn't exist. That's all. You're not causing anything. You're just doing exactly as thousands of others are doing—exactly as circumstances will do with you, never realizing that, in all regarding yourself, you are the main circumstance. An explorer, an artist, a poet, even a prime minister, attempts to cause something that is unnecessary, and that he needn't do except of his own motion—but you!'

'Miss Armstrong,' he said steadily, as I paused for breath, 'you are very excited. Won't you sit down?'

'No,' I almost shouted. 'Don't you see that I have made up my mind not to submit? I won't be experimented on with impunity. I should like to sit down, that's true; but I refuse to yield to such a miserable circumstance. I won't be experimented on with impunity,' I repeated, liking the sound of the sentence, and thinking, with what I suppose I must call feminine inconsistency, that it would have pleased Mr Standard.

William Somers looked very much annoyed—grieved, even. I ought to say that he was a tall man of twenty-three, with reddish beard and hair, and hazel eyes. I had not paid much attention to men up to that time, and did not know how handsome William Somers was. The trouble in his face did put me about; but, again, if paltry circumstances were not to be combated, how was I to challenge and overcome the great ones which hemmed me in on all sides?

'I see some meaning in what you say, Miss Armstrong,' he said; 'but I think it is stated a little wildly.'

I felt on the point of crying, so I laughed. He looked at me inquiringly.

'Do you know,' he said, 'I never heard your age. How old are you?'

'I was seventeen two months ago.' That staggered him. 'I suppose you thought I was thirty?'

'No; but I thought you were twenty until you laughed just now, and then I saw that you must be younger. How precocious you are!' he added.

I laughed again, and he saw what a stupid remark he had made.

'I mean—your figure—' he stammered and stuck.

'Mr Somers,' I said, being now mistress of the situation, 'I will not go with you for a holiday; but you will come with me, and escort me in my first assault on circumstances. Observe that I make a concession in having a squire. It is a bad omen.'

'Your causing a bad omen is just another circumstance for you to overcome,' he said, yielding to my humour.

'I'll be ready in ten minutes. Will you please get a hansom?' I said, as we left the room.

He had not succeeded in saying what he came to say.

Mrs Somers, a very bright, quiet little lady, looked askance at the hansom, but wished us a pleasant holiday as we drove off.

It was my first ride in a hansom, a fact which I concealed from William Somers as long as I could—about one minute, not any more.

'You have never been in a hansom before,' he said, looking at me in a quizzical way.

'How do you know?'

'At first I didn't know, you jumped in so smartly, and told the driver where to go with such aplomb; but then, when we started, in spite of yourself, a half-happy, half-frightened look shot across your face, you sighed, and sank back, and embraced yourself.'

'How dare you!' I said hotly.

My feelings had never been examined to my face before, and I felt outraged, just as I did once when I was posting a letter at a druggist's, and a ruffian laid his dirty hand on my shoulder, and turned me round, saying, 'By Jove! a strapper and a beauty.'

'I dare do all that may become a man,' said William Somers priggishly.

'Don't talk to me any more just now,' I said.

'Very well,' he replied; and, leaning his arms on the door, he tilted back his hat, and looked with unaffected interest at everybody and everything we passed.

I have a great liking for mysteries, and often stop people who begin to explain things to me, because I really don't want to know. A great London mystery of mine is that smooth, elastic, carpet-like roadway along which our hansom glided so stealthily. I admit having thought about its composition, but I have succeeded in overcoming the desire to know of what it is made. It seemed that when we jolted over the stones, we were being wound up in some curious, uncomfortable sort of way; and then, when we reached a stretch of that London turf, I felt as if we had been discharged, and were shooting along through space. (I'm thinking of a crossbow, Mr Standard.) Really, everything appeared to me delightful and interesting. I perceived for the first time what a picturesque city London is—all of it we saw that morning. The fantastic stacks of chimneys, like hieroglyphics wrought in the air; the mellow, antique streets of dwelling-houses; brick, and plaster, and paint; umber, red, and dull gold, splashed with creeping green; the squares, and graveyards, and crescents, with their trees, and sunflowers, and fountains—as if Nature were elbowing a way though the crowded buildings, Mr Standard; and the unknown streets of shops and booths where, even on a Bank Holiday, the butchers and the fishmongers cry their wares, and the little children tumble about among mouldy old furniture on the pavements, like dirty Cupids in the lumber-room of Olympus, Mr Standard; and the parks, with their glades, and avenues, and lakes, where Don Quixote and Sancho Panza lurk, and Robin Hood and Maid Marian, too, Mr Standard, if you had eyes to see; and the Thames—but we didn't see the Thames that morning; and while my thoughts were still revelling in the beauty of the City, we stopped, with a jerk that dislocated my imagination, at the house of Herr Herman Neunzehn, Wellpark Terrace, Bayswater. When I got out, and told the driver to wait, Mr Somers sat very still and attentive. He said nothing to me, and I said nothing to him; but I turned on the steps, and nodded my head encouragingly.

Herr Herman had been my music-master in the boarding-school, and had always had a word of praise for my efforts both in playing and composing. He was dusting his coat with his gloves preparatory to going out when I entered his room, but he received me kindly and said he could afford a few minutes.

'I have come on business,' I said.

'Have you?'

'Yes, Mr Neunzehn. I wish to make a start in life.'

Mr Neunzehn's little bright eyes dashed for a moment close up to his spectacles like silver fish in a miniature aquarium, and then became dim again in the depths as he prepared a cigarette.

'I have brought with me,' I said, displaying a roll I had in my hand, 'two songs, the words and music both by myself.'

Mr Neunzehn's fish darted past his pebbles, and he lit his cigarette.

'Will you oblige me by looking over them? and if you think them good enough, will you give me an introduction to a music-publisher?'

'I will,' said Mr Neunzehn, taking my manuscripts, and opening them out with his diabolical fingers. He was all diabolical, but his fingers were the most diabolical thing about him—long, knotty, sinewy, as if made for strangling.

'Thank you very much,' I said, moving towards the door.

'Wait,' he replied. 'I will do it just now.'

I stood stock still and watched him as he glanced rapidly through my scores. He was much more expeditious than I liked. How could he possibly comprehend in a few seconds the full beauty of my melodies, every individual note of which had been chosen with such care out of the old cottage piano's yellow keyboard, and thumped, and stroked, and listened to, positively for hours, alone, and in conjunction with the others of its phrase, until each separate sound had become so charged with emotion that I couldn't hear one of them without quivering! And my chords! and the counterpoint in my symphonies! He couldn't possibly grasp the full harmony and subtlety of these without at least playing the tunes over once.

The silver fish dashed to and fro behind their glasses, and the smoke curled up through a long, thick, gray moustache as if to cure the fish; but no change in the diabolical expression hinted at a decision one way or other. When he had turned over the last page, he rolled up my manuscripts and handed them back to me, rubbed his shaved cheeks, blew a cloud of smoke that hid his face, and said;

'No, my child.'

'Why?' I faltered.

'Because they are not good enough.'

'Oh, but try them!'

'I have read them through.'

'But let me play them to you;' and I made a dash at the piano.

'No,' he said, closing the instrument. 'It would be of no use. Your music is wrong, and it would not make it right to play it.'

I said to myself: 'The battle has begun; here's a circumstance with a vengeance: don't give in.' Then aloud:

'If you show me the mistakes I will correct them.'

'You couldn't.'

'Will you correct them, then?' I suggested faintly.

'I never correct music except for fools whose money might get into worse pockets than mine.'

I thought I understood now.

'But I will pay you, Mr Neunzehn,' I said sweetly, with a sudden burst of patronage, hope flaming up in my heart.

'You're a stupid little girl'—I was a foot taller than he. 'Listen.' He seized a newspaper and read; 'Some prank them up with oaken leaves; some small-pox hospitals, and banished as far as pos-tribution of articles of clothing to the heads of-pensations to large cities.'

He read slowly, making pauses and inflections as if the matter had been important; then his cigarette glowed and crackled faintly like a squib, and a cloud of smoke enveloped him, from which he emitted hoarsely the terrible sentence:

'That is your music.'

'How?' I whispered, stammering. 'I do not understand. Will you read it again?'

He showed me the newspaper, and with his diabolical finger tracked a line of type, straight across three columns and a half. He read also, but without attempting to make it sound like sense.

'That is your music,' he repeated. 'My dear young lady, amateurs come to me every week with things like that—parts of remembered words and phrases, correctly spelt as a rule, and each phrase or sentence quite grammatical, and sometimes containing bits and bobs of the most unconnected meanings; and they think they have made music. It just needs a little polishing, they know; and that is so easy for me. Look at these words again. See: out of four columns on four different subjects! Would you take that to Mr Standard, and ask him to polish it for you, to make it into one clear sentence? Read it again. "Some prank them up with oaken leaves; some small-pox hospitals, and banished as far as pos-tribution of articles of clothing to the heads of-pensations to large cities." You might by taking a few words and rejecting all the others invent a sentence.

But that won't do for my amateurs. They bring me notes, and I supply the music, the meaning; but it must be with their notes. They select nonsense, and I must make it sense. I never can make it sense; but it pleases them, and I make them pay for it, I can tell you. You are young and sensible, and can learn a lesson.'

The cigarette had gone out; the fish were pressed close to the glasses, and there seemed to be more water in the aquarium than usual. The old man was pitying me, I had turned so white.

'My dear child,' he continued, 'you must not be downcast. I am like a surgeon. You come to me and ask me if you have a disease, and I tell you that you have not; that you are not a musician, and will never be one. You ought to be very glad.'

Here he sighed, and I saw that he was pitying himself. I pronounced with difficulty a heartless 'Thank you,' for I felt he was right. Then a new idea occurred to me in a flash.

'Mr Neunzehn,' I said, 'did you look at the words of my songs?'

'Here and there.'

'What do you think of them?'

'Nothing; I am no judge.'

'Will you look at them, and if you like them set them to music and publish them?'

'No. Look here.'

He opened a press and showed me a pile of manuscript.

'There are fifty songs composed by me—the best music I have written; and I cannot get one of them published. It is not my reputation. My reputation is that of a composer of pianoforte pieces.'

But I didn't give in. I said:

'Can you introduce me to anyone who might buy my songs?'

'I can. Howard Dapper lives three doors from here on the right.'

My heart bounded at the name of the famous composer, and I could have kissed old Neunzehn as he wrote me an introduction.

'My time is more than up,' he said, handing me the letter. 'We will go out together.'

He took no notice of the hansom, and I gave Mr Somers another encouraging nod.

'Dapper may be stiff,' said Mr Neunzehn at the door of the great man's house; 'but never mind. If your songs please him he'll buy them.'

Having knocked and rung, my old music-master left me in a great hurry.

'Courage, you miserable, trembling circumstance!' I said to myself, kicking my heels in the hall till Mr Dapper should have read the letter.

Again a little fellow, less than Mr Neunzehn! I thought of the tall, straight, auburn-haired man waiting in the hansom; but I plunged into business. Mr Dapper had received me stiffly, and I was just as stiff.

'I have with me the songs to which Mr Neunzehn refers,' I said. 'May I read them to you?'

'I prefer to read them myself.'

'Unfortunately, I have them set to music here, and as the music is bad, it might affect your opinion of the verses.'

'It might.'

'I know the words by heart. Shall I repeat them?'

Mr Dapper bowed, and I recited my songs very badly indeed. My auditor's pale, oily face and purple eyes, like a plain suet-pudding into which two raisins had got by mistake, had a dispiriting effect. The songs, which I still think fair productions for a girl of seventeen, were both pathetic: in the one a deserted maiden died; in the other, a mother's only child. When I had done, Mr Dapper coughed, puckered his dumpling face, and delivered a short address in a juicy voice.

'Miss Armstrong'—glancing at the letter to make sure of my name–'your songs, I am sorry to say, do not suit me. I will be glad to look at any other verses you may have, here or elsewhere, suitable for pathetic ballads, with a little story; but death I never like introduced. If you have—a sort of musical duologue, say, to occupy about half an hour, with a good, rather startling, plot, and a little fun, I shall be glad to look at it. Or if you have a cantata for female voices only, I shall be glad to look at that; but, remember, I always like something with a story in it; and one thing I always object to—death, in the broad sense; that is, description.'

'But death is a circumstance,' I said, at my wit's end. 'It happens always.'

'We will not argue the point,' he replied, with a wave of his hand. 'If you are really anxious to succeed as a writer of words for music, you must be guided entirely by the requirements of the composer; but—and you must not take it unkindly—I do not think you will ever succeed in that way. If you wish to try, send me a cantata, or songs, or a duologue to occupy half an hour—these are things I need immediately—but I advise you not to.'

'I will,' I cried; 'I will go and write them at once.'

'I advise you not to. I am almost certain that they wouldn't suit me.'

'Why?'

'I will tell you.'

Mr Dapper had gradually dropped his professional tone and air. Some humanity had slipped into him covertly, wrinkling his brow and softening his mouth. His face looked liker a pudding than ever—a pudding that had been boiled in a cloth and creased; but no longer a plain suet-pudding; rather a plum-pudding, with the graciousness and sweetness of that Christmas delicacy. A light also shone in his eyes, as if the cook had lit some brandy, and I expected every minute to see a sprig of holly appear in his hair. His voice was still juicy, not with the tallowy juiciness of a suet-dumpling, but with the rich and fragrant sap of the Yule haggis—for I must 'derange my epitaphs', Mr Standard.

'Miss Armstrong,' he said, 'however clever you may be, you are much too young to succeed in this kind of work. It takes a very practised writer to make a song, or else a special talent, which I don't think you have. I shall tell you how to graduate in the school of song-making. Write a tragedy, and publish it; write an epic in twelve books, and publish it; write a volume of miscellaneous verse, and publish it; write a great novel, and publish it. The sale of these remarkable works will teach you what not to do; and besides having acquired facility with your pen, you will have expended all your idealism. Then you will be in a condition to write six original songs, which no composer will take. Then you will write one song—about sitting by the river in the moon, or walking in the wood when May is young—and a composer—this composer, possibly—will give you a guinea for it; and while you are dying of consumption and starvation your song will be sung at every concert and in every drawing-room, and be well forgotten before the dandelions have grown on your grave. But'—and here, as if he had been a conjurer who performed culinary tricks with his own head, he shifted his face back into a plain suet-pudding—'but if you have a cantata for female voices only, or a duologue for a lady and gentleman to occupy about half an hour, or songs for pathetic ballads, I will be glad to look at them; only, death I always object to—naked, absolute death, or even a broad hint.'

I don't remember getting out of Mr Dapper's house and getting into the hansom. At seventeen hope is very fierce and reckless, and is always staking happiness against some old song or other. I wakened up out of a blank dream in the midst of the very street where, an hour before, the picturesqueness of London had dawned on me. Prisons the houses seemed, the leprous bricks stained with blood, the scanty creepers striving pitifully to cover up the loathsomeness. The fluent roll of the hansom—was it a hansom, or some dragon-car, sweeping along a pavement of good intentions? '*Facilis*,' I began to myself, when Mr Standard's face in ebony, surmounted by ram's horns, flashed in at the window. My own special, private butt become a demon to torment me! What a war he had waged against quotations from pocket dictionaries! '*Fortiter in re*,' I said aloud, in frantic defiance of the fiend. '*Respice finem, Ad libitum, Cui bono!*' The ebony visage vanished; but another was peering into mine—a fresh face, with wondering hazel eyes. I was frightened at it, and turned away—to think about myself again. Why, I had only had the opinion of two men—the one old and soured with his half-success; the other middle-aged and cynical from prosperity. My music was doubtless as bad as Mr Neunzehn said, and my songs too maudlin for Mr Dapper; but as meaningless music and more lachrymose songs were bought and sold and sung every day, I would visit all the music-publishers in London. I laughed, and, stopping the cab, told the driver to go to one of them.

'Closed, ma'am. Holiday.'

Then I burst into tears, and Mr Somers directed the driver to take us home.

I drew myself together and cried quietly. The first comforting thought that came to me was, that if this had not been a holiday I would have kept

William and myself on the rack for hours yet. I had given in. I did make an attempt to return to my own side. 'Circumstances,' I thought, 'are against you. To-morrow isn't a holiday, and you can resume the fight. You can even post your music.' But, deep down in my own heart, I knew I had made a mistake about myself; and gradually that thought came up, and up, and up, until I writhed and wriggled on it as if I had been impaled. I then perceived this was something very like remorse, and, feeling how unworthy it was of one who had determined to fight circumstances to go on suffering when the thing was over, I looked up at William. He was staring out of the window, with his brows knotted and his mouth set. There was pain in his eyes, and I thought at first he was ill. As I watched him a new idea came stealing on me like some melancholy music, unheard before, but strangely familiar. It filled all my senses like the smell of roses in the evening, and made my body feel as light as my soul. This was the new idea: he, here beside me, was not miserable for himself; he was suffering for me. A great desire seized me to lay my head down on this man's shoulder, to feel his arms about me, and sleep or faint away; and this desire would, I am afraid, have had its course had we not arrived home before it overpowered me.

* * * * *

That night, in the half-furnished room, William said to me what he had failed to say in the morning. How he said it, and how I replied to him, shall never be written down. We said things that men and women say to each other only once—things high and sweet that ink would soil, and an eavesdropper mock. . . .

'Ho-ho, boy!'

I must end now. A little circumstance for which William and I are responsible—I *have* helped to cause something—is shouting in the next room for what nobody can give him but me.

S. R. CROCKETT

1859–1914

'THE HEATHER LINTIE'

BEING A REVIEW OF THE POEMS OF JANET BALCHRYSTIE, OF BARBRAX

Janet Balchrystie lived in a little cottage at the back of the Long Wood of Barbrax. She had been a hard-working woman all her days, for her mother died when she was but young, and she had lived on, keeping her father's house by the side of the single-track railway line. Gavin Balchrystie was a foreman platelayer on the PPR, and, with two men under him, had charge of a section of three miles. He lived just where that distinguished but impecunious line plunges into a moss-covered granite wilderness of moor and bog—where there is not more than a shepherd's hut to the half-dozen miles, and where the passage of a train is the occasion of commotion among scattered groups of black-faced sheep. Gavin Balchrystie's three miles of PPR metals gave him little work, but a good deal of healthy exercise. The black-faced sheep breaking down the fences and straying on the line side, and the torrents coming down the granite gullies, foaming white after a waterspout, and tearing into his embankments, undermining his chairs and plates, were the only troubles of his life. There was, however, a little public-house at 'The Huts', which in the old days of construction had had the licence, and which had lingered alone, licence and all, when its immediate purpose in life had been fulfilled, because there was nobody but the whaups and the railway officials on the passing trains to object to its continuance. Now it is cold and blowy on the westland moors, and neither whaups nor dark green uniforms object to a little refreshment up there. The mischief was that Gavin Balchrystie did not, like the guards and engine-drivers, go on with the passing train. He was always on the spot, and the path through Barbrax Wood to the Railway Inn was as well trodden as that which led over the big moss, where the whaups built, to the great white viaduct of Loch Merrick, where his three miles of parallel gleaming responsibility began.

When his wife was but newly dead, and his Janet just a smart elf-locked lassie running to and from the school, Gavin got too much in the way of 'slippin' doon by'. When Janet grew to be woman-muckle, Gavin kept the habit, and Janet hardly knew that it was not the use-and-wont of all fathers to sidle down to a contiguous Railway Arms, and return some hours later with uncertain step, and face picked out with bright pin-points of red—the sure mark of the confirmed drinker of whisky neat.

They were long days in the cottage at the back of Barbrax Long Wood. The little 'but and ben' was whitewashed till it dazzled the eyes as you came over the brae to it and found it set against the solemn depths of dark-green firwood. From early morn when she saw her father off, till the dusk of the day when he would return for his supper, Janet Balchrystie saw no human being. She heard the muffled roar of the trains through the deep cutting at the back of the wood, but she herself was entirely out of sight of the carriagefuls of travellers whisking past within half a mile of her solitude and meditation.

Janet was what is called a 'through-gaun lass', and her work for the day was often over by eight o'clock in the morning. Janet grew to womanhood without a sweetheart. She was plain, and she looked plainer than she was in the dresses which she made for herself by the light of nature and what she could remember of the current fashions at Merrick Kirk, to which she went every alternate Sunday. Her father and she took day about. Wet or shine, she tramped to Merrick Kirk, even when the rain blattered and the wind raved and bleated alternately among the pines of the Long Wood of Barbrax. Her father had a simpler way of spending his day out. He went down to the Railway Inn and drank 'ginger-beer' all day with the landlord. Ginger-beer is an unsteadying beverage when taken the day by the length. Also the man who drinks it steadily and quietly never enters on any inheritance of length of days.

So it came to pass that one night Gavin Balchrystie did not come home at all, at least not till he was brought lying comfortably on the door of a disused third-class carriage, which was now seeing out its career, anchored under the bank at Loch Merrick, where Gavin had used it as a shelter. The driver of the 'six-fifty up' train had seen him walking soberly along towards the Huts [and the Railway Inn], letting his long surfaceman's hammer fall against the rail keys occasionally as he walked. He saw him bend once, as though his keen ear detected a false ring in a loose length between two plates. This was the last that was seen of him till the driver of the 'nine-thirty-seven down' express—the 'boat-train,' as the employés of the PPR call it, with a touch of respect in their voices—passed Gavin fallen forward on his face, just when he was flying down grade under a full head of steam. It was duskily clear, with a great lake of crimson light dying into purple over the hills of midsummer heather. The driver was John Platt, the Englishman from Crewe, who had been brought from the great London and North-Western Railway, locally known as 'The Ellnen-doubleyou'. In these remote railway circles the talk is as exclusively of matters of the four-foot way as in Crewe or Derby. There is an inspector of traffic whose portly presence now graces Carlisle station, who left the PPR in these sad days of amalgamation, because he could not endure to see so many 'Sou'-West' waggons passing over the sacred metals of the

PPR permanent way. From his youth he had been trained in a creed of two articles—'To swear by the PPR through thick and thin, and hate the apple-green of the "Sou'-West".' It was as much as he could do to put up with the sight of the abominations. To have to hunt for their trucks when they got astray, was more than mortal could stand—so he fled the land.

When they stopped the express for Gavin Balchrystie every man on the line felt that it was an honour to the dead. John Platt sent a 'gurrring' thrill through the train as he put his brakes hard down, and whistled for the guard. He, thinking that the Merrick Viaduct was down at least, twirled his brake to such purpose that the rear car progressed along the metals by a series of convulsive bounds. Then they ran softly back, and there lay Gavin fallen forward on his knees, as though he had been trying to rise, or had knelt down to pray. Let him have 'the benefit of the doubt' in this world. In the next, if all tales be true, there is no such thing.

So Janet Balchrystie dwelt alone in the white 'but-an'-ben' at the back of the Long Wood of Barbrax. The factor gave her notice; but the laird, who was not accounted by his neighbours to be very wise, because he did needlessly kind things, told the factor to let the lassie bide, and delivered to herself, with his own hand, writing to the effect that Janet Balchrystie, in consideration of her lonely condition, was to be allowed the house for her lifetime, a cow's grass, and thirty pound sterling in the year as a charge on the estate. He drove down the cow himself, and having stalled it in the byre, he informed her of the fact over the yard dyke by word of mouth, for he never could be induced to enter her door. He was accounted to be 'gey an' queer' save by those who had tried making a bargain with him. But his farmers liked him, knowing him to be an easy man with those who had been really unfortunate, for he knew to what the year's crops of each had amounted, to a single chalder and head of nowt.

Deep in her heart Janet Balchrystie cherished a great ambition. When the earliest blackbird awoke and began to sing, while it was yet grey twilight, Janet would be up and at her work. She had an ambition to be a great poet. No less than this would serve her. But not even her father had known, and no other had any chance of knowing. In the black leather chest which had been her mother's, upstairs, there was a slowly growing pile of manuscript, and the editor of the local paper received every other week a poem, longer or shorter, for his Poet's Corner, in an envelope with the New Dalry post-mark. He was an obliging editor, and generally gave the closely written manuscript to the senior office-boy, who had passed the sixth standard, to cut down, tinker the rhymes, and lop any superfluity of feet. The senior office-boy 'just spread himself', as he said, and delighted to do the job in style. But there was a woman fading into a grey old-maidishness which had hardly ever been girlhood, who did not at all approve of these corrections. She endured them because over the signature of 'Heather Bell' it was a joy to see in the rich, close luxury of type her own poetry, even though it might be a trifle tattered and tossed about by hands ruthless and alien—those, in fact, of the senior office-boy.

Janet walked every other week to the post-office at New Dalry to post her letters to the editor, but neither that great man nor yet the senior office-boy had any conception that the verses of their 'esteemed correspondent' were written by a woman too early old who dwelt alone at the back of Barbrax Long Wood.

One day Janet took a sudden but long-meditated journey. She went down by rail from the little station of the 'Huts' to the large town of Drum, thirty miles to the east. Here, with the most perfect courage and dignity of bearing, she interviewed a printer and arranged for the publication of her poems in their own original form, no longer staled and clapperclawed by the pencil of the senior office-boy. When the proof-sheets came to Janet, she had no way of indicating the corrections than by again writing the whole poem out in a neat print hand on the edge of the proof, and underscoring the words which were to be altered. This, when you think of it, is a very good way, when the happiest part of your life is to be spent in such concrete pleasures of hope, as were Janet's over the crackly sheets of the printer of Drum. Finally the book was produced, a small, rather thickish octavo, on sufficiently wretched grey paper which had suffered from want of thorough washing in the original paper-mill. It was bound in a peculiarly deadly blue, of a rectified Reckitt tint, which gave you dazzles in the eye at any distance under ten paces. Janet had selected this as the most appropriate of colours. She had also many years ago decided upon the title, so that Reckitt had printed upon it, back and side, 'The Heather Lintie', while inside there was the plain acknowledgment of authorship, which Janet felt to be a solemn duty to the world, 'Poems by Janet Balchrystie, Barbrax Cottage, by New Dalry.' First she had thought of withholding her name and style; but on the whole, after the most prolonged consideration she felt that she was not justified in bringing about such a controversy as had once divided Scotland concerning that 'Great Unknown' who wrote the Waverley Novels.

Almost every second or third day Janet trod that long loch-side road to New Dalry for her proof-sheets, and returned them on the morrow corrected in her own way. Sometimes she got a lift from some farmer or carter, for she had worn herself with anxiety to the shadow of what she had once been, and her dry bleached hair became grey and greyer with the fervour of her devotion to letters.

By April the book was published, and at the end of this month, laid aside by sickness of the vague kind called locally 'a decline', she took to her bed—rising only to lay a few sticks upon the fire from her store gathered in the autumn, or to brew herself a cup of tea—she waited for the tokens of her book's conquests in the great world of thought and men. She had waited so long for her recognition, and now it was coming. She felt that it would not be long before she was recognised as one of the singers of the world. Indeed, had she but known it, her recognition was already on its way.

In a great city of the north a clever young reporter was cutting open the leaves of *The Heather Lintie* with a hand almost feverishly eager.

'This is a perfect treasure. This is a find indeed. Here is my chance ready to my hand.'

His paper was making a specialty of 'exposures'. If there was anything weak and erring, anything particularly helpless and foolish which could make no stand for itself, *The Night Hawk* was on the pounce. Hitherto the Junior Reporter had never had a 'two column chance'. He had read—it was not much that he *had* read—Macaulay's too famous article on 'Satan' Montgomery, and not knowing that Macaulay lived to regret the spirit of that assault, he felt that if he could bring down *The Night Hawk* on *The Heather Lintie*, his fortune was made. So he sat down and he wrote, not knowing and not regarding a lonely woman's heart, to whom his word would be as the word of a God, in the lonely cottage lying in the lee of the Long Wood of Barbrax.

The Junior Reporter turned out a triumph of the New Journalism. 'This is a book which may be a genuine source of pride to every native of the ancient province of Galloway,' he wrote. 'Galloway has been celebrated for black cattle and for wool—as also for a certain bucolic belatedness of temperament, but Galloway has never hitherto produced a poetess. One has arisen in the person of Miss Janet Bal—something or other. We have not an interpreter at hand, and so cannot wrestle with the intricacies of the authoress's name, which appears to be some Galwegian form of Erse or Choctaw. Miss Bal—and so forth—has a true fount of pathos and humour. In what touching language she chronicles the death of two young lambs which fell into one of the puddles they call rivers down there, and were either drowned or choked with the dirt—

> '"They were two bonny, bonny lambs,
> That played upon the daisied lea,
> And loudly mourned their woolly dams
> Above the drumly flowing Dee."'

'How touchingly simple,' continued the Junior Reporter, buckling up his sleeves to enjoy himself, and feeling himself born to be a *Saturday Reviewer*, 'mark the local colour, the wool and the dirty water of the Dee—without doubt a name applied to one of their bigger ditches down there. Mark also the over-fervency of the touching line,

> "And loudly mourned their woolly dams,"

which, but for the sex of the writer and her evident genius, might be taken for an expression of a strength hardly permissible even in the metropolis.'

The Junior Reporter filled his two columns and enjoyed himself in the doing of it. He concluded with the words, 'The authoress will make a great success. If she will come to the capital, where genius is always appreciated, she will, without doubt, make her fortune. Nay, if Miss Bal—, but again we cannot proceed for the want of an interpreter—if Miss B, we say, will only accept a position at Cleary's Waxworks and give readings from her poetry, or exhibit herself in the act of pronouncing her own name, she will be a greater draw in this city than Punch and Judy, or even the latest American advertising evangelist who preaches standing on his head.'

The Junior Reporter ceased here from very admiration at his own cleverness in so exactly hitting the tone of the masters of his craft, and handed his manuscript in to the editor.

It was the gloaming of a long June day when Rob Affleck, the woodman over at Barbrax, having been at New Dalry with a cart of wood, left his horse on the roadside and ran over through Gavin's old short cut, now seldom used, to Janet's cottage with a paper in a yellow wrapper.

'Leave it on the step, and thank you kindly, Rob,' said a weak voice within, and Rob, anxious about his horse and his bed, did so without another word. In a moment or two Janet crawled to the door, listened to make sure that Rob was really gone, opened the door, and protruded a hand wasted to the hard flat bone—an arm that ought for her years to have been of full flesh and noble curves.

When Janet got back to bed it was too dark to see anything except the big printing at the top of the paper.

'Two columns of it!' said Janet, with great thankfulness in her heart, lifting up her soul to God who had given her the power to sing. She strained her prematurely old and weary eyes to make out the sense. 'A genuine source of pride to every native of the ancient province,' she read.

'The Lord be praised!' said Janet, in a rapture of devout thankfulness, 'though I never really doubted it,' she added, as though asking pardon for a moment's distrust. 'But I tried to write these poems to the glory of God and not to my own praise, and He will accept them and keep me humble, under the praise of men as well as under their neglect.'

So clutching the precious paper close to her breast, and letting tears of thankfulness fall on the article—which, had they fallen on the head of the Junior Reporter, would have burnt like fire, she patiently awaited the coming dawn.

'I can wait till the morning now to read the rest,' she said.

So hour after hour, with her eyes wide, staring hard at the grey window squares, she waited the light from the east. About half-past two there was a stirring and a moaning among the pines, and the roar of the sudden gust came with the breaking day through the dark arches. In the whirlwind there came a strange expectancy and tremor into the heart of the poetess, and she pressed the wet sheet of crumpled paper closer to her bosom, and turned to face the light. Through the spaces of the Long Wood of Barbrax there came a shining visitor, the Angel of the Presence, he who comes but once and stands a moment with a beckoning finger. Him she followed up through the wood.

They found Janet on the morning of the second day after, with a look so glad on her face and so natural an expectation in the unclosed eye, that

Rob Affleck spoke to her and expected an answer. *The Night Hawk* was clasped to her breast with a hand that they could not loosen. It went to the grave with her body. The ink had run a little here and there, where the tears had fallen thickest.

God is more merciful than man.

J. M. Barrie
1860–1937

A Literary Club

The ministers in the town did not hold with literature. When the most notorious of the clubs met in the town-house under the presidentship of Gavin Ogilvy, who was no better than a poacher, and was troubled in his mind because writers called Pope a poet, there was frequently a wrangle over the question, Is literature necessarily immoral? It was a fighting club, and on Friday nights the few respectable, god-fearing members dandered to the town-house, as if merely curious to have another look at the building. If Lang Tammas, who was dead against letters, was in sight they wandered off, but when there were no spies abroad they slunk up the stair. The attendance was greatest on dark nights, though Gavin himself and some other characters would have marched straight to the meeting in broad day-light. Tammas Haggart, who did not think much of Milton's devil, had married a gypsy woman for an experiment, and the Coat of Many Colours did not know where his wife was. As a rule, however, the members were wild bachelors. When they married they had to settle down.

Gavin's essay on Will'um Pitt, the Father of the Taxes, led to the club's being bundled out of the town-house, where people said it should never have been allowed to meet. There was a terrible town when Tammas Haggart then disclosed the secret of Mr Byars's supposed approval of the club. Mr Byars was the Auld Licht minister whom Mr Dishart succeeded, and it was well known that he had advised the authorities to grant the use of the little town-house to the club on Friday evenings. As he solemnly warned his congregation against attending the meetings the position he had taken up created talk, and Lang Tammas called at the manse with Sanders Whamond to remonstrate. The minister, however, harangued them on their sinfulness in daring to question the like of him, and they had to retire vanquished though dissatisfied. Then came the disclosures of Tammas Haggart, who was never properly secured by the Auld Lichts until Mr Dishart took him in hand. It was Tammas who wrote anonymous letters to Mr Byars about the scarlet woman, and, strange to say, this led to the club's being allowed to meet in the town-house. The minister, after many days, discovered who his correspondent was, and succeeded in inveigling the stone-breaker to the manse. There, with the door snibbed, he opened out on Tammas, who, after his usual manner when hard pressed, pretended to be deaf. This sudden fit of deafness so exasperated the minister that he flung a book at Tammas. The scene that followed was one that few Auld Licht manses can have witnessed. According to Tammas the book had hardly reached the floor when the minister turned white. Tammas picked up the missile. It was a Bible. The two men looked at each other. Beneath the window Mr Byars's children were prattling. His wife was moving about in the next room, little thinking what had happened. The minister held out his hand for the Bible, but Tammas shook his head, and then Mr Byars shrank into a chair. Finally, it was arranged that if Tammas kept the affair to himself the minister would say a good word to the Bailie about the literary club. After that the stone-breaker used to go from house to house, twisting his mouth to the side and remarking that he could tell such a tale of Mr Byars as would lead to a split in the kirk. When the town-house was locked on the club Tammas spoke out, but though the scandal ran from door to door, as I have seen a pig in a fluster do, the minister did not lose his place. Tammas preserved the Bible, and showed it complacently to visitors as the present he got from Mr Byars. The minister knew this, and it turned his temper sour. Tammas's proud moments, after that, were when he passed the minister.

Driven from the town-house, literature found a table with forms round it in a tavern hard by, where the club, lopped of its most respectable members, kept the blinds down and talked openly of Shakspeare. It was a low-roofed room, with pieces of lime hanging from the ceiling and peeling walls. The floor had a slope that tended to fling the debater forward, and its boards, lying loose on an uneven foundation, rose and looked at you as you crossed the room. In winter, when the meetings were held regularly every fortnight, a fire of peat, sod, and dross lit up the curious company who sat round the table shaking their heads over Shelley's mysticism, or requiring to be called to order because they would not wait their turn to deny an essayist's assertion that Berkeley's style was superior to David Hume's. Davit Hume, they said, and Watty Scott. Burns was simply referred to as Rob or Robbie.

There was little drinking at these meetings, for the members knew what they were talking about, and your mind had to galop to keep up with the flow of reasoning. Thrums is rather a remarkable town. There are scores and scores of houses in it that have sent their sons to college (by what a struggle!), some to make their way to the front in their professions, and others, perhaps, despite their broad-cloth, never to be a patch on their parents. In that literary club there were men of a reading so wide and catholic that it might put some graduates of the universities to shame, and of an intellect so keen that had it not had a crook in it their fame would have crossed the county. Most of them had but a thread-bare existence, for you weave slowly with a Wordsworth open before you, and some were strange Bohemians (which does not do in Thrums), yet others wandered into the world and compelled it to recognize them. There is a London barrister whose father belonged to the club. Not many years ago a man died on the staff of the *Times*, who, when he was a weaver near Thrums, was one of the club's prominent members. He taught himself shorthand by the light of a cruizey, and got a post on a Perth paper, afterwards on the *Scotsman* and the *Witness*, and finally on the *Times*. Several other men of his type had a history worth reading, but it is not for me to write. Yet I may say that there is still at least one of the original members of the club left behind in Thrums to whom some of the literary dandies might lift their hats.

Gavin Ogilvy I only knew as a weaver and a poacher; a lank, long-armed man, much bent from crouching in ditches whence he watched his snares. To the young he was a romantic figure, because they saw him frequently in the fields with his call-birds tempting siskins, yellow yites, and linties to twigs which he had previously smeared with lime. He made the lime from the tough roots of holly; sometimes from linseed oil, which is boiled until thick, when it is taken out of the pot and drawn and stretched with the hands like elastic. Gavin was also a famous hare-snarer at a time when the ploughman looked upon this form of poaching as his perquisite. The snare was of wire, so constructed that the hare entangled itself the more when trying to escape, and it was placed across the little roads through the fields to which hares confine themselves, with a heavy stone attached to it by a string. Once Gavin caught a toad (fox) instead of a hare, and did not discover his mistake until it had him by the teeth. He was not able to weave for two months. The grouse-netting was more lucrative and more exciting, and women engaged in it with their husbands. It is told of Gavin that he was on one occasion chased by a gamekeeper over moor and hill for twenty miles, and that by and by when the one sank down exhausted so did the other. They would sit fifty yards apart, glaring at each other. The poacher eventually escaped. This, curious as it may seem, is the man whose eloquence at the club has not been forgotten in fifty years. 'Thus did he stand,' I have been told recently, 'exclaiming in language sublime that the soul shall bloom in immortal youth through the ruin and wrack of time.'

Another member read to the club an account of his journey to Lochnagar, which was afterwards published in *Chambers's Journal*. He was celebrated for his descriptions of scenery, and was not the only member of the club whose essays got into print. More memorable perhaps was an itinerant match-seller known to Thrums and the surrounding towns as the literary spunk-seller. He was a wizened, shivering old man, often barefooted, wearing at the best a thin ragged coat that had been black but was green-brown with age, and he made his spunks as well as sold them. He brought Bacon and Adam Smith into Thrums, and he loved to recite long screeds from Spenser, with a running commentary on the versification and the luxuriance of the diction. Of Jamie's death I do not care to write. He went without many a dinner in order to buy a book.

The Coat of Many Colours and Silva Robbie were two street preachers who gave the Thrums ministers some work. They occasionally appeared at the club. The Coat of Many Colours was so called because he wore a garment consisting of patches of cloth of various colours sewed together. It hung down to his heels. He may have been cracked rather than inspired, but he was a power in the square where he preached, the women declaring that he was gifted by God. An awe filled even the men, when he admonished them for using strong language, for at such a time he would remind them of the woe which fell upon Tibbie Mason. Tibbie had been notorious in her day for evil-speaking, especially for her free use of the word handless, which she flung a hundred times in a week at her man, and even at her old mother. Her punishment was to have a son born without hands. The Coat of Many Colours also told of the liar who exclaimed, 'If this is not gospel true may I stand here for ever,' and who is standing on that spot still, only nobody knows where it is. George Wishart was the Coat's hero, and often he has told in the Square how Wishart saved Dundee. It was the time when the plague lay over Scotland, and in Dundee they saw it approaching from the West in the form of a great black cloud. They fell on their knees and prayed, crying to the cloud to pass them by, and while they prayed it came nearer. Then they looked around for the most holy man among them, to intervene with God on their behalf. All eyes turned to George Wishart, and he stood up, stretching his arms to the cloud and prayed, and it rolled back. Thus Dundee was saved from the plague, but when Wishart ended his prayer he was alone, for the people had all returned to their homes. Less of a genuine man than the Coat of Many Colours was Silva Robbie, who had horrid fits of laughing in the middle of his prayers, and even fell in a paroxysm of laughter from the chair on which he stood. In the club he said things not to be borne, though logical up to a certain point.

Tammas Haggart was the most sarcastic member of the club, being celebrated for his sarcasm far and wide. It was a remarkable thing about him, often spoken of, that if you went to Tammas with a stranger and asked him to say a sarcastic thing that the man might take away as a specimen, he could not do it. 'Na, na,' Tammas would say, after a few trials, referring to sarcasm, 'she's no a crittur to force. Ye maun lat her tak her ain time. Sometimes she's dry like the pump, an' syne, again, oot she comes in a gush.' The most sarcastic thing the stone-breaker ever said was frequently marvelled over in Thrums, both before and behind his face, but unfortunately no one could ever remember what it was. The subject, however, was Cha Tamson's potato pit. There is little doubt that it was a fit of sarcasm that induced Tammas to marry a gypsy lassie. Mr Byars would not join them, so Tammas had himself married by Jimmy Pawse, the gay little gypsy king, and after that the minister re-married them. The marriage over the tongs is a thing to scandalise any well-brought up person, for before he joined the couple's hands, Jimmy jumped about in a startling way, uttering wild gibberish, and after the ceremony was over there was rough work, with incantations and blowing on pipes. Tammas always held that this marriage turned out better than he had expected though he had his trials like other married men. Among them was Chirsty's way of climbing on to the dresser to get at the higher part of the plate-rack. One evening I called in to have a smoke with the stone-breaker, and while we were talking Chirsty climbed the dresser. The next moment she was on the floor on her back, wailing, but Tammas smoked on imperturbably. 'Do you not see what has happened, man?' I cried. 'Ou,' said Tammas, 'she's aye fa'in aff the dresser.'

Of the schoolmasters who were at times members of the club, Mr Dickie was the ripest scholar, but my predecessor at the school-house, had a way of sneering at him that was as good as sarcasm. When they were on their legs at the same time, asking each other passionately to be calm, and rolling out lines from Homer, that made the inn-keeper look fearfully to the fastenings of the door, their heads very nearly came together although the table was between them. The old dominie had an advantage in being the shorter man, for he could hammer on the table as he spoke, while gaunt Mr Dickie had to stoop to it. Mr McRittie's arguments were a series of nails that he knocked into the table, and he did it in a workmanlike manner. Mr Dickie, though he kept firm on his feet, swayed his body until by and by his head was rotating in a large circle. The mathematical figure he made was a cone revolving on its apex. Gavin's reinstalment in the chair year after year was made by the disappointed dominie the subject of some tart verses which he called an epode, but Gavin crushed him when they were read before the club. 'Satire,' he said, 'is a legitimate weapon, used with michty effect by Swift, Sammy Butler, and others, and I dount object to being made the subject of creeticism. It has often been called a t'nife (knife), but them as is not used to t'nives cuts their hands, and ye'll a' observe that Mr McRittie's fingers is bleedin'. All eyes were turned upon the dominie's hand, and though he pocketed it smartly several members had seen the blood. The dominie was a rare visitor at the club after that, though he outlived poor Mr Dickie by many years. Mr Dickie was a teacher in Tilliedrum, but he was ruined by drink. He wandered from town to town, reciting Greek and Latin poetry to any one who would give him a dram, and sometimes he wept and moaned aloud in the street, crying, 'Poor Mr Dickie! poor Mr Dickie!'

The leading poet in a club of poets was Dite Walls, who kept a school when there were scholars, and weaved when there were none. He had a song that was published in a half-penny leaflet about the famous lawsuit instituted by the farmer of Teuchbusses against the Laird of Drumlee. The laird was alleged to have taken from the land of Teuchbusses sufficient broom to make a besom thereof, and I am not certain that the case is settled to this day. It was Dite or another member of the club, who wrote 'The Wife o' Deeside', of all the songs of the period the one that had the greatest vogue in the county at a time when Lord Jeffrey was cursed at every fireside in Thrums. The wife of Deeside was tried for the murder of her servant who had infatuated the young laird, and had it not been that Jeffrey defended her she would, in the words of the song, have 'hung like a troot'. It is not easy now to conceive the rage against Jeffrey when the woman was acquitted. The song was sung and recited in the streets, at the smiddy, in bothies, and by firesides, to the shaking of fists and the grinding of teeth. It began—

> 'Ye'll a' hae hear tell o' the wife o' Deeside,
> Ye'll a' hae hear tell o' the wife o' Deeside,
> She poisoned her maid for to keep up her pride,
> Ye'll a' hae hear tell o' the wife o' Deeside.'

Before the excitement had abated, Jeffrey was in Tilliedrum for electioneering purposes, and he was mobbed in the streets. Angry crowds pressed close to howl, 'Wife o' Deeside!' at him. A contingent from Thrums was there, and it was long afterwards told of Sam'l Todd, by himself, that he hit Jeffrey on the back of the head with a clod of earth.

Johnny McQuhatty, a brother of the T'nowhead farmer, was the one taciturn member of the club, and you had only to look at him to know that he had a secret. He was a great genius at the handloom, and invented a loom for the weaving of linen such as has not been seen before or since. In the day-time he kept guard over his 'shop', into which no one was allowed to enter, and the fame of his loom was so great that he had to watch over it with a gun. At night he weaved, and when the result at last pleased him he made the linen into shirts, all of which he stitched together with his own hands, even to the buttonholes. He sent one shirt to the Queen, and another to the Duchess of Athole, mentioning a very large price for them, which he got. Then he destroyed his wonderful loom, and how it was made no one will ever know. Johnny only took to literature after he had made his name, and he seldom spoke at the club except when ghosts and the like were the subject of debate, as they tended to be when the farmer of Muckle Haws could get in a word. Muckle Haws was fascinated by Johnny's sneers at superstition, and sometimes on dark nights the inventor had to make his courage good by seeing the farmer past the doulie yates (ghost gates), which Muckle Haws had to go perilously near on his way home. Johnny was a small man, but it was the burly farmer who shook at sight of the gates standing out white in the night. White gates have an evil name still, and Muckle Haws was full of horrors as he drew near them, clinging to Johnny's arm. It was on such a night, he would remember, that he saw the White Lady go through the gates greeting sorely, with a dead bairn in her arms, while water kelpie laughed and splashed in the pools, and the witches danced in a ring round Broken Buss. That very night twelve months ago the packman was murdered at Broken Buss, and Easie Pettie hanged herself on the stump of a tree. Last night there were ugly sounds from the quarry of Croup, where the bairn lies buried, and it's not mous (canny) to be out at such a time. The farmer had seen spectre maidens walking round the ruined castle of Darg, and the Castle all lit up with flaring torches, and dead knights and ladies sitting in the halls at the wine-cup, and the devil himself flapping his wings on the ramparts.

When the debates were political, two members with the gift of song fired the blood with their own poems about taxation and the depopulation of the Highlands, and by selling these songs from door to door they made their livelihood.

Books and pamphlets were brought into the town by the flying stationers, as they were called, who visited the square periodically carrying their wares on their backs, except at the Muckly, when they had their stall and even sold books by auction. The flying stationer best known to Thrums was Sandersy Riach, who was stricken from head to foot with the palsy, and could only speak with a quaver in consequence. Sandersy brought to the members of the club all the great books he could get second hand, but his stock-in-trade was Thrummy Cap and Akenstaff, the Fish-wives of Buckhaven, the Devil upon Two Sticks, Gilderoy, Sir James the Rose, the Brownie of Badenoch, the Ghaist of Firenden, and the like. It was from Sandersy that Tammas Haggart bought his copy of Shakspeare, whom Mr Dishart could never abide. Tammas kept what he had done from his wife, but Chirsty saw a deterioration setting in and told the minister of her suspicions. Mr Dishart was newly placed at the time and very vigorous, and the way he shook the truth out of Tammas was grand. The minister pulled Tammas the one way and Gavin pulled him the other, but Mr Dishart was not the man to be beaten, and he landed Tammas in the Auld Licht kirk before the year was out. Chirsty buried Shakspeare in the yard.

VIOLET JACOB
1863–1946

THE DEBATABLE LAND

Of the birth and origin of Jessie-Mary no one in the parish knew anything definite. Those who passed up the unfrequented cart-road by her grandmother's thatched hovel used to see the shock-headed child among the gooseberry bushes of the old woman's garden, peering at them, like an animal, over the fence.

Whether she were really the granddaughter of the old beldame inside the mud walls no one knew, neither, for that matter, did anybody care. The hovel was the last remaining house of a little settlement which had disappeared from the side of the burn. Just where it stood, a shallow stream ran across the way and plunged into a wood in which Jessie-Mary had many a time feasted on the plentiful wild raspberries, and run, like a little squirrel, among the trees.

It was not until she was left alone in the world that much attention was paid to her existence, and then she presented herself to the parish as a problem; for her life was lived a full half-century before the all-powerful Board School arose to direct rustic parents and guardians, and she had received little education. She had grown into a sturdy girl of twenty, with brown hair which the sun had bleached to a dull yellow, twisted up at the back of her head and hanging heavily over her brows. She was a fierce-looking lass, with her hot grey eyes. The parish turned its mind to the question of how she might earn a living and was presently relieved when Mrs Muirhead, who was looking for an able-bodied servant, hired her in that capacity. She was to have a somewhat meagre wage and her clothes, and was to help her mistress in house and yard. When the matter was settled she packed her few possessions into a bundle and sauntered up the green loaning which ran between the hovel and Mrs Muirhead's decent roof, marking where one fir-wood ended and another began.

Mrs Muirhead was the widow of a joiner, and she inhabited a cottage standing just where the woods and the mouth of the loaning touched the high road that ran north to the hills. She was well to do, for a cottager, and her little yard, besides being stacked with planks which her son, Peter, sawed and planed as his father had done before him, contained a row of hen-coops and a sty enclosing a pig whose proportions waxed as autumn waned. When the laird trotted by, he cast a favourable eye on the place, which was as neat as it befitted the last house on a man's property to be. When he had passed on and was trotting alongside the farther wood he was no longer on his own ground, for the green, whin-choked loaning was debatable land lying between him and his neighbour.

As Jessie-Mary, with her bundle, came through the whins and opened the gate, Peter Muirhead, who was in the yard, heard the latch click and looked up from his work. At sight of the yellow head by the holly bushes he laid down the spokeshave he was using and came round to the front. The girl was looking at him with eyes whose directness a youth of his type is liable to misunderstand. He began to smile.

'Will Mistress Muirhead be ben?' said Jessie-Mary tentatively.

Peter did not answer, but approached, his smile taking meaning.

'Will Mistress Muirhead be ben the hoose?' she inquired, more loudly.

It occurred to her that he might not be in his right senses, for the mile or two of debatable track which separated her old home from her future one might as well have been ten, for all she had seen of the world at the other end of it. She knew very well that Muirhead the joiner had lived where she now stood, and she had seen the old man, but the shambling figure before her was entirely strange. Once, at the edge of the wood, she had listened to the whirr of sawing in the vicinity of the road and had gathered that the work went on, though Muirhead himself had departed.

'She's no here. Ye'll just hae to put up wi' me,' said Peter jocosely. His mother was in the house, but he saw no reason for divulging the fact.

Jessie-Mary stood silent, scarcely knowing what to say.

'Ye're a fine lassie,' observed Peter, still smiling alluringly.

She eyed him with distrust and her heavy brows lowered over her eyes; she began to walk towards the cottage. He sprang forward, as though to intercept her, and, as she knocked, he laid hold of her free hand. Mrs Muirhead, from within, opened the door just in time to see him drop it. She was a short, hard-featured woman, presenting an expanse of white apron to the world; a bunch of turkeys' feathers, in which to stick knitting needles, was secured between her person and the band of this garment, the points of the quills uppermost. She looked from one to the other, then drawing Jessie-Mary over the threshold, she slammed the door.

'Dinna think a didna see ye, ye limmer!' she exclaimed, taking the girl roughly by the shoulder.

And so Jessie-Mary's working life began.

The little room allotted to her, looking over the yard, was no smaller than the corner she had inhabited in the mud cottage, yet it had a stifling effect; and its paper, which bore a small lilac flower on a buff ground, dazzled her eyes and seemed to press on her from all sides. In the cracked looking-glass which hung on it she could see the disturbing background behind her head as she combed and flattened her thick hair in accordance with Mrs Muirhead's ideas. In leisure moments she hemmed at an apron which she was to wear when completed. Mrs Muirhead was annoyed at finding she could hardly use a needle; she was far from being an unkind woman, but her understanding stopped at the limits of her own requirements. Jessie-Mary's equally marked limitations struck her as the result of natural wickedness.

Wherever the yard was unoccupied by the planks or the pigsty, it was set about with hen-coops, whose inmates strayed at will from the enclosure to pervade the nearer parts of the wood in those eternal perambulations which occupy fowls. Just outside, where the trees began, was a pleasant strip of sandy soil in which the hens would settle themselves with much clucking and tail-shaking, to sit blinking, like so many vindictive dowagers, at their kind. Through this, the Dorking cock, self-conscious and gallant, would conduct the ladies of his family to scratch among the tree-roots; and the wood for about twenty yards from the house wore that peculiar scraped and befeathered look which announces the proximity of a hen-roost. At night the lower branches were alive with dark forms and the suppressed gurgling that would escape from them. It was part of Jessie-Mary's duty to attend to the wants of this rabble.

There were times when a longing for flight took the half-civilised girl. Life, for her, had always been a sort of inevitable accident, a state in whose ordering she had no part as a whole, however much choice she might have had in its details. But now there was little choice in these; Mrs Muirhead ordered her day and she tolerated it as best she could. She hardly knew what to do with her small wage when she got it, for the finery dear to the heart of the modern country lass was a thing of which she had no knowledge, and there was no dependent relative who might demand it of her.

The principal trouble of her life was Peter, whose occupations kept him of necessity at home, and whose presence grew more hateful as time went on. There was no peace for her within sight of his leering smile. There was only one day of the week that she was free of him; and on these Sunday afternoons, as he went up the road to join the loitering knot of horsemen from the nearest farm, she would thankfully watch him out of sight from the shelter of the loaning. She hated him with all her heart.

He would lurk about in the evenings, trying to waylay her amongst the trees as she went to gather in the fowls, and once, coming suddenly on her as she turned the corner of the house, he had put his arm about her neck. She had felt his hot breath on her ear, and, in her fury, pushed him from her with such violence that he staggered back against a weak place in the yard fence and fell through, cutting his elbow on a piece of broken glass. She stood staring at him, half terrified at what she had done, but rejoicing to see the blood trickle down his sleeve. She would have liked to kill him. The dreadful combination of his instincts and his shamblingness was what physically revolted her, though she did not realise it; and his meanness had, more than once, got her into trouble with his mother. She had no consideration to expect from Mrs Muirhead, as she knew well. To a more complicated nature the position would have been unendurable, but Jessie-Mary endured stubbornly, vindictively, as an animal endures. She was in a cruel position and her only safeguard lay in the fact that

Peter Muirhead was repulsive to her. But neither morality nor expediency nor the armed panoply of all the cardinal virtues have yet succeeded in inventing for a woman a safeguard so strong as her own taste.

It was on a Sunday afternoon towards the end of September that Peter emerged from the garden and strolled up the road. The sun was high above the woods, his rim as yet clear of the tree-tops, and the long shadow from the young man's feet lay in a dark strip between himself and the fence at his side. He wore his black Sunday suit and a tie bought from a travelling salesman who had visited Montrose fair the year before. In his best clothes he looked more ungainly than usual, and even the group of friends who watched his approach allowed themselves a joke at his expense as he neared them. He could hear their rough laughter, though he was far from guessing its cause. Nature had given him a good conceit of himself.

Jessie-Mary drew a breath of relief as his steps died away and she hailed the blessed time, granted to her but weekly, in which she might go about without risk of meeting him. Everything was quiet. Mrs Muirhead was sitting in the kitchen with her Bible; the door was ajar and the girl could just see a section of her skirt and the self-contained face of the cat which blinked on the hearth beside her. She had accompanied her mistress to the kirk that morning and had thought, as they returned decorously together, that she would go down the loaning again to see the thatched cottage by the burn—perhaps stray a little in the wood among the familiar raspberry-stalks. She had not seen these old haunts since she left them for Mrs Muirhead's service.

She took off her apron and went out bare-headed. On the outskirts of the trees the hens were rustling and fluttering in the dust; as she passed, they all arose and followed her. She had not remembered that their feeding-time was due in half an hour and for a moment she stood irresolute. If she were to go on her intended way there would be no one to give them their food. She determined to make and administer it at once; there would be plenty of time afterwards to do what she wished to do.

She was so little delayed that, when the pail was put away and the water poured into the tin dishes, there was still a long afternoon before her. She threaded her way slowly through the fir-stems, stopping to look at the rabbit-runs or to listen to the cooing of wood pigeons, her path fragrant with the scent of pine. After walking some way she struck across the far end of the loaning into the road which led to the mud hovel.

Autumn was approaching its very zenith, and the debatable land offered gorgeous tribute to the season. Like some outlandish savage ruler, it brought treasures unnumbered in the wealth of the more civilised earth. Here and there a branch of broom stood, like a sceptre, among the black jewels of its hanging pods, and brambles, pushing through the whin-thickets like flames, hung in ragged splashes of carmine and orange and acid yellow. Bushes of that sweetbrier whose little ardent-coloured rose is one of the glories of eastern Scotland were dressed in the scarlet hips succeeding their bloom, and between them and the whin the thrifty spider had woven her net. Underfoot, bracken, escaping from the ditches, had invaded the loaning to clothe it in lemon and russet. Where the ground was marshy, patches of fine rush mixed with the small purple scabious which has its home in the vagabond corners of the land. As Jessie-Mary emerged from the trees her sun-bleached hair seemed the right culmination to this scale of natural colour; had it not been for the dark blue of her cotton gown she might as easily have become absorbed into her surroundings as the roe-deer, which is lost, a brown streak, in the labyrinth of trunks.

The air had the faint scent of coming decay which haunts even the earliest of autumn days, and the pale, high sky wore a blue suggestive of tears; the exhalations of earth were touched with the bitterness of lichen and fungus. Far away under the slope of the fields, and so hidden from sight, Montrose lay between the ocean and the estuary of the South Esk, with, beyond its spire, the sweep of the North Sea.

A few minutes later she found herself standing on the large, flat stone which bridged the burn where the footpath crossed it by her grandmother's hovel. She remained gazing at the walls rising from the unkempt tangle to which months of neglect were reducing the garden. The fence was broken in many places, and clumps of phlox, growing in a corner, had been trodden by the feet of strayed animals. Beneath her, the water sang with the same irresponsible babble which had once been the accompaniment to her life; she turned to follow it with her eyes as it dived under the matted grasses and disappeared into the wood.

All at once, from beyond the cottage, there rose a shout that made her heart jump, and she started to see two figures approaching through the field by the side of the burn; the blood left her face as she recognised one of them as Peter Muirhead. She sprang quickly from the stone and over the rail dividing the wood from the path; it was a foolish action and it produced its natural result. As she did so, a yell came from the field and she saw that Peter and his companion had begun to run.

Through the trees she fled, the derisive voices whooping behind her. She was terrified of her tormentor and the unreasoning animal fear of pursuit was upon her. As she heard the rail crack she knew that he had entered the wood, and instinct turned her towards the loaning, where the cover was thick and where she might turn aside in the tangle and be lost in some hidden nook while they passed her by. It was her best chance.

She plunged out from among the firs into the open track. For a hundred yards ahead the bushes were sparse and there was no obstacle to hinder her flight. She was swift of foot, and the damp earth flew beneath her. Through the whins beyond she went, scratching her hands on protruding brambles and stumbling among the roots. Once her dress caught on a stiff branch and she rent it away, tearing it from knee to hem. The voices behind her rose again and her breath was giving out.

Emerging from the thicket, she almost bounded into a little circle of fire, the smoke of which she had been too much excited to notice, though it was rising, blue and fine, from the clearing she had reached. A small tent was before her, made of tattered sail-cloth stretched over some dry branches, and beside it a light cart reposed, empty, upon its tailboard, the shafts to the sky.

In front of the tent stood a tall, lean man. His look was fixed upon her as she appeared and he had evidently been listening to the sound of her approaching feet. His face was as brown as the fir-stems that closed him in on either side of the loaning, and his eyes, brown also, had a peculiar, watchful light that was almost startling. He stood as still as though he were an image, and he wore a gold ring in either ear.

To Jessie-Mary, a living creature at this moment represented salvation, and before the man had time to turn his head she had leaped into the tent. Inside, by a little heap of brushwood, lay a tarpaulin, evidently used in wet weather to supplement its shelter, and she flung herself down on the ground and dragged the thing over her. The man stood immovable, looking fixedly at the bushes, from the other side of which came the noise of jeering voices.

As Peter Muirhead and his friend pushed into the open space, red and panting, they came upon the unexpected apparition with some astonishment. Tinkers and gipsies were far from uncommon in the debatable land, but the tall, still figure, with its intent eyes, brought them to a standstill. Peter mopped his forehead.

'Did ye see a lassie gae by yon way?' he inquired, halting dishevelled from his race through the undergrowth, the sensational tie under one ear.

The brown man nodded, and, without a word, pointed his thumb over his shoulder in the direction in which they were going.

Peter and his companion glanced at each other; the former was rather blown, for he was not naturally active.

'Huts! a've had eneuch o' yon damned tawpie!' he exclaimed, throwing his cap on the ground.

The brown man looked him carefully over and smiled; there was a kind of primitive subtlety in his face.

Like many ill-favoured persons, Peter was vain and the look displeased him, for its faint ridicule was sharpened by the silence that accompanied it.

'A'll awa' to Montrose an' get the pollis tae ye the nicht,' he said, with as much superiority as he could muster; 'the like o' you's better oot o' this.'

'Ye'll no can rin sae far,' replied the other.

The answer was a mere burst of abuse.

'Come awa' noo, come awa',' said Peter's friend, scenting difficulties and unwilling to embroil himself.

But Peter was in a quarrelsome humour, and it was some time before the two young men disappeared down the track and Jessie-Mary could crawl from her hiding-place. She came out from under the sail-cloth, holding together the rent in her gown. The brown man smiled a different smile from the one with which he had regarded Peter; then he stepped up on a high tussock of rush to look after the pursuers.

'Are they awa'?' she asked, her eyes still dilated.

'Aye,' he replied. 'A didna tell on ye, ye see.'

'A'd like fine tae bide a bit,' said the girl nervously, 'they michtna be far yet.'

'Just sit ye doon there,' said he, pointing to his tattered apology for a dwelling.

She re-entered the tent and he seated himself before her on the threshold. For some minutes neither spoke and he considered her from head to foot. It was plain he was one chary of words. He took a short pipe from his pocket and, stuffing in some tobacco, lit it deliberately.

'A saw yon lad last time a was this way,' he said, jerking his head in the direction in which Peter had disappeared.

As she opened her mouth to reply the snort of a horse came through the bushes a few yards from where they sat. She started violently. There was a sudden gleam in his face which seemed to be his nearest approach to a laugh. 'Dod, ye needna be feared,' he said. 'Naebody'll touch ye wi' me.'

'A was fine an' glad tae see ye,' broke out the girl. 'Yon Muirhead's an ill lad tae hae i' the hoose—a bide wi' his mither, ye ken.'

As she spoke the tears welled up in her eyes and rolled over. She was by no means given to weeping, but she was a good deal shaken by her flight, and it was months since she had spoken to anyone whose point of view could approach her own. Not that she had any conscious point of view, but in common with us all she had a subconscious one. She brushed her sleeve across her eyes.

He sat silent, pulling at his pipe. From the trees came the long-drawn note of a wood-pigeon.

'A'll need tae be awa' hame and see tae the hens,' said the girl, at last.

The man sat still as she rose, watching her till the whins closed behind her; then he got up slowly and went to water the pony which was hobbled a few yards off. When evening fell on the debatable land, it found him sitting at his transitory threshold, smoking as he mended the rabbit-snare in his hand.

For Jessie-Mary, the days that followed these events were troublous enough. The tear in her gown was badly mended, and Mrs Muirhead, who had provided the clothes her servant wore, scolded her angrily. Peter was sulky, and, though he left her alone, he vented his anger in small ways which made domestic life intolerable to the women. Added to this, the young Black Spanish hen was missing.

The search ranged far and near over the wood. The bird, an incorrigible strayer, had repaid previous effort by being found in some outlying tangle with a 'stolen nest' and an air of irritated surprise at interruption. But hens were not clucking at this season, and Mrs Muirhead, in the dusk of one evening, announced her certainty that some cat or trap had removed the truant from her reach for ever.

'There's mony wad put a lazy cutty like you oot o' the place for this!' she exclaimed, as she and Jessie-Mary met outside the yard after their fruitless search. 'A'm fair disgustit wi' ye. Awa' ye gang ben the hoose an' get the kitchen reddit up—just awa in-by wi' ye, d'ye hear?'

Jessie-Mary obeyed sullenly. The kitchen window was half open and she paused beside it before beginning to clear the table and set out the evening meal. A cupboard close to her hand held the cheese and bannocks but she did not turn its key. Her listless look fell upon the planet that was coming out of the approaching twilight and taking definiteness above a mass of dark tree-tops framed in by the window sash. She had small conscious joy in such sights, for the pleasures given by these are the outcome of a higher civilisation than she had yet attained. But even to her, the point of serene silver, hung in the translucent field of sky, had a remote, wordless peace. She stood staring, her arms dropped at her sides.

The shrill tones of her mistress came to her ear; she was telling Peter, who stood outside, the history of her loss. Lamentation for the Black Spanish hen mingled with the recital of Jessie-Mary's carelessness, the villainy of serving-lasses as a body, the height in price of young poultry stock. Like many more valuable beings, the froward bird was assuming after death an importance she had never known in life.

A high-pitched exclamation came from Peter's lips.

'Ye needna speir owre muckle for her,' he said, 'she's roastit by this time. There's a lad doon the loan kens mair aboot her nor ony ither body!'

'Michty-me!' cried Mrs Muirhead.

'Aye, a'm tellin' ye,' continued he, 'the warst-lookin' great deevil that iver ye saw yet. He gie'd me impidence, aye did he, but a didna tak' muckle o' that. "Anither word," says I, "an' ye'll get the best thrashin' that iver ye got." He hadna vera muckle tae say after that, I warrant ye!'

Seldom had Mrs Muirhead been so much disturbed. Her voice rose to unusual heights as she discussed the matter; the local policeman must be fetched at once, she declared; and, as she adjured her son to start for his house without delay, Jessie-Mary could hear the young man's refusal to move a step before he had had his tea. She was recalled to her work by this and began hurriedly to set out the meal.

As she sat, a few minutes later, taking her own share at the farther end of the table, the subject was still uppermost, and by the time she rose mother and son were fiercely divided; for Peter, who had taken off his boots and was comfortable, refused to stir till the following morning. The hen had been missing three days, he said, and the thief was still in his place; it was not likely he would run that night. And the constable's cottage was over a mile off. The household dispersed in wrath.

In the hour when midnight grew into morning, Jessie-Mary closed the cottage door behind her and stole out among the silent trees. The pine-scent came up from under her feet as she trod and down from the blackness overhead. The moon, which had risen late, was near her setting, and the light of the little sickle just showed her the direction in which she should go. In and out of the shadows she went, her goal the clearing among the whins in the debatable land. As the steeple of distant Montrose, slumbering calmly between the marshes and the sea, rang one, she slipped out of the bushes and, going into the tent, awakened the sleeping man.

It was some time before the two came out of the shelter, and the first cock was crowing as the pony was roused and led from his tether under the tilted shafts. The sail-cloth was taken down and a medley of pots and pans and odd-looking implements thrown into the cart; the wheels were noiseless on the soft sod of the loaning as, by twists and turns, they thrust their way along the overgrown path.

Day broke on the figures of a man and woman who descended the slope of the fields towards the road. The man walked first.

And, in the debatable land among the brambles, a few black feathers blew on the morning wind.

John Buchan

1875–1940

The Loathly Opposite

Oliver Pugh's Story

> How loathly opposite I stood
> To his unnatural purpose.
>
> King Lear

Burminster had been to a Guildhall dinner the night before, which had been attended by many—to him—unfamiliar celebrities. He had seen for the first time in the flesh people whom he had long known by reputation, and he declared that in every case the picture he had formed of them had been cruelly shattered. An eminent poet, he said, had looked like a starting-price bookmaker, and a financier of world-wide fame had been exactly like the music-master at his preparatory school. Wherefore Burminster made the profound deduction that things were never what they seemed.

'That's only because you have a feeble imagination,' said Sandy Arbuthnot. 'If you had really understood Timson's poetry you would have realised that it went with close-cropped red hair and a fat body, and you should have known that Macintyre (this was the financier) had the music-and-metaphysics type of mind. That's why he puzzles the City so. If you understand a man's work well enough you can guess pretty accurately what he'll look like. I don't mean the colour of his eyes and his hair, but the general atmosphere of him.'

It was Sandy's agreeable habit to fling an occasional paradox at the table with the view of starting an argument. This time he stirred up Pugh, who had come to the War Office from the Indian Staff Corps. Pugh had been a great figure in Secret Service work in the East, but he did not look the part, for he had the air of a polo-playing cavalry subaltern. The skin was stretched as tight over his cheek-bones as over the knuckles of a clenched fist, and was so dark that it had the appearance of beaten bronze. He had black hair, rather beady black eyes, and the hooky nose which in the Celt often goes with that colouring. He was himself a very good refutation of Sandy's theory.

'I don't agree,' Pugh said. 'At least not as a general principle. One piece of humanity whose work I studied with the microscope for two aching years upset all my notions when I came to meet it.'

Then he told us this story.

'When I was brought to England in November '17 and given a "hush" department on three floors of an eighteenth-century house in a back street, I had a good deal to learn about my business. That I learned it in reasonable time was due to the extraordinarily fine staff that I found provided for me. Not one of them was a regular soldier. They were all educated men—they had to be in that job—but they came out of every sort of environment. One of the best was a Shetland laird, another was an Admiralty Court KC, and I had besides a metallurgical chemist, a golf champion, a leader-writer, a popular dramatist, several actuaries, and an East-end curate. None of them thought of anything but his job, and at the end of the War, when some ass proposed to make them OBEs, there was a very fair imitation of a riot. A more loyal crowd never existed, and they accepted me as their chief as unquestioningly as if I had been with them since 1914.

'To the War in the ordinary sense they scarcely gave a thought. You found the same thing in a lot of other behind-the-lines departments, and I daresay it was a good thing—it kept their nerves quiet and their minds concentrated. After all our business was only to decode and decypher German messages; we had nothing to do with the use which was made of them. It was a curious little nest, and when the Armistice came my people were flabbergasted—they hadn't realised that their job was bound up with the War.

'The one who most interested me was my second-in-command, Philip Channell. He was a man of forty-three, about five-foot-four in height, weighing, I fancy, under nine stone, and almost as blind as an owl. He was good enough at papers with his double glasses, but he could hardly recognise you three yards off. He had been a professor at some Midland college—mathematics or physics, I think—and as soon as the War began he had tried to enlist. Of course they wouldn't have him—he was about E5 in any physical classification, besides being well over age—but he would take no refusal, and presently he worried his way into the Government service. Fortunately he found a job which he could do superlatively well, for I do not believe there was a man alive with more natural genius for cryptography.

'I don't know if any of you have ever given your mind to that heart-breaking subject. Anyhow you know that secret writing falls under two heads—codes and cyphers, and that codes are combinations of words and cyphers of numerals. I remember how one used to be told that no code or cypher which was practically useful was really undiscoverable, and in a sense that is true, especially of codes. A system of communication which is in constant use must obviously not be too intricate, and a working code, if you get long enough for the job, can generally be read. That is why a code is periodically changed by the users. There are rules in worrying out the permutations and combinations of letters in most codes, for human ingenuity seems to run in certain channels, and a man who has been a long time at the business gets surprisingly clever at it. You begin by finding out a little bit, and then empirically building up the rules of decoding, till in a week or two you get the whole thing. Then, when you are happily engaged in reading enemy messages, the code is changed suddenly, and you have to start again from the beginning. . . . You can make a code, of course, that it is simply impossible to read except by accident—the key to which is a page of some book, for example—but fortunately that kind is not of much general use.

'Well, we got on pretty well with the codes, and read the intercepted enemy messages, cables and wireless, with considerable ease and precision. It was mostly diplomatic stuff, and not very important. The more valuable stuff was in cypher, and that was another pair of shoes. With a code you can build up the interpretation by degrees, but with a cypher you either know it or you don't—there are no half-way houses. A cypher, since it deals with numbers, is a horrible field for mathematical ingenuity. Once you have written out the letters of a message in numerals there are many means by which you can lock it and double-lock it. The two main devices, as you know, are transposition and substitution, and there is no limit to the ways one or other or both can be used. There is nothing to prevent a cypher having a double meaning, produced by two different methods, and, as a practical question, you have to decide which meaning is intended. By way of an extra complication, too, the message, when decyphered, may turn out to be itself in a difficult code. I can tell you our job wasn't exactly a rest cure.'

Burminster, looking puzzled, inquired as to the locking of cyphers.

'It would take too long to explain. Roughly, you write out a message horizontally in numerals; then you pour it into vertical columns, the number and order of which are determined by a keyword; then you write out the contents of the columns horizontally, following the lines across. To unlock it you have to have the key word, so as to put it back into the vertical columns, and then into the original horizontal form.'

Burminster cried out like one in pain. 'It can't be done. Don't tell me that any human brain could solve such an acrostic.'

'It was frequently done,' said Pugh.

'By you?'

'Lord bless you, not by me. I can't do a simple cross-word puzzle. By my people.'

'Give me the trenches,' said Burminster in a hollow voice. 'Give me the trenches any day. Do you seriously mean to tell me that you could sit down before a muddle of numbers and travel back the way they had been muddled to an original that made sense?'

'I couldn't, but Channell could—in most cases. You see, we didn't begin entirely in the dark. We already knew the kind of intricacies that the enemy favoured, and the way we worked was by trying a variety of clues till we lit on the right one.'

'Well, I'm blessed! Go on about your man Channell.'

'This isn't Channell's story,' said Pugh. 'He only comes into it accidentally. . . . There was one cypher which always defeated us, a cypher used between the German General Staff and their forces in the East. It was a locked cypher, and Channell had given more time to it than to any dozen of the others, for it put him on his mettle. But he confessed himself absolutely beaten. He wouldn't admit that it was insoluble, but he declared that he would need a bit of real luck to solve it. I asked him what kind of luck, and he said a mistake and a repetition. That, he said, might give him a chance of establishing equations.

'We called this particular cypher "PY", and we hated it poisonously. We felt like pygmies battering at the base of a high stone tower. Dislike of the thing soon became dislike of the man who had conceived it. Channell and I used to—I won't say amuse, for it was too dashed serious—but torment ourselves by trying to picture the fellow who owned the brain that was responsible for PY. We had a pretty complete *dossier* of the German Intelligence Staff, but of course we couldn't know who was responsible for this particular cypher. We knew no more than his code name, Reinmar, with which he signed the simpler messages to the East, and Channell, who was a romantic little chap for all his science, had got it into his head that it was a woman. He used to describe her to me as if he had seen her—a she-devil, young, beautiful, with a much-painted white face, and eyes like a cobra's. I fancy he read a rather low class of novel in his off-time.

'My picture was different. At first I thought of the histrionic type of scientist, the "ruthless brain" type, with a high forehead and a jaw puckered like a chimpanzee. But that didn't seem to work, and I settled on a picture of a first-class *Generalstaboffizier*, as handsome as Falkenhayn, trained to the last decimal, absolutely passionless, with a mind that worked with the relentless precision of a fine machine. We all of us at the time suffered from the bogy of this kind of German, and, when things were going badly, as in March '18, I couldn't sleep for hating him. The infernal fellow was so water-tight and armour-plated, a Goliath who scorned the pebbles from our feeble slings.

'Well, to make a long story short, there came a moment in September '18 when PY was about the most important thing in the world. It mattered enormously what Germany was doing in Syria, and we knew that it was all in PY. Every morning a pile of the intercepted German wireless messages lay on Channell's table, which were as meaningless to him as a child's scrawl. I was prodded by my chiefs and in turn I prodded Channell. We had a week to find the key to the cypher, after which things must go on without us, and if we had failed to make anything of it in eighteen months of quiet work, it didn't seem likely that we would succeed in seven feverish days. Channell nearly went off his head with overwork and anxiety. I used to visit his dingy little room and find him fairly grizzled and shrunken with fatigue.

'This isn't a story about him, though there is a good story which I may tell you another time. As a matter of fact we won on the post. PY made a mistake. One morning we got a long message dated *en clair*, then a very short message, and then a third message almost the same as the first. The second must mean "Your message of to-day's date unintelligible, please repeat," the regular formula. This gave us a translation of a bit of the cypher. Even that would not have brought it out, and for twelve hours Channell was on the verge of lunacy, till it occurred to him that Reinmar might have signed the long message with his name, as we used to do sometimes in cases of extreme urgency. He was right, and, within three hours of the last moment Operations could give us, we had the whole thing pat. As I have said, that is a story worth telling, but it is not this one.

'We both finished the War too tired to think of much except that the darned thing was over. But Reinmar had been so long our unseen but constantly pictured opponent that we kept up a certain interest in him. We would like to have seen how he took the licking, for he must have known that we had licked him. Mostly when you lick a man at a game you rather like him, but I didn't like Reinmar. In fact I made him a sort of compost of everything I had ever disliked in a German. Channell stuck to his she-devil theory, but I was pretty certain that he was a youngish man with an intellectual arrogance which his country's *débâcle* would in no way lessen. He would never acknowledge defeat. It was highly improbable that I should ever find out who he was, but I felt that if I did, and met him face to face, my dislike would be abundantly justified.

'As you know, for a year or two after the Armistice I was a pretty sick man. Most of us were. We hadn't the fillip of getting back to civilised comforts, like the men in the trenches. We had always been comfortable enough in body, but our minds were fagged out, and there is no easy cure for that. My digestion went nobly to pieces, and I endured a miserable space of lying in bed and living on milk and olive-oil. After that I went back to work, but the darned thing always returned, and every leech had a different regime to advise. I tried them all—dry meals, a snack every two hours, lemon juice, sour milk, starvation, knocking off tobacco—but nothing got me more than half-way out of the trough. I was a burden to myself and a nuisance to others, dragging my wing through life, with a constant pain in my tummy.

'More than one doctor advised an operation, but I was chary about that, for I had seen several of my friends operated on for the same mischief and left as sick as before. Then a man told me about a German fellow called Christoph, who was said to be very good at handling my trouble. The best hand at diagnosis in the world, my informant said—no fads—treated every case on its merits—a really original mind. Dr Christoph had a modest kurhaus at a place called Rosensee in the Sächischen Sweitz.

By this time I was getting pretty desperate, so I packed a bag and set off for Rosensee.

'It was a quiet little town at the mouth of a narrow valley, tucked in under wooded hills, a clean fresh place with open channels of running water in the streets. There was a big church with an onion spire, a Catholic seminary, and a small tanning industry. The kurhaus was half-way up a hill, and I felt better as soon as I saw my bedroom, with its bare scrubbed floors and its wide verandah looking up into a forest glade. I felt still better when I saw Dr Christoph. He was a small man with a grizzled beard, a high forehead, and a limp, rather like what I imagine the Apostle Paul must have been. He looked wise, as wise as an old owl. His English was atrocious, but even when he found that I talked German fairly well he didn't expand in speech. He would deliver no opinion of any kind until he had had me at least a week under observation; but somehow I felt comforted, for I concluded that a first-class brain had got to work on me.

'The other patients were mostly Germans with a sprinkling of Spaniards, but to my delight I found Channell. He also had been having a thin time since we parted. Nerves were his trouble—general nervous debility and perpetual insomnia, and his college had given him six months' leave of absence to try to get well. The poor chap was as lean as a sparrow, and he had the large dull eyes and the dry lips of the sleepless. He had arrived a week before me, and like me was under observation. But his vetting was different from mine, for he was a mental case, and Dr Christoph used to devote hours to trying to unriddle his nervous tangles. "He is a good man for a German," said Channell, "but he is on the wrong tack. There's nothing wrong with my mind. I wish he'd stick to violet rays and massage, instead of asking me silly questions about my great-grandmother."

'Channell and I used to go for invalidish walks in the woods, and we naturally talked about the years we had worked together. He was living mainly in the past, for the War had been the great thing in his life, and his professorial duties seemed trivial by comparison. As we tramped among the withered bracken and heather his mind was always harking back to the dingy little room where he had smoked cheap cigarettes and worked fourteen hours out of the twenty-four. In particular he was as eagerly curious about our old antagonist, Reinmar, as he had been in 1918. He was more positive than ever that she was a woman, and I believe that one of the reasons that had induced him to try a cure in Germany was a vague hope that he might get on her track. I had almost forgotten about the thing, and I was amused by Channell in the part of the untiring sleuth-hound.

'"You won't find her in the Kurhaus," I said. "Perhaps she is in some old schloss in the neighbourhood, waiting for you like the Sleeping Beauty."

'"I'm serious," he said plaintively. "It is purely a matter of intellectual curiosity, but I confess I would give a great deal to see her face to face. After I leave here, I thought of going to Berlin to make some inquiries. But I'm handicapped, for I know nobody and I have no credentials. Why don't you, who have a large acquaintance and far more authority, take the thing up?"

'I told him that my interest in the matter had flagged and that I wasn't keen on digging into the past, but I promised to give him a line to our Military Attaché if he thought of going to Berlin. I rather discouraged him from letting his mind dwell too much on events in the War. I said that he ought to try to bolt the door on all that had contributed to his present breakdown.

'"That is not Dr Christoph's opinion," he said emphatically. "He encourages me to talk about it. You see, with me it is a purely intellectual interest. I have no emotion in the matter. I feel quite friendly towards Reinmar, whoever she may be. It is, if you like, a piece of romance. I haven't had so many romantic events in my life that I want to forget this."

'"Have you told Dr Christoph about Reinmar?" I asked.

'"Yes," he said, "and he was mildly interested. You know the way he looks at you with his solemn grey eyes. I doubt if he quite understood what I meant, for a little provincial doctor, even though he is a genius in his own line, is not likely to know much about the ways of the Great General Staff. . . . I had to tell him, for I have to tell him all my dreams, and lately I have taken to dreaming about Reinmar."

'"What's she like?" I asked.

'"Oh, a most remarkable figure. Very beautiful, but uncanny. She has long fair hair down to her knees."

'Of course I laughed. "You're mixing her up with the Valkyries," I said. "Lord, it would be an awkward business if you met that she-dragon in the flesh."

'But he was quite solemn about it, and declared that his waking picture of her was not in the least like his dreams. He rather agreed with my nonsense about the old schloss. He thought that she was probably some penniless grandee, living solitary in a moated grange, with nothing now to exercise her marvellous brain on, and eating her heart out with regret and shame. He drew so attractive a character of her that I began to think that Channell was in love with a being of his own creation, till he ended with, "But all the same she's utterly damnable. She must be, you know."

'After a fortnight I began to feel a different man. Dr Christoph thought that he had got on the track of the mischief, and certainly, with his deep massage and a few simple drugs, I had more internal comfort than I had known for three years. He was so pleased with my progress that he refused to treat me as an invalid. He encouraged me to take long walks into the hills, and presently he arranged for me to go out roebuck-shooting with some of the local junkers.

'I used to start before daybreak on the chilly November mornings and drive to the top of one of the ridges, where I would meet a collection of sportsmen and beaters, shepherded by a fellow in a green uniform. We lined out along the ridge, and the beaters, assisted by a marvellous collection of dogs, including the sporting dachshund, drove the roe towards us.

It wasn't very cleverly managed, for the deer generally broke back, and it was chilly waiting in the first hours with a powdering of snow on the ground and the fir boughs heavy with frost crystals. But later, when the sun grew stronger, it was a very pleasant mode of spending a day. There was not much of a bag, but whenever a roe or a capercailzie fell all the guns would assemble and drink little glasses of *kirschwasser*. I had been lent a rifle, one of those appalling contraptions which are double-barrelled shot-guns and rifles in one, and to transpose from one form to the other requires a mathematical calculation. The rifle had a hair trigger too, and when I first used it I was nearly the death of a respectable Saxon peasant.

'We all ate our midday meal together and in the evening, before going home, we had coffee and cakes in one or other of the farms. The party was an odd mixture, big farmers and small squires, an hotel-keeper or two, a local doctor, and a couple of lawyers from the town. At first they were a little shy of me, but presently they thawed, and after the first day we were good friends. They spoke quite frankly about the War, in which every one of them had had a share, and with a great deal of dignity and good sense.

'I learned to walk in Sikkim, and the little Saxon hills seemed to me inconsiderable. But they were too much for most of the guns, and instead of going straight up or down a slope they always chose a circuit, which gave them an easy gradient. One evening, when we were separating as usual, the beaters taking a short cut and the guns a circuit, I felt that I wanted exercise, so I raced the beaters downhill, beat them soundly, and had the better part of an hour to wait for my companions, before we adjourned to the farm for refreshment. The beaters must have talked about my pace, for as we walked away one of the guns, a lawyer called Meissen, asked me why I was visiting Rosensee at a time of year when few foreigners came. I said I was staying with Dr Christoph.

'"Is he then a private friend of yours?" he asked.

'I told him No, that I had come to his kurhaus for treatment, being sick. His eyes expressed polite scepticism. He was not prepared to regard as an invalid a man who went down a hill like an avalanche.

'But, as we walked in the frosty dusk, he was led to speak of Dr Christoph, of whom he had no personal knowledge, and I learned how little honour a prophet may have in his own country. Rosensee scarcely knew him, except as a doctor who had an inexplicable attraction for foreign patients. Meissen was curious about his methods and the exact diseases in which he specialised. "Perhaps he may yet save me a journey to Homburg?" he laughed. "It is well to have a skilled physician at one's doorstep. The doctor is something of a hermit, and except for his patients does not appear to welcome his kind. Yet he is a good man, beyond doubt, and there are those who say that in the War he was a hero."

'This surprised me, for I could not imagine Dr Christoph in any fighting capacity, apart from the fact that he must have been too old. I thought that Meissen might refer to work in the base hospitals. But he was positive; Dr Christoph had been in the trenches; the limping leg was a war wound.

'I had had very little talk with the doctor, owing to my case being free from nervous complications. He would say a word to me morning and evening about my diet, and pass the time of day when we met, but it was not till the very eve of my departure that we had anything like a real conversation. He sent a message that he wanted to see me for not less than one hour, and he arrived with a batch of notes from which he delivered a kind of lecture on my case. Then I realised what an immense amount of care and solid thought he had expended on me. He had decided that his diagnosis was right—my rapid improvement suggested that—but it was necessary for some time to observe a simple regime, and to keep an eye on certain symptoms. So he took a sheet of note-paper from the table and in his small precise hand wrote down for me a few plain commandments.

'There was something about him, the honest eyes, the mouth which looked as if it had been often compressed in suffering, the air of grave good-will, which I found curiously attractive. I wished that I had been a mental case like Channell, and had had more of his society. I detained him in talk, and he seemed not unwilling. By and by we drifted to the War and it turned out that Meissen was right.

'Dr Christoph had gone as medical officer in November '14 to the Ypres Salient with a Saxon regiment, and had spent the winter there. In '15 he had been in Champagne, and in the early months of '16 at Verdun, till he was invalided with rheumatic fever. That is to say, he had had about seventeen months of consecutive fighting in the worst areas with scarcely a holiday. A pretty good record for a frail little middle-aged man!

'His family was then at Stuttgart, his wife and one little boy. He took a long time to recover from the fever, and after that was put on home duty. "Till the War was almost over," he said, "almost over, but not quite. There was just time for me to go back to the front and get my foolish leg hurt." I must tell you that whenever he mentioned his war experience it was with a comical deprecating smile, as if he agreed with anyone who might think that gravity like his should have remained in bed.

'I assumed that this home duty was medical, until he said something about getting rusty in his professional work. Then it appeared that it had been some job connected with Intelligence. "I am reputed to have a little talent for mathematics," he said. "No. I am no mathematical scholar, but, if you understand me, I have a certain mathematical aptitude. My mind has always moved happily among numbers. Therefore I was set to construct and to interpret cyphers, a strange interlude in the noise of war. I sat in a little room and excluded the world, and for a little I was happy."

'He went on to speak of the *enclave* of peace in which he had found himself, and as I listened to his gentle monotonous voice, I had a sudden inspiration.

'I took a sheet of note-paper from the stand, scribbled the word *Reinmar* on it, and shoved it towards him. I had a notion, you see, that I might surprise him into helping Channell's researches.

'But it was I who got the big surprise. He stopped thunderstruck, as soon as his eye caught the word, blushed scarlet over every inch of face and bald forehead, seemed to have difficulty in swallowing, and then gasped. "How did you know?"

'I hadn't known, and now that I did, the knowledge left me speechless. This was the loathly opposite for which Channell and I had nursed our hatred. When I came out of my stupefaction I found that he had recovered his balance and was speaking slowly and distinctly, as if he were making a formal confession.

'"You were among my opponents . . . that interests me deeply. . . . I often wondered. . . . You beat me in the end. You are aware of that?"

'I nodded. "Only because you made a slip," I said.

'"Yes, I made a slip. I was to blame—very gravely to blame, for I let my private grief cloud my mind."

'He seemed to hesitate, as if he were loath to stir something very tragic in his memory.

'"I think I will tell you," he said at last. "I have often wished—it is a childish wish—to justify my failure to those who profited by it. My chiefs understood, of course, but my opponents could not. In that month when I failed I was in deep sorrow. I had a little son—his name was Reinmar—you remember that I took that name for my code signature?"

'His eyes were looking beyond me into some vision of the past.

'"He was, as you say, my mascot. He was all my family, and I adored him. But in those days food was not plentiful. We were no worse off than many million Germans, but the child was frail. In the last summer of the War he developed phthisis due to malnutrition, and in September he died. Then I failed my country, for with him some virtue seemed to depart from my mind. You see, my work was, so to speak, his also, as my name was his, and when he left me he took my power with him. . . . So I stumbled. The rest is known to you."

'He sat staring beyond me, so small and lonely, that I could have howled. I remember putting my hand on his shoulder, and stammering some platitude about being sorry. We sat quite still for a minute or two, and then I remembered Channell. Channell must have poured his views of Reinmar into Dr Christoph's ear. I asked him if Channell knew.

'A flicker of a smile crossed his face.

'"Indeed no. And I will exact from you a promise never to breathe to him what I have told you. He is my patient, and I must first consider his case. At present he thinks that Reinmar is a wicked and beautiful lady whom he may some day meet. That is romance, and it is good for him to think so. . . . If he were told the truth, he would be pitiful, and in Herr Channell's condition it is important that he should not be vexed with such emotions as pity."'

Lorna Moon

1886–1930

Silk Both Sides

'And two and a half yards o' four-inch black satin ribbon.'

Jessie MacLean added this last fatal item with an upward jerk of her head lest Mistress MacKenty, at the other side of the counter, should think she was ashamed of her purchase. But Mistress MacKenty had a nose for news rather than an instinct for tragedy, and by the suppressed eagerness in her voice as she asked, 'And ye'll want it silk on both sides I'll warrant?' you could see that she was already half-way down the road to the smithy to spread the news that: 'Jessie MacLean had lost heart and would be out in a bonnet in the morn, so help her Davey.'

'Aye, silk both sides,' Jessie answered, letting her eyes range the shelves carelessly to prove that there was nothing momentous in her buying bonnet strings.

Silk both sides proved it! A satin-faced ribbon might have many uses, but silk both sides was a bonnet string by all the laws of millinery known to Drumorty.

Telling about it five minutes later, Mistress MacKenty said, 'I might hae been wrang when she bought the silk geraniums, and I may hae been over-hasty when she said "half a yard o' black lace"—but silk both sides is as good as swearing it on the Bible.'

In Drumorty, a bonnet with strings tied below the chin means that youth is over. About the time the second baby is born, the good wife abandons her hat—forever—and appears in a bonnet with ties. She may be any age from eighteen to twenty-five; for matrimony, and motherhood, and age, come early in Drumorty. The spinster clings longer to her hat, for while she wears it, any bachelor may take heart and 'speer'* her; and if she be 'keeping company' she may cling to her hat until she be thirty 'and a bittock'; but after that—if she would hold the respect of her community, she must cease to 'gallivant' about 'wi' a hat' and dress like a decent woman, in a bonnet with ties.

And Jessie MacLean was six and thirty as Baldie Tocher could tell you, for did he no' have the pleasure of burying the excise-man the very morning that Jessie first saw the light of day, and was it no' the very next year that Nancy MacFarland's cow got mired in the moss?

Drumorty had been very lenient with her. Many a good wife thought

* Ask, particularly ask in marriage; proposal.

it was high time that Jessie laid aside her hat, but always she held her peace, remembering her bridal gown and the care with which Jessie had made it, for Jessie was the village seamstress, and it was a secret, whispered, that she charged only half price for making wedding gowns, because she liked to make them so much.

Another reason for their lenience was Jock Sclessor. For fifteen years Jock had 'kept company' with her; not one Sunday morning had he missed 'crying by' for Jessie to go to morning service. He would come round the bend of the road from Skilly's farm just as the sexton gave the bell that first introductory ring which meant 'bide a wee till I get her goin' full swing and then bide at hame frae the kirk if ye dare'; and Jessie would come out of her door and mince down the sanded walk between the rows of boxwood to the gate, and affect surprise at seeing Jock just as if she had not been watching for him behind her window curtain this past five minutes.

Jock was the cotter* on Skilly's farm. Every year he intended to 'speer' Jessie when thrashing was over. Tammas, his dog, will tell you how many times he had been on the very point of asking her the very next day, but—always the question of adding another room came up and not for the life of him could he decide whether to level the rowan tree and build it on the east—or to move the peats and build it on the west; and by the time he had made up his mind to cart the peats down behind the byre and build it on the west, lambing was round again and he let it go by for another year.

And every year Jessie was in a flutter as thrashing was nearing the finish. One year she had been so sure that he would 'speer' her that she bought a new scraper; for Jock could make your very heart stand still, he was that careless about scraping the mud off his boots. Often when she was alone, she would practise ways of telling Jock that he must clean his boots before he came in: 'Good man, hae ye forgotten the scraper?' was abandoned because it wasn't strictly honest, for Jessie knew full well that he always forgot the scraper. 'Gang back and clean your feet,' was set aside also, because it was too commanding and 'Dinna forget the scraper' was also discarded because it isn't good to nag a man before he has set foot in the door. But none of the expressions had been tried out yet—for when Jock dropped in with her, after service, Jessie hurried him by the scraper as if it might shout at him, 'She expects you to speer her!' and so put her to shame.

But now thrashing had been over for weeks, and every Sunday since then, she had looked for the white gowan† in Jock's coat, for what swain 'worth his ears full of cold water' ever asked the question without that emblem of courage in his button-hole? It is a signal to the world that he

* One who rents a small portion of a farm from the farmer and gives his services in exchange. The cotter's wife also assists in the farm kitchen and in the fields at harvest time, so a cotter's wife must be strong.

† Field daisy.

means to propose, that he is going in cold blood to do it—and forever after his good wife can remind him of that, should he suggest that he was inveigled into it by some female wile.

Last Sunday, on their way to church, Jessie cleared her throat nervously and grasped her new testament, bracing herself as she asked in a thin voice that was much too offhand: 'Would your peats no' be better sheltered in the lea o' the byre?' and Jock replied: 'Na, they are better where they are.'

And so a hope, nourished fifteen years, died, and through the service she sat gazing straight ahead, with her eyes wide open, for the wider eyes are open, the more tears they can hold without spilling over. And next Saturday she bought black satin ribbon 'silk both sides' and the world, meaning Drumorty, knew that Jessie's tombstone would not read 'Beloved wife of—'

In Jessie's cottage the blinds were drawn on Saturday evening, and you who have suffered will not ask me to pull them aside and show you how a faded spinster looks when she weeps; nor how her fingers tremble when she sews upon bonnet strings; but let me tell you how bravely she stepped out next morning wearing her bonnet, with never a look through the curtain to see if Jock was on his way, nor a glance to see if the neighbours were watching. Her step was just as firm upon the sanded path, and her head just as high; perhaps she grasped her testament more tightly than usual, but what soul on the rack would not do that?

Jock came round the bend as she reached the gate. She clung to the latch to keep herself from tearing the bonnet from her head. Oh, Fate, that sits high and laughs, have you the heart to laugh now? Jock was wearing a white gowan. It was just a dozen steps or so back to the house, and a hat, and happiness; but the world knew that she had bought bonnet strings and was that not Mistress MacKenty watching from behind her curtain? Go forward, Jessie—there is no turning back, and go proudly! Open the latch and answer his 'guid morning' and smile—and don't, don't, let your hands tremble so, or he will surely guess!

Look your fill from behind the curtain, Mistress MacKenty! You can not see a heartache when it is hidden by a black alpaca gown and when the heart belongs to Jessie MacLean!

Jock Sclessor, your one chance of happiness is now! Lead her back into the house and take the bonnet from her head! No, laggard and fool that you are, you are wondering if she has noticed the gowan! Has she not! She has watched for it for fifteen years! Speak, you fool! Don't keep staring at her bonnet! You dullard, Jessie must come to your rescue, and she does, 'Is the sexton no' late this morning?' Jessie turned out of the gate as she spoke, snapping the latch with the right amount of care.

'Aye, later than usual,' Jock agreed. (The sexton was never late in his life, and at that moment the bell rang out to give Jock the lie.) But Jock was so dazed, he would have agreed if she had said, 'Let us choke the minister.'

As they walked to the church, he meditated, 'Evidently she never expected me to speer her, and she's never so much as glimpsed the gowan. I'll slip it out when we kneel in the kirk. But maybe I better sound her out first. It's gey* and lonesome for a man biding by himself.'

They were just turning round by the town hall where the rowan trees are red, when he said, 'I had been thinkin' o' levellin' my rowan tree.' Jessie's heart thumped. Here it came! (That was why he wouldn't move the peats.) But she wouldn't help him a foot of the road—she had waited too long—he must come every step himself.

'Oh, that would be a pity—it's a bonnie tree,' she answered.

Not much help here, but he would try again. 'I was thinking o' building.'

'Building? My certies! What could ye be building so near the house?' this with some malice for all the fifteen long years. But you have gone too far, Jessie; he needs help.

'I—oh—I thought I'd build a shed for peats.'

Thud! That was Jessie's heart you heard and that queer thin voice is Jessie's, saying, 'I thought—ye were minded to leave the peats the other side o' the house.'

'Aye, I am minded to leave them there—but a body canna hae too many peats.'

And as they knelt in kirk, he slipped the gowan out—and Jessie did not need to widen her eyes to hold the tears this time. That sorrow was past, she would never weep over it again.

At home, she brewed her tea, looking round at her rag rugs and white tidies with pleasure. The tidy on the big chair was as white and smooth as when she pinned it there in the morning, and there was no mud to be carefully washed off the rug by the door. There was a certain contentment in knowing that it would never be; a certain exhilaration in knowing that next Sunday she could not be disappointed, because next Sunday she would not hope. She sipped her tea peacefully and smiled at the bonnet sitting so restful-like on the dresser, and at the tidy on the big chair so spotless and smooth, and thought, 'Jock Sclessor would have been a mussy man to have about a house. I'm thinkin' his mother was over-lenient when she brought him up.'

And Jock, at Skilly's, was thinking, 'I wouldna had time to build it anyway. Lambing is here—and that is too bonnie a tree to be cut down.'

* Very.

NEIL GUNN

1891–1973

THE TAX-GATHERER

'Blast it,' he muttered angrily. 'Where is the accursed place?'

He looked at the map again spread before him on the steering-wheel. Yes, it should be just here. There was the cross-roads. He threw a glance round the glass of his small saloon car and saw a man's head bobbing beyond the hedge. At once he got out and walked along the side of the road.

'Excuse me,' he cried. The face looked at him over the hedge. 'Excuse me, but can you tell me where Mrs Martha Williamson stays?'

'Mrs Who?'

'Mrs Martha Williamson.'

'No,' said the face slowly, and moved away. He followed it for a few paces to a gap in the hedge. 'No,' said the man again, and turned to call a spaniel out of the turnips. He had a gun under his arm and was obviously a gamekeeper.

'Well, she lives about here, at Ivy Cottage.'

'Ivy Cottage? Do you mean the tinkers?' And the gamekeeper regarded him thoughtfully.

'Yes. I suppose so.'

'I see,' said the gamekeeper, looking away. 'Turn up to your right at the cross-roads there and you'll see it standing back from the road.'

He thanked the gamekeeper and set off, walking quickly so that he needn't think too much about his task, for it was new to him.

When he saw the cottage, over amongst some bushes with a rank growth of nettles at one end, he thought it a miserable place, but when he came close to the peeling limewash, the torn-down ivy, the sagging roof, the broken stone doorstep thick with trampled mud, he saw that it was a wretched hovel.

The door stood half-open, stuck. He knocked on it and listened to the acute silence. He knocked again firmly and thought he heard thin whisperings. He did not like the hushed fear in the sounds, and was just about to knock peremptorily when there was a shuffling, and, quietly as an apparition, a woman was there.

She stood twisted, lax, a slim, rather tall figure, with a face the colour of the old limewash. She clung to the edge of the door in a manner unhumanly pathetic, and looked at him out of dark, soft eyes.

'Are you Mrs Williamson?'

After a moment she said, 'Yes.'

'Well, I've come about that dog. Have you taken out the licence yet?'

'No.'

'Well, it's like this,' he said, glancing away from her. 'We don't want to get you into trouble. But the police reported to us that you had the dog. Now, you can't have a dog without paying a licence. You know that. So, in all the circumstances, the authorities decided that if you paid a compromise fine of seven-and-six, and took out the licence, no more would be said about it. You would not be taken to court.' He looked at her again, and saw no less than five small heads poking round her ragged dark skirt. 'We don't want you to get into trouble,' he said. 'But you've got to pay by Friday—or you'll be summonsed. There's no way out.'

She did not speak, stood there unmoving, clinging to the door, a feminine creature waiting dumbly for the blow.

'Have you a husband?' he asked.

'Yes,' she said, after a moment.

'Where is he?'

'I don't know,' she answered, in her soft, hopeless voice. He wanted to ask her if he had left her for good, but could not, and this irritated him, so he said calmly, 'Well, that's the position, as you know. I was passing, and, seeing we had got no word of your payment, I thought I'd drop in and warn you. We don't want to take you to court. So my advice to you is to pay up—and at once, or it will be too late.'

She did not answer. As he was about to turn away the dregs of his irritation got the better of him. 'Why on earth did you want to keep the dog, anyway?'

'We tried to put him away, but he wouldn't go,' she said.

His brows gathered. 'Oh, well, it's up to you,' he replied coldly, and he turned and strode back to his car. Slamming the door after him, he gripped the wheel, but could not, at the last instant, press the self-starter. He swore to himself in a furious rage. Damn it all, what concern was it of his? None at all. As a public official he had to do his job. It was nothing to him. If a person wanted to enjoy the luxury of keeping a dog, he or she had to pay for it. That's all. And he looked for the self-starter, but, with his finger on the button, again could not press it. He twisted in his seat. Fifteen bob! he thought. Go back and slip her fifteen bob? Am I mad? He pressed the self-starter and set the engine off in an unnecessary roar. As he turned at the cross-roads he hesitated before shoving the gear lever into first, then shoved it and set off. If a fellow was to start paying public fines where would it end? Sentimental? Absolutely.

By the following Tuesday it was clear she had not paid.

'The case will go on,' said his chief in the office.

'It's a hard case,' he answered. 'She won't be able to pay.' His voice was calm and official.

'She'll have to pay—one way or the other,' answered his chief, with the usual trace of official satire in his voice.

'She's got a lot of kids,' said the young man.

'Has she?' said the chief. 'Perhaps she could not help having them—but the dog is another matter.' He smiled, and glanced at the young man, who awkwardly smiled back.

There was nothing unkindly in the chief's attitude, merely a complete absence of feeling. He was dealing with 'a file', and had no sympathy for anyone who tried to evade the law. He prosecuted with lucid care, and back in his office smiled with satisfaction when he got a conviction. For to fail in getting a conviction was to be inept in his duty. Those above him frowned upon such ineptitude.

All the same, the young man felt miserable. If he hadn't gone to the cottage it would have been all right. But the chief had had no unnecessary desire for a court case—particularly one of those hard cases that might get into the press. Not that that mattered really, for the law had to be carried out. Than false sentiment against the law of the land there could, properly regarded, be nothing more reprehensible—because it was so easy to indulge.

'By the way,' said the chief, as he was turning away, 'I see the dog has been shot. You didn't mention that?'

'No. I—' He had forgotten to ask the woman if the dog was still with her. 'I—as a matter of fact, I didn't think about it, seeing it was a police report, and therefore no evidence from us needed.'

'Quite so,' said his chief reticently, as he turned to his file.

'Who shot it?' the young man could not help asking.

'A gamekeeper, apparently.'

The young man withdrew, bit on his embarrassment at evoking the chief's 'reticence', and thought of the gamekeeper who might believe that if the dog was shot nothing more could be done about the case. As if the liability would thereby be wiped out! As if it would make the slightest difference to the case!

In his own room he remembered the gamekeeper and his curious look. Decent of him all the same to have tried to help. If the children's faces had been sallow and hollow from under-feeding, what could the dog have got? Nothing, unless—The thought dawned: the gamekeeper had probably shot the brute without being asked. Poaching rabbits and game? Perhaps the mainstay of the family? He laughed in his nostrils. When you're down you're right down, down and out. Absolutely. With a final snort of satire, he took some papers from the 'pending' cover and tried to concentrate on an old woman's application for a pension. It seemed quite straightforward, though he would have to investigate her circumstances. Then he saw the children's faces again.

He had hardly been conscious of looking at them at the time. In fact, after that first glance of surprise he had very definitely not looked at them. The oldest was a girl of nine or ten, thin and watery, fragile, with her mother's incredible pallor and black eyes. The stare from those considering eyes, blank and dumb, and yet wary. They didn't appeal: trust could not touch them; they waited, just waited, for—the only hope—something less than the worst.

And the little fellow of seven or eight—sandy hair, inflamed eyelids, and that something about the expression, the thick, half-open mouth, suggesting the mental deficient. Obviously from the father's side, physically. The father had deserted them. Was perhaps in quod somewhere else, for they had only recently returned from their travels to the cottage. How did they manage even to live in it without being turned out? But the police would have that in hand as well! There was something too soft about the woman. She would never face up to her husband. When he was drunk, her softness would irritate him; he would clout her one. She was feckless. Her body had slumped into a pliant line, utterly hopeless, against the door. All at once he saw the line as graceful, and this unexpected vision added the last touch to derision.

The young man had observed in his life already that if his mind was keen on some subject he would come across references to it in the oddest places, in books dealing with quite other matters, from the most unlikely people. But this, carefully considered, was not altogether fortuitous. For example, when the old woman who had applied for the Old Age Pension asked him if he would have a cup of tea, he hesitated, not because he particularly wanted to have a cup of tea, but because he vaguely wanted to speak to her about the ways of tinkers, for Ivy Cottage was little more than a mile away.

His hesitation, however, the hesitation of an important official who had arrived in a motor car and upon whom the granting of her pension depended (as she thought), excited her so much that before she quite knew what she was doing she was on her knees before the fire, flapping the dull peat embers with her apron, for she did not like, in front of him, to bend her old grey head and blow the embers to a flame. As he was watching her she suddenly stopped flapping, with an expression of almost ludicrous dismay, and mumbled something about not having meant to do that. At once he was interested, for clearly there was something involved beyond mere politeness. The old folk in this northern land, he had found, were usually very polite, and he liked their ways and curious beliefs. The fact that they had a Gaelic language of their own attracted him, for he was himself a student of French, and, he believed, somewhat of an authority on Balzac.

Fortunately for the old lady, a sprightly tongue of flame ran up the dry peat at that moment, and she swung the kettle over it. 'Now it won't be long,' she said, carefully backing up the flame with more peat.

She was a quick-witted, bright-eyed old woman, and as she hurried to and fro getting the tea things on the table they chatted pleasantly. Presently, when she seated herself and began to pour the tea, he asked her in the friendliest way why she had stopped flapping her apron.

She glanced at him and then said, 'Och, just an old woman's way.'

But he would not have that, and rallied her. 'Come, now, there was something more to it than that.'

And at last she said: 'It's just an old story in this part of the world and likely it will not be true. But I will tell it to you, seeing you say you like stories of the kind, but you will have to take it as you get it, for that's the way I got it myself, more years ago now than I can remember. It is a story about our Lord at the time of His crucifixion. You will remember that when our Lord was being crucified they nailed Him to the Cross. But before they could do that they needed the nails, and the nails were not in it. So they tried to get the nails made, but no one would make them. They asked the Roman soldiers to make them, but they would not. Perhaps it was not their business to make the nails. Anyway, they would not make them. So they asked the Jews to make the nails, but they would not make them either. No, they would not make them. No one would make the nails that were needed to crucify our Lord. And when they were stuck now, and did not know what to do, who should they see coming along but a tinker with his little leather apron on him. So they asked him if he would make the nails. And he said yes, he would make them. And to make the nails he needed a fire. So a fire was made, but it would not go very well, so he bent down in front of it and flapped it with his leather apron. In that way the fire went and the nails were made. And so it came about that the tinkers became wanderers, and were never liked by the people of the world anywhere. And that's the story.'

When at last he drove away from the old woman's house he came to the cross-roads and, a few yards beyond, drew up. This is the place, he thought, and he felt it about him, gripped the wheel hard, and sat still. Irritation began to get the better of him. Anyone could see he was a fool. He got out and stretched his legs and lit a cigarette. There was no one in the turnip field, no one anywhere. All at once he walked back quickly to the cross-roads, turned right, and again saw the cottage. It was looking at him with a still, lopsided, idiotic expression. His flesh quickened and drew taut in cool anger. He threw the cigarette away, emptied his mind, and came to the door, which was exactly as it had been before, half open and stuck.

When he had knocked once, and no one answered, he felt like retreating, so he knocked very loudly the second time, and the woman materialised. There was no other word for it. There she was, with the graceful twist in her dejected body, attached to the edge of the door. Was she expecting the blow? Was there something not so much antagonistic as withdrawn, prepared to endure, in the pathos of her attitude? She knew how to wait, in any case.

'I see you didn't pay,' he said.

She did not answer. She could not have been more than thirty.

'Well, you have got to appear before the court now,' he asserted, and added, with a lighthearted brutality, 'or the police will come and fetch you. Hadn't you the money to pay?' And he looked at her.

'No,' she said, looking back at him.

'So you hadn't the money,' he said, glancing away with the smile of official satire. 'And what are you going to do now—go to prison?'

The children were poking their heads round her skirts again. Their fragility appeared extreme, possibly because they were unwashed. Obviously they were famished.

She did not answer.

'Look here,' he said, 'this is no business of mine.' He took out his pocket-book. His hands shook as he extracted a pound note. 'Here's something for you. That'll pay for everything. The only thing I want you *never* to do is to mention that I gave it to you. Do you understand?'

She could not answer for looking at the pound note. If he had been afraid of a rush of gratitude he might have saved himself the worry. She took it stupidly and glanced at him as if there might be a trick in all this somewhere. Then he saw a stirring in her eyes, a woman's divination of character, a slow welling of understanding in the black deeps. It was pathetic.

'That's all right,' he said, and turned away as if she had thanked him.

When he got back into his car he felt better. That was all over, anyway. She was just stupid, a weak, stupid woman who had got trodden down. Tough luck on her. But she certainly wouldn't give him away. Perhaps he ought to have emphasised that part of it more? By God, I would never live it down in the office! Never! He began to laugh as he bowled along. He felt he could trust her. She was not the sort to give anything away. Too frightened. Experience had taught her how to hold her tongue before the all-important males of the world—not to mention the all-important females! She knew the old conspiracy all right and then some! His mirth increased. That he had felt he could not afford the pound—a pound is a pound, by heavens!—added now to the fun of the whole affair.

He did not go to court. After all, he might feel embarrassed; and the silly woman might, if she saw him, turn to him or depend on him or something. Moreover, he did not know how these affairs were conducted. So far it had not been his business. Besides, he disliked the whole idea of court proceedings. Time enough for that when he *had* to turn up.

Before lunch the chief came into his room for some papers. The young man repressed his excitement, for he had been wondering how the case had gone, having, only a few minutes before, remembered the possibility of court expenses. He could not bring himself to ask the result, but the chief, as he was going out, paused and said: 'That woman from Ivy Cottage, the dog case, she was convicted.'

'Oh. I'm glad you got the case through.'

'Yes. A silly woman. The bench was very considerate in the circumstances. Didn't put on any extra fine. No expenses. Take out the licence and pay the seven-and-six compromise fine—or five days. She was asked if she could pay. She said no. So they gave her time to pay.'

The young man regarded the point of his pen. 'So she's off again,' he said, with official humour.

'No. She elected to go to prison. She put the bench rather in a difficulty, but she was obdurate.'

'You mean—they've put her in prison?'

'Presumably. There was no other course at the time.'

'But the children—what'll happen to them?'

'No doubt the police will give the facts to the Inspector of Poor. It's up to the local authorities now. We wash our hands of it. If people will keep dogs they must know what to expect!' He smiled drily and withdrew.

The young man sat back in his chair and licked his dry lips. She had cheated him. She had . . . she preferred . . . let him think. Clearly the pound mattered more than the five days. His money she would have left with the eldest girl, or some of it, with instructions what to buy and how to feed the children. She would have said to the eldest girl, 'I'll be away for a few days, but don't worry, I'll be back. And meantime . . .'

But no, she would tell the eldest everything, by the pressure of instinct, of reality. That would bring the eldest into it; make her feel responsible for the young ones. And food . . . food . . . the overriding avid interest in food. Food—it was everything. The picture formed in his mind of the mother taking leave of her children.

It was pretty hellish, really. By God, he thought, we're as hard as nails. He threw his pen down, shoved his chair back, and strolled to the window.

The people passed on the pavement, each for himself or herself, upright, straight as nails, straight as spikes.

He turned from them, looked at his watch, feeling weary and gloomy, and decided he might as well go home for lunch, though it was not yet ten minutes to one. Automatically taking the white towel from its nail on the far side of the cupboard, he went out to wash his hands.

EDWARD GAITENS

1897–1966

DANCE OF THE APPRENTICES

'Workers Of The World Unite! Ye Have Nothing To Lose But Your Chains! Ye Have A World To Win!' Eddy Macdonnel read the world-famous slogan of Karl Marx, blazed in big white letters on the long red streamer hung across the drop-scene of the Rivoli Music Hall. Long ago he had learned it by heart from Socialist pamphlets and he had read the scarlet streamer many times. Here among crowded humanity it urged him like a battle-cry. His heart flared with enthusiasm and sentimental tears glimmered in his eyes, which he kept intently forward lest the two companions he was crushed between should notice his emotion.

But they were unaware of him. On his right, a stoutish raw-boned youth of nineteen puffed a clay pipe, emitting bitter fumes, while he concentrated on a column of propaganda facts and figures under the heading 'Grapeshot' in the *Socialist Banner*; to his left a rosy-cheeked young man of twenty, short, stocky and of clerkish aspect, perused with a conscious intellectual frown, intensified by pince-nez, the arguments of *Man and Superman* in a heavy volume of Shavian plays and prefaces. Eddy Macdonnel held a *Labour Leader* and two new pamphlets he had bought from the bookstall at the entrance, but his elation at the feel of life surpassed his desire to read; the storm of voices, the cries of paper-sellers at the exits and down the sides of the gallery, stirred him like the beat of the sea.

The occasion was one of a winter series of ILP Sunday evening lectures which offered an open platform to all exponents of progressive political thought. Tom Mann was billed to speak and the three youths congratulated themselves on getting such a good seat. They had anticipated this moment with feverish impatience, for they all knew something of the speaker's dynamic history, of his leadership of a great strike that brought him imprisonment, then his triumphant organization of the dock labourers into a powerful union.

The comfortable little Rivoli was packed to capacity, though it wanted a good half-hour before the meeting opened and late comers were shouldering for standing room in the gallery promenade. Into the electric blaze from an enormous, ungainly chandelier, that revealed a roof crowded with flying cupids chasing naked nymphs, tobacco smoke coiled a thick, blue haze through which the sound of voices mounted like the drone of insects. A few sober handclaps greeted the arrival of the deputy chairman and the usual thirty minutes' musical prelude began. It was announced that a Miss Gunn would beguile the interval with the Celtic songs of Kennedy Fraser, and a small, obese soprano toddled on to the stage, posed, beaming, by the piano and sang in a sweet, spirited Scotch, 'The Road to the Isles'.

Young Macdonnel was immediately sped to islands misted in purple heather and shimmering in opalescent seas; Jimmy Hamilton ceased scrutinizing 'Facts and Figures' and sucked his pipe less furiously, and Willie Mudge closed *Man and Superman*. After two more songs of a sadder strain the vocalist was followed by a violinist, a sallow young man with long dark hair, who looked like a genius in his velvet coat and Bohemian tie. Then the deputy chairman apologized for the speaker's lateness, explaining that he had travelled from London and had already addressed two meetings in other parts of Scotland. The violinist had received applause enough for a Kreisler, but the handclapping became thunderous as all heads on the stage turned simultaneously to the wings. Tom Mann appeared and Eddy Macdonnel committed one of the proudest boldnesses of his young life by springing to his feet and impetuously singing the first notes of the 'Red Flag'. Tom Mann waved the whole audience to its feet and swung it into the anthem and Eddy's confusion was drowned in the roar of a thousand voices.

But he was as hot as a caldron with pride throughout the speech. He felt he had inspired the whole audience and when Willie Mudge dug him in the ribs, saying: 'That was fine, Eddy! My, ye're coming out!' he visioned himself for a moment as a great working-class leader, as great as Tom Mann! Fancy being praised like that by Willie Mudge, who had written essays—he had actually delivered two at the Study Circle of the Unitarian Church—who could discuss philosophers and the plays of Shaw so easily! He wanted to cap his triumph by asking a question at discussion time. Trembling, he began to formulate one. Yes, he would ask: 'Will the working-class ever be free while it is priest-ridden and doped by Catholicism?'

Already he saw himself standing up, felt the gaze of the audience, and then he remembered when he had risen to ask a question at another meeting, his heart almost suffocating him, and the words had scattered from his head like frightened birds. The speaker had been kind, waited, signed and smiled encouragement, and the audience smiled at his youth and embarrassment. The man beside him laughed: 'Go ahead, laddie! Dinna be frichtit!', but he sat down, in a hell of confusion, and for a week after went about with the shame on him.

Mr Mann signed an end to the applause and got to business without the formality of introduction, punching out his theme, 'Syndicalism and Socialism', with two-fisted belligerence as he strode from end to end of the stage in his big, bluff style, waving his arms and exhaling prodigious energy and confidence. Infected by his vigour the three youths leant excitedly over the gallery rail. 'By juv, he's great!' exclaimed Jimmy Hamilton, using his favourite expletive and forgetting, in his excitement, his affected adult manner. As the propagandist ranged the stage, smashing his thesis point by point with his right fist into his left palm, as though his incendiary ideas were concrete, tangible objects he was shaping there before them all, like a sculptor chiselling marble, Willie Mudge remarked in a tone of generous patronage: 'Aye, he's a rare oarator!', then settled back in a judicious calm lest his appreciation of oratory should be mistaken for agreement with extremist views.

Willie was a 'Constitutional Socialist', who regarded anti-parliamentarians with kindly aloofness, but Jimmy Hamilton and young Macdonnel yearned for 'Blood-red Revolution' and they plunged into Tom Mann's ardour like healthy swimmers into a riotous sea. Eddy Macdonnel studied feverishly the agitator's slightest movement, his tossing leonine head and blazing eyes that flashed into every corner of the auditorium. He loved him and the tall, dark, intelligent chairman, thin and keen, who occasionally smiled sedately or lightly clapped his long, slim hands. He loved all this gathering of Clydeside workers, boilermakers, hefty riveters, pale clerks, railwaymen and miners. They were all intelligent and fearless, and the 'others', the great host of the city's 'wage-slaves', who put their 'joabs' before the 'Cause'; the 'prood' middle-class that preferred property and comfort to ideals—were craven and ignorant.

Eddy's feeling was like a desire to embrace the world. Those 'others', the Capitalists, his parents, were blind. 'They have not seen the Light. They listen, but they do not understand!' he quoted silently, thrilled by a condescending pity for misled human kind. Stuffed, as he was, with sentimental tags from the pamphlets, he believed men would one day march in friendship 'Sunwards'—the Capitalist converted to sharing the 'Fruits of the Earth' which provided 'Plenty for All'—to end for ever the fester of slums and the beastliness of war. 'Burns is right!' he thought. 'It's comin' yet for a' that! Man an' man the warld ower shall brithers be an' a' that!' He would have to be more courageous, he decided, more wholehearted, and he felt ashamed of lying to his mother this morning that he had attended Mass, even inventing a sermon to erase her suspicion, though he had skulked miserably around the streets in the Mass-time hour. He admired the audacity of Jimmy Hamilton, who recently had renounced his religion and openly professed himself an atheist. Yes, he would have to outface even his parents for the 'Cause'!

Beside him Jimmy stirred, eager to shoot his question across that intimidating abyss of the auditorium. He was up. 'Mister Chairman! I should like to ask the Speaker "What is the present attitude of the International Trade Union Movement to war; also, if war broke out, would the International Strike make a decisive weapon for Peace?"' Without a flicker of embarrassment, his clay pipe steadily poised, pronouncing clearly every word, Jimmy delivered his question, then sat down coolly and leaned forward, pipe in mouth, elbow on knee, to consider like a sage adult the reply. Mr Mann replied that despite dangerous war clouds piling up, Trades Unionism persisted in a very parlous state of disunity and roared mightily that the International Strike would be a very decisive weapon indeed! At the close of deafening applause, Jimmy rose easily and said 'Thank you, Mister Speaker!' quite unnecessarily.

Eddy Macdonnel was mystified by Jimmy's sublime self-confidence and despaired of ever attaining it. Nerved by example he resolved to speak and the intention surged hot waves of daring and timidity through him; he became alert and had cast inhibition to the winds when Willie Mudge moved violently and exclaimed 'Damn!' and Eddy turned to see him, extremely flushed and annoyed, groping among their feet for his notebook which he had dropped as he snatched at his pince-nez that slipped from his nose when he stood up. The chairman's 'Any further questions?' sailed up from the stage, but Willie's poise was destroyed, and Eddy, affected by his disaster, lost his nerve. The little green bag for the collection came their way; Willie snappishly passed it along and Eddy nervously dropped in three pennies.

The meeting was over, their opportunities to shine in discussion gone, and they both looked with envy at Jimmy Hamilton's irritating mien of self-satisfaction.

In the street they walked along silently, bracing to a snell November wind. Jimmy Hamilton paused in a dark archway to re-light his pipe and pounded after them with his big steps, shouting the beginning of a conversation. He was met with silence. Willie Mudge was still smarting with frustration and Eddy reverberated with enthusiasm, his mind a gorgeous entanglement of words, music and vague ideas about Syndicalism, Organization, Revolution. He wanted to hear them talk, being too shy to open discussion. He invariably listened, sometimes dropped in a word or two and flushed with pleasure when Willie complimented him, surprised by an original phrase. 'My, that was damn well expressed, Eddy!' he would say. 'Ye oaght tae write something. Ye'll be giving us an essay up at Ross Street next!' But he quailed at the image of himself calmly reading a paper to the 'Men's Study Circle' of the little Unitarian Church in the Calton slums, coolly handling discussion like Jimmy Hamilton, when he delivered his 'Essay on Milton', but he would often vision that accomplishment, enjoying the effort and victory, till some other fancied triumph replaced it or he was exhausted by emotional imagery.

They walked sharply for twenty minutes in profound Scotch quiet, Jimmy Hamilton holding his solid, Roman nose very aloof, his determined jaw invincibly rigid, as he swung along mannishly, imposing the pace from his mood, hands deeply plunged in a rough-napped brown overcoat, long thin lips clenching his pipe, his carriage expressing adult disdain of Willie's huff. They were passing Jail Square, opposite the unimposing façade of the Law Courts and the Public Mortuary, a wide, plaza-like gateway to Glasgow Green, the ancient, smoky riverside park, the city's public rostrum, where political and religious zealots harangued and the Salvation Army recruited souls; where lively racing tipsters offered unlimited wealth on threepenny and sixpenny slips of paper, and quack doctors, for a mere trifle, supplied all-comers with infallible cures for every known disease; where amateur philosophers coiled endlessly throughout the labyrinths of Free Will and Determinism.

Lustily swinging the walking-stick he affected, Willie Mudge ended the twenty minutes' silence, looking straight ahead, with a faint smile, as he said: 'My, it's damn cold! It's no sae warrm as it was in the Rivoli.'

Jimmy Hamilton plucked his pipe from his lips. 'What?' he said, and Willie repeated his remark less gruffly, slightly turning his head.

'No, by juv; it isn't, by juv! It wad be a gran' night tae be hung, by juv!' cried Jimmy, his stony expression demolished by uproarious laughter at his reminder of the Jail Square's ancient function as a place of public execution. 'Ye'd be blawn aff the gallows!' he shouted, stopping to guffaw helplessly while his pals paused and laughed.

'My, ye're a soambre divil, Beefy!' said Willie, poking him with his stick.

Jimmy hated the nickname which had stuck to him since his schooldays, but he was so immoderately gratified with his wit that he took no offence.

Two sparse groups, like hangers-on of Learning, were gathered round disputants in the Square, whose contending voices struck feebly against the wind. They would linger there till the small hours in the bitter cold, drifting away in ones and twos, probably leaving the wranglers still at variance, quoting chapter and verse. Jimmy suggested biding awhile to listen. In this way they frequently enlivened the suicidal dullness of Glasgow Sundays, when Socialist meetings were few or they considered the speakers of those advertised as too 'moderate' for their recognition. But Willie pushed on. 'No, thanks!' he said. 'Philosophy won't keep ye warrm. I want to get home!'

The wind stung their cheeks like arrows as they crossed Albert Bridge. Up the river to their left the two lights of the weir gleamed dim as glow-worms where the weir fall thrashed the darkness. Fascinated, Eddy Macdonnel watched the beams of the parapet lamps thrill the black water, writing romantic signs on swirl and flow. His blue eyes laughed at the tortured shine as they hurried across, happier at his companions' resumption of friendliness, for he lacked the skill to keep the conversation going. In eightpenny woollen gloves his hands tingled fervently, thrust deep in his flimsy overcoat pockets and the clink of his ironshod heels pealed through his body. Cars and trams crossing the bridge flashed cheer at him, and as he looked happily heavenward he fancied that the keen stars of this brilliant November night glowed like a jewelled crown around his head and sang like a choir in his heart: 'Life is glorious! Glorious at seventeen!'

They turned left into Clyde Walk and crossed diagonally right into Lily Street, a long, monotonously straight thoroughfare, fouled at this end on one side with slums only fit for demolition, on the other, harsh with the stony silence of a school playground and the blind gables of factories. Eddy listened to his friend's arguing: 'Well, anyway, Tom Mann's nut afraid tae speak out!' Jimmy was saying. 'It's a pity we haven't mair like him! The Labour Movement wants more red blood in its veins. We've too many namby-pamby reformers crawlin' like snails tae the Millennyum!'

'Syndicalism presupposes violence; violence is anti-social,' said Willie Mudge, quietly authoritative. 'Revolution creates Revolution *and* Revolution! Consistent constitutional progress will eliminate Capitalism without bloodshed.'

Jimmy lifted his head, laughed pityingly and blew out a long, audacious spire of smoke. 'A foolish consistency is the hobgoblin o' little minds!' he sneered, quoting Ralph Waldo Emerson without acknowledgement. 'Capitalism bears within itself the seeds o' its own destruction,' he added with dogmatic finality, passing off the dictum of Marx as his own original phrase.

'But, my dear chap!' exclaimed Willie, in a tone that implied charitable long-suffering with faulty reasoning, 'if capitalism is self-destructive, whence the necessity for revolution?' He clutched the rim of his bowler as they turned a corner against the wind and coughed a little cough replete with dialectic satisfaction.

'Revolution is the historic mission o' the Workers!' retorted Jimmy, primed with a Marxian slogan for every turn. He flicked a dead match from his fingers; it flashed a white parabola in the darkness and vanished in a gutter stank. 'Damn guid shot that, eh? Did ye see it? Right across the pavement and doon the stank, by juv!' he cried, delighted more with his dexterity than with his skill in debate.

Gowan Street surrounded them, a dreary main street which changes its name every other mile and runs from east to west of the town. Great blocks of tenements loured with their dark maws of entries and hundreds of windows dimly aglow with gaslight, seeming to crush the life from the low-browed shops beneath, all shut and blinded, except for the rare gleam of an Italian ice-cream saloon.

'We swear by the beard of the Prophet, Karl Marx!' sneered Willie, with a theatrical gesture at his chin.

'There is nae Goad but Compromise an' Ramsay MacDonald is his prophet!' counter-sneered Jimmy.

Tramcars swayed past them, galleons of light and humanity, their steps echoed by the wide gateway of a timber yard, gorged with deals of sweet-smelling pine, adjacent to a fur factory that exhaled the odour of dead rabbit. 'That beastly place always stinks enough tae knock ye doon!' exclaimed Willie with a refined shudder, apparently deprived of the power to argue by the loathsome smell.

'Ah don't smell anything,' said Jimmy blithely.

'I'm not surprised—wi' that filthy pipe o' yours!' gibed Willie, as they crossed the road and walked alongside Saint Peter's School, where young Macdonnel had spent his boyhood. A square, four-storied building of red sandstone, blackened by industrial mirk, it reared huge at the farther side of its playground, open to the road; the whole lower floor was the parish chapel, the three upper floors serving as school. Unaware of its ugliness, concerned only whether he should bow his head and lift his cap like a good Catholic in respect to the chapel, he struggled between the influences of religion and his new beliefs, and in guilty anxiety lost his companions' argument, heard only their voices. He knew they would have respected the gesture, but armed with the teaching of the pamphlets he successfully resisted making it. He felt it would be a cowardly submission to superstition which, according to the pamphlets and books, was the 'dope of the medicine-man, the Priest, in league with the Capitalist to keep the Working Class in ignorance and subjection'. He was relieved when they got beyond the chapel-house, which stood in the left corner of the playground, narrowly railed from the pavement, as if priestly eyes were spying on him. As they tramped under the shadow of a hideous cabinet-making factory he breathed more easily and thrilled anew to the contention of his friends. 'Compromise,' Willie was saying, 'is a social necessity. Without it there could be no progress. Society and Compromise are synonymous.'

Jimmy Hamilton plucked his pipe from his mouth in manly disgust. 'That's the atteetude of your ILP, the skulking-place o' lily-livered Democrats, all theory an' no guts! We want more action! Action is the very life-blood o' the Movement!'

'What about the Vote?' rejoined Willie.

Jimmy puffed twice prodigiously, his smoke ascending like amazement. 'The Vote!' he cried. 'Well, what about the Vote?'

'Used intelligently it will eventually emancipate the proletariat,' coolly advanced Willie.

'Evenchilly! Evenchilly!' sarcastically returned Jimmy. 'An' when wud that be? Has the worker ever used it intelligently? Doesn't the Capitalist Press instruct him how tae use it? The Vote is the red-herrin' drawn across the Worker's path b' the Powers That Be tae seeduce him from his Historic Mission!'

At the corner of Commerce Road, another turning of slums and small factories stretching from the river and ending hereabout in an evil, narrow lane, Jimmy, who affected to be bored with the argument, suggested, 'Shall we call in at the Tallie's?' and they wheeled left into a parallel main road of tenements, tramlines and shops and entered a small ice-cream saloon, their chosen place of retirement from wandering round the blocks. A highly-coloured sign entitled the place 'THE THISTLE SALOON', and on the lintel of the door minute letters said: 'Joe Boganny. Prop.' In the narrow window, curtained at the back, a few bottles of wrapped and bare sweets reposed among several dummy chocolate-boxes covered with ribbons and impossibly charming ladies. A sturdy Italian of forty, with ornate moustaches beautifully curled, bright floral waistcoat, gold albert and fob, was leaning on the counter and gazing stolidly into the street as the youths entered. He greeted them in broad Glaswegian: 'Hullo, boays. Hoos things? Ah hivnae seen ye fur a week.'

'How do, Joe?' said Willie, brightly, 'ye're lookin' proasperous. Taking all the money? How's business?'

'Och, no sae bad!' said the Italian. 'No sae bad! It's the weather, ye ken. There's no sae much ice-cream takkin' this cauld weather.'

'My, that's a rare fine waiskit ye've goat, Joe,' said Jimmy. He saw Joe in the garment every Sunday evening, but he enjoyed the fun of titillating his memory, sometimes being rewarded with a long, enthusiastic description of his life in Naples. Joe stuck out his chest, proudly exhibiting the waistcoat, which was thick and furry, with a design like a Brussels carpet. 'It's twenty year auld that waiskit an' as guid as the day it was boaght,' he said. 'It was ma faither's in Naples. Whit's yer oarder?'

The youths gave their order and sidled into one of the four stalls which furnished the accommodation of the saloon. Each stall had a table of imitation marble and the partition panels were gaudily painted; above their heads Italian battle-pictures crowded the walls, where the central place of honour was given to brilliant lithographs of Italian royalty, draped with the national flags. Joe Boganny looked more Italian than his king and talked broader Scotch than his customers. He called through to the kitchen: 'A Macallum, plate o' hoat peas, plenty bray, an' a hoat raspberry', and immediately the shop echoed with a babel of female voices speaking in rapid Neapolitan and apparently raised in furious indignation. Young Macdonnel, who could never overcome a feeling of strangeness among these saloon-keepers, became simultaneously alarmed and interested, filled with violent visions of flashing knives and bloody struggle, convinced that they were murderously quarrelling. The apparent battle continued for five minutes, then Mrs Boganny, still talking backwards excitedly, sailed into view with a steaming tray and smiling, as if she had joyfully murdered someone, presented the refreshments to her husband, who lifted the counter-flap, came through and served.

Mrs Boganny, a dark little woman in black, with bright scarf round her neck and heavy, brilliant earrings, leant her plentiful breasts on the rounded counter that swung out at the entrance and twinkled her eyes and earrings at the youths.

'Colda nighta!' she said.

'Aye, it's enough tae freeze the ears off a brass monkey!' said Jimmy Hamilton, as he took an enormous bite from his 'Macallum'—two circular sponges sandwiching a colossal amount of ice-cream and a bar of Fry's cream chocolate.

Willie Mudge winced. 'My Goad, I don't know how ye can eat that stuff in this cauld!' he said.

'Ye want good teeth for the job,' said Jimmy boastfully, taking another prodigious bite. 'Look!' He gaped his long mouth and showed his large, strong teeth, tainted with nicotine. 'Have ye ever tried bitin' through a lump o' ice?'

Willie Mudge took a long drink of hot raspberry to rid himself of the chilly suggestion. 'Jimmy, ye're disgusting!' he said emphatically.

'This pea bray's champion!' exclaimed Eddy Macdonnel, spooning up the green peas and scalding gravy on which he had liberally showered salt, pepper and vinegar. 'I say!' he said, 'I hear Tom Mann's speaking away out at Partick next Sunday morning but one. I'd like to hear him again,' and he looked hopefully at his friends.

'Good! We'll go!' said Jimmy, then he stood at the counter and called through to the kitchen where the two Bogannys had retired for warmth, 'a Tallie's Blood, Joe, an' a small Woodbine!' When his order was brought he presented the cigarettes to Eddy and attacked with unvitiated appetite the 'Tallie's Blood', a glass of lurid ice-cream, scarlet with raspberry essence.

Willie Mudge toyed with his hot cordial, coughed, flushed and began fingering his pince-nez, his usual sign of embarrassment or grave delivery. 'Hem!' he exclaimed. 'I'll be giving my treetiz on "Individualism" to the Men's Study Circle on that morning. I thought you chaps would like to hear it,' and he looked directly at Eddy, his pince-nez seeming to contract and jerk violently up his nose as if in sympathy with his acute frown of mortification. He had attended the reading of Jimmy's 'Essay on Milton', and he expected a return of the courtesy. He coughed pettishly again, removed his glasses, polished them agitatedly with a small square of chamois leather and frowned up at the King of Italy.

Eddy reddened with confusion. He regarded Willie as far superior to himself and Jimmy in intellect and culture. Jimmy was a glazier's apprentice; he himself was a plater's apprentice in the shipyards; while Willie was a bookkeeper in a big city office, went to business every morning togged up in his best and was studying to become a chartered accountant, and his pince-nez gave him an air of refined authority which intimidated Eddy. He was hoping that Willie wouldn't press him to attend his reading because he wouldn't have the courage to refuse and the prospect of hearing Tom Mann tugged at his heart. He loved the great mass meetings vibrating with human excitement and was always painfully shy at the intimate Study Circle.

Jimmy let his spoon rattle into his finished glass and remarked in Willie's averted ear: 'Did ye speak?'

'Och, it doesn't matter,' said Willie, still glaring at Italy's king. 'I was just mentioning my treetiz.'

'Aw, that thing o' yours on Indiveedulism,' said Jimmy. 'Sure ye could read it tae us next Saturday. I wouldn't miss hearin' Tom Mann for you or anybody.'

'It's a very difficult theeziz an' I'm putting my best into it,' Willie said impressively. 'It's sure tae raise treemendous discussion. Will you be at the Circle, Eddy?' he asked, and Eddy, amazed at himself, found the courage to say, 'I think I'd rather hear Tom Mann,' saying it almost naturally because he had become aware for the first time that Willie, who had not resumed his glasses, was less impressive without them. About his gaze lay that shade of tiredness that dims the eyes of most short-sighted people, and suddenly this devitalized look made Eddy feel regret at having refused him. Joe Boganny reappeared with the warmth of the kitchen clinging to him and dissipated their confusion by lolling lazily on the counter and commencing a discussion on Scottish League Football which they all entered with enthusiasm, forgetting their feelings of a moment before.

During the following fortnight Eddy and Jimmy saw little of Willie, who was giving every spare moment to the creation of his treatise, but on the Saturday afternoon they met by chance and decided to call at Willie's home and find if his supreme literary effort was finished.

'I was just speaking' to Jeannie Gordon,' said Eddy, as they walked down Calder Street. 'She was lookin' lovely the day.'

'Och, Ah spoke tae her masel,' confessed Jimmy, whose magnificence had evidently failed to startle the young lady. 'Wimmin are fast becomin' ma beet noyer,' he said, proud of his piddle of French and his nodding acquaintance with Nietzsche. 'It's nae wunner that some o' the world's greatest men hiv been missohgynists. Wimmin hiv nae souls; their only purpose in creation is tae breed the race an' when they've done that they're aboot as interestin' as stripped bean-puds. There's nut wan o' thim is fit mate for a superman. They don't know the meanin' o' the words, "Freedom" an' "Independence".'

Eddy listened silently, studying Jimmy's supermanly expression, envying him his arrogant way with girls. Women were a mystery to Eddy. He never could find the things to say that flattered or amused them, though he was always dreaming of himself in triumphant situations with maidens of fancy and the few girls he knew, situations in which he delighted them with brilliant conversation, holding them spellbound by his masterful personality; but when he was actually with them he stood wordless, burning with dumb desire.

As they crossed the well-burnished tramway-lines of Gowan Road and turned into Mathieson Street that led down to the Clyde, back-street life surged and shouted round them. It was unnaturally mild and close for a November day, with blue-black heavy clouds creeping low above the high tenements. Two distinct games of football were in progress in this last portion of the street, with about forty rough lads and youths wildly shouting as they kicked two huge footballs about improvised of paper and string.

'This is life with the lid off,' remarked Eddy as they crossed the street, threading among the rushing footballers.

'It's the Mob at play, the slaves o' ceevilization seethin' on the dunghill o' Capitalism,' said Jimmy, relishing his phrases, which he believed he had just invented, utterly forgetting that he had collected them from the lips of a soap-box orator.

They regarded the 'mob' with tragic superiority, then suddenly they both laughed heartily as a little bow-legged goalkeeper in a desperate dash at the ball tripped headlong over one of the goal posts, which were composed of heaped jackets, overcoats and caps.

'It must be hard for wee Rabbie tae save the ba' wi' they bandy legs o' his,' said Jimmy. 'Ye could run a tramway through them!'

They turned left into Clyde Walk, entered one of the closes that were cleaner and quieter than those in Gowan Road and Calder Street, and, mounting to the third story, knocked at one of the three doors, which was opened by Mrs Mudge, an extremely small, slight Scotchwoman of the size locally referred to as a 'nice wee body'. She led them into the infinitesimal square entrance hall and called through to Willie, who could be heard rustling papers in his bedroom, that his friends had arrived.

Young Macdonnel was hoping that the six Mudge sisters were out, because he always perspired with embarrassment in their presence. They were all studious, all prize-winners at school and disconcertingly intelligent. But the whole sextette was at home as he dragged in, hiding himself behind Jimmy. Janet, the eldest, a fat young woman of thirty, was washing her hair at the sink and she whipped a towel from a nail and pinned it round her bare shoulders, but not before they saw her robust breasts, urgent against her stays. She greeted them with a bright, flushed smile, and began drying her brown hair, her big, confident eyes, lively and intelligent, considering them. Charmed by that glimpse of nakedness, Eddy, glancing at her russet, homely complexion, thought her quite pretty; then his old, agonizing confusion in this female kitchen overcame him while Jimmy was quite coolly putting his pipe away.

'Och, ye can smoke awa, Jimmy,' said Mrs Mudge. 'We're used tae smoke here. Sure Mister Mudge smokes like a lum!' and she offered to bring them chairs, but Jimmy said they wouldn't be staying.

At the window sat Flora Mudge, a dark young woman of twenty-four, knitting and reading a novel; two sisters, respectively sixteen and seventeen, sat reading at opposite sides of the fireplace; the two youngest sisters monopolized both ends of the kitchen table, their childish heads bent over homework, with copybooks and volumes opened before them. The six sisters were bespectacled and all were bantam size, as though destiny had strictly forbidden them to overreach their mother. Suddenly the five who were reading raised their heads and smiled at the visitors, and the gaslight glancing on their glasses gave them a queer, owlish look, then they immediately applied their eyes to their books again.

Jimmy and Janet were political opponents, and Jimmy asked her if her essay on 'Feminism' which had been set as a subject for a literary competition by the *New Statesman* had been accepted. Janet answered with proud pleasure that it was being published as a prize winner and they began a lively argument on 'Feminism'. Janet dried her hair and was easily triumphing in the argument when Willie entered and saved his friend from shameful defeat.

Willie's hair, which sprouted like lawn grass, was tousled and implied that he had just emerged from a terrific struggle with thought. His mother asked him if he had finished his essay. 'Yer eyes look gey tire't,' she said. 'Ye'd best gie them a wee bit rest.'

Willie's stocky figure never at any time suggested intellectual asceticism, but he loved to act the man fatigued by thought. He put his shirt-sleeved arm round his mother's shoulder and uttered with fond condescension: 'Ay, I've just finished the pairoration. Puir wee mither, of the making of books there's nae end, an' much study is a weariness o' the flesh. We Thinkers plum' the very depths o' misery an' scale the heights o' joy!' He sighed heavily, passing his hand across his brow, removed his pince-nez, dangled them for a moment, as though plunged in unfathomable pre-occupation, slowly put them in a case, slowly inserted that in his waistcoat pocket and shook his head, muttering audibly: 'Ay Thoaght! It leads ye oan—tae what?' then, staring at space with a Faustlike stare, he recited in a tragic voice:

> Introspection's cancer,
> Baffling surgeon's knife,
> Hoping, groping, blinking, thinking
> Some men call it 'Life'!

Five sisters lifted heads to look at him with simultaneous pride, and Janet, while she wrapped a towel turban-wise round her damp hair, smiled with maternal possessiveness at the back of his head.

'Yer hair's toozl't, son,' said his mother. 'Ye'll hae tae brush it doon afore ye gang oot.'

'He's sicklied ower wi' the pale cast o' thoaght,' remarked Jimmy Hamilton impressively.

'What?' asked the little woman, who was faintly deaf.

'It's fae *Hamlit*, ye ken,' explained Jimmy, repeating the quotation.

Mrs Mudge nodded her head. '*Hamlit*', she echoed musingly, as if it was the name of an optical disease, 'aye, *Hamlit*. Willie reads too much. It's no' guid for his een, d'ye think it is?' she inquired, peering anxiously at the youths.

'Och, there's nothing wrong wi' ma hair nor ma een, mother,' Willie exclaimed irritably. He was very proud of his hair's literary disorder and her suggestion that it was untidy always irritated him. In a state of great annoyance he walked into his bedroom to get his jacket, his head slightly inclined to the right, a posture that, beginning in affectation, had become a habit.

He came out again, still looking intellectually weary and pre-occupied.

'Have ye goat yer treetiz?' asked Jimmy.

Willie struck a modest air. 'Och, I'll not bring it the night,' he said. 'I don't feel I'd do it justice. Composin' the pairoration has left me exhausted.'

His pals proffered the expected encouragement and he threw off the mask of modesty and produced his essay. All its power and glory were concealed between the glossy blue covers of a twopenny copybook which he stuffed into his pocket as they went downstairs. Suddenly, beaming innocent enthusiasm as they turned into the street, he said: 'My, they damn pairorations always give me trouble, but, by jings, I think this is the finest I've ever done! Y'know, I sat starin' at the blank page till I thought my reason had gone. It was like "Waitin' for the star from Heavin' tae fall"—now, who was it said that?' he mused, momentarily delaying his account—'then it just rushed oan me like a sudden stoarm and words poured through me like a toarrent. By jings, inspiration's a wonderful thing!'

Willie mentally patted Eddy on the shoulder, smugly gratified by his look of awe while thinking that Jimmy was just a bit on the coarse side and incapable of appreciating his finer flights. 'Aye, there's something in young Macdonnel, though he's so quiet,' he thought, deciding to present him with a volume of 'Keats', judging, after 'mature reflection' that 'Coleridge' would be too intellectual for him.

'Och, Ah never find essays difficult, once Ah get past the start,' said Jimmy nonchalantly. 'It's the first sentence that jiggers me; once Ah've done that Ah carry oan like a hoose oan fire. Noo, Ah remember when Ah wrote ma study o' Milton; Ah began it at twelve o'cloack at night efter trying fur three hoors tae get the op'nin' words an' Ah feenished it at three o'cloack in the mornin' an' was up at six as fresh as a daisy,' he concluded, taking three deep, gratified puffs at his pipe and drawing himself up to his full stature while Willie frowned, nettled by his cock-sureness.

The three youths went through the gaslit streets for an hour's walk over the town saying little, for Willie was conning over his essay, finding it flawless in composition and devoid of fallacies, while Jimmy was collecting all his well-used arguments, quotations and slogans to riddle it with criticism, convinced that Willie wouldn't produce anything original, and Eddy darted ardent futile glances at all the prettiest girls in the Saturday night crowds of workers parading brilliantly-lit Argyle Street, wondering if they were smitten with his amazing wealth of hair. So they meandered round to the close in Calder Street where they always foregathered and where Jeannie Gordon lived. Jimmy Hamilton looked eagerly ahead as they approached, hoping to see her there and was disappointed at beholding only Paddy Maguire and Bobby Logan, their unintellectual pals, who made no pretensions to politics, poetry or culture, but who always listened respectfully when the other three debated, exclaiming in amusement at immensely long words, with which their arguments were invariably well sprinkled.

Bobby Logan, a youth of nineteen, with thick pasty features and silvery light hair, almost albino, was a counter-hand in the local Co-operative Stores. The one obsessing ambition of his life was to become its manager, but he reckoned he would have to wait about twenty years for that choice promotion. Happily he was a patient youth and did not goad Time on but dutifully served the Co-operative ladies of the South Side with viands and dividends, knowing well that the grave would open up and swallow a manager or two and that he would one day stand forth resplendent in the dead one's shoes as sole director of the Butchery, the Bakery and Grocery, the Dairy and Provisions.

Paddy Maguire was a dark, sturdy young riveter of twenty, Glasgow born of an Irish farmer. His father's dairy, which conjured a living from a portion of Calder Street, competing with the other nine dairies in the thoroughfare, and where rancid cheese, sour milk, cats and fly-paper caused an outstanding smell, stood opposite. Paddy worked in the shipyards and helped to eke out the shop's income which was sadly depleted by defaulting credit customers.

The five friends met with cheerful greetings and stood about humming snatches of ragtime songs, then unlovely Glasgow rain began to fall, loaded with smoke and grime, and they drew further into the close where Willie Mudge, acutely longing to be asked to deliver the bundle of great thoughts that were bulging out his pocket, assented eagerly to Jimmy's request that he should read his essay.

'"Essay?" What's that?' asked Paddy Maguire.

'Och, ye'll hear what it is in a minnit,' said Jimmy. 'Come intae the foot o' the stairs an' listen,' and they all grouped beneath the broken entry lamp, where Willie produced the glossy, blue exercise-book, folded back the covers with trembling fingers and revealed the word 'INDIVIDUALISM' written in large block letters at the top of the first page.

'It's no' very long,' he said modestly.

'Och, it disnae matter,' said Jimmy. 'We've goat all night.'

Willie coughed and uttered 'Individualism' in a small, nervous voice and at that moment six children rushed downstairs screaming in play, skeltered back into the close, alarmed by the rain, and gathered curious round the youths, till Jimmy drove them away, when they ran and sat staring, half-way up the stairs.

'Damn they kids!' exclaimed Willie, proceeding with his introduction. His voice was acquiring power as he turned the third page, when a happy drunkard rolled into the close, shaking off rain like a dog and stopping to leer fatuously at them all.

'My, it's wat enough tae droon the nine lives o' a cat!' he solemnly assured them, bowing and raising a battered, mouldy bowler, then he staggered upstairs singing, 'Ah luv a lassie, A bonny Heilan' lassie!' and the children ran before him amused and afraid, while the essayist sighed, frowned and resumed and read for ten minutes uninterrupted. Paddy Maguire and Bobby Logan were becoming restive, eyeing each other with low-browed sympathy; Eddy Macdonnel listened engrossed, and Jimmy Hamilton folded his arms and puffed his pipe, looking very judicial. Willie's 'treetiz' was concerned with the relations of the individual to the

State, and it threatened every kind of State, Communist, Socialist, Conservative or Ecclesiastical, with dire consequences if it dared to make unreasonable inroads on personal liberty, and Jimmy condemned its logic with god-like scorn, for it was evident that Willie was refusing the State any right to live at all.

Jimmy became utterly aghast at the essay's wild theories and was about to interrupt the entranced reader when a stout hawker with a vast packsheet bundle of cast-off garments, her day's collection from middle-class houses, waddled into the close, her burden hanging from her shoulders down her back and swinging from side to side. The listeners made way for her, but Willie, transported by the glory of his composition, read on unaware, and the hawker's bundle grazed the copybook and swept it from his hand, while she puffed on upstairs, unconscious of her interference with poetry and wisdom.

Paddy and Bobby laughed, but Eddy looked with sincere concern at Willie, thus rudely wakened from his self-imposed hypnotism. He snatched, startled, at his pince-nez, and as Jimmy handed him the copybook he stuffed it indignantly into his pocket, walked rapidly to the closemouth and stood quivering with wrath, feeling that he had suffered the final indignity and toying with the suspicion that Jimmy had listened as indifferently as Paddy and Bobby, despite his appearance of intense attention.

Jimmy strode after him, loudly asking him if he wasn't going to finish the essay, for he didn't want to lose this glorious opportunity to lam into Willie's conglomeration of fallacious ideas. Willie glowered, silent, then exclaimed: 'Ach, it's rotten. It's not worth reading!' and obstinately refused to continue till Eddy, whose mind had flown along elated in the procession of words, happily indifferent to association of ideas, expressed his genuine appreciation of Willie's prose, and Willie, flattered by his praise and incited by a desire to rise on the wings of his 'pairoration', consented to read the remaining pages.

They all returned to their stand under the lamp; Paddy and Bobby hovered near, out of politeness, though they were pale and feverish with boredom; Jimmy resumed his sage attitude; Eddy, his look of eager attention. Willie began reading, at first with diminished ardour, but as he reached the concluding page his zeal was in full blaze and he threw his whole body into the utterance of the last phrase, the crest of his 'pairoration'; 'Individualism!' he almost shouted, 'we await Thy Glorious Dawn!' Tears glinted against his glasses as he stood for a moment with bent head and arms hung, like an exhausted orator.

'Ha, that was damn guid!' cried Paddy Maguire with terrific relief.

'My, that was great, Willie! Great!' cried Eddy, aglow with enjoyment, trembling with Socialistic emotion and wanting to shake 'comrade' Willie by the hand, while Bobby Logan smiled his Punch-like smile, his long chin tilting up to meet his long curved nose.

As if he was forcing himself with fearful effort back to reality, Willie shrugged his shoulders dramatically, pocketed the exercise book, faced

Jimmy with a challenging look and said quietly, apparently indifferent to Jimmy's or the world's opinion:

'Well, what did ye think of it?'

Jimmy slowly gestured with his pipe. 'Ay, it was good,' he admitted in a hesitating, reflective voice. 'It didnae lose interest onywhere. Ah think it was an improvement oan yer treetiz oan Free Wull an' Determinism.' He suspected that the bulk of the essay had been contributed by Schopenhauer, Emerson, Thomas Carlyle and Bernard Shaw, though he had to allow that the plagiarisms had been well organized. 'Mind ye,' he hastened to add, 'Ah don't agree wi' the theeziz in wan particular!' and he shook his pipe at Willie like an elderly philosopher.

'I didn't expect ye would!' said Willie, resenting his knowing manner and thinking, 'Who is he tae pass literary judgments?'

Jimmy then made a scornful reference to one of the essay's major points and they began an argument on 'The Individual and the State', with Willie vigorously attacking, heartened by his literary success. Eddy Macdonnel lit a Woodbine and stood close to them with a broad, amiable smile, drinking in their erudition; Paddy Maguire and Bobby Logan strayed to the middle of the close and commenced singing 'Alexander's Ragtime Band' in execrable harmony, Paddy contributing a tuneless base, Bobby's Punch-like features moving up and down in unison with his nasal tenor.

At that moment Jeannie Gordon danced down the stairs. She was carrying a very gay, blue umbrella for the time of the year and beneath it her neat blonde hair gleamed tantalizingly; she wore a shimmery rubber rainproof of an emerald green and held in her hand a highly-coloured quart jug, as she was going to fetch her parents a quart of bedtime stout from the 'Jug and Bottle' of The Shamrock Bar at the end of the street. She reached the foot of the stairs, gave Jimmy an injurious look, threw to Eddy an inviting smile—in the light of which his face assumed the hue of a boiled carrot—and responded with a small, lyrical laugh to a naughty remark of Paddy Maguire's, then her tiny, seductive steps bore her into the pouring rain.

Thunderstruck, Jimmy realized that loveliness had passed by, and, thrusting his glowing pipe in his pocket, he shouted: 'Hi! Jeannie! What the hell are ye scootin' away like that for? Hi, wait a minute. Ah want tae talk tae ye!' and dashed furiously into the downpour like a hot satyr pursuing a desirable nymph.

'Well! The ignorant bugger!' cried Willie, swallowing the torrent of eloquence that was gushing from his lips. 'What bloody manners!' and he looked at Eddy with indignation inflaming his pince-nez.

'It wasn't exactly polite,' agreed Eddy, not wishing to be hard on a pal who shared his political outlook.

'Polite!' cried Willie. 'It was atavistic. It was uncouth!' Then, giving a theatrical shudder, he made a gesture that entirely discarded Jimmy. 'Och, what else could ye expect from a Materialist an' a Communist?' he said, and went on talking about higher things, his young voice mincing in the tones of all Scotchmen who constrict their native vocal breadth to imitate upper-class English. Across the street an enormous natural voice was booming: 'Hi! Paddy! Will ye come an' look afthur the shop?'—a command that Paddy Maguire immediately obeyed by racing across the street to deputize for his father, who stood at his shop door, a tall, massive, black-bearded man, stooping and peering under a streaming umbrella, impatient to be going to have his evening dram in the 'Snug' of the Rob Roy Arms.

Bobby Logan was still singing, crooning unctuously:

> You made me love you,
> I didn't want to do it. I didn't want to do it!
> You made me happy, sometimes
> You made me sad,
> But there were times, dear, you made me simply mad!
> I want some love that's true,
> Yes I do. 'Deed I do! You know I do!
> So give me, give me, what I sigh for,
> You know you've got the sort of kisses that I'd die for!
> You know you made me love you!

Eddy Macdonnel was suddenly overcome by a great weariness with learning and culture and Willie's voice. A deep longing for love and ragtime troubled his heart. He wanted to be standing also at the closemouth, shuffling his feet to syncopation and humming 'Itchy Koo' and 'Casey Jones', and he glanced impatiently from Willie to the end of the close that held the echoes of song and debate and the drum of relentless rain, while the broken mantle of the lamp overhead shed ghastly light.

Eric Linklater

1899–1974

Country-Born

Sometimes when it was growing late in the Club bar, little O'Driscoll's voice would be heard in the richness of its native brogue. At ordinary times he spoke like ordinary mortals who have been educated in the ordinary English way, and during business hours—he was in the Bombay office of the West of India Bank—his sentences were sedulously clipped and his intonation was as smooth as a pebble. But with one peg over the statutory eight a rhythm would creep into his voice and a luxuriance into his syntax such as are familiar to students of the Irish theatre. There was almost certainly a literary quality in the accent which fell on him like a flame (will-o'-the-wisp or pentecostal), at that hour when well-behaved people have dressed for dinner, and the others are coming to the conclusion that after another short one a steak at Green's would suit them better than anything else. And yet with a name like Jerry O'Driscoll can anyone blame a man for encouraging the gentle growth of an Irish accent? It lent point to his stories, apparent subtlety to his repartee, and a warmth to all his jokes.

O'Driscoll played inside-right for the Gymkhana hockey eleven with the characteristic dash and brilliance of an Irish forward, and on the day they defeated the Poona team in the final of the Presidency Cup Tournament it was his goal, cleverly taken from a short corner, that won the match five minutes from time.

There was a dinner afterwards—a hilarious dinner—and O'Driscoll sang 'The Wearing o' the Green' and 'The Old Side-Car', and then—by request, as restaurant orchestras complacently put it—he told his famous and shameless story of the fabulous Duchess of Kilkenny and the mythical Major O'Gorman. 'Sure, and is it a man ye are at all, Major O'Gorman?' the Duchess asks at a critical stage of the history; and there was the inherited wealth of all histrionic disdain in O'Driscoll's voice as he put the question to his other hand. And when the story had reached its appointed end a roar of laughter swelled and spread through the room, and shook the echoing glass before it died away to chuckling ripples of mirth and the comparative silence of general talk.

But India is an unchancy land where you may drink cheerfully with a man one week and write letters of condolence to his widow the next, and on the Tuesday following the dinner, which was on a Saturday, O'Driscoll went to hospital, where the Civil Surgeon removed, evacuated, or did whatever was appropriate to an abscess in his liver.

Jerry took his anaesthetic badly, and he had to be given more chloroform and ether than a doctor cares to give. The operation was successful, but there was anxiety in the hospital after it, and a nurse was continually inside the barrier of tall blue screens that surrounded the little man's bed. He was very weak when the chloroform wore off, and when Manderson, who lived with him and who had been waiting at the hospital most of the morning, tried to see him about midday he was told uneasy things about saline injections and the danger of collapse. But the saline did its work, and later, when Manderson asked him how he felt, he said, 'Not half as bad as I did the morning after the Hockey Dinner, old man', and grinned feebly beneath the ice-pack balanced on his head.

At night Jerry seemed stronger, and in the morning he was cheerful and comparatively comfortable. But the following night his temperature ran up suddenly, his breathing grew heavy and fast, and his face was flushed and puffy. What the doctors feared had happened, and pneumonia had set in.

They telephoned to Manderson at four o'clock in the morning, and when he arrived Jerry was breathing quickly from the mouth of a glass funnel which a length of rubber tubing connected to an oxygen cylinder.

'And what are ye doin' here at this time o' night?' said Jerry excitedly, his eyes bright with fever.

'It's nearly breakfast time, old fellow,' said Manderson pacifically.

'The devil it is,' said Jerry, 'I've never seen such a rotten *bandobast* as this since I came to—since I came to India,' he shouted with unreasonable defiance; or tried to shout, for his voice broke and he whispered, 'O God!' very low, and then he seemed to sleep a little.

Manderson tip-toed out and sat in the waiting-room at the end of the ward. Across the veranda the blackness of the night and the brightness of the stars were fading together; the sky was growing pale at the first distant threat of dawn. He rang up O'Driscoll's Burra Sahib, and Atkinson and another man or two who were his friends, more because of loneliness than for anything they could do. They said they would come down to the hospital very soon.

Gradually the sky lightened, and grey that would turn in a minute to saffron took the place of darkness. Manderson watched the day coming with something of that feeling with which one waited for dawn in France during the war years. There was something like fear and something like hope in his waiting. He thought at times that with the sun strength would come to Jerry, and that if he lived for one day he might live for another. When the sun was just below the horizon the buildings in the middle distance grew startlingly black and hard, as though they were made of gaunt black girders, and the sky looked more and more immeasurably remote and fragile, like very thin muslin over an enormous greenish bowl that was quite empty. Then, as the light became suddenly certain, a white-faced Sister came hurrying and Manderson's heart sank.

It has been said—though most men die without asking for anybody—that a dying man will most often call for his mother, and Manderson thought of this as he followed the nurse, and tried to remember whether it was Cork or Kerry that O'Driscoll came from; one or the other, he thought, but he was not certain.

Jerry was quieter and the fever seemed to have gone. His face was drawn and there were hard lines on it.

'Manderson,' he said, 'for God's sake go and get my mother, for I'm dying and she doesn't know.'

He spoke sensibly enough, but Cork is six thousand miles and more from Bombay, and Manderson choked and stammered as he tried to tell him so.

'Don't be a fool,' said O'Driscoll desperately. 'She's here, within a mile of me, and O Christ! I want her.' His hand caught wildly at the sheet which covered him.

Manderson said, 'There, there, old man,' and felt foolish, and O'Driscoll, coughing weakly, replied, 'Damn you, she's living here, I tell you.' And he gave an address in Byculla, which is a district of Bombay where few Europeans live, and very few of the privileged class known as Club members.

Manderson, bewildered, made some remark about Cork, but O'Driscoll said again with an effort, 'She's there, I tell you, an' it's nearly too late.'

It was the Sister who said at last, 'All right, Mr O'Driscoll, we'll send a taxi along now and she'll be here in no time,' and beckoned Manderson out into the corridor.

'Write a note and give it to your driver, if you've got a car, and send him to that address,' she said. 'It'll do no good, but if his mother's there she's got a right to see him die.'

'But he's Irish, and his mother's in Cork,' replied Manderson limply.

The Sister looked at him and considered for a moment. 'Are you sure?' she asked.

'He always said so,' said Manderson.

'You'll find his mother at that address unless I'm very far mistaken,' said the Sister a little grimly. 'And I think you'll find that he's country-born.' She was a widow, and maybe knew more about the country-born—English people born in India with an anna in the rupee or so of native blood in their veins—than Manderson did. Or maybe Jerry's voice, when he was raving a little, had lost its Irish accent and taken on the sing-song *chi-chi* tone of the country instead.

Anyway, Manderson wrote his note and his car went to the address in Byculla, and Manderson was left to ponder and rake his memory as deep as he could for all that O'Driscoll had told him of his people and his birth. He found that it was not very much. Once O'Driscoll had said something about Trinity College, Dublin, and Powell, who had spent many years there getting his Rugger blue, and, in the end, a somewhat gratuitous MB, asked in astonishment, 'Why, when were you up?'

'I wasn't,' O'Driscoll answered very gravely. 'Borstal was my Alma

Mater, and it's brought me more friends than you'd credit—especially in business circles in Bombay,' and he winked portentously.

'Old Borstalians!' someone had shouted, and the toast was drunk noisily.

'You can talk of your Eton and the House, and the playing-fields that Waterloo was won on, but bedad!'—O'Driscoll had taken his cue and was holding the stage—'Bedad! It's the playing fields of Borstal that Second Ypres was won on—or lost, for I'm damned if I can remember which it was.'

He had gone home sometime in 1915 and been given a commission in the—Munster Fusiliers, was it?—Manderson could not remember. He had been wounded, once at any rate, and he had come back to Bombay shortly after the Armistice. That was certain; but his earlier life, like the Dark Ages, was lighted only by hearsay and uncertain reference. There were many people you knew whose histories went back to the war, but no further. A temporary commission had been a sort of social birth certificate, and the necessary gestation previous to it was often enough forgotten. The twice-born castes have their privileges everywhere.

'God knows!' Manderson said to himself, and watched the sun, now insolently radiant above the white buildings. A crow, lean, black, and beggarly, perched clumsily on the window-sill and eyed him furtively.

The Sister came in to him. 'They're here,' she said, and her lips tightened.

He got up.

A fat, ungainly woman, her dark sallow cheeks quivering as she walked, came down the corridor. A girl, slim and strangely like Jerry, was at her side; her face was heavily powdered and she was trying to muffle her sobs. Behind them were another woman, fat too, with her hat astray over black, untidy hair, and a small beady-eyed boy.

'Oh, Gerald,' said the first woman, 'where is my Gerald? It was nice of you, indeed, Mr Manderson, to help him as you could, but why did you not tell me before about him?'

Manderson cleared his throat, and the Sister said, 'I think perhaps you had better see him alone, Mrs O'Driscoll. Too many people would be bad for him. He is very ill, you know.'

The girl started to cry loudly, but the mother said, 'Hush, Tessie', and went with the Sister. The other fat woman said, 'It is so sad, Mr Manderson, is it not? For Gerald was so popular, and now he is going to die.'

'I think you had better wait here a little,' replied Manderson uncertainly, and led them into the waiting-room.

The sight of them, so obviously country-born—Anglo-Indians, half-castes, he thought, and shuddered at the ugly name—made him feel cold and numb. Poor old Jerry; he remembered the slighting things that were said about such people as these, the jokes that were made about them, the heartless jokes that were, some of them, so terribly true to facts. There was that one about the country-born girl who, from the depths of illicit embraces on an agitated couch, said resignedly, 'Oh, Mr James, indeed you are such a flirt!' Surely Jerry had told him that one himself. And it was as though it was his sister of whom he had told it. Manderson considered her unhappily. She was pretty, short and pliant, with big dark eyes and the sort of helpless allure of her kind. Her lips, half-open, were darkish red against the faintly yellow pallor of her face . . . The small boy stared at him. His forehead bulged and his mouth was stupidly open. These were Jerry's blood relations.

He would have been eligible for none of the clubs had it been suspected that he was country-born, and the mere suggestion of relationship with one of these people would have damned him for ever. It was a cruel system, this social philosophy of Englishmen in the East, but looking at the woman and the disgusting small boy Manderson could not honestly condemn it. It was this racial fastidiousness which had saved England from the fate of Portugal. Portuguese India, that gorgeous crown of a proud and gallant nation, had become the home of cooks and menials; Goa, the city of magnificent cathedrals and unimaginable wealth, was a dilapidated ruin. And all because the Portuguese had married into India after invading it, had consummated their victory in lawful wedlock and begotten their etiolated kind on the women they had captured and converted. But England, avoiding conscious procreation and devoting her energies to trade and conquest, had won supremacy as the reward of continence—or better, perhaps, of discretion.

The woman tearfully began to rehearse the virtues of Gerald. He had always given his mother plenty of money—she herself was his mother's sister and had shared in the benefits. The ivory necklace which Tessie wore was a present from him on her last birthday. Of course they had not seen very much of him lately, but 'He was so very popular, was he not, Mr Manderson? We used to be very proud when he talked about people who were rather wealthy and famous in Bombay, and whose names we saw in the papers.'

'We were all very fond of Jerry,' said Manderson uncomfortably.

'There, Tessie, indeed you should be proud of your brother and not cry so much before Mr Manderson. Do you think he will have a very large funeral, Mr Manderson?'

'Oh, Auntie,' sobbed the girl, and clutched a small soiled handkerchief to her eyes. A fringe of petticoat showed above her thin legs, and her narrow shoulders shook with weeping. The small boy stood by the window and solemnly picked his nose.

They probably encouraged him to desert them, thought Manderson, so that they could talk proudly of his achievements to their neighbours in that queer, half-light society of the Anglo-Indian. Not that he had really deserted them, apparently, for he had given them money—a lot of money, according to the fat aunt—and that was decent of Jerry, for he was always hard up and living on February's salary in December. It was easy to blame him, perhaps; but Manderson pictured the house in Byculla, the noisy throng of relatives, the uneasy proud assertions of a white ancestry, the constant defiance of Indian associations, the pathetic loyalty to an indifferent and unsympathetic rule, the squalor and dust and smell of a teeming neighbourhood. And then he thought of the clubs, their coolness and ease; their members might be vulgar and commonplace and ignorant, but they were white, the dominant race, and they could talk naturally of England as Home—they did not say 'Going home to England', as subconscious fear of misunderstanding made the country-born say; they said 'Going Home', secure in the knowledge that everyone recognized what they meant. And Jerry had been accepted as one of them without hesitation; he looked white enough; he had been a good fellow, a good companion; he had undoubtedly been popular even as that woman his aunt (a greasy strand of hair hung over her fat sallow cheek, and her exuberant breast strained the shiny fabric of her blouse) had said he was.

He should be blamed, possibly; he had not honoured his father and his mother. But he had seemingly supported his mother, and heaven knew how many of her relations, and he had flattered her maternity by his acceptance in the clubs and the bungalows of successful Europeans on Malabar Hill. Perhaps that absolved him. 'I wonder,' thought Manderson. 'Service is supposed to count; perhaps he's established a precedent, and saved his soul in the Yacht Club.'

He started to his feet. They all did, the woman and the girl with renewed wailing, as Mrs O'Driscoll came in, her face distorted with weeping.

'Oh, Tessie,' she cried, 'Tessie, he's dead.' A broken harmony of grief went up, thrown into piercing discord by the unexpected screaming of the boy. The aunt, sobbing with her mouth open and a stream of tears running down either cheek, began to pray confusedly, and crossed herself mechanically and without unction. Manderson, his grief for Jerry's death almost driven out of his mind by the acute discomfort of the scene at which he was assisting, crossed to the door where the Sister stood, silent and stern of mouth.

'Was he conscious?' he asked.

'He laughed,' she answered. 'He sat up and laughed, and lay back and was dead. I don't know whether he was properly conscious or not.'

'What did he say, mother?' sobbed the girl, her lips twisted. 'Could he speak to you before he died?'

Mrs O'Driscoll's grief quietened to sniffing and an occasional hiccup. Nobody took any notice of the small boy, who clawed at her dress, ripped open her placket, and screamed hoarsely. 'I don't know, Tessie,' she answered. 'Oh, that is honest, I don't know what he said. And now he is dead—oh, blessed Mary, he is dead, and we had no time to get the Father, and—oh! he will never say anything again!'

'Come, Mrs O'Driscoll,' said the Sister. 'Try and control yourself, and do make the little boy stop crying. I'll get you some tea, and then you'll feel better.'

'He said something about a Major Gorland. I think it was Major

Gorland. Do you know him, Mr Manderson? Indeed, I knew so few of his friends, and he had so many, had he not?'

'He was so popular, Tessie,' said the aunt.

'He said—you know how Irish he was, Mr Manderson; we are all Irish, you know; indeed, we were all going Home to Ireland next year—he said, "Sure, Major Gorland, is it a man I am at all, at all?" and then he sat up and laughed. And then he died. Oh, Tessie!'

'It was rather a funny thing for a dying man to say,' remarked the aunt, 'but perhaps Major Gorland was a very great friend of his, Mr Manderson, was he?'

'I think he was,' answered Manderson unsteadily.

The thought of that last night at the Gymkhana came to him: Jerry with one foot on the table, Hibernian hybris on his face, the art of the histrionic Gael in his finger-tips, the mockery of the playboy in his soul. At any rate he died as he had chosen to live. If he were not true diamond he was amazingly good paste; there were even such things as synthetic diamonds, Manderson's sympathy suggested. . . .

'I'm afraid I must go,' he said abruptly. 'I'm so sorry, Mrs O'Driscoll. Jerry was a dear soul. I shall see you afterwards, of course.'

Mrs O'Driscoll sobbed faintly and took his hand in a soft, moist embrace.

'He was so very popular, wasn't he?' said the aunt pathetically as Manderson turned and left them.

LEWIS GRASSIC GIBBON

1901–1935

SMEDDUM

She'd had nine of a family in her time, Mistress Menzies, and brought the nine of them up, forbye—some near by the scruff of the neck, you would say. They were sniftering and weakly, two-three of the bairns, sniftering in their cradles to get into their coffins; but she'd shake them to life, and dose them with salts and feed them up till they couldn't but live. And she'd plonk one down—finishing the wiping of the creature's neb or the unco dosing of an ill bit stomach or the binding of a broken head—with a look on her face as much as to say *Die on me now and see what you'll get!*

Big-boned she was by her fortieth year, like a big roan mare, and *If ever she was bonny 'twas in Noah's time*, Jock Menzies, her eldest son would say. She'd reddish hair and a high, skeugh nose, and a hand that skelped her way through life; and if ever a soul had seen her at rest when the dark was done and the day was come he'd died of the shock and never let on.

For from morn till night she was at it, work, work, on that ill bit croft that sloped to the sea. When there wasn't a mist on the cold, stone parks there was more than likely the wheep of the rain, wheeling and dripping in from the sea that soughed and plashed by the land's stiff edge. Kinneff lay north, and at night in the south, if the sky was clear on the gloaming's edge, you'd see in that sky the Bervie lights come suddenly lit, far and away, with the quiet about you as you stood and looked, nothing to hear but a sea-bird's cry.

But feint the much time to look or to listen had Margaret Menzies of Tocherty toun. Day blinked and Meg did the same, and was out, up out of her bed, and about the house, making the porridge and rousting the bairns, and out to the byre to milk the three kye, the morning growing out in the east and a wind like a hail of knives from the hills. Syne back to the kitchen again she would be, and catch Jock, her eldest, a clour in the lug that he hadn't roused up his sisters and brothers; and rouse them herself, and feed them and scold, pull up their breeks and straighten their frocks, and polish their shoes and set their caps straight. *Off you get and see you're not late*, she would cry, *and see you behave yourselves at the school. And tell the Dominie I'll be down the night to ask him what the mischief he meant by leathering Jeannie and her not well.*

They'd cry *Ay, Mother*, and go trotting away, a fair flock of the creatures, their faces red-scoured. Her own as red, like a meikle roan mare's, Meg'd turn at the door and go prancing in; and then at last, by the closet-bed, lean over and shake her man half-awake. *Come on, then, Willie, it's time you were up.*

And he'd groan and say *Is't?* and crawl out at last, a little bit thing like a weasel, Will Menzies, though some said that weasels were decent beside him. He was drinking himself into the grave, folk said, as coarse a little brute as you'd meet, bone-lazy forbye, and as sly as sin. Rampageous and ill with her tongue though she was, you couldn't but pity a woman like Meg tied up for life to a thing like *that*. But she'd more than a soft side still to the creature, she'd half-skelp the backside from any of the bairns she found in the telling of a small bit lie; but when Menzies would come paiching in of a noon and groan that he fair was tashed with his work, he'd mended all the ley fence that day and he doubted he'd need to be off to his bed—when he'd told her that and had ta'en to the blankets, and maybe in less than the space of an hour she'd hold out for the kye and see that he'd lied, the fence neither mended nor letten a-be, she'd just purse up her meikle wide mouth and say nothing, her eyes with a glint as though she half-laughed. And when he came drunken home from a mart she'd shoo the children out of the room, and take off his clothes and put him to bed, with an extra nip to keep off a chill.

She did half his work in the Tocherty parks, she'd yoke up the horse and the sholtie together, and kilt up her skirts till you'd see her great legs, and cry *Wissh!* like a man and turn a fair drill, the sea-gulls cawing in a cloud behind, the wind in her hair and the sea beyond. And Menzies with his sly-like eyes would be off on some drunken ploy to Kineff or Stonehive. Man, you couldn't but think as you saw that steer it was well that there was a thing like marriage, folk held together and couldn't get apart; else a black look-out it well would be for the fusionless creature of Tocherty toun.

Well, he drank himself to his grave at last, less smell on the earth if maybe more in it. But she broke down and wept, it was awful to see, Meg Menzies weeping like a stricken horse, her eyes on the dead, quiet face of her man. And she ran from the house, she was gone all that night, though the bairns cried and cried her name up and down the parks in the sound of the sea. But next morning they found her back in their midst, brisk as ever, like a great-boned mare, ordering here and directing there, and a fine feed set the next day for the folk that came to the funeral of her orra man.

She'd four of the bairns at home when he died, the rest were in kitchen-service or fee'd, she'd seen to the settling of the queans herself; and twice when two of them had come home, complaining-like of their mistresses' ways, she'd thrashen the queans and taken them back—near scared the life from the doctor's wife, her that was mistress to young Jean Menzies. *I've skelped the lassie and brought you her back. But don't you ill-use her, or I'll skelp you as well.*

There was a fair speak about that at the time, Meg Menzies and the vulgar words she had used, folk told that she'd even said what was the place where she'd skelp the bit doctor's wife. And faith! that fair must have been a sore shock to the doctor's wife that was that genteel she'd never believed she'd a place like that.

Be that as it might, her man new dead, Meg wouldn't hear of leaving the toun. It was harvest then and she drove the reaper up and down the long, clanging clay rigs by the sea, she'd jump down smart at the head of a bout and go gathering and binding swift as the wind, syne wheel in the horse to the cutting again. She led the stooks with her bairns to help, you'd see them at night a drowsing cluster under the moon on the harvesting cart.

And through that year and into the next and so till the speak died down in the Howe Meg Menzies worked the Tocherty toun; and faith, her crops came none so ill. She rode to the mart at Stonehive when she must, on the old box-cart, the old horse in the shafts, the cart behind with a sheep for sale or a birn of old hens that had finished with laying. And a butcher once tried to make a bit joke. *That's a sheep like yourself, fell long in the tooth.* And Meg answered up, neighing like a horse, and all heard: *Faith, then, if you've got a spite against teeth I've a clucking hen in the cart outbye. It's as toothless and senseless as you are, near.*

Then word got about of her eldest son, Jock Menzies that was fee'd up Allardyce way. The creature of a loon had had fair a conceit since he'd won a prize at a ploughing match—not for his ploughing, but for good looks; and the queans about were as daft as himself, he'd only to nod and they came to his heel; and the stories told they came further than that. Well, Meg'd heard the stories and paid no heed, till the last one came, she was fell quick then.

Soon's she heard it she hove out the old bit bike that her daughter Kathie had bought for herself, and got on the thing and went cycling away down through the Bervie braes in that Spring, the sun was out and the land lay green with a blink of mist that was blue on the hills, as she came to the toun where Jock was fee'd she saw him out in a park by the road, ploughing, the black loam smooth like a ribbon turning and wheeling at the tail of the plough. Another billy came ploughing behind, Meg Menzies watched till they reached the rig-end, her great chest heaving like a meikle roan's, her eyes on the shape of the furrows they made. And they drew to the end and drew the horse out, and Jock cried *Ay,* and she answered back *Ay,* and looked at the drill, and gave a bit snort, *If your looks win prizes, your ploughing never will.*

Jock laughed, *Fegs, then, I'll not greet for that*, and chirked to his horses and turned them about. But she cried him *Just bide a minute, my lad. What's this I hear about you and Ag Grant?*

He drew up short then, and turned right red, the other childe as well, and they both gave a laugh, as plough-childes do when you mention a quean they've known over-well in more ways than one. And Meg snapped *It's an answer I want, not a cockerel's cackle: I can hear that at home on my own dunghill. What are you to do about Ag and her pleiter?*

And Jock said *Nothing,* impudent as you like, and next minute Meg was in over the dyke and had hold of his lug and shook him and it till the other childe ran and caught at her nieve. *Faith, mistress, you'll have his lug off!* he cried. But Meg Menzies turned like a mare on new grass, *Keep off or I'll have yours off as well!*

So he kept off and watched, fair a story he'd to tell when he rode out that night to go courting his quean. For Meg held to the lug till it near came off and Jock swore that he'd put things right with Ag Grant. She let go the lug then and looked at him grim: *See that you do and get married right quick, you're the like that needs loaded with a birn of bairns—to keep you out of the jail, I jaloose. It needs smeddum to be either right coarse or right kind.*

They were wed before the month was well out, Meg found them a cottar house to settle and gave them a bed and a press she had, and two-three more sticks from Tocherty toun. And she herself led the wedding dance, the minister in her arms, a small bit childe; and 'twas then as she whirled him about the room, he looked like a rat in the teeth of a tyke, that he thanked her for seeing Ag out of her soss, *There's nothing like a marriage for redding things up.* And Meg Menzies said *EH?* and then she said *Ay,* but queer-like, he supposed she'd no thought of the thing. Syne she slipped off to sprinkle thorns in the bed and to hang below it the great hand-bell that the bothy-billies took them to every bit marriage.

Well, that was Jock married and at last off her hands. But she'd plenty left still, Dod, Kathleen and Jim that were still at school, Kathie a limner that alone tongued her mother, Jeannie that next led trouble to her door. She'd been found at her place, the doctor's it was, stealing some money and they sent her home. Syne news of the thing got into Stonehive, the police came out and tormented her sore, she swore she never had stolen a meck, and Meg swore with her, she was black with rage. And folk laughed right hearty, fegs! that was a clour for meikle Meg Menzies, her daughter a thief!

But it didn't last long, it was only three days when folk saw the doctor drive up in his car. And out he jumped and went striding through the close and met face to face with Meg at the door. And he cried *Well, mistress, I've come over for Jeannie.* And she glared at him over her high, skeugh nose, *Ay, have you so then? And why, may I speir?*

So he told her why, the money they'd missed had been found at last in a press by the door; somebody or other had left it there, when paying a grocer or such at the door. And Jeannie—he'd come over to take Jean back.

But Meg glared *Ay, well, you've made another mistake. Out of this, you and your thieving suspicions together!* The doctor turned red, *You're making a miserable error*—and Meg said *I'll make you mince-meat in a minute.*

So he didn't wait that, she didn't watch him go, but went ben to the kitchen where Jeannie was sitting, her face chalk-white as she'd heard them speak. And what happened then a story went round, Jim carried it to school, and it soon spread out, Meg sank in a chair, they thought she was greeting; syne she raised up her head and they saw she was laughing, near as fearsome the one as the other, they thought. *Have you any cigarettes?* she snapped sudden at Jean, and Jean quavered *No,* and Meg glowered at her cold. *Don't sit there and lie. Gang bring them to me.* And Jean brought them, her mother took the pack in her hand. *Give's hold of a match till I light up the thing. Maybe smoke'll do good for the crow that I got in the throat last night by the doctor's house.*

Well, in less than a month she'd got rid of Jean—packed off to Brechin the quean was, and soon got married to a creature there—some clerk that would have left her sore in the lurch but that Meg went down to the place on her bike, and there, so the story went, kicked the childe so that he couldn't sit down for a fortnight, near. No doubt that was just a bit lie that they told, but faith! Meg Menzies had herself to blame, the reputation she'd gotten in the Howe, folk said, *She'll meet with a sore heart yet.* But devil a sore was there to be seen, Jeannie was married and was fair genteel.

Kathleen was next to leave home at the term. She was tall, like Meg, and with red hair as well, but a thin fine face, long eyes blue-grey like the hills on a hot day, and a mouth with lips you thought over thick. And she cried *Ah well, I'm off then, mother.* And Meg cried *See you behave yourself.* And Kathleen cried *Maybe; I'm not at school now.*

Meg stood and stared after the slip of a quean, you'd have thought her half-angry, half near to laughing, as she watched that figure, so slender and trig, with its shoulders square-set, slide down the hill on the wheeling bike, swallows were dipping and flying by Kinneff, she looked light and free as a swallow herself, the quean, as she biked away from her home, she turned at the bend and waved and whistled, she whistled like a loon and as loud, did Kath.

Jim was the next to leave from the school, he bided at home and he took no fee, a quiet-like loon, and he worked the toun, and, wonder of wonders, Meg took a rest. Folk said that age was telling a bit on even Meg Menzies at last. The grocer made hints at that one night, and Meg answered up smart as ever of old: *Damn the age! But I've finished the trauchle of the bairns at last, the most of them married or still over young. I'm as swack as ever I was, my lad. But I've just got the notion to be a bit sweir.*

Well, she'd hardly begun on that notion when faith! ill the news that came up to the place from Segget. Kathleen her quean that was fee'd down there, she'd ta'en up with some coarse old childe in a bank, he'd left his wife, they were off together, and she but a bare sixteen years old.

And that proved the truth of what folk were saying, Meg Menzies she hardly paid heed to the news, just gave a bit laugh like a neighing horse and went on with the work of park and byre, cool as you please—ay, getting fell old.

No more was heard of the quean or the man till a two years or more had passed and then word came up to the Tocherty someone had seen her—and where do you think? Out on a boat that was coming from Australia. She was working as stewardess on that bit boat, and the childe that saw her was young John Robb, an emigrant back from his uncle's farm, near starved to death he had been down there. She hadn't met in with him near till the end, the boat close to Southampton the evening they met. And she'd known him at once, though he not her, she'd cried *John Robb?* and he'd answered back *Ay?* and looked at her canny in case it might be the creature was looking for a tip from him. Syne she'd laughed *Don't you know me, then, you gowk? I'm Kathie Menzies you knew long syne—it was me ran off with the banker from Segget!*

He was clean dumbfounded, young Robb, and he gaped, and then they shook hands and she spoke some more, though she hadn't much time, they were serving up dinner for the first-class folk, aye dirt that are ready to eat and to drink. *If ever you get near to Tocherty toun tell Meg I'll get home and see her some time. Ta-ta!* And then she was off with a smile, young Robb he stood and he stared where she'd been, he thought her the bonniest thing that he'd seen all the weary weeks that he'd been from home.

And this was the tale that he brought to Tocherty, Meg sat and listened and smoked like a tink, forbye herself there was young Jim there, and Jock and his wife and their three bit bairns, he'd fair changed with marriage, had young Jock Menzies. For no sooner had he taken Ag Grant to his bed than he'd starved to save, grown mean as dirt, in a three-four years he's finished with feeing, now he rented a fell big farm himself, well stocked it was, and he fee'd two men. Jock himself had grown thin in a way, like his father but worse his bothy childes said, old Menzies at least could take a bit dram and get lost to the world but the son was that mean he might drink rat-poison and take no harm, 'twould feel at home in a stomach like his.

Well, that was Jock, and he sat and heard the story of Kath and her say on the boat. *Ay, still a coarse bitch, I have not a doubt. Well if she never comes back to the Mearns, in Segget you cannot but redden with shame when a body will ask 'Was Kath Menzies your sister?'*

And Ag, she'd grown a great sumph of a woman, she nodded to that, it was only too true, a sore thing it was on decent bit folks that they should have any relations like Kath.

But Meg just sat there and smoked and said never a word, as though she thought nothing worth a yea or a nay. Young Robb had fair ta'en a fancy to Kath and he near boiled up when he heard Jock speak, him and the wife that he'd married from her shame. So he left them short and went raging home, and wished for one that Kath would come back, a Summer noon as he cycled home, snipe were calling in the Auchindreich moor where the cattle stood with their tails a-switch, the Grampians rising far and behind, Kinraddie spread like a map for show, its ledges veiled in a mist from the sun. You felt on that day a wild, daft unease, man, beast and bird: as though something were missing and lost from the world, and Kath was the thing that John Robb missed, she'd something in her that minded a man of a house that was builded upon a hill.

Folk thought that maybe the last they would hear of young Kath Menzies and her ill-gettèd ways. So fair stammy-gastered they were with the news she'd come back to the Mearns, she was down in Stonehive, in a grocer's shop, as calm as could be, selling out tea and cheese and suchlike with no blush of shame on her face at all, to decent women that were properly wed and had never looked on men but their own, and only on them with their braces buttoned.

It just showed you the way that the world was going to allow an ill quean like that in a shop, some folk protested to the creature that owned it, but he just shook his head, *Ah well, she works fine; and what else she does is no business of mine.* So you well might guess there was more than business between the man and Kath Menzies, like.

And Meg heard the news and went into Stonehive, driving her sholtie, and stopped at the shop. And some in the shop knew who she was and minded the things she had done long syne to other bit bairns of hers that went wrong; and they waited with their breaths held up with delight. But all that Meg did was to nod to Kath *Ay, well, then, it's you—Ay, mother, just that—Two pounds of syrup and see that it's good.*

And not another word passed between them, Meg Menzies that once would have ta'en such a quean and skelped her to rights before you could wink. Going home from Stonehive she stopped by the farm where young Robb was fee'd, he was out in the hayfield coling the hay, and she nodded to him grim, with her high horse face. *What's this that I hear about you and Kath Menzies?*

He turned right red, but he wasn't ashamed. *I've no idea—though I hope it's the worse—It fell near is—Then I wish it was true, she might marry me, then, as I've prigged her to do.*

Oh, have you so, then? said Meg, and drove home, as though the whole matter was a nothing to her.

But next Tuesday the postman brought a bit note, from Kathie it was to her mother at Tocherty. *Dear mother, John Robb's going out to Canada and wants me to marry him and go with him. I've told him instead I'll go with him and see what he's like as a man—and then marry him at leisure, if I feel in the mood. But he's hardly any money, and we want to borrow some, so he and I are coming over on Sunday. I hope that you'll have dumpling for tea. Your own daughter, Kath.*

Well, Meg passed that letter over to Jim, he glowered at it dour, *I know—near all the Howe's heard. What are you going to do, now, mother?*

But Meg just lighted a cigarette and said nothing, she'd smoked like a tink since that steer with Jean. There was promise of strange on-goings at Tocherty by the time that the Sabbath day was come. For Jock came there on a visit as well, him and his wife, and besides him was Jeannie, her that had married the clerk down in Brechin, and she brought the bit creature, he fair was a toff; and he stepped like a cat through the sharn in the close; and when he had heard the story of Kath, her and her plan and John Robb and all, he was shocked near to death, and so was his wife. And Jock Menzies gaped and gave a mean laugh. *Ay, coarse to the bone, ill-gettèd I'd say if it wasn't that we come of the same bit stock. Ah well, she'll fair have to tramp to Canada, eh mother?—if she's looking for money from you.*

And Meg answered quiet *No, I wouldn't say that. I've the money all ready for them when they come.*

You could hear the sea plashing down soft on the rocks, there was such a dead silence in Tocherty house. And then Jock habbered like a cock with fits *What, give silver to one who does as she likes, and won't marry as you made the rest of us marry? Give silver to one who's no more than a—*

And he called his sister an ill name enough, and Meg sat and smoked looking over the parks. *Ay, just that. You see, she takes after myself.*

And Jeannie squeaked *How?* and Meg answered her quiet: *She's fit to be free and to make her own choice the same as myself and the same kind of choice. There was none of the rest of you fit to do that, you'd to marry or burn, so I married you quick. But Kath and me could afford to find out. It all depends if you've smeddum or not.*

She stood up then and put her cigarette out, and looked at the gaping gowks she had mothered. *I never married your father, you see. I could never make up my mind about Will. But maybe our Kath will find something surer. . . . Here's her and her man coming up the road.*

ROBERT MCLELLAN
1907–1985

THE DONEGALS

In the auld horse days, afore the fruit-growers in Clydeside had motor lorries to send to Larkhaa and siclike places for loads o keelie tounsfolk to pou their strawberries, fetchin them ilka mornin and takin them back at nicht, aa the berry-pouin was dune by weemen and laddies frae Kirkfieldbank and ither bits of places roun aboot, and by squads o gangrel Irish, caaed Donegals, that bade at the ferms in barns and bothies till the season was ower, whan they moved awa to the upland ferms for the hey and the tatties.

Ae sic squad cam to Linmill ilka year. The mairrit anes had the auld bothy abune the milk-hoose that had been biggit for the byre lassies, but the single weemen bade at the tae end o the big barn, and the single men at the tither, wi a bit raw o auld blankets hung up for a waa atween them.To get into the milk-hoose bothy ye had to gang into the barn and up a lether, sae whan the Donegals were aa beddit doun at nicht they were aa ahint the ae barn door, and that was lockit aye by my grandfaither, at ten o'clock, whan he lockit the muckle yett at the closs mou.

He had whiles a gey job to get them aa in for the nicht, wi the aulder anes gey aften taiglet on their wey back hame frae the Kirkfield Inn, and the younger anes oot amang the hedges or in the ditches daffin, wi nae sense o time. But he was weill respeckit amang them, and though they whiles blarneyed him in the maist droll fashion, the mair whan he had a bit dram taen himsell, they did his biddin aye whan he grew douce wi them, and nane eir daured to lift a haund against him, ein whan he gat crabbit wi them, as he did whiles, and ordert them to their beds like a pack o bairns.

Whan they were aa in and the barn door lockit, and the closs mou yett, he wad come into the hoose and mak ready for his bed, and nae maitter what unco steer he micht hear frae the barn as the Donegals made ready for theirs, he wadna set fute ootbye in the closs again that nicht. Let them howl and skrech for aa they were worth, and they gaed their dinger whiles, I can tell ye, he gaed intil his bed aside my grannie in the kitchen closet, poued the claes ower his heid, and snored like a grumpie.

Save ance, on a Setterday, and it was a dreidfou nicht that, as ye shall hear.

But first ye maun ken that amang the Donegals that cam to Linmill ilka year, like the peesweeps to the tap parks, there was an auld couple caaed

Paddy and Kate O'Brien. Auld they maun hae been, for they had poued strawberries at Linmill afore my minnie was born, and aye whan they cam, and had settled doun in the best corner o the milk-hoose bothy, neist to the fire and awa frae the door, they wad seek oot my minnie, to hinnie and dawtie her as if she was a bairn o their ain, no forgettin to tell her what a bonnie laddie she had gotten, kecklin around me like clockin hens, and clappin my heid and cuddlin me till I could smell the smeik aff them, for they were aye beylin tea in a wee black can.

This pair, Paddy and Kate, for aa their great age, behaved aye when they were sober like a trystit lad and lass, and ein whan they were thrang on the berry beds, whaur they were aye pairtners, they wad rookety-coo in ilk ither's lugs like cushie doos. But whan they were fou, and they were baith deils for the drink, they seemed to hate the grun ilk ither walkit, and it was a diversion to watch them gaun aff at ein to the Kirkfield Inn, wi arms linkit and heids thegither, to come back at nicht pairtit, Kate walkin aheid dour and grumlie, and Paddy stacherin efter her, a lang way ahint, cursin and sweirin at her, and flingin stanes.

Ae Setterday nicht, the wild ane I hae mentioned, they had gotten byordinar fou, for they had gane to the Lanark races and won some siller on a horse, and a sair job my grandfaither had to get them into the barn. But get them in he did, and lockit the door, and syne the closs mou yett, and in the end he won to his ain bed, weary for sleep, for he had a guid dram in him that nicht himsell.

A while efter, whan the steer in the barn had deed doun, and the rummle o Stanebyres Linn had come into its ain, there was a skrech frae the milk-hoose bothy that wad hae curdlet yer bluid. The haill barn bizzed again like a bees' byke, and the dug on the chain at the stable door lat oot a yowl.

Syne there was sic a fleechin and flytin, and duntin and dingin, as we had neir heard in the barn afore in aa oor days. In the end the barn door itsell was dung and daddit, and there were cries o murder.

My grandfaither juist had to rise to see whit was whaat.

He turnt the big key in the barn door and poued it open, and oot stachert auld Kate O'Brien, wi her hair doun her back, her claes torn gey near aff her, and bluid rinnin doun her face and breist. And she was cursin her man Paddy wi sic dreidfou spleen that it was terrible to hear.

My grandfaither waitit wi his muckle neive liftit for Paddy to come oot tae, but a wheen o the younger Donegals were haudin him doun inbye, and they cried to my grandfaither to fetch a raip, for they thocht that Paddy had gane wud athegither, and suld be weill tied up.

My grandfaither peyed nae heed to their clavers, but gaed into the barn, took Paddy frae them by the scruff o the neck, and mairched him oot into the closs. In the closs he had to let him be for a while, to fin the key o the closs mou yett, for he was gaun to pitch him aff the ferm athegither, and as sune as he lowsed his grip he gat a clowt on the lug. Paddy had gane for him.

That wasna to be tholed, sae my grandfaither gaed for Paddy, and a bonnie fecht it micht hae been, hadna Kate, whan she saw her man haein the warst o it, turnt her coat athegither, and gane for my grandfaither like a cat wi kittlins. Atween the twa he was in a fair wey to bein torn to daith.

The ither Donegals lookit on frae the barn door, feart to interfere, and though I was fain mysell to fling a stane or twa I didna daur, for I micht hae missed the pair I was aimin at and hit my grandfaither.

I had reckoned withoot my grannie. She ran to the kitchen for the big claes beetle, gethert up her goun, and laid aboot her.

That settled it. Kate and Paddy ran to the barn and poued the door tae ahint them.

That micht hae been the end o the haill affair, but my minnie, whan she had seen my grandfaither in a fair wey to bein murdert, had taen fricht, and no waitin to see my grannie's pairt in the affair, had let hersell oot at the Linmill front door and run for Airchie Naismith, and Airchie had yokit his gig at ance and driven helter-skelter for the polis.

Whan the polis cam, a muckle weill-fed lazy craitur by the name o Gilfillan, he and my grandfaither gaed into the barn, and ye could hae heard a preen drap, the Donegals were sae awed. In a wee while oot they cam again, Gilfillan leadin Paddy wi the ae haund and Kate wi the tither, and my grandfaither cairryin their bundle o claes and their auld smeikie can.

My grandfaither convoyed them oot o the closs, lockit the closs mou yett ahint them, and came into the hoose. And afore lang there wasna a soun to be heard bune the rummle o Stanebyres Linn.

I had thocht that Gilfillan wad tak Paddy and Kate to the jeyl, and nae dout that was what he ettlet whan he left Linmill, but it seems that, whan the three had won doun the road a wee, the twa auld Donegals had stertit to blarney, and priggit sair to be lowsed, and whan Gilfillan, gey angert at bein trailed frae his warm bed sae late at nicht, tried to quaiten them and gar them hurry, they stertit to taigle him wi aa the wiles in their pouer. The upshot was that he grew hairt seik o them afore he gat them the length o the Black Bog, and in the end he telt them no to gang near Linmill again, and left them skaithless.

On the Saubbath we saw naething o them, though we heard they had been beylin their black can in the wuid on the Hinnie Muir road, and thocht they maun be makin for the upland ferms, and we wad be weill redd o them, but on the Monday mornin, whan my grandfaither was leadin his squad o warkers across the Clyde road to the field aside the waal orchard, he fand Paddy and Kate wi their bundle aside the waal yett.

They rase as he cam forrit and stude in his wey, and their blarney wad hae saftent the hairt o the Lanark factor. It was Kate that stertit.

'Sure now, master, and ye wouldn't be after giving us the sack, a poor owld couple the like of us, that has worked our fingers to the bone for ye, year after year, with never a word of complaint. Sure ye wouldn't be after sending us away from ye, and we never wishing ye any ill at all. He hit ye, master, but ye wouldn't be blamin him for the like of that, and ye with a fondness for a drop yerself. And didn't ye give him as good as ye got, ye owld warrior, for be looking at the eye he has on him. It's as black as a tinker's pot.'

And syne Paddy.

'Look at the poor owld sinner, master, and have pity in yer heart. She's not fit to be travelling the roads and sleeping outbye. She would die on me, so she would, for her breast's black and blue, master, with the lamming I gave her when my head was fuddled, God help me, and my five senses dulled with the drink. Take us back, master, like the kind man ye are, for it's sorry we are for all the trouble we gave you, and that's the solemn truth. Take us back, master, for the love of God, and not be sending us both out to die on the roads.'

My grandfaither felt gey sorry for them, I hae nae dout, but he daurtna tak them back again against my grannie's will.

'I canna dae it, Paddy. Ye'd better baith gang awa up and speir at the hoose.'

They lookit gey taen aback whan he said that, for weill they kent my grannie was anither nuit to crack. But there was nae help for them, for my grandfaither shouthert his wey past them and led his squad doun to the field.

The pair sat for a while and argle-barglet, and syne maun hae made up their minds, for in the end they cam to the hoose door, timrous like, and gied a blate wee chap.

My grannie cried oot frae the kitchen.

'Wha's that?'

'Sure it's meself, mistress, and me poor wife Kate, come to beg yer pardon.'

'What! Ye'll get nae paurdon here. Awa wi the pair o ye, or I'll lowse the dug. Awa I tell ye or I'll send for the polis again. The thowless lump suld hae putten ye baith in jeyl.'

And no anither word wad she say, though they stude at the door, disjaskit lookin, for hauf the mornin.

They had their denner by the hedge at the Falls road-end, and they were there still, beikin in the sun, whan I gaed doun the Falls road in the efternune, on my wey to the shop to buy sweeties.

Auld Kate saw me comin and sat up.

'Sit up, Paddy dear, and just look here. Isn't he the little darlint with his curly hair, and the freckles on the nose of him? The Lord bless ye, honey boy, and yer lovely ma, for she's the prettiest lady in the broad land, and it's the truth I'm telling. Isn't she, now, Paddy dear, and isn't he the living image of her?'

I kent it was aa blarney, for my minnie had black hair, and mine was reid. But there was mair to come.

'Where would ye be after going, now, on a day like this? Is it for Clyde ye are?'

'Na.'

'For the Falls, then, maybe?'

'Ay.'

'He'll be for buying sweeties, the little treasure, at Martha Baxter's shop. Is that what's in it?'

'Ay.'

'There now, and I after saying it. Sure, and that'll be a penny ye have, shut tight in yer hand?'

'Ay.'

'A penny. He couldn't be buying much with a penny. Could he now, Paddy dear?'

'A penny. No. A few sweeties, maybe, or a lucky bag, or maybe a little box of sherbert, but what's sweeties, or a lucky bag, or sherbert itself, on a hot thirsty day the like of this?'

'It's a bottle of lemonade he should be buying, to quench his thirst.'

'Yes, indade.'

'And a swate biscuit or two.'

'Ah sure, a swate biscuit or two, for drink should never be taken on an empty stomach, and there's nobody in the wide world knows that better than meself.'

'Ach wheesht now, Paddy, and give the boy a sixpence.'

'A sixpence, is it? Sure now, and haven't ye a sixpence yerself in yer petticoat pocket?'

'I wonder now. Ah yes, indade I have. Come here, little swateheart, and be holding your hand out.'

I kent I suldna tak the sixpence, but the temptation was mair nor I could staun, and though I held back, blate like, I didna rin awa till she pat the sixpence in my haund and tried to kiss me.

I ran then.

Whan I had gotten the sixpence, though, I was feart to ware it. My heid gaed roun like a peerie whan I thocht o aa I could buy, but I was shair that if I gaed hame wi my pooches fou my minnie wad speir, and wad think shame o me whan she fand whaur I had gotten the siller. I thocht for a while o hidin it to ware some ither day, but that wad hae made things waur. In the end, for I was ower greedy to gang and gie it back, I gaed to my minnie to show it to her, and tell her they had forced me to tak it.

I fand her by her lane in the front gairden.

'Kate O'Brien gied me a sixpence, minnie.'

'Dear me, a haill sixpence. Let me see.'

'She forced me to tak it.'

'She forced ye, did she? Hou that?'

'She stude on the Falls road and blarneyed me, and wadna let me bye, syne she pat the sixpence in my haund and tried to kiss me and I ran awa.'

'She blarneyed ye, did she? What did she say?'

'She said ye were the bonniest leddy in Clydeside.'

'Did she? And what else did she say?'

'That I suld buy some lemonade. But I dinna want lemonade. I want sweeties, and a luckie bag, and a box o sherbert. Minnie, can I ware the sixpence, or shall I gie it back?'

'Na, na, ware it, but dinna mak yersell no weill.'

Whan I had been to the Falls shop I gaed back to the Linmill kitchen for a tumbler of watter. My grannie saw me wi the sherbert and strauchtent her back, for she was bendin ower the girdle.

'Whaur did ye get that trash?'

'At the Falls shop.'

'Wha gied ye the siller?'

My minnie was tappin and tailin some grossets.

'Kate O'Brien gied him a sixpence.'

'A sixpence. Tryin to win favour.'

'Ay.'

'It's a woner she had sixpence to gie him, efter Setterday nicht.'

'Ay.'

Nae mair was said till my grandfaither came in for his supper. By that time Paddy and Kate were sittin wi their bundle on the dyke fornent the closs mou yett.

My minnie spak first.

'Paddy and Kate O'Brien are at the yett, faither.'

'Ay. They want back into the bothy, puir sowls.'

My grannie flared up.

'Puir sowls! Did they no try to kill ye on Setterday nicht?'

'They didna ken what they were daein. They were baith fou.'

'They're fou ower aften.'

'Ye canna blame them, wi the life they hae. And I missed them on the field the day.'

'Ye missed auld Kate's flaitterin tongue, nae dout, but ye didna miss them for ony wark they wad hae dune.'

'Oh but I did, for they're gey guid warkers. The best in the field.'

My minnie spak then.

'Ye canna gainsay that, mither, for ye hae said it gey aften yersell.'

'Oh ye're as bad as yer faither. They hae saft southert ye and aa, wi the sixpence they gied the bairn.'

My grandfather cockit his lugs.

'Did they gie the bairn a sixpence?'

'Ay, this efternune.'

'Weill, I declare. It maun hae been gey near their last.'

'Nae dout.'

'Then damn it, wumman, they're gaun back into the bothy!'

'They'll gang back to the bothy ower my corp!'

And she stude in my grandfaither's wey.

He pat his twa haunds to her waist and liftit her aff the flair.

'Ye're for the closet, then.'

He had lockit her in the box-bed closet ae Burns' nicht, when he was fou efter a spree, and had left her there till she had gien ower her flytin and stertit to greit. And nou whan he had made up his mind on a thing, and she wadna gie in, he wad threaten her wi the closet.

She lauched.

'Aa richt, hae it yer ain wey, ye big saft sumph.'

And she gied him a dad on the lug.

He gaed awa oot to the closs mou, and in a wee while there was a great cheer frae the Donegals, as Paddy and Kate gaed forrit to the barn door.

George Friel

1910–1975

I'm Leaving You

They were both in their early thirties. They had no children, and their parents were dead. She was the daughter of a surgeon in the Royal Infirmary, and he was the son of a lawyer. When they were orphaned they each inherited a parental house and some six or seven thousand pounds. They had no brother or sister to share the legacies.

She was educated at Laurel Bank, and he had gone to Glasgow Academy. They met at the university. Their first date was the night he took her to the Rugby Club dance. After a wasted year he stopped idling with beer and billiards and scraped a degree in Economics the year she got an honours degree in French and German.

They married when he began to find his way in the world of business, and she did occasional teaching in her old school. She didn't want a regular job. She didn't mind staying at home. She managed to get a woman to come in once a week to do the rougher chores, but that was all. The rest she did herself, playing the part of a good wife who looked after the comforts of her husband.

He agreed she didn't need to work. They weren't rich, he said, but they weren't poor. By the time he was thirty he had a good post in a big imports-and-exports firm, with his own office and a secretary. They were living in a flat in Kersland Street, but they wanted out of it.

When her father died they sold his house in Giffnock because it was on the wrong side of the river for them. But when his father died too they moved into his house in Bearsden. They liked the place. It had a two-car garage, and they could afford two cars. They thought it would be silly to sell another house just to put more money in the bank, and then go on living in a flat they didn't like.

They had been married seven years when it happened. He was dumb when she told him.

She came out with it one night in April after she had cleared away their evening meal. She marched from the kitchen to the sitting-room and thumped down on her chair. She kept staring at him as if her eyes could send a laser beam through his *Financial Times*.

Her penetrating silence made him peep over his paper. He never said much to her after their first couple of years together, and she never said much to him. Her silence shouldn't have disturbed him. But that night it came over as a demand to sit up and listen. He put his paper on his lap and looked at her patiently. He always paid her the courtesy of listening whenever she had a passing mood to talk to him.

The moment he put his paper down she spoke abruptly.

'I'm leaving you,' she said.

She wasn't smiling and she was quite calm. He couldn't take it as a joke or hysteria. It was a cold statement of fact.

He didn't know what to say or how to behave. He gaped at her, and waited. His reaction angered her. She spoke rather loudly.

'You're not the least bit bothered, are you? You sit there, and I might as well be telling you I'm going shopping in town tomorrow, for all you care.'

'That's not fair,' he said. 'It's just I don't understand. What do you mean, you're leaving me?'

'I mean I'm leaving you,' she repeated. 'Packing up and going.'

'But why on earth,' he started to ask.

He felt she was trying to be dramatic when their life hadn't a script with any drama in it.

'Because I'm fed up,' she said fiercely.

'I see,' he said. But he didn't.

'Is that all you've got to say?' she cried. 'But of course you never have anything to say, have you? My God, you're dull! Dull, dull, dull!'

'I'm sorry,' he said.

He was too shocked to say any more, and she was furious at his humility. She had expected him to make frantic appeals, to argue and fight about it, to be angry and try to talk her out of it. She was as baffled as he was.

He picked up his paper and used it as a shield against her till he thought of something to say. The idea that she would leave him was absurd, yet it terrified him. His retreat provoked her to another attack.

'I'm fed up living here,' she raised her voice again. 'And I'm fed up with you! Your laziness, and your selfishness. The way you sit there night after night and fall asleep on your chair. My God! I daren't contemplate life with you when you're middle-aged.'

'You'll be middle-aged too then,' he said behind his paper, very sour.

'My mind is made up,' she said. 'I'm going. In fact, my case is packed and in my car. I'm going tonight.'

'If that's what you want,' he said.

He saw no use arguing if her mind was made up. He wouldn't go on his knees to her and make a fool of himself.

'My God! Will nothing move you?' she screamed. 'Good God Almighty, are you made of wood?'

'I wish you'd leave God out of this,' he said.

'You sit there like a bloody turnip,' she said.

The remark hurt him. He knew he was putting on weight. He laughed it off by saying all rugby players put on weight when they gave up the game. But he thought it unkind of her to call him a turnip. Worse, a bloody turnip.

'There's no need to swear,' he said.

'You'd make a saint swear,' she retorted. 'A woman tells her husband she's leaving him, and the husband reads his paper and says all right, if that's what you want. What kind of a man are you?'

'I hope I'm too much of a gentleman . . .' he began to answer.

'Oh yes, always the gentleman, that's you!' she interrupted him. 'Never raises his voice or his hand to a woman. All manners and no matter.'

'I was going to say,' he continued, proud to be calm when she wasn't, 'I hope I'm too much of a gentleman to demand obedience from any woman. I respect you—'

'Thanks very much,' she said.

'I have never regarded you as my slave or my property,' he kept going.

'Aren't we noble!' she jeered.

She bounced out of her chair and walked round the room, pulling her fingers. His head swivelled to watch her as he went on talking.

'I've always respected the freedom of women. You know that. If you want to go, and your mind is made up, I can't stop you. You're a free person.'

'You mean I can go and you don't care?' she asked over her shoulder.

'That's not what I said,' he replied. 'I care very much. You've never given me any reason to think you felt this way. But what I feel, what I may suffer, doesn't concern you if your mind is made up.'

She went back to her chair and tried again to make him understand.

'I'm in a rut,' she said. 'I haven't had a holiday abroad since the day we were married. It has always been your fishing and golfing holidays. I'm fed up with it.'

'You've never complained,' he said.

'I'm complaining now,' she told him. 'It's eight years since I was last in Germany, and I've never been anywhere in Austria. The places I've never seen! I'm stuck here in this house all day, and there's not enough in it for an intelligent woman.'

'You could get a job if you want to,' he said.

'And look after you as well?' she challenged him. 'You want me to do two jobs?'

'Other women do two jobs,' he said.

'That's not the point,' she snapped. 'I want away from you, and that's all.'

'Well, if that's what you want,' he said again.

She nearly apologised before she left, as if at the last minute she felt she was treating him rather harshly. She said there didn't need to be anything final about it, she didn't want a divorce or anything like that. But she must have a change or she would go off her head. She would think about it again after a month or two. Then she told him she had got herself a job in a translation bureau, and she had arranged to share a service-flat with Shona McGregor. He was amazed at her duplicity in planning it all without ever giving him a word of warning.

'I'll get in touch,' she said. 'Not at once, of course. But later on. And we can discuss how it's working out.'

'Yes, that's fine,' he told her. 'We'll do that.'

When she left him he went about his work in a state of anesthesia. Nothing seemed quite real. He didn't sleep well.

He told nobody he was living a bachelor's life again, and he didn't think his wife would go around telling people she had left him because she was bored. Yet it was soon common knowledge. Perhaps he helped to make it so, for he was clumsy in deceit the first time he was asked about her. He was brooding over a drink at the bar before going for lunch in the Malmaison when he was slapped hard on the shoulder. He turned irritably. It was Bob Ramsay, a genial fellow who was scrum-half with him for a season in the university fifteen.

'Hullo there!' he welcomed the intruder.

'Hiya, Jack?' said Ramsay with a grin of manly affection. 'Long time no see, eh?'

They shook hands, agreed it was over a year since they met. When their chat rambled on they bought each other a drink, and then another drink.

'And how's Jean these days?' Ramsay asked at the fag-end of their conversation.

'Oh, Jean?' he said cautiously. 'Well now—Jean—she's gone to her mother for a week or two. Just for a change of air, you know. She's been off colour lately.'

Ramsay frowned at the limp falsehood.

'Back to her mother? I thought her mother was dead.'

'Oh God, so she is!' said Jack. 'That's right. I forgot.'

He splayed his fingertips across his temple, his elbow on the bar. He had meant to have one short drink only before lunch, and now he was on his fourth. It wasn't that he couldn't carry his liquor, but it was the wrong time of day. He felt silly.

Ramsay squeezed his arm, shook it gently.

'Come on now,' he wheedled. 'Tell the truth. You're not looking yourself at all, Jack. You've lost weight. What's going on? Tell me.'

Jack told him.

'I know what's the matter with her,' said Ramsay.

'Yes?' said Jack.

He was eager to listen to anybody who would talk about Jean.

'She's had things too damned easy,' said Ramsay. 'You've been too soft with her. If she had a couple of kids to look after, and no money of her own, she wouldn't act so high and mighty. She's a spoiled girl. Always was. Too much money behind her.'

'Money has nothing to do with it,' said Jack. 'Jean and I were never hard-up, but we were never well-off.'

'That's what you think,' said Ramsay.

'She was never a spoiled girl with me,' said Jack. 'There was nothing she wouldn't do for me.'

'Except live with you,' said Ramsay.

'Well, that's the problem, isn't it?' said Jack. 'It doesn't make sense. I never expected it. I can't think what came over her.'

'You take my Kath,' said Ramsay. He too was affected by extra drinks at midday, and he spoke with foolish pride. 'That girl didn't bring me a penny when I married her. And you know the old man left me a lot less than I thought he had. So Kath goes out to work mornings in a prep school and looks after me and our wee girl and runs the house, and she's too busy to be bored.'

'Good for her,' said Jack.

'Yes, she's a good woman, my Kath,' said Ramsay. 'We pull together.'

'Good for you,' said Jack, but it was more of a snub than a compliment.

'All right,' Ramsay apologised. 'I always say the wrong things, don't I? I'm sorry, Jack. But I do feel for you.'

He wanted to help. He told Jack there was no point going about looking miserable and feeling sorry for himself. He coaxed him to come to a club where four or five old rugby players met every Friday night for a drinking-session. He said they often spoke of him. They still remembered the great try he scored after a forty yards run when he played in a select fifteen against the London Scottish.

The welcome he got when Ramsay took him there was like the kiss of life to a man rescued from drowning. He moved from the dark of loneliness to the light of company. His memorable try was mentioned in the course of the evening and he felt he was a person of some standing among old friends.

He went back fuddled to his empty house. Somewhere in a bureau-bookcase there was an envelope with presscuttings from *The Scotsman* and the *Herald*. His reborn ego was confirmed when he read again the report that said his try against the London Scottish was 'a thrilling performance'.

'And she called me dull!' he said, swaying. 'Me? Dull? It's not me, it's her.'

His moping days were over. He began to look at the many girls the firm employed. It was, he believed, a purely aesthetic interest in the walk and figure of certain females. Then his secretary went away to look after an ailing father. He was given the smartest girl in the typing-pool as a stand-in. She was young and pretty, and she became very congenial to him. From the way she always hovered at the door before she would leave him, the way she looked at him tenderly, he guessed she knew his wife had left him. Her fond young eyes were silently saying she was sorry for the wrong done to him.

She wasn't the only one who made him feel better. All the girls went out of their way to be nice to him. It gave him a twisted amusement to see people being sorry for him when he was trying to stop being sorry for himself.

He was coping quietly with his new life when a senior colleague's secretary left to get married. It was a surprise. She had been with the firm for years, and she was turned thirty. Nobody ever thought she had a life outside the office. And because she had given such long and excellent service there was an office party and a wedding presentation, with plenty of drink and a buffet. He was stuck in a corner most of the time with one girl after another.

His temporary secretary came very close, shoulder to shoulder, thigh against thigh. He cuddled her discreetly, his hand squeezing her waist, then under her arm to fondle a breast. She was flushed with sherry followed by gin, and he was carefree with whisky.

The incident made him ambitious, but he didn't start an affair until his new secretary came. She was a beautiful slim brunette, efficient and attentive. He meant nothing by it when, looking at the *Herald* over mid-morning coffee, he remarked there seemed to be a lot of hotels opening round about the city. She read the full-page advert he showed her, and commented on the picture of the luxurious lounge-bar and the dining room.

'Looks super,' she said. 'I do love a drink and a meal in a place like that.'

To prove he wasn't so dull that he couldn't take a hint he asked her to go there with him and see how the reality compared with the advert.

He behaved very prudently when he took her out. He was no excited schoolboy, he kept telling himself. He could wait and see. He didn't even attempt a good-night kiss when they parted. It was three weeks before he took her at the end of an evening to the bleak house where he lived alone. He started to explain once more about his wife, but she said it didn't matter. She had heard it all already, and she wasn't bothered.

'It's her own fault if you—' she said, and stopped.

'A man like you needs a woman,' she tried again.

She made it easy for him, but he didn't often take her into his house overnight. He didn't think it wise to get involved with a woman working in the same firm, and he had no complaints when, as calmly as she started the liaison, she told him it would have to end. The man she was engaged to marry was coming home from an eight months tour of duty with an oil company in the Middle East. She had never mentioned any man before. He was surprised again how secretive women could be when it suited them.

'I was just as lonely as you', she explained. 'But I didn't want to say I was engaged in case it sort of inhibited you. You're so moral really. Still, I think we helped each other through a difficult time.'

'Yes, indeed,' he said.

'I don't think I took more than I gave,' she said.

'Oh no,' he said.

When it was all over he was left with a feeling of gratitude rather than affection, and with that experience behind him he was confident he could find another woman whenever he liked.

But he didn't particularly want to start another affair. He drifted back to his bachelor's habits. Every week he had at least one drinking-session with Ramsay and others. Sometimes two or three of them went out on a pub-crawl for the sake of variety, wearing an old suit and raincoat.

There was a touch of daring in it, a quest for adventure. They drank pints from the city centre to the south side or east end, trying pubs that catered for queer types or rough customers.

They knew it was madness, but they enjoyed their pub-crawls and made them a weekly habit.

On those nights he left his car at home and travelled by bus. He was very strict about not driving when he had been drinking. And it was on one of those nights that Jean saw him from her car when he was waiting for a bus home. She was held up at the lights, and there he was, loitering at the kerb.

If he had gone to the bus station she wouldn't have seen him, but his company broke up at a corner where it was easier for him to walk on to the next stop than to go a long way back to the terminus where the Bearsden bus came in. He was so confused after an evening's heavy drinking that he didn't notice he was waiting at a Corporation bus stop instead of the stop for a country bus.

When the lights changed Jean reacted quickly. She made a left hand turn into a sidestreet, left her car smartly, locked the door, and hurried back. She was unhappy at what she had seen. He was rocking there blind to the world, round-shouldered and talking to himself. He looked wretched and neglected.

'What's he standing there for?' she wondered as she ran. 'And that old coat! I could have sworn I gave it away to a jumble sale. My God, he had let himself go. Oh, the fool! And why hasn't he his car?'

She was unhappy enough before she saw him. She was on her way back from Pollokshaws after a visit to an old girlfriend who had a husband and two little boys. Her visit was a flop. The husband disappeared five minutes after she arrived, as if she was a bore he couldn't be expected to endure. And the two little boys clamoured so much for attention that her conversation with their mother was a series of interruptions. She didn't like it. She saw it as proof that even from childhood the male insists on women giving him priority.

She was still in a bad mood on her way back to her service-flat, and she wasn't comforted to think what she would have to put up with when she got there. Shona McGregor never stopped talking, and she always had to have the radio or television on. It had given her a recurrent headache over the months, and she longed for some domestic peace and quiet. She missed her own corner and the chance to sit down with a book.

Jack was still rocking at the bus stop when she came running round the corner.

'What are you doing here?' she demanded, very strict with him.

'Waiting for a bus,' he said.

He didn't say it as a rude answer to a daft question. He was, as always, polite. He said it with a smile, patiently explaining what might not be obvious, even to a person of her intelligence.

'You're drunk,' she said.

'Not me,' he said, and raised a palm in protest.

'And you need a shave,' she said.

He rubbed his chin with trembling fingers.

'You could be right,' he said.

'Where's your car?' she asked.

'Car?' he said. 'Oh yes, car. Well, you see—'

He couldn't think. She was so severe she frightened him. He saw his drinking-sessions banned, his next affair stopped before it started, and an end to his free and easy hours of coming and going when he pleased.

'I've been phoning you every night for the past month,' she said. 'You're never in.'

'That's right,' he mumbled. 'I'm always out.'

'Oh, stop your nonsense,' she said. 'It's time you—'

'Here's my bus,' he interrupted her, and moved from the kerb to the road, his arm up in a signal.

'That's not your bus!' she called after him. 'That's a Corporation bus.'

He was on it and away, and he raised a hand in a parting salute from the platform.

She ran back to her car. She would drive on at speed and be home before him. He would have a long walk after he left the Corporation bus. She would go in and wait for him and talk frankly. Then she wasn't so sure.

'What if he's not going straight home?' she faltered as she doubled back to the main road. 'And that's why he took that bus. But then, where can he be going at this time of night?'

She drove on, arguing with herself.

'It doesn't matter,' she said in the end. 'He must come in sooner or later. And I'll be there. I'm going back home, back to my own house and my own things, and I'll wait for him. And I'll tell him something. I'll shake him.'

It was only then she remembered that when she left she had forgotten to take her key.

FRED URQUHART

1912–

THE LAST SISTER

For over twenty years Hutcheon the grieve's daughters, one after the other, had been maids at the Mill of Burnhill. First there had been Mary, who had come as a gawky child of fourteen. She was there six years, and by the time she had left to get married she had her sister Sarah well trained to take her place. In her turn Sarah left to get married, but Bella stepped into her shoes. And then after Bella came Nell and Nanny and Jean and Agnes. Mrs Crichton, the farmer's wife, had always hoped that Agnes would not follow her sisters, but after seven years Agnes decided to get married, too. And now Norma, the last sister, had come to be maid.

Mrs Crichton had often wondered what she would do after the 'last' sister. Who would she get so well trained, so dependable? For she had needed to train only the first sister. The following sisters had been trained by the sister immediately in front of her. As children each of them had had the run of the house and had seen how things were done and knew everything long before it was her turn to pack her kist and bring it from the grieve's cottage and cart it up the narrow wooden stairs to the small room above the scullery.

Mrs Crichton had had misgivings about Norma ever since she was a small, snottery-nosed child peering fearfully in the back door. Mrs Hutcheon had been well into middle-age when she'd had her, and there was a space of nine years between her and Agnes. Norma's birth had been a surprise to everybody, even to Mrs Hutcheon herself. At the time, Bella, the third sister, was maid in the farm-house, and Mrs Crichton had said: 'And what'll yer mother be callin' this one, I wonder? I doubt she'll have run out o' names by this time!' But Mrs Hutcheon, who had become an inveterate film fan since the building of a cinema in Auchencairn, had christened the child after her favourite actress. A most unsuitable name, Mrs Crichton had thought that day she had gone to see the baby for the first time, and she had had misgivings even then when Mrs Hutcheon grinned up at her and said: 'This ane'll be a maidie some time, I hope. The last sister. . . .'

All through Norma's childhood Mrs Crichton had thought about this, wondering if the day would ever come when she would have the sleekit little brat as a domestic. For Norma was sleekit. There was no other word for it. The other sisters, whatever their various and individual faults, were honest and above-board. But Norma was underhand and a thief.

Knowing this, Mrs Crichton had done her utmost to keep Agnes from marrying. She had successfully put a spoke in the wheel of all the maid's suitors, and for a long time she had thought she had managed, but Joe McIntyre had been stronger than all Mrs Crichton's hints and warnings, and after seven years Agnes had handed in her notice. 'I dinna ken what we're to do,' Mrs Crichton lamented to her husband. 'I've priggit and priggit wi' her, but no matter what I say she'll nae bide.'

'Well, well, if the lassie wants to up tail and get married we canna do anythin' about it,' Mr Crichton said. 'We maun make the best o' it, wife.'

Mrs Crichton sighed. She knew that the best would have to be Norma. Even though other maids had been available, she knew that she could not go past the last of Hutcheon's daughters. Norma was nineteen now, and she had been in several jobs already. She had been maid to young Mrs Moyes of the Barns of Dallow, but she had stayed there only a term. She had told people that the work was too heavy and that Mrs Moyes expected too much, but there was a rumour that Mrs Moyes had missed too many pairs of silk stockings and other articles. Then Norma had joined the ATS, but she had been put out of that. She said it was because of a weak heart, but different other stories went round the countryside. 'If she'd said because o' a weak head,' Mrs Stormont, one of the cottar women said, 'begod, I'd ha'e believed that quick enough!'

For, added to her other faults, Norma was simple. She was quite a pretty girl, but she had a vacant look. She was always giggling and putting up her hand to her mouth. And she was a dirty slut. She never bothered to tidy herself, except when she went to Auchencairn to the pictures. In the house she shambled about all day in her dust-cap and a huge pair of felt slippers. At eleven o'clock at night she would be found crouching in front of the dying kitchen fire, still wearing her dust-cap, writing letters.

She wrote and received a lot of letters. 'Norma's fan mail,' Mr Crichton called them. 'Who the hell does she get to write to?' he often asked his wife.

'Oh, she's got dozens of boy friends. God knows where she gets them all.' Mrs Crichton puffed with exasperation. 'I whiles wish some o' them would come here and see her sittin' like a craw nancy in front o' that fire. I wonder if they'd be sae keen on writin' to her then?'

Norma's amours were a continual source of interest not only to Mrs Crichton but to many others on the farm. Norma could not keep from telling everybody and asking their opinion about the merits of this swain or that. This infuriated Mrs Crichton, and she said: 'The quaen's a feckless jade. It's a wonder to me that the lads put up wi' her. I ken if I was a lad I widna be havin' my name bandied about like that by her or any other glaikit limmer.'

But Norma had no sense. 'Tammy doon at the Mains is after me noo,' Mrs Crichton heard her say one morning to two of the cottar women, who were waiting for milk at the byre door. 'I wonder what I should dae? I'm nae terrible keen on him, but . . . he's a nice kind o' childe. . . .'

'Has he speired ye yet, Norma?' asked Mrs Stormont.

'No, nae yet,' Norma giggled. 'But he's goin' to. I ken by the way he carries on.'

'Mercy, quaen, ye'd think ye'd had a lot o' experience,' Mrs Dickie said, winking at Mrs Stormont.

'Well, so I ha'e!' Norma giggled again.

'What aboot yon soldier laddie you met last week at the dance?' Mrs Stormont said. 'I thought you were keen on him.'

'Och, him!' Norma gave her head a toss. 'I'm for nae mair ado wi' him. He was tryin' to set me off against Biddy Smith at the Mains, and I'm for nane o' that.'

'Well, what aboot the sailor ye were goin' to get engaged to?'

'I dinna ken.' Norma put her hand up coyly to her weak mouth. 'He's a bit ower serious for me.'

'But they're better to be serious than the other way,' Mrs Dickie said.

'Och, I dinna ken,' Norma tittered.

'I dinna ken what to make o' that maidie, John,' Mrs Crichton said to her husband that evening. 'She's really nae wice. There she was standin' gawpin' and gigglin' while thae two cottar wifies took a loan o' her. I fairly had to bawl at her to make her shift. She was leanin' against the byre door as though fire widna lift her.'

'I wish fire would lift her in the mornin's,' Mr Crichton grumbled. 'She's that damned slow. She widna run even though a mole was howkin' it's way up her arse.'

'Really, John! There's nae need to be sae vulgar aboot it,' Mrs Crichton said. She clicked her knitting-needles viciously. 'I'm fair worried aboot her. She's the last sister, mind. I dinna ken what we're to do for a maidie if she gings.'

'I whiles think she'd be better to ging,' Crichton said, puffing his pipe moodily. 'The other Hutcheon lassies were nae bother ava. But this ane! I dinna ken what's come ower the young lassies nowadays. They have respect for neither man nor beast. This mornin' when I asked that limmer, Norma, why my breakfast wasna ready in time she had the cheek to say: "Well, why do ye nae come and boil the bloody kettle yersel'?"'

'Agnes would never have said that,' he said morosely.

'None o' them would ha'e said it,' his wife said. 'They kenned better than talk like that to their betters. I canna see why Agnes had to go and get married after a' thae years.'

'Well, well, I suppose the lassie felt the urge like everybody else,' Mr Crichton said, switching on the wireless.

'I never thought she was the marryin' kind,' Mrs Crichton said. 'If I had kenned she was for leavin' I would ha'e lookit for another maidie long syne; I widna ha'e put up wi' that Norma and her daft capers.'

Her needles clicked irritably in time with Oscar Rabin and his band. 'It would be awfa to have to train a new maidie after all thae years,' she said. 'And she'd be sure to follow the rest and get married afore long.'

'Ach, there's time enough to worry aboot that when Norma ups and gives notice,' Mr Crichton said.

But despite all her boy friends, Norma never seemed to get any nearer to matrimony. Mrs Crichton soon lost count of her sweethearts. There were always letters from new ones. 'Ye canna blame the lads for stoppin' writin' to her,' Mrs Crichton said. 'That slow, soft way she has o' speakin' would drive ony man to drink. And that face o' hers—faith, I ken it's bonnie enough, but it's got as much expression as a cow's hint end.'

'A cow's hint end has often more,' Mr Crichton said.

When she wasn't crouched in front of the fire, writing letters, Norma would spread all her photographs on the large kitchen table and pore over them. She wrote to all her favourite male film stars, enclosing postage, for their photographs; and she had photos of all her boy friends. 'I dinna ken which is worse,' Mrs Crichton lamented. 'Whether it's waur to see her moonin' like a sick cow over thae photos or whether it's better to see her wi' that writing-pad on her knee. Every time I see her gettin' it out, I feel like screamin'. I only wish she was as keen on scrubbin' as she is on writin'. Ye should see the blethers she writes, too!'

Norma was not content with writing the letters; she had to show them to somebody else before she sent them off. And Mrs Crichton was usually that somebody. Night after night Norma would insist upon reading over what she had written. 'Now do ye think that's all right?' she would say. 'If I say that, do ye think he'll ken what I mean? Or do ye think I should say this?'

'As if it mattered what she said!' Mrs Crichton said to her husband or whoever else would listen to her. 'She just gets an answer frae them once or twice, and then they stop writin'. Not that Norma seems to care. She's got new fowk to write to by that time. Faith, it beats me what the men see in her. They've never seen her in that dust-cap or else the game would be up.'

But apparently Norma did not think this. Slatternly dressed though she was at all hours of the day, she was continually going out of her way to ogle any man. She waved and shouted out of the windows to the farm-hands, and they were always chaffing her. If she had not been so exasperated, Mrs Crichton would have been amused at all the trouble Norma took. She would slither frantically in her large slippers from the front room where she was making a bed to the stair-window to lean out and shout: 'Ay, there, Wullie!' or 'Ay, ay, Geordie, what a cauld mornin'!'

Most of the farm-hands were safely married, so nothing was thought of their harmless chaffing of Norma. But things took a different turn when three hefty young Irish labourers came to the Mill of Burnhill at the May term. By this time Norma had been with the Crichtons for a year, and Mrs Crichton's fears about her leaving them had died down. She realized that there was nothing in Norma's letters and photographs, and so she worried more about her other faults, which were many. And she worried especially about her habit of petty pilfering.

But the coming of the Irishmen restarted all the old fears about Norma being the last sister. The Irishmen were bachelors, or said they were, and they lived in the bothy at the back of the steading. This bothy had been a stable, but some years ago Mr Crichton had converted it into a dwelling-house for his unmarried men. A regular pigsty of a place at the best of times, it rapidly became worse by the time the Irishmen had been in it for a few weeks. Every time she passed the open door, Mrs Crichton's nose wrinkled with disgust, and she complained bitterly to her husband. 'Can ye nae make them clean it up, John?' she said. 'It's a fair scunner. I'm sure they could easy take a brush to the floor and gi'e the windows a bit dicht now and then.'

'It's nane o' my business, wife,' he said. 'I pay them for workin' in my fields. They can live like pigs for all I care, so long as they do their work.'

'If ye're sae keen on cleanliness,' he said, 'why do ye nae send that limmer Norma ower wi' a brush and pail? She could easy gi'e the place a bit red up.'

'Ay, and maybe get red up hersel' into the bargain,' Mrs Crichton said. 'No, no, she's nae settin' foot inside that bothy. If she ever does, I'll have to know the reason why. I widna trust thae Irish loons as far as I could throw them. And I certainly widna trust Norma wi' them.'

'I widna trust Norma wi' anything,' she added.

The bothy-lads gave Mrs Crichton a sad time of it. They borrowed her towels and crockery, they demanded sheets and blankets, and they calmly proceeded either to break or dirty them all. 'My guid bath-towels that I got frae yer sister, Bess, for our Silver Wedding!' she lamented. 'My faith, ye should see the colour o' them now! I was past that bothy this afternoon and there they were, lyin' at the door, draped a' ower the auld horse-trough, and the colour o' a dirty kitchen grate. When I think that I've kept them bye a' thae years in the cupboard, to let thae filthy Irishers make them look like dishcloots. . . .'

The Irishmen did not care what they said or did, because they had no fixed roots in the neighbourhood. Knowing they were transient, they went to great lengths in larking with the local girls, and simple Norma was easy bait for them. Every night when they came to the back door to collect their cans of milk Norma nearly broke her neck rushing to greet them.

'Sure now, you're the loveliest girl this side of the Shannon, Norma me darlin',' one of them was saying one evening when Mrs Crichton came into the scullery.

'Och, now, Paddy McShane, ye're just kiddin' me,' Norma giggled.

'Kiddin' ye, is it!' Paddy made a mournful face. 'Sure now, you would be knowin' me better than that, surely. It's destroyed I am entirely for the love of ye.'

'Well, destroy yourself quick and get awa' hame out o' here,' Mrs Crichton cut in. 'Takin' a rise out o' a decent girl! Ye should be ashamed o' yourself.'

Paddy glowered, but he said nothing. He had already felt the rough edge of Mrs Crichton's tongue. He was a tall, loose-limbed blond youth, and he might have been termed handsome except that he shambled rather than walked, and he was always dirty.

'Come now, Norma, get on wi' the tatties and stop standin' there like a sick heifer,' Mrs Crichton ordered.

'Isn't Paddy McShane a smasher,' Norma said, picking up a knife and starting to peel a potato in a desultory fashion. 'I could go for that fella in a big way.'

'If ye do, I'll go for you in an even bigger way,' Mrs Crichton snapped. 'Wi' Mr Crichton's big walking-stick!'

As the weeks went past Mrs Crichton sensed rather than saw that an intimacy had sprung up between Paddy and Norma. Try though she did, she could never catch them together, unless in the most innocent way, and so she could not pin them down to anything. And then one night she found Paddy in the kitchen. And not only Paddy. The other two bothy-louts were standing in the doorway, and they grinned uneasily and backed outside when they saw her. But Paddy held his ground. He shifted his thighs lazily from their half-seat on the table and said: 'Good evening to ye, Mrs Crichton, we were after askin' Norma here if we could be gettin' the use of a couple more pots. Sure now, the last ones ye gave us are so full of holes that they can be holdin' nothing in them.'

'So it's pots you want now, is it?' cried Mrs Crichton. 'I thought that a tink like yourself would be able to mend the ones ye've got. Do ye think I keep an ironmongery here to be supplying ye all the time for nothing?'

After they had gone with two old pots that Mrs Crichton had used to boil her hens' meat in, she turned on Norma and said: 'If those stinkin' tinks set one foot inside this kitchen again, Norma Hutcheon, I'll give ye such a flytin' that your ears 'll ring for weeks afterwards.'

'I didna ask them in,' Norma whined. 'I'm nae keen on their company. I'd rather be on my lone wi' the cat than listen to their daft Irish tongues.'

But a few nights later Mrs Crichton found Paddy in the kitchen again. And then she knew what everybody else on the farm but herself had known for weeks. Paddy and Norma were courting. 'Ay, the Irish loon is sleevin' Norma,' Mr Crichton said when tackled. 'I hear tell that they're gettin' marriet in the spring.'

'And who told ye that?' Mrs Crichton cried.

'Geordie the cattler. He heard tell from one o' the other Irish childes.'

'Geordie the cattler should mind his own business,' Mrs Crichton snapped. 'Startin' up a rumour like that! Before we ken where we are that glaikit limmer will be givin' in her notice.'

'And what'll I do then?' she wailed.

She determined to put a spoke in Norma's wheel in exactly the same way as she had tried with Agnes. And to do this she attempted to enlist the aid of Mrs Hutcheon and of Agnes herself. If they had not lived so far away, she would have tried to get the sympathy and help of the other sisters, too. In fact, she did think of going the twenty-three miles to Inverbervie to visit Bella, but she couldn't persuade Mr Crichton to drive her there.

Mrs Hutcheon was not much help. 'Och, I dinna see what harm there is in it,' she said. 'If Norma wants to ha'e the loon, let her ha'e him. We've a' got to get marriet sometime.'

'But he's a filthy orra tink,' Mrs Crichton cried. 'Ye widna want your daughter to marry a pig o' an Irisher, would ye? Have ye seen the stir him and the others ha'e that bothy in? The stink would knock ye doon a mile away.'

'That'll nae worry Norma ava,' Mrs Hutcheon said placidly.

'Faith, I know that,' Mrs Crichton said. 'She and that Irish tink are a well-met pair. I dinna ken how decent fowk like yersel' and Hutcheon ever managed to ha'e such a gormless goose. And surely you and he widna want a Papish for a son-in-law?'

'Och, me and Hutcheon widna worry ower much aboot that,' said Mrs Hutcheon. 'We're nae what ye might ca' kirk-goers ourselves.'

'Ay, I ken that,' Mrs Crichton said.

She did not get any more sympathy when she called upon Agnes. That young woman still remembered Mrs Crichton's campaign to prevent her from marrying her own Joe. She said: 'I doot the best thing ye can dae, Mrs Crichton, is to start lookin' for another maidie. I hear that wee Jessie Stormont 'll be leavin' the school soon. If I wis you, I'd put in a word for her.'

'Ay, and ha'e that mother o' hers aye sittin' in ma kitchen, drinkin' tea and gossipin'!' Mrs Crichton cried. 'No, no, before it comes to that I'll apply to an agency.'

But she did not need to do that, because a few days later something happened to stop the intimacy between Norma and Paddy.

For months Mrs Crichton had been pestering her husband to build a fence near the back door so that it would keep her hens from straying into the kitchen and leaving their droppings on the floor. He had remained stolidly impervious to her cajolings and pleadings, and she had reconciled herself more or less to being bothered by the hens in the same way as she had reconciled herself to the idiosyncrasies of Norma. And then one morning the bothy-lads appeared with a tractor-load of wood and, directed by Mr Crichton with a measuring-tape, they started to hammer posts into the ground.

'Mercy on us!' Mrs Crichton cried. 'Dinna tell me this is to be my fence at last!'

Norma giggled and leaned over her shoulder to watch. 'Don't fence me in!' she sang, waving to Paddy. 'Let me ride through the open country that I lu-uve! Don't *fence* me in. . . .'

'You ride up to the front bedroom and get that bed made,' Mrs Crichton ordered. 'Or I'll give ye boots and saddles where ye'll feel them the most!'

Though she was glad to get her fence at last, Mrs Crichton wondered, as the day wore on, whether it was worth all the bother. Norma did scarcely any work. She leaned out of the kitchen window, shouting and giggling at the men, and whenever possible she would be outside, leaning against a post, hugging her breasts, gawping and ogling. By late afternoon Mrs Crichton was exasperated almost beyond words. And when from an upstairs window she saw Norma take out tea to them she prepared to give battle.

'I'll warm that quaen's ears for her,' she muttered, hastening down to the kitchen, lips pressed tightly. And she folded her arms and stood at the window, waiting until Norma would return with the tray and empty cups; for she was determined not to upbraid her before the men, knowing that her own reputation and not Norma's would be the one to suffer.

Norma sat on a barrow while the bothy-lads drank the tea. They had started to paint the fence with tar, otherwise she would probably have leaned against it.

'Sure now, and that was grand tea, Norma me darlin',' Paddy said, handing her his empty cup. 'Now, off about your business, girl, while we're after gettin' on with the good work.'

He picked up his brush and belaboured the virgin palings. Norma stood, hugging the tray against her waist.

'Och, what's a' the hurry?' she pouted. 'I've nothin' else to do. I'll just bide here and watch ye slappin' on the tar.'

'Sure and if you do, it's over yourself we might be tempted to slap it,' Paddy grinned.

'I'd like to see ye try,' she giggled.

'Would ye now!' he said, and he winked at the others.

'Ay, would I!' Norma tossed her head, confident of her own invulnerability. 'Ye'd never hear the end o' it, Paddy McShane, if one drop o' that tar went on me.'

Paddy laughed and dipped his brush into the tar-pot again. 'Ye never know what might be comin' over you, me darlin',' he said. 'We might not be after stoppin' at tar. We might put a few feathers, too, to make ye upsides with the other ould hens.'

'Just you dare!' Norma giggled.

Mrs Crichton, straining her ears at the window, could not hear what one of the other bothy-lads said, but she heard the roar of laughter. And then she saw Paddy flip his brush playfully at the maid.

There was a scream of rage from Norma. 'Look what ye've done, ye filthy brute! Look at my clean overall!'

'Clean! It was never clean, me darlin'.' Paddy laughed again. 'Sure now, and you wouldn't be after tellin' us a bare-faced lie, would ye?'

The youngest bothy-lad guffawed, and emboldened by Paddy's example, he flipped his brush at Norma. And then in a second they were all flipping their brushes at her, and she rushed screaming into the house.

Mrs Crichton was so overjoyed that she did not give Norma the row she had intended; instead she sympathized with her over 'thae filthy Irish tinks'. And that night when Norma spread her photographs and writing-pad over the kitchen table, she smirked with satisfaction. She was glad to see them again. There was no need to worry any longer about the last sister leaving. A Hutcheon had aye been maid at the Mill of Burnhill, and a Hutcheon would go on being a maid there. Photographs and all!

A week later Norma announced that she was going to marry a Pole.

Mrs Crichton was preparing to make jam. 'A Pole!' she said, pouring sugar into the jelly-pan. 'What Pole's this?'

'Och, it's a fella I met at a dance a while syne,' Norma said. 'I'd forgotten all aboot him, but he turned up at the pictures last night, and he's askit me to marry him. I might as well.'

'Norma Hutcheon, are ye in yer right senses?' Mrs Crichton cried. 'Do you mean to stand there and tell me in cold blood that you're goin' to marry *a Pole*?'

'Ay, and what for no?' Norma said. 'He's a smashin' fella.'

'A Pole!' Mrs Crichton cried. 'A Pole! Goad Almighty, quaen, ye winna be able to understand what he's talkin' aboot.'

'Ach, I ken fine what he's talkin' aboot,' Norma said. 'And, onywye, we widna aye be talkin'.' And she sniggered lewdly.

All morning Mrs Crichton argued with her while they made the jam, but Norma said her mind was made up. 'I'm for leavin',' she said.

'And when may I ask is the great event to take place?' Mrs Crichton said, thinking that perhaps sarcasm might work where all other methods had failed.

'Well, Alexei wants me to marry him next week,' Norma said.

'Next week!' Mrs Crichton almost spilled the jam she was pouring into glass jars. 'Next week! But ye winna be able to get marriet by that time. It's ower short notice. What am I to do? Who's to help me here in this big hoose on my lone? Who's to help me wi' Mr Crichton?'

'I dinna ken,' Norma said. 'And what's more, I dinna care.'

'But the ceremony 'll be in Polish,' Mrs Crichton said. 'Ye'll nae be able to understand one word o' it. Not one word.'

'I dinna care,' Norma said. 'I'm for marryin' him, and it would take more than you to stop me, ye interferin' auld bitch.'

There was a long silence.

Mrs Crichton replaced the jelly-pan on the range. 'Norma Hutcheon,' she said slowly, 'I've put up wi' a lot for your father's and mother's sakes and for the sakes o' yer sisters—nice lassies—decent quaens every one o' them—but I winna stand any more o' yer impiddence. Ging upstairs and pack yer kist. You leave this house this very night.'

'Even if I've to do without a maidie,' she said, turning her back.

Half an hour later Norma pranced downstairs, dressed in her Sunday clothes. 'I'll awa' and get ane o' the tractor boys to lift my kist,' she said. 'I'm damned if I'm humphin' it along to our house.'

Mrs Crichton did not answer. She was stirring another panful of jam. And other things had been stirring in her mind during the time Norma had been upstairs. Already Mrs Crichton had counted and recounted the rows of jars of newly made jam. She was sure that two were missing.

She watched Norma mince in her too-high heels across the steading, walking in a roundabout way to avoid the tractor-wheel ruts. Then she pressed her lips together and went upstairs to Norma's room.

The maid's trunk stood in the middle of the floor. Mrs Crichton gave a quick look out of the window before she tried the handle of the trunk. It was unlocked. She lifted the lid.

The missing two pots of jam were standing on top of the clothes.

Mrs Crichton bared her upper teeth for a second, then she tore off the paper-coverings and lifted the jars upside down so that the jam fell in a thick sticky mass over the entire contents of the trunk. Then she replaced the lid and went downstairs.

ROBIN JENKINS

1912–

A FAR CRY FROM BOWMORE

Macpherson wished the Reverend Donald Dougary would take his hand off his: not because it was clammy and feeble, or because in the big airport hall small brown-skinned men were showing their gold teeth in impertinent goodwill at such a show in public of affection between one white tuan and another, but simply because it confused and sullied his reactions to what the old minister was saying, or dribbling rather, for out of the corner of his wrinkled tired mouth saliva kept trickling.

'You could do it, Hugh. I've no doubt of that. You've got a natural piety. You were brought up to it: on that Skye croft long ago. It's still marked on your face. Isn't that so, Mary?'

Mary Macpherson, also uneasy about that old sweaty hand—it had tried to hold hers in the car, for no reason she could think of—nodded and tried to smile. She knew how offended Hugh would be by that much too intimate gesture. Though proud of her husband's solid presbyterian worth in this hot lush land of dissipated expatriates, simple-minded pagans, and sinisterly successful Roman Catholic missionaries, she knew he had weaknesses; these she tried harder than he to remedy, since she suffered from them more than he did.

With his other hand the old clergyman took a sip of the sweet locally made orange that only increased his thirst. He wished it was time to go on the plane. At his home in Singapore, a thousand miles away over the South China Sea, he would have drunk refreshing beer. Here, in Kalimantan, on his annual visit to perform christenings and give communion, he had had to be very careful not to offend his hosts, the Macphersons, who allowed no alcohol into their house, far less their mouths. In a hot thirsty land such self-denial was undoubtedly Christian, but, alas, also in this day and age a bit priggish. The truth unfortunately was that Hugh Macpherson, gaunt in the face as his native Coolins, was a good man and an expert engineer, but as diligent a prig as Mr Dougary had ever met. His affections, even for his wife and children, were never warm, or at any rate warmly expressed. Moral calculations entered into everything he did and said.

What they were now talking about, in the airport hall, his holding weekly religious meetings in his house, was really Macpherson's own suggestion, but since it was his habit to avoid giving himself credit for anything, preferring to let others do it, he had wanted Mr Dougary to put it forward as his idea. The effort to humour so humourless a man was making the old minister sweat, more than the heat.

Macpherson laughed. 'I wouldn't want to make too much of it,' he said. 'Just an extension of Mary's Sunday school.'

'Exactly.' The old man longed for the announcement that would release him. He had a headache. During the past three days he had been obliged to discuss man's religious duties more deeply than he cared nowadays to do. He was far too old for talk: he just wanted to lie and float on faith, like the children he'd seen yesterday on a lilo in the warm sea.

'All the same,' said Macpherson, 'if I did it I'd have to take it seriously.'

'Of course, Hugh. But don't forget there's joy in Christ too.'

A plane came roaring in to land, with a roar. The old minister, with a secret smile, imagined it was the Lord angry at being saddled with so many conscientious fools, like Hugh Macpherson.

'That's true,' said Macpherson. 'But you should know, Mr Dougary, a country like this gives opportunity to so many empty pleasures it can rot a man's soul. I've seen it happen in a dozen cases.'

'I'm sure you're right, Hugh.'

'Don't misunderstand me. I'm not saying it's wrong to enjoy yourself.'

You are, but you shouldn't, thought the old minister who'd seen the Macpherson's launch, their big white car, their beautifully appointed house, their three servants, their gold Rolex watches, and their well-stocked table. By comparison he himself, with his congregation of two hundred prudish, abstemious Chinese, lived like an ascetic, despite his beers. His wife too, poor Peggy, however she tried, and no woman had tried harder, had never been in pew or bed the pleasure that Mrs Macpherson with her fair hair, ample breasts, and sonsy buttocks must be. Had King David got his eyes on a woman like her, messengers would have been sent in the night.

Over the loudspeaker came some blurred Chinese English.

Mr Dougary rose. 'What's she saying?' he cried.

'It's not for you,' said Mrs Macpherson. 'She was just saying the plane from Keningau has arrived.'

He sat down again. 'Keningau? Where's that?'

'In the interior. No road into it, so you've to fly, or go by river. It's on the Pensiangan.'

'I hope you won't mind, Mr Dougary,' said Macpherson, 'if I write to you now and then for advice.'

'By no means. I should be only too pleased to help, if I can.'

Inwardly the old minister asked himself why a man in confident charge of over a hundred natives should make so much fuss about weekly prayer meetings with no more than half a dozen present, if he was lucky. Taking a charitable view, it must be because Macpherson, accustomed to making certain that the bridges he built from one bank of a river to the other were strong, was similarly thorough about bridges intended to carry a man's soul to God. Mr Dougary himself nowadays was inclined to risk it with any old frayed rope he could lay hands on.

Passengers off the Keningau plane were now passing through the hall. Some were natives, in coloured robes. One was a brisk spectacled Englishman, dapper in a white cotton suit; he carried a black suitcase.

He hesitated and then stopped at their table.

'Hello, Macpherson,' he said. 'The very man I wanted to see. How extraordinarily convenient. How are you, Mrs Macpherson?'

She smiled. 'This is Mr Dougary, from Singapore. Mr Dougary, meet Dr Willard of Api hospital. He operated on Morag when she had appendicitis.'

The doctor smiled: an Anglican himself, with a cathedral in the town to go to if he wished, he found amusing this once-a-year service of the dour Scotch Presbyterians.

'I'm just back from Keningau,' he said. 'With a message for you, Macpherson, as a matter of fact.'

'For me?'

'Yes. Sent on by a colleague of mine, Dr Lall, an Indian, attached to Keningau hospital.'

'Can't say I know him, doctor.'

'And he doesn't know you. Lall goes out on up-river calls. He's just back from one, and he brought this message: from a fellow countryman of yours, a planter at Pensiangan, called McArthur.'

'McArthur? I don't know any McArthurs here.'

'Well, he thinks he knows you. His description was accurate enough: divisional engineer, PWD Api, Scotsman, tall, lean, called Macpherson, with a fair-haired wife.'

'That's you all right, Hugh,' said the fair-haired wife.

Macpherson frowned and shook his head.

'According to Lall,' said the doctor, 'he's a man of substance. Beautiful house. Owns the estate, as well as manages it. Well, his wife does. She's a Dusun, daughter of the chief that owned the whole territory. They've got a couple of children.'

Macpherson went on frowning. It was none of his business, but he did not like white men marrying native women, amahs or princesses, and having half-breed children. Others did not like it either; but his reasons were religious. Instructed by Christ as well as by Jehovah, he knew that compassion must be shown in addition to disapproval.

'He's dying,' said Dr Willard, quietly. 'Of cancer. In considerable pain, I'm afraid.'

Mrs Macpherson gasped with pity. 'Poor man,' she whispered.

'He wants to see you, Macpherson. He asked Lall to pass on the message.'

'Me? But I don't know him. I've never set eyes on the man.'

'He must have heard of you; must have been impressed. You should be flattered. That was three days ago. Lall thought he had a week at most left. You'd have to hurry to get there in time.'

'I assure you,' said Macpherson, 'he's made a mistake. Perhaps he was delirious.'

'No. According to Lall he's remarkably clear-headed, considering. Perhaps because you're a Scotsman too?'

'There are at least a dozen Scotsmen in Kalimantan.'

'Well, you're the one he's asked for. Whether or not you go is up to you. I've done my bit; passed on the baton, so to speak. Cheerio. Glad to have met you, Mr Dougary. Hope you have a pleasant flight.'

They watched him stride through the crowd in the hall. At the exit he stopped, turned, hesitated, and came back.

'Sorry,' he said. 'Just remembered. This damned climate rots the memory. It seems McArthur was born in Islay, one of the Hebrides. So were you, weren't you, Macpherson?'

'I was born in Skye.'

'Well, isn't that next door as islands go?'

'Far from it. They're hundreds of miles apart.'

'We went on holiday to Islay about five or six years ago,' said Mrs Macpherson.

'Perhaps you met him there then?'

'No. No, we didn't. Do you remember, Hugh?'

'We met nobody from Kalimantan there.'

'That can't be it then,' said the doctor. 'The mystery remains. Well, goodbye again.'

Again they watched him go.

If he had turned a second time he would have seen Macpherson snapping his fingers in triumph.

'Got it,' cried Macpherson. 'My memory's usually good. Years ago, in the Sports Club. Some of us were chatting about unusual churches we'd seen. I mentioned the Round Kirk at Bowmore, in Islay. This man was sitting at the bar. He must have been listening. He was pretty drunk, if I remember. He got up and came over. 'I was born in Bowmore,' he said. That was all. Then he went. I said nothing. I hardly got a good look at him. None of us knew who he was, but somebody at the bar called over that he was a planter from Pensiangan. I can't even remember what he looked like. Can you, Mary?'

'I didn't even notice him, Hugh.'

'Well,' murmured Mr Dougary, 'whatever he looked like then, he looks different now, poor soul.'

Then came the announcement they were waiting for. Passengers for Singapore were to go to the departure gate.

The Macphersons escorted the old man. He quavered thanks for their hospitality: perhaps they would see him again next year, if he was spared.

Then at the very last moment he muttered, 'Go and comfort that poor man.'

Macpherson had no time to reply. He gazed after the old dotard in astonishment.

'Did you hear that?' he asked. 'Did he say I should go to Pensiangan?'

'That's what he said, Hugh.'

'I thought so.' Still astonished, Macpherson watched the old man staggering across the hot bright runway; a hostess helped him up the steps into the plane. He had been too long in the tropics, thought Macpherson: he had lost that fair but tight grip on essentials which every effective Christian must have.

'I'm afraid he's past it,' he said, as they made for their car.

'He does his best, Hugh.'

'Yes, but half the time he's not listening to what you're saying. Then out of the blue to say a thing like that!'

He shook his head, still astonished: she kept hers very still; what astonished her was his astonishment.

II

For the rest of that day Macpherson never once thought of Mr Dougary or of McArthur. Until five o'clock he was much too busy with his work where, indeed, his Christian forbearance was put at strain. This wasn't because the Chinese drivers of the big earth-moving machines, or the numerous native labourers, some of them women, provoked him with laziness and inefficiency. On the contrary, they all worked with their usual honesty and diligence, almost as if, he had often remarked to Mary, they were doing it, not for money—not that they got much—but for pride in themselves. Pagans or Buddhists of a peculiar sort, they gave many Christians a showing up.

No, what vexed him that day was what had been vexing him for weeks: this was the stupid stubbornness of an old native whose miserable ramshackle hovel happened to be right in the path of the new double carriage-way. Other much better houses had already been bull-dozed out of the way, their owners having eagerly accepted compensation. This old nuisance had refused. Work would soon be held up. The trouble was, owing to the political situation in the country, not long after Merdeka or freedom from British rule, with each of the three parties aspiring to be the champion of the aboriginal natives, compulsion was not to be used. Moreover, though the road would be to the whole country's benefit, those native politicians, with childish perversity, seemed to enjoy watching the mighty white tuans of the PWD being thwarted by one old man with a face like a coconut. When it suited them they would very quickly throw the old fool aside, and reduce his compensation for having made a pest of himself.

In the meantime, however, work would soon be halted.

As an engineer, Macpherson had hitherto left the negotiations to the PWD administrators; but that day, accompanied by his assistant Jock Neilson, he went to the house and tried to persuade the old man to give up. Obstructive senile obduracy, exhibited with imbecilic cackles, is never easy to oppose with patience or reasonableness or Christian tolerance. Macpherson tried hard, standing in that stinking little house with flies on his lips; but when he looked out of the window and saw not far off the great orange hungry machines that in a day or two would be baulked by this small, brown, skinny, 'ragged-arsed', to use Jock's word, semi-imbecilic native, he had the greatest difficulty in reminding himself that in spite of everything this was a human being, a soul capable of worshipping God, and not a decrepit animal to be kicked out of the way. He had been harsher than he should.

When he got home he said nothing to Mary about his failure. As much as possible he liked to protect her from his professional troubles. After a shower and a cup of tea, he took the two children, Donald aged seven and a half and Morag aged nine, a stroll along the beach near their house. Soon the sky, and sea, and the sand, and their faces grew pink with the usual magnificent sunset. Better still, their young souls were peaceful with it.

An observant onlooker, such as his wife from her verandah, would have noticed that though he was obviously proud and fond of his children, and they of him, still there appeared to be a strangeness or awkwardness between them. Mary of course was most keenly aware of it. She knew it was because of his own upbringing, when affection might have been felt but never was spontaneously shown: as a result his had grown stiff and awkward.

More intimately still, she knew that he made love as though there was something impure in it; it made no difference that they had God's permission, having been married by a Church of Scotland minister in one of the best churches in Glasgow. With a similar puritanic upbringing, she had started out with the same foolish feeling of guilt, but she could easily have overcome it if Hugh just once had given joy full rein. Instead he had curbed it resolutely. Their love-making was always done in the dark, even the moon being shut out; and there were no preliminaries of affectionate nonsense. When he was finished he went to sleep as fast as he could, as if fleeing from something.

Often she lay in the bed next to his wondering what she could do to save him, though she could not name what it was he needed saving from.

That night, after Mr Dougary's departure, suddenly reckless, for the first time in her marital life, before he had time to escape into sleep, she said: 'Hugh, that poor man in Pensiangan, who's dying of cancer, who asked you to go and see him, don't you think you should, as Mr Dougary said?'

He said nothing, did not move. She was sure he was only pretending to be asleep. She did not have the necessary courage or vindictiveness or hope to say it twice. After four or five minutes of increasing disappointment she sighed, and prepared herself for sleep.

Suddenly he heaved round to face her.

'Did I hear right?' he asked. 'Did you say I should go to Pensiangan?'

'Yes, Hugh, I did.'

'To see a man I wouldn't even recognise?'

'Is that so important? He's dying.'

'Mary, this is Asia. There are millions dying.'

'But he's one of us, Hugh, a Scotsman.'

'Does that mean he's any better than all those millions of others?'

'No, but it makes him closer to us.'

He shook his head. 'Let's get this clear, Mary. I'm as sorry for him as you are. But pity's cheap, too cheap. We've got to keep up standards. Pity can lower them, you know. You're speaking as if he was alone. He's married, isn't he? He's got his wife to comfort him. Whose fault is it she's a native? He's got children, and in-laws. He's far from being alone. He made his choice. Good luck to him. But why should we lower our standards to accommodate him, or anyone else?'

She did not understand. His anger was as unintelligible as his meaning. He had no imagination. She had little, he had none. She remembered his mother, the devout Skye woman, telling her that proudly: her Hugh always gave things their true value.

'This is something that goes beyond nationality or even principle,' he said.

'Yes, Hugh. I'm sorry. Good-night.'

'Good-night.'

He was asleep inside five minutes. She lay awake for much longer, tears in her eyes, only now and then thinking of the unhappy man at Pensiangan.

III

Next morning Macpherson had forgotten their post-coital disagreement and the man who had caused it. He was very patient with the children at breakfast. Usually they were noisy and talkative, asking each other and their parents half a dozen shrill questions a minute, without waiting for answers. Previously their father, when asked something had insisted on answering and on being listened to. Recently though he had decided it was a mistake to attempt to discipline his children as rigorously as he had been disciplined at their age. Without anyone telling him, he had become aware that perhaps he was too slow and awkward at expressing his feelings. He must not imprison his children's feelings in the same long enforced silences.

Therefore this morning, when he was in the midst of an explanation as to why chichaks were able to walk on ceilings, he just smiled and shook his head when the children began to chatter about school.

He did not find it easy to see as harmless childish illogicality what he thought was downright impudence. He still rebuked them, but not as sternly as before. Morag seldom wept now when he scolded her. She kissed him good-night more fondly.

He was on his way down to the car when the telephone rang. Mary answered it.

'It's for you,' she called down. 'Dave Sloan.'

He came hurrying up. Sloan was his boss, the Director of the PWD. About to retire in less than a year, he had more or less promised to recommend Macpherson as his successor. His qualifications as an engineer were not as good as Macpherson's.

'Hello, Dave. An early call surely. You just about missed me.'

'Sorry, Hugh. Something's come up. Another of these damned conferences. I just heard about it ten minutes ago, by phone. What you have to expect now that Jack's the master. I expect you know I was going to Keningau today, to consult with the Aussies about that damned village they want to knock down.'

At Keningau the Australian army, as a Commonwealth gesture, was building a road to open up part of the interior. A village stood in their way: they wanted to knock it down and build a better one a hundred yards or so away. The trouble was some holy trees would have to be chopped down too, and this the villagers objected to. Pagans though they were, Macpherson was inclined to sympathise with them: men willing to forego new houses for the sake of ancestral gods however heathenish were, in his view, to be complimented in an age so greedy for material benefits.

'Could you go in my place, Hugh? It's very short notice, I know, but it's really an engineering problem. You know more about building roads than I do.'

'Pleased to go, Dave. No bother.'

'I don't want to influence your decision, but the word is here at the Secretariat that the Aussies are to have their way.'

'I must say I've got a lot more sympathy for these villagers than for that old fellow that's holding up my own road.'

'Sure. Well, the plane's at ten. Your ticket will be waiting for you, and you can have my room at the Rest House. All right?'

'All right, Dave.'

Macpherson put the telephone down. He felt pleased. It certainly looked as if the succession was his. Being Director for a few years would make a big difference to his pension.

'What did Dave want?' asked Mary.

He told her, and was puzzled by the way she put her hand to her mouth as if to stifle a cry of amazement.

'I'll be back tomorrow,' he said, smiling.

'It's not that. I just thought, how funny. First you happened to be at the airport when Dr Willard arrived, and now you've been asked to go to Keningau.'

Having again forgotten McArthur, he saw no connection. 'What's funny about that?'

She saw that he was genuinely puzzled. 'Nothing. I'm being silly. I'd better get your bag packed.'

There were tears in her eyes again. He thought it was because he would be away from home for a night. He was touched, yet puzzled. She had never been a woman given to easy tears. He would never have married her if she had. It must be that her period was due. Perhaps that was the reason why last night she had behaved so oddly. Her nails had dug into his back. He had had to tell her not to.

Two hours later, flying high above the Pensiangan valley, with the jungle below like a vast garden of curly kale, he thought about McArthur. To die in that land of thick jungle, primitive villages, and hot, steamy climate was sad for any white man but especially so for one born in Islay, that island of delightful breezes, invigorating sea, and fresh green hills. Still, what difference did it make where a man died? One place was as near to God as another. For a few minutes of luxurious supposing, Macpherson pretended it was him dying in the Pensiangan jungle, far from Mary and the children. Thank God they were safe home in Scotland. He hoped he would not be sorry for himself: he would not make appeals to strangers. If he was too weak to read his Bible he would recite passages learnt in childhood and never forgotten. He would be an example to the heathens watching him die: they would talk about his Christian fortitude afterwards for years, and to that degree at least would be converted.

Just before they were over Keningau the plane lurched and dropped suddenly: a box fell off a shelf. Some of the other passengers, natives, looked as apprehensive as their dark thick faces could. Macpherson knew there was no danger, it had been only a freak current of air; already the plane was steady again. But for a second or two he, like the rest, had tasted the possibility of imminent death, and had been, for that fraction of time, terrified: Christian consolation had not been quite so swift as terror. This happened in his mind so quickly, and was over so soon, that he was never fully aware of it. He did not either notice any contrast between his moment of real panic and his long, slow, patient, faithful, agonised, pretended dying in McArthur's place.

Before the plane landed on the grassy runway he had assumed the role of deputy of the Director, first in command in that area. In this capacity he greeted Phil Barnes, Superintendent at Keningau, who was waiting with a Land Rover. On the way into town he asked what arrangements had been made.

Barnes was a quiet red-haired man with an enormous number of freckles on his brow and his knees. He was known to enjoy conviviality. The coming of the drouthy Australians had been a godsend to him. He did not like Macpherson, but thought he concealed it very well. An honest man, he could not help feeling a little embarrassed as he remembered how, last night in the soldiers' mess, exuberant on whisky and Foster's beer, he had described the divisional engineer as 'good at his job, better than anybody else to be sure, and a dour hater of strong drink'. He had said it in a Scotch accent, which he could imitate very well.

'We're meeting the Aussies at two at the village,' he said.

'What's your own opinion?'

'They're lucky to get offered a new village. The present one's filthy and falling to bits.'

'It's the trees they're worried about. They're sacred.'

'As I was saying, why can't they carry their ancestors' spirits with them, and instal them in new trees?'

'It's not funny to them, Phil.'

Barnes replied in a high-pitched sing-song querulous voice: 'To chop down their trees will be sacrilege, no less.'

'Who's that supposed to be? Not Maluku?'

'No. A doctor at the hospital, Lall, an Indian. Passionate little character, especially after a couple of pints.'

'I thought Indians didn't touch alcohol.'

'He does. He thinks it shows he's a man of the world.'

The name Lall had meant nothing to Macpherson. 'Shows he's a fool, more like,' he said. 'What business is it of his?'

'None. One dark skin supporting other dark skins against white skins. He's a lot darker than they are, almost black.'

'His colour doesn't matter. But we can ignore him. What about Maluku, the assembly-man?'

'Sniffing around for baksheesh.'

'A bribe, do you mean?'

'Sure. Mind you, what he gets for being an assembly-man wouldn't keep you in beer.'

'Very little would keep me in beer.'

'Sorry. You know what I mean. So he thinks he's got a right to add to it any chance he gets.'

'Well, he'll find this is no chance.'

'All the same, in these parts a little baksheesh can make things go a lot more smoothly.'

'I don't believe in bribes.'

'Oh well, if you call it bribes.'

'What else is it?'

Macpherson did not approve of bribes because, as he had more than once explained to Mary, when he was offered one himself he felt insulted; and, as he also explained to her, not noticing her slight incredulity, he always applied to other people, whatever their colour or station in life, the same standards by which he judged himself.

One bribe offered him had taken the form of a headman's bare-bosomed daughter with lice in her hair. She had been sent to his bed in the guest hut. He had received her courteously and asked her to get rid of some of the cockroaches with which the place was infested. She had done so, picking them up in her hands and throwing them out of the door. Then, with her hands unwashed, she had slipped off her one garment and lain down on the thin mat that was his bed. He had had difficulty in making her go. She could not understand, just as her old father hadn't understood why Macpherson had refused to take any tapai or rice beer. Even if she had been beautiful and clean he would have sent her away. When he saw her next morning, looking dejected, he had been surprised at the feeling of disgust that had surged up in him. He had hated her very nipples.

At the conference held that afternoon in the village under sentence he took an instant aversion to Maluku, who kept on talking with all the nasty ingratiation of the brown man in power jealous of the white man his servant but also overwhelmingly his superior. The villagers, on the other hand, won his sympathy. They were not only humble and respectful, they were also genuinely concerned about their sacred trees: if these were cut down they believed their own souls would be snatched away. Macpherson reassured them this was not so. He tried gently to remove their heathenish fears. He praised the trees as magnificent.

In the end of course he had to agree with the Australian engineers that to make the necessary detour would involve too much delay and extra work. He broke the bad news to the villagers as humanely as he could. He pointed out that the new village which would be built for them—it would take the sappers a couple of days—would be clean, free from the filth of generations which had brought so much disease. (Too many were consumptive and syphilitic.)

It would have been too much to expect them to be convinced, but they were obviously grateful to him for the trouble he had taken to make the position clear. It was him they kept clamouring round. When they asked him what would happen to them if they did not consent to their homes and their shrines being demolished he had to answer truthfully and regretfully that they would be shifted by force. Progress could not be held up by stupid or superstitious people. He took trouble to explain to them what progress meant.

To the whining Maluku he was outwardly most affable. As a person the man was a sneaky self-seeker, deserving contempt; but as an assembly-man and therefore a representative of the country he deserved respect. Macpherson did not hesitate to show it. The way to the Directorship was lined with Malukus.

He would have been angry but not ashamed if he had been in the Australians' mess that evening and heard Barnes mocking the diplomatic way he had spoken to Maluku. He would have reminded Barnes that in the UK he, who had only a mechanic's qualifications, would be lucky to be foreman of a garage.

However, since he declined the invitation to the mess, and never for a moment suspected he would be the subject there of drunken and ribald mockery, he felt content with himself in the Rest House that evening. He had it to himself, which pleased him. The last time he had been there two educationists, one Canadian and one from New Zealand, were his fellow guests. They were on an inspection tour of interior schools. Not only had they drunk too much whisky, they had also gone out on the prowl for whores. At breakfast they had expected him to laugh at their exploits.

Though the chef was Chinese, noted for his bamboo shoots, Macpherson asked for a plain meal of tomato soup, steak, and cherry pie. After eating he read an adventure story while waiting for his call to Api.

The line was bad. It blurred Mary's voice and gave the absurd impression she was reluctant to talk. He mentioned this to her.

'You sound cheery enough,' she said.

'Well, I've got reason to be. I think I handled it well. I'll be back tomorrow.'

'So you're not going to Pensiangan?'

He did not know what she meant. He had forgotten McArthur. 'I said nothing about going to Pensiangan.'

'I thought you might have changed your mind.'

'It never was on my mind. What are you talking about?'

'I'm talking about Mr McArthur, that's dying.'

He remembered at once. 'I thought I'd explained that.'

'I haven't been able to get him out of my mind.'

She seemed to be weeping.

Trying to be fair to her, he forced himself to think again about McArthur, and assess his own responsibility. He could not understand how his own wife, who knew how conscientious he was, could ever expect him to drop his work and go off into the wilds to see a man he'd never even spoken to.

'He's a Scotsman, Hugh. He was born in Bowmore. Remember you said you might retire to Islay? You could never be happy there if you turned your back on him.'

Yes, she was weeping.

Compassion for her, and anger against her, grew in him equally. He loved her for her kind heart but resented her hypocrisy. He had seen her in India in tears of vexation because some maimed and blind beggars wouldn't give her peace. For the past eight years she had lived in Api in great comfort, oblivious to the hungry, dying millions of Asia.

He let her weep. She struggled to subdue it. He loved her and was proud of her. At bottom she was a sensible, loyal, trustworthy woman. She was now honestly realising that the emotion she had just shown was extravagant and, in the circumstances, more of a hindrance than a help.

'Never mind,' she said, dully. 'I'm sorry I spoke. You'll be home tomorrow then?'

'Yes. The plane's in about ten. Will you be there?'

'Yes, I'll be there. Good-night.'

Surprised, and rather hurt, by the abruptness with which she had ended the conversation, he went into the big dimly lit lounge that hummed with mosquitoes, and found there, waiting for him, a small fat bald very black Indian with a smell fragrant but nasty. He bounced up, hand held out, and announced himself, almost ecstatically, to be Dr Lall.

'And you are Mr Macpherson, of the PWD, Api. Tall, lean, with a hard face, and a soft fair-haired wife: just as poor McArthur described you. Pity the said wife is unfortunately absent. How do you do, Mr Macpherson? You are a kind man, a stern man as I can see but also a kind one. So my good friend Dr Willard conveyed to you my message, and here you are, en route to Pensiangan. How wonderful is humanity.'

'Just a minute, Dr Lall. You've got it wrong, I'm afraid. I'm here in Keningau on PWD business. Tomorrow I return to Api. I'm going nowhere near Pensiangan.'

Lall, an exaggerating fellow if ever there was one, shrank back as if a cobra had just struck at him.

'For God's sake, man,' he cried, 'you must go to Pensiangan. A hand has been held out, you must take it.'

'Mr McArthur seems to have made a mistake. I don't know him.'

Lall had been drinking: as well as that unpleasant perfume, there was a stink of beer off him. The tears that now came into his brown bloodshot eyes were beery. He was trying to make a tragedy out of what was an unhappy but very frequent occurrence, a man dying in pain. It was a fault of his race. Small wonder a handful of level-headed British had held countless millions of them in subjection for over a hundred years.

'He has his wife and family with him,' he said, reasonably. 'And all his wife's relations. I should think strangers wouldn't be welcome at such a time.'

Lall rose on his toes; even so he hardly came up to Macpherson's chin.

'Mr Macpherson,' he said solemnly, 'you are a man without vision.'

Macpherson took no offence. It would have been ridiculous letting himself be provoked by a fat half-drunk overheated Indian doctor with pretensions to be a poet or prophet. India, after all, was the land of fakirs.

'I see what's there,' he said. 'That's vision enough.'

'If you do not go to hold that dying man's hand, do you know what you will have done? You will have stopped the stars in their courses.'

Macpherson grinned. 'I'll take that risk.'

'You make us all take that risk. Imagine, please, it is your brother who is dying yonder in the jungle.'

'As I said, Dr Lall, I only see what's there. He isn't my brother.'

'Is not every man your brother?'

Macpherson shook his head. He was not to be enticed by that nonsense.

He had too often been in companies in Scotland when that brotherhood of man had been proclaimed with far more passionate tears than this little black man was capable of. Next day he had seen those fervent proclaimers of 'a man's a man for a' that', sober again, with dry eyes and a hard-faced acceptance of man's selfishness and the limitations it would impose forever.

'That's too easy to say,' he said.

'Forgive me,' whispered Lall. 'I see now you are a most unhappy man. You are a desert, not a green field. You, not Mr McArthur, is to be pitied. He has been very happy in his day. If I was a man of religion, sir, if I believed in God, I would pray to Him for you; but since I do not, since I believe that man must achieve his own destiny, all I can offer you is my pity.'

Then out he walked, or rather waddled: his very gait showed up his words as ludicrous.

From the verandah Macpherson watched and listened. The stars were bright and normal. The cicadas' hum was very loud. It appeared that the odd little man was either talking to himself or weeping. Once again it became obvious why the British, with so many Scots among them, had been able to conquer India and hold it so long.

Perhaps it was because this superiority of the Scot over the Indian was in his mind, as he was cleaning his teeth and looking at his face in the bathroom mirror, that the resolution, though it was nothing so dramatic at first, occurred to him. He was not merely cleaning his teeth, he was also saying his prayers. This he had done all his life. As a boy in Skye he had knelt on the cold waxcloth and shut his eyes tightly, but since then, when he could please himself, he had sat on the edge of his bed, or stood in front of the bathroom mirror as at present, or even, with no sense of impropriety, perched on the lavatory. In Skye too there had been a fixed form of words, to be spoken aloud, with grim humility. This also had changed. Adult prayer was best silent: nothing now about sins repented or favours begged; only a flicker or two of gratitude for another day safely and prosperously fulfilled. At some times the gratitude was warmer and more spontaneous than at others.

Suddenly then, with the toothbrush arrested, into his mind to join the thankfulness came the suggestion, the proposal, the challenge. It was not in any way inspired or provoked by Mary or by Dr Lall, for he hadn't been thinking of them at all. It simply rose out of the depths of his prayerful mind.

He found himself asking why not go to Pensiangan and see McArthur. It would be against his nature, against his common-sense, against his lifelong habit of avoiding extravagance in spending and living; and therefore it would be proved a mistake, in some way he could not foresee. His objections would be vindicated, Mary would have to admit his judgment was better than hers.

Next morning, when he awoke, he was rather surprised to find that what after all had been only a vote of self-confidence had, during sleep, hardened into a determination. If anyone had suggested the Lord had been at work in him he would not have denied it, though he might have tried to describe the process in more modern phraseology, for what had been adequate in Skye forty years ago would not quite do nowadays.

Before breakfast he telephoned Barnes and astonished him by saying he wanted to use the PWD launch to go up river to Pensiangan.

'What have we got going up there?' asked Barnes.

'This is private. I want to visit a planter called McArthur. He's dying.'

'I didn't know you knew him. You never mentioned him.'

'I don't know him. For some reason he sent a message through Lall that he wanted to see me.'

'My God! I mean, why?—if you don't know him.'

It was clear that Barnes, who seemed to know McArthur, could see little in common between him and Macpherson. Macpherson felt vindicated already.

'I don't know why. Perhaps I'll find out. Do you know him?'

'I've had a drink with him once or twice here at the club. I've been in his house. Look, Macpherson, would you like me to go with you? I'd like to see him, before he pushes off.'

'If you don't mind, I think I'd rather go myself. It's a kind of personal thing.'

'Please yourself. I wouldn't have thought Mac was your sort.'

'What's my sort?'

'Well, you don't drink, to begin with; and he does. You don't care much for coloured people; he married one. He's always been a great man for the women.'

Macpherson was angry at being accused of not caring much for coloured people. It was unfair and untrue. He had spoken to the villagers yesterday far more sympathetically than Barnes had done.

'I'd be obliged if you could have the boat at the quay by ten,' he said.

'It'll be there,' replied Barnes, after a long pause. He had been tempted to say it was under repair. Why should poor McArthur be plagued on his death-bed by this arrogant pussy-foot?

'Since this isn't PWD business I'll pay for the petrol and the men's wages.'

'No need for that.'

'I prefer it.'

'Suit yourself. You'll not manage back today, you know. Mac will put you up. He's got a beautiful house, and his wife's a charmer. Give them my regards.'

It was nearer eleven before the launch set out. Macpherson had been delayed trying to get through to Mary in Api to tell of his change of plan.

Her exclamations of joy, and of pride in him, had been strangely excessive; and there had been an element of forgiveness in them. He had been curter than he should in reminding her that all he was doing was travelling sixty miles or so up a tedious but not dangerous river, in a fairly comfortable launch, to humour a man either delirious or dead. There was nothing heroic or big-hearted in that. Lall did it fairly often, sometimes with an additional trek through sweltering jungle; and Lall, whom he'd met, was fat and conceited, nobody's idea of a hero.

She could not be dissuaded from her praises. She had even wept. Her present state must be menopausal. Yet she was only forty-two.

From early childhood Macpherson had been conditioned to look upon all of life as a duty, even the parts he enjoyed. Mary once, in the early days of their marriage, in half-humorous half-frightened expostulation, had accused him of making love dutifully. He had rather huffishly denied it, but later, after some private rumination, he had seen that she was right. He had not been ashamed as she seemed to think he should. On the contrary, he felt pleased and relieved, that even his sexual instincts were under the authority of the Lord.

Most men would have looked on that trip up the river as an adventure, a holiday not only from work, but also from home and family; especially those who, like him, were fond of boats. Macpherson could never feel like that: responsibility was closer to him than his shadow, for this in the dark when he was asleep took time off.

The two native boatmen, on the other hand, always felt on holiday. Dressed in the briefest shorts, and with coloured rags round their foreheads, they had squat-nosed piratical faces. As they gazed ahead, on the look-out for logs or crocodiles or, as they said gleefully, drowned women, or as they frowned at the wall of jungle on either side, they looked very like their great-grandfathers not so long ago as these raided villages along this river, raping, burning, killing, and hacking off heads.

Whenever they caught Macpherson's eye they would instantly grin in acceptance of their role as menial hirelings of the PWD, but when they didn't know he was looking at them they played at pirates. In their belts were knives, and razor-sharp parangs were to hand, to chop away obstructing branches or slaughter imaginary foes.

Now and then, for the first hour or so, he would call out something to them, harshly, to remind them of their true position.

For a few miles the river was the colour of cocoa made with milk, then it grew as green as crocodiles' backs, then suddenly blue in open spaces, and once as clear as a Skye burn. Crocodiles slept on banks. A python lay on sand, guarding its eggs. Monkeys swore from the tops of trees. Butterflies as big as birds, and birds as brightly coloured as butterflies, flew past. Sometimes jungle gave way to plantations of sago palms, and villages would be seen, with bamboo stairs leading down to the water's edge. Bare-breasted women washed clothes, or themselves; they waved to the boatmen. Big-bellied children slept on their feet.

The constant heat, the hypnotic swish of the water, the glimpses of people as remote as the stone age, the dazzle of sun seen through millions of leaves, all combined to give Macpherson the feeling that he was in the midst of a dream, especially as his purpose had become no more urgent or comprehensible, but less and less meaningful. He felt that this journey would never end, but would keep going round bends of the river and of time for all eternity.

He even wondered if he was ill, if the sweat pouring off him was that of fever. He had already taken the day's dose of anti-malaria tablets, and was too characteristically cautious to take more than was stated on the box.

He only noticed the quay ahead when one of the boatmen shouted and pointed to it. It was like a dream to see there, after so much jungle, a substantial pier, with sheds, and boats, and men dressed in shirts and shorts, and a Land Rover. Beyond were many thousands of rubber trees, healthy and well-kept. Among the trunks could be glimpsed tappers as devoted as priests in a huge pillared church.

On the quay was a small dapper Malay wearing a wide-brimmed white hat. He introduced himself, in prim self-satisfied English, as Razak the under-manager. He explained that Mr Barnes had sent a message by radio telephone. So Mr McArthur was expecting him.

'Is he well enough to expect anybody?' asked Macpherson. 'I was told he is dying.'

'Please listen.'

Macpherson now heard, some distance off, the sound of gongs, not beaten merrily as at a wedding, but slowly and sonorously.

'They are frightening off the evil spirits,' said Razak, with a Muslim's smile. 'They believe that these spirits gather round to snatch the soul as it leaves the body and carry it off to the mountain. Therefore they must be scared off.'

'Who believes such nonsense?' asked Macpherson, with a Christian's frown. 'Surely not McArthur?'

'Indeed, not. His wife's people. They are all pagans.'

'Why does he allow it? It would get on my nerves.'

'Mr McArthur says he has the tolerance of the man who believes in no-one's god.'

'That isn't tolerance,' said Macpherson. He was about to add that it was arrogance, but remembered in time he was speaking to a brown man.

They climbed into the Land Rover. The driver was told to take it easy. Even so, they had to hold on for the road was rough. Macpherson could smell its coral foundation. A lot of money and labour had been spent on this estate.

'I understand the estate belongs to McArthur,' said Macpherson.

'More properly to his wife. According to the law only natives can own land. Mrs McArthur is the daughter of the king of all this district.'

'You mean she's a kind of princess?'

'That is so.'

Not that she would look like a princess. None of the native women were beautiful. Their lips thick, their noses flat, their bellies big, and their legs lumpy.

'I've been told the house is beautiful,' he said.

'See for yourself.'

There it was, in a vast clearing, high, round, and made of wood. The roof was conical; on the point sat a great carved bird with wings outspread. The house itself was built of teak and ebony and other handsome woods. Surrounded by flowering trees and bushes, it too was like a tree, gigantic and glorious. Macpherson was astounded. In front of it was the most magnificent frangipani he had ever seen. The grass beneath was white with blossom. He was reminded of foam, and of the great strand at Laggan in Islay, near Bowmore, where it had been blown across the sand like white balls.

Here, though, was no sea breeze. It was very hot and still. No birds chirped. The gongs, louder now, emphasised the stillness.

Macpherson felt dirty and hot. He resented too an enticement he could not understand.

In the shade of a majestic tree, with white flowers that reminded him of the horse chestnuts of home, except that these were larger and heavily perfumed, was waiting a woman, his hostess, mistress of this extraordinary house, McArthur's wife, the princess. As he had thought, she was stout and heavy, both in the face and body. But even at rest she had what he was forced to acknowledge as dignity. She wore a native dress of black, red, and gold, with a head-band to match.

His Mary was considered to be a fine-looking woman. So she was, but she did not have anything of this woman's impressive presence. The admission shook the very foundations of his soul.

He began to feel a strange deep sadness. Here, in this marvellous place, in this house which had so much character, McArthur, the man from far-off Bowmore, must have lived a happy life. Into the sadness that Macpherson felt came pity, not only for McArthur, soon to be cut off from this happiness, but also somehow for himself, although his wife was waiting for him in Api and they would live together for many more years.

It was as though his mind, with its ballast of principles and beliefs, had broken loose and was beginning to drift.

The Land Rover with Razak in it drove away. The little Malay had been very respectful towards her.

Macpherson greeted her. His mind drifted still further from the shore of lifelong habit. He had thought that when he met McArthur's native wife he would naturally be sorry for her because her husband was dying, and because too she was a native woman. He had taken for granted that she would be protected from grief by the animal-like stoicism peculiar to primitives. After all, her ancestors less than fifty years ago had hunted for heads, and had been hunted for their own heads. Sudden bloody death must have been a commonplace for hundreds of years. The inhabitants must therefore have acquired a protective thickness of soul.

How wrong he had been. Here she was greeting him with a dignity that the Governor's wife, in the British days, would have been proud of. She shed no tears, but no one, not even Macpherson with his prejudices loose within him, could have doubted that her husband was dying and she loved him.

She thanked him for coming and said her husband was expecting him; but first he must wash and rest and refresh himself.

The outside of the house had impressed him, the inside had him gaping in wonder and admiration. The great round room was like a church, panelled and floored with exotic woods. He had never been in a house that smelled so pleasantly. Everywhere were delicate green plants, and bright flowers. All round the wall was carved a frieze of leaves. There was a profusion of carvings, the largest and most astonishing being two life-sized figures, in a shining wood, one male and the other female, as naked as Adam and Eve, and looking, as Macpherson had to admit, just as innocent, though their private parts were boldly and meticulously fashioned.

Someone—could it have been a man from Bowmore?—had made beautiful and inspiring use of primitive skills.

Through a doorway came, like a creature in his dream, a young girl, a servant, feet and bosom bare; in her black hair was frangipani. If she had dropped her bright red sarong she could have been mistaken for the life-size statue.

In other expatriate houses Macpherson had seen amahs much too comely for mere dishwashing. They had evidently slept with their masters. He had been disgusted. This girl, though, lacked utterly that smugness which in a coloured woman indicated sexual claim on a white man. When she smiled, showing beautiful teeth, she meant only friendliness. It was a part of the welcome and the delight of that house.

Anxiously he warned himself not to exaggerate. McArthur had a large estate, a magnificent unusual house, a dignified native wife, and servants who knew their place. That was a great deal, worthy of wonder; but did it justify this feeling of light-headedness, worse of light-heartedness, that kept growing in him? He had always believed that his life of duty, obedience to God, thrift and caution had been leading towards some revelation. Could this be it? This was not his triumph, but McArthur's.

He remembered that McArthur was dying.

He followed the girl up a staircase with a balustrade that deserved hours of study, so numerous and exquisite were its carvings of faces, birds, flowers, and fish. Under the thin red cloth the girl's buttocks were plump: sweat glistened faintly on her smooth back. For a moment or two he let himself imagine that this was his house, and this girl was his wife: they were going up to bed. He felt he knew her more intimately than he had ever known poor Mary.

She showed him into a bedroom with open french windows and a view of red and orange flowers. She opened a door and showed him a bathroom. Everything was ready. He felt deeply moved. A man in mortal pain, whom he did not know, was treating him with flawless consideration.

He had asked Mrs McArthur how her husband was. With a smile that not even he could have misinterpreted as congenital callousness, she had shaken her head. He was asleep, she said. Macpherson knew she meant drugged, deliberately no doubt to escape pain. She said there was no hurry. After he had washed and rested and refreshed himself, he would be taken to see her husband.

The gongs were still sounding. He could also hear the humming of a big generator.

He asked the girl where the children were. At their grandfather's house in the village, she replied.

Before taking a bath, he stood on the terrace gazing down at the garden and smelling its many perfumes. He was curiously conscious of the great bird on the roof above, though he could not see it from there. He wondered again why McArthur had asked to see him. Previously he had thought of the planter as living in squalid conditions with his native wife and children; therefore his asking Macpherson to come had been a kind of whine for help. Now he saw he had been honoured. Whatever he had McArthur did not need it. Why then had he been asked to come?

About two hours later—it was now after six and almost dark—he went downstairs, much refreshed. The house was lit by electricity. In the large round room he studied the carvings more carefully. On a shelf were several dancing figures, about a foot high, in the shape of headhunters, each with a severed head clutched to his breast. He had heard that this taking of heads had been part of their religion: their gods could only be propitiated by such offerings. He had been disgusted. Now he gazed in uneasy wonder at the expression of fierce rapture depicted on all of these faces. He was still horrified, but he looked more deeply, and more humbly, at his horror.

Dimly he perceived that there were aspects and areas of faith that he had not known existed.

Mrs McArthur appeared. She looked tired. She wore the same dress. He thought she had been watching by her husband's bed.

'Can I see him now?' he asked.

'He is still asleep. Perhaps you would like to eat now?'

Coming up in the launch, he had swallowed two Enterovioform tablets. He always did this when in danger of having to accept native food. But in the dining-room he found prepared for him a meal he could have ordered in any good Scottish hotel: chilled pineapple juice, fish, gammon steak with peas, brussels sprouts and potatoes, and for dessert ice-cream with tinned pears. He was able to eat the lot with enjoyment and without a qualm. Again he felt grateful to McArthur who, in spite of his severe pain, had taken thought to what his guest would like.

The girl with the frangipani in her hair served him. She wore a red blouse. He was asked if he wanted wine or beer. His preference for water caused no surprise.

He was just finished, drinking his coffee, when his hostess appeared at the door. She was as calm as ever.

'He is awake now,' she said. 'He wants to see you.'

He noticed that, though she did not seem to hurry, she went quickly. As he followed her the thought that came into his mind, not altogether incongruously, was that he had never shown his love for Mary as he should.

The bedroom into which he was led was almost in darkness: the only light was a small lamp with a red shade. McArthur lay facing the door. He opened his eyes with a great effort as Macpherson came in, smiled very faintly, and closed them again. Macpherson had no recollection of what the man from Bowmore had looked like in the Sports Club: he had not looked at him attentively, or interestedly, enough. But he knew that in this dying man a terrible change had taken place. The face was hollow, the neck thin, the hair grey and wet.

Mrs McArthur wiped the sick man's face and head with a damp cloth. She bent low and whispered to him in her own language. He replied with hoarse painful slowness.

On a table was a tray with the means of keeping him drugged, safe from pain.

The gongs still sounded, louder now that it was dark outside. In Macpherson's hip pocket was the small New Testament he always carried with him on his travels. He had thought that even if he did not read from it or even take it out of his pocket, its presence would comfort the dying man. Now he saw that it was to him what the gongs were to the natives. He felt for them a respect he had never been capable of before, and for himself a pity that he had never thought he needed.

'You will pardon him if he cannot talk to you,' Mrs McArthur whispered. 'Perhaps a word or two. He wants you to talk about the place where he was born.'

'Of course.' Macpherson wished Mary was with him. She was so much better at finding something to talk about, and she had got to know more Islay folk than he during their holiday.

'I must tell you,' whispered Mrs McArthur. 'He may die at any moment.'

'Yes.' But he did not think he would notice any difference. Poor McArthur already looked dead.

'Only five minutes,' she said, with a glance at the syringe on the table.

'I understand.'

Despite his stillness the agonies were again gathering in the sick man.

In any case, thought Macpherson, five minutes would be as long as he could stand. Again he wished Mary was there.

Mrs McArthur went out. He sat down by the bed. For over half a minute, listening to the gongs and to McArthur's faint breathing, he could think of nothing to say.

'Well,' he said at last, 'this is a far cry from Bowmore.'

McArthur's hand moved an inch closer. Macpherson put his on it, lightly. Its touch reminded him of old Mr Dougary's, and he realised that the reason why the old minister was so often lost in thought and didn't listen was because he too was in pain.

'When my family and I were on holiday in Bowmore, five or six years ago, we went to the Round Kirk twice. It's a very bonny little church.'

It wasn't what he wanted to say, and he hadn't said it as he had wanted to say it. He had been too stiff, too formal, too much himself.

McArthur seemed to be trying to say something. Macpherson bent down and listened. But, after almost a minute, nothing was said.

'We took a car with us,' he went on, 'so we explored pretty nearly the whole island. I thought I wouldn't mind retiring to Islay. It's milder than Skye, where I was born. My wife liked Port Charlotte: all the white houses.'

Furtively he glanced at his watch. Three minutes had passed already.

'We went to a ceilidh in the Masonic Hall,' he said. 'Songs in Gaelic. We enjoyed it, though we hardly understood a word. I used to know it, but I let it go. Mary, that's my wife, said it would soon come back, if I was prepared to make the effort. There was one old man who'd been singing Gaelic songs for forty years, they said. Perhaps you'd have known him. But I forget his name.'

Again McArthur tried to speak, and again failed.

'Our children, one was about three and the other five at the time, liked the beaches best. Marvellous beaches. Particularly Laggan. We used to watch the planes coming in. A Land Rover drove the sheep off the runways.'

The thumb moved against his hand, with the force of a butterfly; yet it was trying to say something. What, Macpherson had no idea. It might have to do with why he had been asked to come.

Then words at last, one at a time. The eyes stayed closed. The fingers tensed and formed a claw. He should not have been speaking. It was too great an effort: it could kill him.

Macpherson had to put his ear close.

'Your . . . wife's . . . a . . . bonny . . . woman.'

At least that was what it sounded like. Macpherson could hardly ask for it to be said again. He felt dismayed, and yet happier than he had ever been in his life before.

McArthur was now tense all over and shaking; and whimpering.

Mrs McArthur came in, impassive as ever.

Macpherson got up. She nodded, and he went out, tears in his eyes. He felt a great need for Mary, and a great fear that he might never see her again; but above all he felt a great pride that she was his. McArthur had sent for him because of her; and there had been many previous unacknowledged occasions when he had been welcomed, accepted, tolerated, or even liked, for her sake.

If he had gone to his room he could not have rested; so he went downstairs, where he found Razak seated in a chair, his hat on his lap.

He got to his feet when he saw Macpherson.

'How is he?' he asked, in Malay.

He spoke defensively, as if he believed that Macpherson disapproved of him; and so Macpherson had, that afternoon, for no reason except that his way of life, his appearance, his religion, his people, his food, were different from Macpherson's. Mary, had she seen, would have said nothing, out of loyalty, but she would have been hurt.

He had a great longing to tell Razak about her.

'Very ill, I'm afraid,' he answered, also in Malay. 'I don't think Mrs McArthur expects him to last the night.'

'She will miss him very much. They are very devoted to each other.'

Could a man who thought his Mary bonny think Mrs McArthur bonny too? Yes, yes, he could.

'How long have you been here, Mr Razak?' he asked.

'Three years.'

'You must know them very well then.'

The little brown man nodded, and then hung his head. As if it was a signal the gongs began to sound much more loudly and wildly; those striking them seemed to have gone mad. Almost at the same time from upstairs, from McArthur's room, came a curiously deliberate shrieking; and from somewhere else in the house it was echoed.

'He is dead,' whispered Razak. 'They think the evil spirits are very close.'

But how had those beating the gongs known that McArthur was dead?

Macpherson touched his New Testament, not to counteract the pagan shrieks and gongs, but to assist them.

Neil Paterson
1915–

The Life and Death of George Wilson

who was an Extraordinary Fine Specimen

It was Madame Gollatz who brought it all back to me. She came into the office sideways, being the only way she could get through the door, eased herself on to the chair that I had had specially strengthened for her, and sat panting and nodding and smiling till she had enough breath to speak. Then she said: 'Did you see about the Princess Circassy?'

I hadn't, and I said so.

'She died. It was in the Sunday paper. Aged twenty-nine. That kind don't live long, do they?'

'No,' I said. Ma Gollatz is thirty-six stones and only sixty-one inches high, and as a rule her kind do not live long either, but I did not think of that at the time, because of course I was thinking only of George; and Ma knew that, and began a fine chatter of words to make me think of other things.

'And the second George?' she said. 'Well, well, the *liebchen*. What news do you have of the little man, I wonder?'

'My boy? Oh, he's fine. He's doing pretty good at this new school I sent him to. Do you know what it costs me a year?'

'Yes, you told me, Joe. It must be a very high-tone establishment.'

'The best,' I said. 'The best. Nothing but the best is good enough for young George. He is going to be a doctor, you know.'

'Yes, I know. Old Max and me, we are very proud of it. We have all come a long way, Joe.'

I looked round the office. I could see my name on the frosted glass door. 'J. Harrap & Co., Entertainments.' It had an important look. 'Yes, I suppose we have,' I said. 'Did it say anything else in the paper about the Princess Circassy?'

'No, just that she was dead. It didn't even give her measurements.'

'Well, that's the way it goes,' I said. I drew a couple of lines on my account-sheet, and when I looked up I saw that Ma was crying.

'George was a great gentleman, though undersized,' she said. The tears were sliding along the creases in her cheeks. There was nobody had tears like Ma for size, and she cried easy of course, like most of these fat ones. 'Been thinking of him all day. I can't get him out of my mind. Wasn't he a great little gentleman, Joe?'

'He was all that,' I said.

II

I bought George Wilson in 1928. I bought him from old Sol Goldstein, who was in the same line of business as me, although at that time Sol's side-shows were of a higher class than mine, Sol being established, and long past one-night stands. I knew that Sol would sell George to me, because Sol was beholden to my Dad, and Sol never forgot a favour nor an insult, though he paid both in his own time and in his own way, and he did not like to be rushed. If I had asked for George Wilson I would not have got him. I asked for the lightweight out of the boxing booth.

Sol raised his hands and screamed at me. That boy was the best he'd got. That boy meant the difference between life and death to Sol Goldstein. Without that boy Sol Goldstein would starve.

I thought for a bit and tried again. 'The indiarubber man,' I said.

'My God,' Sol said. 'Hark at him now! The indiarubber man. Do you know what I pay that indiarubber man in wage! More then your ha'penny shows draw in a month. Look, Joe, it's you I think of when I refuse. Can Joe afford it, I say to myself? Can he support the burden? A thousand times no. With a show like yours you cannot carry the expense of them important artistes. Be reasonable, Joe. Be prudent. Now, was there anything else you saw?'

'No, Sol, there was nothing else,' I said. I got up like I was going. 'I don't mind telling you I'm disappointed, because I need new blood, and need it bad, and if I don't get it now it's curtains for the Harrap shows.'

'To tell you the truth,' Sol said, 'as I would tell it only to the son of your father, I cannot afford to give you the Murder Kid or the Rubber Man, and that is the honest truth of the matter, Joe.'

I reckoned that I could have them if I wanted them. But I didn't want them. I wanted George Wilson. I had wanted him from the first time that I saw him. I put my hand on the doorknob.

'Wait, Joe,' Sol said.

'All right, Sol,' I said. 'Let's see what else you have. How about? . . . No. I'll tell you what, Sol. I don't much want him, but I'm willing to take the midge, provided you let me have him cheap.'

'He's a great artist, that midge,' Sol said. 'Frankly, I would not like to part with that midge. Just listen, Joe.'

I had to listen to the list of things the midge could do. He could walk the trapeze. He could sing and dance. He could swim. He had a good memory. He could learn a part. He could even read and write and he lived clean. He was no trouble. Any showman that had that midge had gold in his hands.

'I'd rather have the fighter,' I said.

'All right, Joe,' Sol said. 'How much you bid?'

'For the fighter?'

'For the midge.'

'Ten pounds.'

We talked for the rest of the morning, and in the end I got George Wilson for twenty-eight guineas. He was cheap at the price, and I was very pleased about it. I knew it was one of my best bargains.

III

I have seen a lot of midges in this business, but I have never seen one like George Wilson. The fact is, there is something wrong with most midges. Sometimes it is their heads. Sometimes it is inside their heads. Sometimes it is their bodies. Their heads are too big for their bodies or their bodies are too big for their legs, or their voices, if they have voices, aren't like the voices of human beings at all. You might think that a midge was like you and me, only on a small scale, but, bar one, I never saw any midge like that, that was just right, good-looking in the face and with small perfect bones and all its muscles working and its actions natural and no loose flesh round its neck. I never saw one like that yet except for just this one midge George Wilson that I'm telling you about. Mind you, I'm not saying that George was perfect, because he wasn't. George had faults like anybody else.

At the time I bought George he was twenty-eight inches high and weighed nineteen pounds. This was because he was out of condition. He wasn't being given enough exercise. He wasn't being watched, and all the time he was supposed to be taking his exercise he was sitting around studying for some examination at London University. Can you beat that! A midge studying out of books sent him by a university. Nobody would credit it that didn't know George, but that was the kind of midge he was. Different. Just plumb different.

He told me about it the first day. 'I want to be frank with you,' he said. 'This thing is important to me. I must have a degree of privacy.'

And then he told me about this university business.

I suppose I must have gaped at him with my mouth open.

'I don't want to be forced into evasion and disloyalties,' he said. 'I want to be perfectly honest with you.'

I noticed that he was pale and that his hands were trembling, and I guessed that he was frightened for all his fine manner, and I spoke up to him like he was a person. 'That's okay with me,' I said. 'I suppose a midge has a right to a private life like anybody else.' I thought about that after I heard my voice saying it, and I reckoned it was fair enough, although it was not a point that had struck me before. 'Provided,' I added, 'you don't stray out of your compound, of course.'

Looking back, it seems crazy that I ever thought of keeping George in a compound, but you have to remember that when I first got him I did not understand George very well and he did not understand me either.

Take that first day when we got to the station, for instance.

'Where's the shawl?' George said.

'What shawl?' I asked.

'For wrapping me in. Aren't you going to pretend I'm a baby?'

'Hell, no,' I said.

'Pity,' George said. 'I much prefer the shawl to the basket. The shawl is *infra dig.*, but the basket is dehumanising and also damned uncomfortable. I see you haven't got a basket. You're not going to put me in that attaché-case, are you?'

'I'm not going to put you in anything,' I said.

'Then you'll have to pay for me.'

'I'll get you for a half.'

'My God,' George said, grinning. 'Don't tell me there's a man in your trade with morals!'

I didn't want any cracks from him or anybody else about my morals, and I told him so. 'Everything's on the up and up in my stalls,' I said, 'as you'll find out for yourself, and I don't hold for no looseness any more than my father did before me.'

I got him through for a half.

'What's your age?' I asked in the train.

'Eighteen.'

'I thought you were older. That's what Sol said. That you were eighteen. Are you really only eighteen?'

'Yes,' he said. 'I don't tell lies.'

I looked at him and knew he was speaking the truth.

'And you, Mr Harrap? How old are you, if I may ask?'

'I'm twenty-five.'

'You look younger, you know.'

That got me on the raw. After all, I had been through a lot, and I was sole proprietor of Harrap's Side-shows and Stalls. I was a man of substance and I owned several people. I was touchy about looking so young. I had, I remember, even grown a moustache to look older. I was pretty foolish in these days.

'Personally, Mr Harrap,' George said, 'I should have put you at about twenty-one.'

'Well, I'm twenty-five,' I snapped, 'and *I* don't tell lies either.'

'I believe you,' George said. He sat looking up at me for a long time, then he said: 'You know, I think you and I have struck rather lucky in each other.'

Until you got used to George he knocked the breath out of you.

'What do you think, Mr Harrap?'

'I think you talk too much,' I said.

IV

The spring I got George I was working in with J. V. Green. It was his fair, and I hired the space off him for my tents, and when he moved on

I moved too. I had four shows at this time, five if you count the coco-nut shy. The shows were Madame Gollatz, the Fattest Woman in the World, Max Brown (Ma Gollatz's husband) and his performing fleas, Vendini the Strong Man, and the boxing booth with two pugs, both old. I was working very close to the bone. I remember I was so hard up that I had to offer Madame Gollatz and Max as securities to Entertainments Ltd., before I could get the new stuff I needed for George. Anyway, after a lot of letters and paper-signing, I got the gear, and I set George up in the new tent as King Minos VII, the Smallest Man in the World. He did a five-minute work-out on his trapeze, his tight-rope act, and a trick-cycle stunt, and you could shake hands with him as you filed out at tuppence a time. The weather was good that year and I did all right, and George was a success. I paid off my debts, including the one to Entertainments Ltd., and hired a new tent and platform and scenery, and started a ventriloquist act, in which I had George dressed up in a doll's mask with a flap lip and a jointed wooden suit that covered him completely, including his hands and feet, and by the end of autumn this act was doing so good that I raised the price from threepence to sixpence and was still packed out, most performances.

In the winter I got an offer to put the act on the stage, but I turned it down, because I did not see any future in it, and then at Christmas time I hired George out at great profit to play a character known as Tinker Bell in a London show called *Peter Pan*; his name was in the papers, and I got an offer for him from a London theatrical agent. It was a good offer. It was a mighty good offer when you think what I paid for him—£250. And I think he would have gone to three hundred.

I spoke to George about it. By this time George was living in my room and calling me Joe and going everywhere with me.

'What do you think, George?' I said. 'It's a good offer and I'll give you a decent split.'

'If you need the money, Joe.'

'It's not a matter of needing the money. It's just good business. I'll give you a fair share.'

'I don't care about the money, Joe.'

'For Pete's sake,' I said. 'If you don't care about money, what do you care about?'

George just looked at me. I can see him now—his mouth opening to say something and then closing tight and his face getting red, and his eyes—those small grey steady eyes of his—hooked right into mine. He was sitting on the edge of the table with the white silk rope across his knees. He had these ropes fixed all over the place so that he could climb on to anything without being helped. He had his hands clenched on the rope, and I saw the whites of his knuckles, and I got angry and shouted at him. 'You'll never amount to anything if you don't care for money,' I said. 'The sooner you start caring about money the better, my lad, see!'

'All right, Joe,' George said. 'I will, I'll start caring.'

I didn't like the quiet way he said it.

'After all, business is business,' I said.

'It's all right, Joe. I understand. When'll I have to go?'

'Go where?'

'To London.'

'You're not going to London,' I said. 'Two hundred and fifty pounds is chicken-feed. I'll get a lot more than that for you yet.'

I never talked or thought of parting with George again. I never even told him of the offers I got. As far as I was concerned he was off the market. It just wouldn't have been good business to sell him. He was worth too much to me.

The next year I got three new fighters, and I started a Hall of Mirrors and two shooting galleries. It was about this time that I first began to have money in the bank, but I never left it there long. I used it to buy up new acts and new stalls. I had one idea only. To expand. To have a fair of my own. To be the boss.

George had given up the university business.

'I don't need it now,' he said. 'I'm fulfilling myself. The old show business is good enough for you and me, Joe.'

'You're damned tootin, pardner!' I would say.

I was on the way up then. Nothing could stop me. I had spent my life in the business, and I knew all the people that counted. I always got on well with everybody, and even men I didn't know trusted me because I was Honest Joe Harrap's son, and around this time it seemed that everybody was falling over everybody else to put something good my way. I was always travelling, always buying, always working. George worked hard too. He was interested in everything about the business, and he had a kind of gift for it. I talked my deals over with him and I listened to what he had to say. He became my right-hand man, and when I went to look over a turn or a site I used to leave him in charge and he would run the show and be the boss for maybe two or three days at a stretch. The queer thing was that nobody ever tried to make a monkey out of him.

'He's got dignity,' Madame Gollatz used to say, winking and wheezing and smiling. 'Dignified, that's what he is. The little one.'

Maybe that was it. Maybe it was because of this dignity of his that nobody teased George. I don't know. I do know that George could speak for me to everybody, and they would do just like it was me talking, that is, all except Vendini the Strong Man.

Vendini and George never struck it off. Vendini had the strength and the brain of an ox, and he could not get it into his skull that George was a person. He used to follow George around just to stare and laugh at him. He did not mean any harm. He liked George, but the way he liked George was the way Max liked his fleas and he never got wise to the fact that George did not like him or that George had feelings.

'You clod!' I've heard George say. 'I have more matter in my little finger than you have in your whole body!'

'Listen!' Vendini would say, delighted. 'Listen to 'm talking! The words!'

'You big ape, what does size count! You may not know it, but the greatest men who've ever lived have been little men. Take Napoleon.'

'Mother o' God, listen! Just listen. Such speech! Say more, dwarf. The words that he has, in a voice!'

I think George hated Vendini. He used to say that he didn't care what Vendini said or did. Vendini was a moron, beneath a man's notice, but I think that Vendini always hurt him very much and that really he hated Vendini.

I tried to explain George to Vendini myself, but he could not understand that it hurt George to be treated like a freak out of business hours; he just could not understand that George was human.

'I say sorry,' Vendini said, nodding. 'You get him. I say sorry now.' And he began to laugh again, just thinking of George, looking at the bottom of the door where George would appear. 'You get him. I say sorry. Yes. Ha, ha, ha!'

It was hopeless.

V

I remember it was soon after George's twenty-first birthday that I took him to see the Princess Circassy. I had found her some months before when I was in Newcastle to sign up a middle-weight, and I had it all fixed with Iggy Pullitzer who owned her that if George liked her he would sell for £100. The Princess Circassy couldn't do much—she was just an exhibit—and she wasn't worth £100 or anything like it to me; but I didn't mind spending the money if it made George happy. I didn't tell George what we were going to see, I wanted it to be a surprise to him, but I wired Iggy Pullitzer that we were coming and to have the Princess all primed up.

Well, it wasn't a success.

The Princess Circassy was the prettiest female midge I ever did see—she was only one inch taller than George—and she looked just fine when you saw her by herself, but alongside George she had a kind of coarse look, and I knew right off that they'd never make a match of it.

'Hiya, big boy,' the Princess Circassy said, looking George over.

'Gives you a turn, don't it,' Iggy said, 'to hear them midges talkin' up like folks! B'jeez, if they ain't just a pair now!'

'C'mon in,' the Princess Circassy said to George, standing at the door of her compound and touching her hair with her hand. 'We can be kinda private in here. C'mon, big boy.'

'No, thanks,' George said.

'The little runt's scared,' Iggy said, laughing. 'Go on you, George! Go on, Lofty. Go to it!'

I knew I had made a bad mistake. I got George away as quickly as I could. I reckon I felt almost as bad about it as he did himself. I didn't know where to look.

'I'm sorry, George,' I said.

'It's okay, Joe.'

'I meant it for the best. I didn't know it would be like that.'

'It's okay. I understand, Joe.'

We didn't say much in the train. George was white as a sheet, and I hadn't any use for myself and didn't want to hear my voice talking.

'Joe,' George said after a long time.

'Yes, George.'

'Did you see that place she lived? It was no better than a . . . manger. And, Joe—she smelled! Did you notice?'

'Yes, George,' I said, though I hadn't noticed. 'I'm sorry. I'll make it up to you.'

'It's all right, Joe.'

We never spoke of it again, but later that night, after we got home and George was turning in—he needed a lot of sleep: twelve hours was his usual—he said something to me that I never forgot. 'Joe,' he said, 'do you know something? I hate midgets.'

VI

The following year George took ill with the 'flu, and I thought that I would lose him. I got three doctors in. The doctors said positively there was nothing any of the others could do that they each of them couldn't do on his own, but I reckoned there was bound to be one of them smarter than the others, same as in any kind of business, and I wanted only the best for George. Two days and two nights I never left his bed, and then, on the morning of the third day, his temperature went down, they said he was out of danger, and he started moving his lips to speak.

'I'm sorry to be such a trouble, Joe,' he said.

'You shut up,' I said. 'You lie quiet like the nurse says, or I'll tan the hide off you.' I had a temperature myself then and I could of wept like a woman, I was so relieved. After all, midgets that are a commercial proposition like George do not grow on trees.

When he was better I sent him down to Eastbourne with Max Brown to take a holiday and get strong. He didn't want to go. He argued this and that. You never heard such arguing. He kept on at it. His place was with me, he said.

'Listen,' I said; 'what kind of mug am I and what kind of master-mind are you that I cannot get on very well without you for a bit! You'll do what you're damn well told.'

'All right, Joe,' George said at last. 'I'll go. But you'll send for me if you want me, won't you?'

'I won't want you,' I said.

I missed him though. The place was kind of empty and the evenings seemed long with nobody to tell about the day's work and nobody to help count the money. I looked forward to his letters—he wrote every day—and I spent time fixing up some gadgets that I knew he would like, including a little rope ladder up to the safe in the wall which nobody had ever touched before except me.

Max and George were to stay in Eastbourne for three weeks but before the end of the third week I got a letter from George that ended:

'Will you please come down, Joe? There is somebody that I want you to see. It is very important, so please come, Joe. Max will add something to this letter too.'

Max had added:

'Yes, you better come, Joe.'

I went down to Eastbourne the next day. I couldn't think who it was that George wanted me to see, and of course I never guessed the truth.

Max got in a word to me at the station. 'It's a girl, Joe,' he said, behind his hand. But even then I didn't cotton on.

They took me to see her that evening. She was working behind the bar in the *Duke of Buccleugh*, a house that the police have now closed. She had a white face and very black hair, and she was about eighteen years old. Her name was Mary Guiseppi.

'Well, there she is,' George said. He kept glancing at me nervously, waiting for me to speak. 'Isn't she beautiful, Joe?'

'I'll not deny it,' I said. I watched her. I could see she was ill-treated like George had said. It was in her manner all the time, and when the woman in the black dress, the manager's wife, spoke to her, her eyes turned over like a frightened darkie's. She was scared all right, and she was moreover not a girl who knew her way around. It was plain that she had not been in the trade for long, because when the men at the bar laughed out loud, teasing her, she got red in the face and acted kind of paralysed. I was not impressed.

'Beautiful maybe,' I said. 'But dumb.'

'You should hear her sing, Joe,' George said. 'One night, when there was hardly anybody in and the manager's wife was at the pictures, she sang to us. She has a wonderful voice, hasn't she, Max?'

'She sings real pretty,' Max said. Max never took sides. 'Yes, sir, real pretty, Joe.'

'She's got talent,' George said. 'She's just what we want, Joe, isn't she? The glamour angle. I knew you would see that. Wait till Joe sees her, I said to Max. Joe can spot talent a mile off. Joe is the greatest discoverer of talent in the business. . . .'

'All right, George,' I said. 'All right, take it easy. You can't talk me into buying something I don't want, and I don't want that girl, and that's final.'

'You don't understand, Joe,' George said. He started to tell me all over again how badly she was treated, and then he stopped suddenly, his body stiffened, and his face set like a mask, and he said in a new hard voice: 'Look! There he is. The manager.'

The manager was a little man with a waxed moustache. He had a part in the story George had been telling me about this girl Mary Guiseppi. It was the usual part. Middle-aged man, ageing wife, pretty young girl. You know the kind of thing. And the wife knowing it too and knocking sweet hell out of the girl.

'Joe, she's got nowhere at all to go, and nobody in the world, and she's good, Joe, and sweet, and she has a beautiful voice. She really has. Listen, Joe, she's got glamour. She's what we need. She would brighten up any act.'

'Be reasonable, George,' I said. 'What could I use a girl like that for?'

'For nearly anything.'

'For instance?'

He thought about that. 'Well, at least she could work in with the fleas. Please, Joe, listen to me. . . .'

'Sit down, George,' Max said. 'Everybody's looking at you.'

George got down off the table. He had got so excited that he had climbed up in order to look me in the eyes. He was very pale, and I guessed he had not yet got over being ill. He was trembling.

'Please, Joe,' he said. 'To please me!'

'No,' I said.

'Joe, I'll never ask you anything else. . . .'

'No,' I said. 'For the last time. No.'

The rest of the night I hardly got a word out of him and the next day it was the same. I didn't like the look of him, and I wanted to get a doctor.

'There's nothing wrong with me,' he said.

'Sulking, are you, then?'

'I'm not meaning to, Joe.'

I stayed on another day because I was worried about him, and he saw that, and tried to be cheerful, and that was worse than if he'd just sulked. At night I heard him tossing and turning in his bed, and in the morning I knew that he had not slept.

Max puffed away at his pipe and read the paper like it was none of his business and he didn't care anyway. 'I reckon he's pining away,' he said. 'Yes, sir, I reckon his little heart's broke.'

I stood it till dinner-time, and then I lost my rag watching George pick at his food, pretending to eat.

'Oh, for God's sake!' I said. 'All right, I give in. I'll take your girl.'

Max was smiling as if he had known all along that this would happen.

'On one condition only,' I snapped. 'Provided I get her cheap.'

VII

Everybody thought it was very funny, George having a girl. I must say I thought it a pretty good joke myself, at first, and one way or another George had to put up with a lot of chaff and sniggering that was meant kindly enough, but which must have hurt him considerably, as I realised later.

The girl Mary Guiseppi was no trouble to anybody. Ma Gollatz took her into her tent—Ma was too big for a caravan, and she and Max did not live together, although they were as devoted a couple as ever I saw—and Mary acted as a kind of housekeeper to Ma and made herself useful doing odd jobs about the stalls. I was busy—I was expanding all this time—and I might have entirely forgotten the girl if George hadn't kept harping on about giving her a chance.

George had taken to books again, and any time he wasn't rehearsing or doing a job for me you'd find him in a partitioned-off corner of Ma Gollatz's tent poring over books with Mary Guiseppi.

'All she needs is polish,' he said. 'You wait. She'll be a great lady one day. A great star. You'll see, Joe.'

After George had been on at me for about a month I asked Max Brown how he would feel if the girl sang outside his fleas. Max didn't stand on his dignity. He said he would feel fine, he didn't see what harm it could do, and neither did I, and so it was fixed that Mary Guiseppi would do a short number to draw the crowd before each of Max's shows. We threw up a platform, and George used his own money to hire a microphone and buy some of the latest records.

Mary herself came to thank me.

'It's George you want to thank,' I said.

'I know. But I wanted to thank you too.'

I asked her how she was getting on. Was she settling down all right?

'Oh yes,' she said. 'I'm the happiest I've ever been in my life. I'm very, very happy, Mr Harrap.'

She was too. You could see it shining out of her, and it puzzled me, because I couldn't think for the life of me what she had to be so happy about.

Well, Max's takings got good. They got better than they had ever been before, and I started paying Mary a wage, and I bought her a couple of dresses and sent her into town to have a good doing over at a beauty place, hair and nails and all that, and George was on top of the world, and strutted about like a cocky boxer, saying, I told you so.

I do not think that I ever regarded Mary Guiseppi as a good proposition, however, until I teamed her up with Gaspar and Lily Freud, the lions. It was George that gave me the idea, saying that Lily appreciated Mary's singing, she was always quiet when Mary was singing, and when Mary spoke to her she would scratch her ears on the bars of the cage and do a kind of soft growl. I had seen this happen before, a lion cottoning on to a person for no special reason that you could see, and this time it gave me the idea for the Singing Beauty and the Beasts turn that I later showed in all parts of the civilised world.

George did not like the idea of Mary going into the cage with the lions. He did not like it at all at first, and Signor Blanchino (Jack Evans, the trainer) and me had to talk him into it. It was quite safe. Gaspar and Lily were circus lions of the third generation, and there was no viciousness in them, although Lily got kind of contrary at times. But, as I say, she started off liking Mary, and she soon got wise to the set-up, namely, that she got fed after Mary finished singing and she didn't get fed at all if she hollered. That year I raised Mary's wage three times and George started calling it her salary.

At first I had it figured out that George would get over the craze he had for this Mary Guiseppi. I thought that George had just too much intelligence to let himself fall seriously for a full-sized girl, but I am wiser now, and I know that intelligence does not count much in a man where women are concerned. Anyway, George got worse. They called him Mary's lamb. He followed her everywhere, and he talked about nothing else. She was never out of his mind and hardly ever out of his mouth. It was Mary this and Mary that till you were sick to the teeth of it. At first when she did her turn with the lions he got so white you would have thought there wasn't a thimbleful of blood in his whole body, and though later he got used to the act, he never liked it, and was always in a blue funk when Mary was in the cage. If ever a man was crazy about a girl, that man was George.

It was only after Mary joined us that I began to notice how touchy George was about his size. He was always telling you about little men. Napoleon and Nelson and J. M. Barrie—it was J. M. Barrie who wrote the play *Peter Pan* that George did so good in—and Lloyd George and Louis XIV, a foreign king that invented Louis heels, and many other characters out of history that I had never heard of. According to George, they were all little men and he kept telling you about them. He got Bill Dekker, the carpenter, to build him a rack for stretching exercises and he sent secretly for booklets on How to be Six Feet Tall. He got thin and pinched-looking, and I did not like it because he had started on a trapeze act with the Stein Brothers in which, hanging by their legs, they tossed him to and fro like a ball, George turning a somersault in mid-air in the finale. He needed all his strength and his wits for a dangerous stunt like that, and I finally told him I wasn't going to stand for any more nonsense. I told him he had to give up stretching himself on this rack of Bill Dekker's.

'All right, Joe,' he said. 'Anyway, it was a mistake. I have been thinking a great deal about it, and I have come to the conclusion that the physical side of things is not important to Mary. For instance, it never occurs to you that I'm a midget, does it, Joe? I mean, you regard me as just a normal human being, don't you?'

I nodded.

'Well, it's the same with Mary. As a matter of fact, I don't think she's the physical type at all.'

'She looks physical enough to me,' I said.

'That's because you don't know the side of her that I know. You don't know the real Mary, Joe. It's things of the mind—yes, and of the spirit, that interest Mary, Joe.'

'Maybe you're right, George,' I said. But I did not think he was. It had seemed to me that I had a problem on my hands with Mary Guiseppi. Now that she was well-fed and no longer scared, she was getting to be a good-looking girl. She was shining with health; she had that kind of ripe look you see in girls that need marrying off, and a lot of the men around the place were watching her. But naturally I didn't want to say anything about that to George.

'In eighteen months' time,' George said, 'I should have five hundred pounds saved. That's a lot of money, Joe.'

I waited. I knew there was more to come, and I knew it was something unusual from the way George was flipping and scratching at the pages of his book. I could not see his face, but his neck was red. He said in an ordinary voice: 'I'm going to ask her to marry me then.'

I knew that I was the only person he would tell, that he was trying the idea out on me, and I made my voice as ordinary as his.

'That'll be fine, George,' I said.

I didn't like it at all, but I reckoned a lot could happen in eighteen months.

One of the things that happened was that I got robbed. It was at the time of the first of my big deals. I was buying up Solomon's show, lock, stock and barrel. Seven thousand quids' worth. I drew the money from the bank in the morning, put it in the safe in the caravan, and left George in charge while I went uptown to the lawyer's office where Solomon had said to meet him. We sat all day getting the paper work squared up, and then we went across the road to an hotel for a drink, and the lawyer bought an evening paper, and there was the story on the front page.

ROBBERY AT HARRAP'S SHOWGROUND

We read it.

'Bad luck, Joe,' they said.

'This is not bad luck, Solly,' I said. 'This is the end of Joe Harrap.'

On the way back to the ground I read the report half a dozen times. George had been assaulted, the safe forced, and a considerable sum of money stolen! The police wished to interrogate a man in connection with the affair. This man was known, the paper said, as Henry Spicer, showman, and it gave his description.

Harry the Spice! He was no showman. He was a roundabout engineer. He had been with me two years, and I had always thought he was a steady kind of chap. Seven thousand quid—most of it borrowed on security—plus the week's takings, plus the week's wages. Harry the Spice had made a pretty job out of me. I was finished with a capital F.

George wasn't in the caravan and nobody knew where he was. Nobody knew anything. They were all calling me Mr Harrap now. When I was alone I put my head down on my arms on the table and cried.

Freddy, the Kid from the Boxing Booth, came in without knocking and said: 'Do you know what? Harry the Spice ran off with old Dave Green's daughter as well as your dough. The youngest.' He stood with his hands in his pockets smiling at me. He was after Mary Guiseppi, and she had told him where to get off, but he hadn't, and George and me had had to tell him too. He hated the lot of us. 'The old Spice, eh!' he said, laughing.

'Get the hell out of here,' I said.

George came in then.

'Where the hell have you been?' I shouted.

George didn't say anything. He reached up and laid an envelope on the table.

'Answer me! Where have you been, you miserable little runt?' I said, and I hit him on the side of the head with my open hand. George went down like a clay duck and lay on the floor at my feet staring up at me with an awful look on his face.

'You damned useless pigmy!' I said.

He pulled himself up then and scuttled out like a rat.

After a long time I saw the envelope and opened it. There were four cheques inside, all payable to Mr Joseph Harrap: £200 from Max Brown, £100 from Hermione Gollatz, £75 from Mary Guiseppi, and £407 15s. 6*d*. from George Wilson.

Even his shillings and pence. His last bit.

I went out after him. I went out running. Folks who heard me calling George thought I had gone off my head.

I asked Mary.

'No, Joe,' she said. 'I haven't seen him at all. I'm sorry about the money, Joe.'

'That doesn't matter now,' I said.

I searched the whole ground. I tried all his usual places first and then I tried every place I could think of. There was no sign of him at all, and nobody had seen him.

'I ain't seen him since the afternoon,' Ma Gollatz said. 'He called a meeting. You should of heard him speak, Joe. "Joe's always done right by us," he said. "Joe is the finest man in the world. He'd never let any of youse down if you was in a jam." You should just of heard him. My, he had the tears pouring down my cheeks. Didn't he speak good, Max?'

'Eloquent,' Max said. 'Eloquent indeed.'

'We gave all we could, Joe.'

'You won't be sorry,' I said. 'When I get on my feet there'll be bonuses for all of you, you Max and you Ma and Mary and George, and I'm doubling all your wages as from now. Only for Pete's sake let's find George.'

When it got too dark to look for him any longer I went back to the caravan, and there he was in bed, undressed, with his eyes closed, pretending to be asleep. He wouldn't let me say I was sorry and he wouldn't let me thank him. He wanted the whole thing forgotten.

'After all, we're partners, Joe,' he said.

I told him about doubling the wages. I thought that would please him, but it didn't. He said I paid them all plenty. He said it wasn't good business. But he was wrong there, of course, though I couldn't get him to admit it. Good friends are always worth good money, and something more besides.

Later I asked him where he had been. 'I looked everywhere for you,' I said. 'Where were you, George?'

'Underneath,' he said, pointing.

'I don't understand.'

'I was underneath the caravan.'

I stared at him. 'That's a hell of a place to go.'

'It's private,' George said.

In the middle of the night I was wakened by the police hammering on the door. They had caught Harry the Spice, and they thought that I would be glad to know that they had recovered the money. They were quite right, of course. I was very glad.

VIII

Soon afterwards I started in Big Top. This had been my father's ambition before me, and I wished he could of been there the night I opened Harrap's Continental Circus, featuring an Amazing Collection of World-famous Artistes and Savage Beasts, a Thousand Thrills and a Thousand Laughs. George and me moved into a new big caravan coated all over in a highly artistic way with gilt, and we gave a party to all the old hands, including Mary Guiseppi, to celebrate our success. The takings went from good to better. In the trade they said I had the golden touch, and it was around this time they started calling me Lucky Joe Harrap. Myself, I think Hard-working Joe Harrap would have been a better name, because I never had my nose off the grindstone, but I'll not deny that I did have luck.

It never struck me now that there was anything funny about George's craze for Mary. I thought about it often, but I could not find a way to stop George from being hurt and hurt bad. For all his size he had the biggest heart I ever saw in any man, and there was a fair chance that Mary Guiseppi would break his heart, and if George's heart broke there was more than a fair chance that he would die. It wasn't very funny. Mary Guiseppi meant the whole world to him, and he showed it in a hundred ways every day of his life.

I soon reckoned that George had more than his £500 saved and I kept expecting the fireworks to start. I watched them both, but George was the same as always, and it wasn't him but Mary Guiseppi who began to act like she was under a strain. George noticed, of course. George always noticed everything about Mary, and he came to me about it. He said she was working too hard. He said he thought she was scared of the lions. He said she needed a holiday. He asked me to have a talk with her.

'And, Joe,' he said, 'maybe at the same time you could sound her out about me, eh?'

'What do you mean?'

'About marrying me.'

I didn't like it at all.

'I would be extremely obliged to you, Joe.' You should of seen his serious face. 'Please, Joe.'

'Oh, all right,' I said at last.

I spoke to Mary Guiseppi. I beat all around the bush and didn't say much at first. She looked ill. She had a white face at the best of times, and now there were blue shadows under her eyes and her breasts kept heaving all the time she spoke. She had kind of noticeable breasts.

'There's only one thing wrong with me, Joe,' she said. 'I want a change.'

'Maybe we could arrange a holiday.'

'No, Joe, I want more than a holiday. I want to leave.'

'Quit, you mean!'

'Yes. Quit,' she said.

I didn't need to think about that. Mary Guiseppi was one of my best acts. She was worth money to me, and I didn't want to lose her, and besides, there was George to consider.

'Is it money?' I asked.

'No.'

'You won't get better money anywhere, you know. You won't get as good.'

'It's not money.'

'What is it, then?'

'It's just that I want a change, Joe. You know how it is.'

'I don't know how it is. People that work for me just don't walk out on me.' I hesitated. I was so sharp around this time that I could of caught a weasel asleep, and I had a sudden hunch of the thing that was troubling her. It was George, of course. She was tired of seeing George on the door-step. He had got on her nerves. I had never thought of her angle before, but I thought of it now, and I saw it might not be very nice for a normal dumb sort of girl to be courted by a midget. I asked her point-blank. 'George couldn't be getting on your nerves, could he?' I said.

'No, of course not. I like George.'

'George loves you.'

'And I love George. I think he's wonderful. You know that, Joe.'

'Yes, yes,' I said. 'But what I want to know is, do you love George the way a woman loves a man?'

I saw the colour rush up her throat to her cheeks, but I was paying more attention to getting the words out than anything else. I wanted to clear up this thing once and for all and I said bluntly: 'Would you marry him?'

She jerked her arm as if she was going to strike me, but her hand finished up on her throat. 'That's a pretty bloody rotten thing to say, Joe Harrap!' she said, eyes blazing, and flew out of the door. George hadn't quite finished making a lady of her, you see.

I'd made a mess of it. I scratched my head and tried to think what I had said to make her fly off the handle. It wasn't like her. She was usually such a good-natured girl and, though she was a pretty singer, she had never come over temperamental before. In the end I sent for Ma Gollatz, because I reckoned Ma was sure to know what the trouble was.

I told her what had happened.

'What's wrong, Ma?' I said. 'Why has she been off her stroke these past months? Why does she want to leave?'

Ma held up her hands and spoke the foreign way she always does when she gets angry. 'Clever the word for you is,' she said. 'You make money. You understand reading and writing. You fill the big tent with people. Oh, clever the word for you is, Joe Harrap. Extreme clever.'

'Oh, come off it, Ma,' I said. 'I'm not so clever, and you know it.'

'No, you are a great fool. The girl is in love with you. Simply, that is all.'

I couldn't believe my ears.

'But, Ma,' I said. 'Ma, she's George's girl. George loves her.'

'So do you,' Ma said. 'You are really a great fool, Joe, although the boss.'

IX

I didn't sleep much that night.

A few days later George came to me and told me he'd asked Mary to marry him, and though she hadn't said yes she hadn't exactly said no. The way George spoke I reckoned Mary had said no all right, but she'd wrapped it up so soft he'd never felt it hit him. George was full of hope. I put my head down on my arms on the table so that he couldn't see my face.

'What's wrong?'

'I'm thinking,' I said. 'There must be some way out of this.'

'If you would speak to her again, Joe.'

'You think I can fix pretty near anything, don't you?'

'Pretty near, Joe,' George said.

Well, I spoke to Mary. I went into the tent and gave Ma the thumb and she cleared out. Mary was sitting by the door darning stockings. She had on a blue shiny frock and she looked very pretty. I cleared my throat.

'Listen,' I said. 'Ma says you're in love with me.'

'Yes, Joe.' She never even looked up from her darning. 'Ma says you're in love with me too.'

'Well, that's right. I reckon I am.'

'Oh, Joe!' she said, and her face was beautiful. She came and put her arms round me and her body pressed everywhere against mine and I shook like a frightened foal.

'I don't know much about this kind of thing, Mary,' I said. 'I wouldn't fool you. I've always been too busy.'

'Oh, Joe, you're so beautiful,' she said. 'And I love you so much. I've always loved you. Put your arms round me, darling.' And she began to cry. She was shaking just the same as me.

After that I couldn't face George, and that was why I went up to Liverpool next day to look over a couple of acts I had had in mind since the previous winter.

'You'll write, won't you, Joe?' George asked before I left. He always asked me that. 'Even if it's only a post-card.'

'Sure I'll write, George,' I said.

I had planned to stay away for a week. Mary and I talked it over, and we reckoned a week would be long enough for us to find some way of breaking the news to George.

'It's got to be some way that won't hurt him, Mary,' I said.

'I wouldn't hurt George. You know that, Joe. I wouldn't hurt him for anything. But there must be some way, and we'll find it between us.'

Well, I didn't find a way, and I knew that I never would. There are times when thinking just does not help, and this was one of them. George reckoned Mary was his girl. You couldn't think past that, and after three days I gave up trying. I couldn't stay away from Mary any longer. I went back.

I took the morning train from Liverpool and was in London in the afternoon. I had to wait for a connection, and I did not get to the town where we were pitched till after the show had begun.

The clowns were in the ring and the Stein Brothers, trapezists, were just running in to take their first bow when I arrived. I couldn't see Mary, and I waited for a minute, watching, to get the laugh that always came when Stefan Stein took George out of the pocket of his big coat.

Whenever I saw George I knew that there was something wrong. At first I thought it was the make-up, but when I saw him do his preliminary work-out I knew it was more than that. I had a kind of premonition then.

'My God, something's going to happen,' I said.

Mary had heard that I was back, and the next thing I knew she was tugging at my sleeve.

'Well, hullo,' she said. I didn't hear the rest.

'There's something wrong with George,' I said.

'I've fixed all that,' Mary said. 'That's what I'm telling you, darling. Do come and see.' She kept pulling at me, and I wanted to look at her, but couldn't take my eyes off George. 'In Ma's tent,' she said. 'Come on, Joe. You've seen this act a hundred times before.'

Gee, she was pretty! We held hands going across the Common, and she asked me if I had missed her and I told her I had, and for a few minutes I forgot all about George. Then the tightness came back to my stomach.

'You haven't done anything to hurt George, have you, Mary?'

She laughed at me. Course she hadn't, she said. But she had fixed everything. She had had a brainwave and fixed everything, and now she just couldn't wait for me to see. She was giggly with happiness.

We went into Ma's tent. It was dark inside.

'Yoo-hoo, where are you?' Mary said.

Something moved in the back of the tent, and I had to jerk my eyes down to see what it was.

'Hiya, big boy,' the Princess Circassy said.

I stared at her. Mary was watching me, smiling. 'Well, what do you think of her? Isn't she wonderful, Joe? She's George's. I bought her for George, and he's crazy about her. . . .'

I didn't listen to what she was saying. I was thinking of a day several years ago, and I was hearing George's voice. 'She smelled, Joe,' George was saying. And then: 'Joe, do you know something? I hate midgets.'

'Well, say something, Joe!' Mary demanded. 'Don't you think she's wonderful?'

'Oh, my God!'

'What's the big idea! Who are you my Godding at?' the Princess Circassy said.

I turned on my heel and started to run back towards the Big Top, but I knew all along that I wouldn't be in time. I heard the scream from a hundred voices and I heard the music stop. When I got there they were carrying something into the dressing tent and I knew it was George. There was a doctor with them.

I pushed my way to the front.

'Will he live, doctor?' I said. 'You! Doctor!'

'Back's broken,' the doctor said.

George opened his eyes and looked at me then. 'Hullo, Joe,' he said.

'George,' I said. I couldn't say anything more, and when he smiled I couldn't even look at him.

'Don't take it so hard, Joe,' he said.

Mary was beside me now, and she was crying without trying to hide it and Stefan Stein was crying with great dry sobs in his brother's arms at the back of the tent. It was Stefan who had dropped him, although it wasn't Stefan's fault, of course.

'Do you know what she did, Joe?' George said. 'She gave the Princess Circassy four hundred pounds. Every penny she had. All her savings. She did that for me, Joe.'

I stared at him.

'Imagine her doing that for me, Joe. She must have loved me in a sort of way, mustn't she, Joe?'

'Of course I love you,' Mary said, crying.

'Only a few minutes now,' the doctor said.

'He'll want to see the Princess Circassy,' Mary said. 'You want to see the Princess, don't you, George?'

'Yes.'

'Come on, Joe,' Mary said to me. 'He'll want to be alone with her.'

'No, no!' I said. I tried to explain. 'It's us that George wants. We're his people.'

'I want the Princess Circassy,' George said. 'Mary's quite right. Leave us alone together.'

'George,' I said.

Mary took my arm.

'Take care of her, Joe,' George said.

I nodded. I couldn't speak.

'We'll both take care of her, George,' Mary said. She thought George was speaking about the Princess. She didn't understand at all.

I couldn't take my eyes off George.

'George,' I said.

He just smiled at me.

'Oh, come on, Joe,' Mary said, crying. She pulled me out of the tent. 'Where is she? Oh, there you are. Quick!' she said to the Princess Circassy. 'He wants to see you.'

I lifted the Princess Circassy off the barrel that somebody had put her on because she was nervous about being left on the ground alone since a pony had trampled on her, and gave her a push towards the tent. She took a couple of steps and looked round doubtfully, and Mary said, 'Go on,' and she went in, walking very slowly.

Mary hid her face in my chest, and after a minute I put my arms round her and held her. There was no use blaming her. She didn't know what she had done, and I would see to it that she never did know. That was the way George wanted it. That was the thing he had been saying to me.

'They were in love, Joe,' she said. 'It was love at first sight. You should of seen how pleased George was. Oh, poor, poor George!'

I held her very tight and stroked her hair. I made my voice come. 'It's all right, baby,' I said. 'It's all right.'

Later, I tried to settle up with the doctor. He wouldn't take anything, but he kept talking. He had never had his hands on a midget before. He had found it very interesting, he said.

'Amazing, you know,' he said. 'Turned his head away when his girl came in. And she wouldn't go near him. No human feelings at all. I suppose they're all like that, aren't they?'

I would have liked to of told him, but I couldn't. I could only look at George.

The doctor pulled on his gloves and nodded at the table.

'Extraordinary fine specimen, though,' he said.

Jessie Kesson

1916–1994

The Gowk

You'd felt pity for the Gowk, when yourself was young. And he was a boy—debarred. Clutching the school gates. Engrossed in the rough and tumble of the playground.

In manhood, this on-looking compulsion was still with him. But you had outgrown pity. Revulsion, tinged with apprehension, had taken its place. Until you thought about it, and realised that maybe, maybe the half-witted dribbling boy was now imprisoned grotesquely in the flesh of manhood.

But you didn't often think about that. And certainly the boys on the inside of the school playground never thought about it at all. Ettling always to get out, and within taunting distance of the Gowk.

We saw Gowkit Jockie
We saw him run awa
We saw Gowkit Jockie
And his nakit Bum and a'!

'Come *inside*, Rob! *And* you, Peter!' Jean Aitken shouted from her kitchen window.

'And stop tormenting the life out of that poor bloody Gowk!' her father admonished, over her shoulder.

'That "Poor Gowk", as you call him,' Jean Aitken shrugged, 'should have been lockit up and away a long time ago. Terrifying the life out of the bairns.'

'Jockie's harmless enough,' the old man defended. 'He wouldna mind *them*. If they didna keep tormenting *him*!'

'You try telling Kate Riddrie that, Father. She's had her bellyful of the Gowk!'

'That's true enough,' her Father agreed. 'But not until the Gowk's father put the wedding ring on her hand!'

'And she's living to regret *that*!' Jean Aitken pointed out. 'Forbye, the Gowk was but a bairn, then. He's a man. Full-grown, now.'

'. . . and the older he grows, the worse he grows,' Kate Riddrie was complaining. 'He's started to abuse himself again. In broad daylight now! You'd think he hadn't got the wit for *that* even!'

'Maybe it's the instinct he's gotten,' Hugh Riddrie said. 'Even the brute beasts have gotten *that*.'

He had long since found that words failed to justify to himself the existence of his idiot son. And was beginning to discover that they failed even to protect him.

'I could *cope*,' his wife claimed. 'I could *cope* when he was young. But he's getting beyond me now.'

'You could never cope, Kate.' Hugh Riddrie reached above the dresser for his bonnet. 'You could only pretend he wasna there at all.'

'*Better*!' Kate Riddrie flared. 'Than pretending he wasn't an idiot *born*! But then, of course, he's *your* son.'

'So you aye keep reminding me, Kate.'

'And you *need* reminding! Do you know something?'

Hugh Riddrie shook his head. 'No. But I know you're just about to *tell* me something.'

'High time somebody did! You puzzle me,' Kate Riddrie admitted. 'Where other folk would try to keep a Gowk out of sight, *you* seem to like flaunting him in the face of the world.'

'Letting everybody share the *shame*, like, Kate?'

'I don't know what you'd call it!' Kate Riddrie snapped. 'But Nell Crombie was saying that she gets a red face, every time she puts her foot across this door!'

'She would,' Hugh Riddrie agreed. 'A very modest woman, Nell. Forever bragging that her man has never seen her nakit. In his life. Come to think of it,' he reflected, 'neither have you! What the hell is it makes you all so feared to *look*!'

'Decency!' Kate Riddrie said. 'Just plain *decency*!'

'Is *that* the name they've gotten for't? Ah well. I'm aye learning.'

'Not fast enough!' Kate Riddrie shouted, as he made for the door. '*Something's* got to be done. About the Gowk!'

'*Jockie*! *You mean*. Don't you, Kate?' Hugh Riddrie spun round on his heel. '*Jockie*!'

'I meant *Jockie*.' She flustered. 'It was just . . . it was just that everybody else calls him . . .'

'*THE GOWK*!' Hugh Riddrie finished the sentence for her. His quiet anger rising loud. Out of control. 'What do you suggest I do with him, Kate? *Lib* the poor bugger! The way I'd lib a young calf! Or would you have rather I had thrappled him at *birth*! With my bare hands! . . . I've killed a calf for less. For just being shargered . . .'

He could hear the school bairns taunting in the distance. Forcing his forefingers between his teeth, the shrillness of his whistle brought the taunting to a halt. And evoked the memories of the workers on their way home from the farm.

Old Riddrie. Whistling his Gowk again. Poor bugger. Other men had dogs to whistle for. Still. The man himself could be more sociable. Oh, but they minded on Riddrie, young. Another man then. Another man, altogether. That, of course, was before the Gowk was born. They themselves found little enough wrong with the Gowk! A pat on the head. A word in his ear, in the passing. A chew of tobacco slippit into his hand. And God! The Gowk was as happy as if he was in his right mind!

The shrill whistle halted their wives on their way back from the baker's van. Myth and memory blending in a confusion of truth.

The *minute* the Gowk was born. The *instant* the doctor set eyes on him . . . 'Poultice Jimmy', as he was known. For he believed that a bread poultice could cure anything from a blind boil on your bottom to a broken heart. Though a poultice was of little use to the Gowk. But at least the doctor knew *something* was far wrong.

It was the midwife, of course, that had let the cat out of the bag. In *confidence*, mind you! Though she should never have done the like. Not in a job like hers. According to her, the doctor cursed and swore like a tinker when he set eyes on the Gowk. Roaring away at the midwife. To pay heed to the *mother* . . . The midwife swore to the day she died, that Poultice Jimmy *knew*. That he hopit, if they paid no attention to the bairn, it might just dwine away. But the Gowk had survived. Never a day's illness in his life. To the great regret of Mistress Riddrie the Second.

Still. There was nothing on the *women's* consciences. The Gowk, young, had never been debarred from *their* games as girls. Always willing to be 'poor Gracie' lying dead and in her grave. While they circled mournfully around him . . .

We planted an Apple-tree
Over his head
Over his head
We planted an Apple-tree . . .

. . . 'Did you not hear me the *first* time? I'll comb your hair for you!' Jean Aitken threatened. 'If you don't come inside. And stop crying after that Gowk!'

'The Gowk was following our Liz.' Young Rob dodged his mother's upraised hand. 'Liz didn't see him. That's why Peter and me was shouting. They were going down Sue Tatt's road.'

'Sue Tatt's road!' The information halted Jean Aitken's enraged intention.

'There you are, then!' Dod Aitken laughed. '*There's* something for you to pick your teeth on! We know Sue's not all that particular. But even Sue Tatt would draw the line at the Gowk!'

'Are you *sure*, Rob?' His mother demanded.

'Positive!'

'*Certain*!' Peter added. Enjoying the effect the information had produced. 'We was trying to warn Liz. That's why we was shouting.'

'Our Liz,' Jean Aitken remembered. 'Should have been home by *this* time! The school bus gets in at the back of five. What on earth would Liz be seeking down Sue Tatt's road?'

Liz Aitken, herself, knew what she was seeking. But was not sure whether it was to be found.

'Sue Tatt will know what to do,' Chris Forbes had informed Liz. 'They say she's had more men than we've had suppers.'

That had sounded reassuring enough, last night. But then night had always brought reassurance to Liz. Expecting its very privacy to produce the dark, quiet miracle. And herself waking up. To confirm it, in the morning.

'I've *often* been late,' Chris had said. Sounding it like some special privilege rather than a comfort. 'Sometimes a whole *week* late.'

But then Chris Forbes had never been enticed up into the woods. How glad Liz had always been that she was herself. And not Chris Forbes. Never Chris Forbes. Now, she could have torn Chris right out of her skin. And gone inside it. To be safe. Like Chris was safe.

The rumours surrounding Sue Tatt were such that her house, itself, should have imparted an aura. Secret. Erotic. Its ordinariness disappointed Liz. But then the ordinariness of familiar things had begun to confuse her.

They should *know*. They should look *different*. The thing that had happened to herself should lie distorted, reflected in everything she set eyes on. The skeleton of Rob's bike, stripped of essentials, lying out in the shed. The handles of her father's old plough, curving high above the nettles.

But it was her landscape that was the ultimate traitor. Lochnagar couldn't *stand* there. The Dee it should flood . . .

> The sky it should fall
> Since I am with bairn
> Unwedded and all . . .

'This *friend* . . . this friend of yours, Liz,' Sue Tatt asked. 'About how old would she be, then?'

'Sixteen-and-a-half. Nearly seventeen!' Liz extended her age, thinking somehow that it might advance her cause. 'Chris Forbes said you could help!'

'Oh she *did*, did she? It could be nothing, Liz,' Sue Tatt concluded, transforming her irritation with Chris Forbes into an attempt to reassure Liz Aitken. 'That whiles happens to young lassies. Till they become regular, like.'

'But I *am* regular!' Liz protested. 'I've always been regular. Till *now*.'

'Oh Liz! Liz Aitken. Not *You*!'

The roof at home would have fallen in, under such an admission. It was the echo of its fall that sounded in Sue Tatt's voice.

'But you could *help*!' Liz urged. 'Everybody says . . .'

'Everything except their prayers, Liz. The thing is,' Sue stood pondering the paradox, 'Everybody knows the cure. Till the ailment happens. Syne, they know nothing. For *myself*,' Sue recollected, 'I just fell back on the old Penny Royal. Quinine. And the skin of my legs peeling off in a pail of hot water and mustard. Knowing they were all useless. But always just . . . hoping. Nothing ever budged mine an inch! Not until they were good and ready to be born. But cheer *up*, Liz! It *could* be a "wrong spy"! And I've had my share of them! You might just waken up the morn's morning to find that everything's just fine, again. And oh, whatten a fine feeling *that* is, Liz, stroking yourself under the sheet. As if your hands loved your body again. And the sweat pouring out of you. With relief, just. And thanking God. Even though you're not a Christian. Because you cannot think of anybody else to thank. And promising never to do it again. Not as long as you live . . . But of course you'll do it again, Liz!' Sue Tatt bent towards her, laughing. Pressing her hands on Liz's shoulders, as if they might leap up, and dance together, to a bright reel of Sue's composing. 'Again. And again, Liz! And it will be *right* then. And fine. For some lad will have *wedded* you! There's no chance of *this* lad wedding you?' Sue asked, as if the music itself had ended, and the bright bubble of hope drifted high up. Out of mind's reach.

'*None*!' That was a certainty. And Liz merely confirmed it. 'He's sitting his Highers,' she explained, 'and I'm trying for a Bursary. I'm going to the University. My mother's *set* on that. And my father will kill me. You'll not tell!' she urged. For, although hope had gone, secrecy still seemed essential. 'You'll never tell.'

'I'll not tell,' Sue Tatt promised. 'But you should, Liz. Tell your mother. And tell quick! Before other folk get in there first. That's what "gets" mothers. Not having the time to get their faces ready. To look on the world again.'

'It's my *father*!' Liz rose to go. 'He'll kill me. When he finds out.'

'I doubt that, Liz. I very much doubt that!'

'You don't know my father.'

Liz Aitken could well be right, Sue Tatt thought, as she watched Liz turn the bend of the road. But still Sue doubted. It was with the *mothers* of the parish that she had a mere 'nodding acquaintance'.

'A fine night, again, Jockie!' Sue Tatt cried out to the Gowk, as he shambled past her gate. Poor silly creature, he wouldn't understand a word she was saying. But he might just know that somebody was speaking to him. 'Another fine night again, Jockie!'

The brambles down in the King's Howe were always the first to ripen. Liz Aitken stood amongst the bushes, caught up once more in a deceptive sense of security. The taste of childhood on her tongue. The colour of it staining her mouth. Savouring a fallacy.

The reeshling in the bushes behind her didn't disturb her peace of mind. It was the unseen hands that gripped her shoulders, that sent her cry rising across the Howe.

Such cries breaking the silence of the quiet Howe were common enough. Easily enough analysed by listeners in the passing. A screaming rabbit cornered at last by the watchful weasel. A bleating ewe worried by a stray dog. The black sweep of the Hoodie Crow. And the rising protest of its victim. Distress traced easily enough to its unseen source. It was the source, itself, that could always momentarily stop the listening heart.

The Gowk, was no solitary. Hugh Riddrie nearly always knew where to find him. The smiddy, the general shop, the bus stop. For Jockie liked to be amongst folk. A pity, that. For folk either ignored his presence. Or acknowledged it, the way they acknowledged old Moss, the shepherd's dog. With a pat on the head.

In all the years, Hugh Riddrie had never got rid of the ache that caught at him at the sight of his son, standing with, but not of, normal men. It was rare. But easier, at times like now, when they came upon each other alone. In the nakedness of their relationship. When communication, though primitive, was natural. When tone of voice transcended interpretation. And monologue, comprehended by the listener, gave release to the speaker.

'*There* you are, Mannie! I've been whistling on you all night.
What have you been up to, Jockie?
Riving head first amongst the bushes!
Steady! Steady now! Till I get you wipit down.
Let's see your mouth now! You've been dribbling again!
The moustache of you's all slavers!
Steady now! Steady *on*!
Your flies are wide open again! Will you *never* learn to button yourself up!
You know fine that drives her clean mad.
She's gotten such a spite to flies.
Especially open flies.
STILL!
You're *fine*, now!
In you go, then, Jockie.
Up the stairs. As nippit as you can!
Hold it! Hold it, Jockie. Till I get the boots off you.
That's *it*! *That's* it!
She'll not hear you, now.
We'll better her, this time!
Eh, Jockie? Eh, Mannie!
In you go, then! You're fine, now.
All present and correct!
NO! Jockie, *NO*!
Let my hand *go*.
I'm *coming* in! Right *behind* you.
Let my hand *go*!
Do you not *see*, Jockie?
You've got to go in *first*!
As if you'd been a *good* mannie!

And come all the way home. By *yourself*!
It's easier, that way, Jockie.
In you go then. We'll be all right!'

'It's all *wrong*! All wrong, I tell you!' Jean Aitken insisted. 'That Gowk should never be allowed to roam the countryside. Just look at the state Liz has come home in. Are you all *right*, Liz? What did that mad bugger of a Gowk *do* to you?'

'We tried to warn Liz,' Young Rob remembered. '*And* me!' Peter confirmed. 'We was shouting after the Gowk.'

'Off up to your beds! The pair of you!' Jean Aitken commanded. 'Are you *sure* you're all right, Liz? Are you *sure*!'

'Liz will be all right,' Rob Aitken said. 'She got a fleg just.'

'She's gotten more than a fleg! She's looking *terrible*.'

'He grabbed me,' Liz explained. 'And I didn't see him. *That's* what it was. I didn't *see* him. I ran all the way from the King's Howe. But I thought I'd never get out of the spot.'

'What took you down to the King's Howe, like?' her father asked. 'That's bit out of your road, isn't it?'

'My homework. I forgot to take it down. I went over to get it from Chris Forbes.'

'I wouldn't bother about homework the night,' Jean Aitken advised. 'You should hold straight on up to your bed. You've had a gey shake-up.'

'I'll be all right. I couldn't sleep if I went to my bed.'

'Liz is right!' her Father agreed. 'Stop fussing her, woman!'

'Well then!' Jean Aitken turned in attack on both of them. 'If she's all right, and can't go up to her bed, and can't sleep, she's *not* going to sit molloching here all night! She can just take herself through to the sink. And make a start to the washing-up!'

She would 'tell them on Saturday'. The decision taken, Liz leant against the sink, comforted by the postponement of time that lately she had begun to allow herself, when days could seem almost normal.

'I could have sworn I put preserving ginger down on the grocer's list.' Her mother's voice drifted through to the scullery. 'I'm sure I noticed some at the back of the press, the other day . . .'

There *couldn't* be anything wrong with Liz! Her mother would *know*. She would never be worrying about preserving ginger, if there *was* something wrong . . .

'But I think it's *last* year's preserving ginger that's in the press. The strength will have gone out of it . . .'

If there *was* something wrong, her mother would stop going on about preserving ginger forever . . .

'I could be speaking to *myself*!' her mother was complaining to her father. 'I *told* you she would be better off in her bed! Standing through there in a dwam. She's had a bigger upset than she'll admit. And if it's the last thing I do, I'll make Hugh Riddrie's ears blister! Him and that Gowk of his.'

The Gowk, himself, was beginning to take on a subtle new dimension in the eyes of the Howe. A curious kind of normality. An ability to share in the venial sins of ordinary men. It was Liz Aitken who began to lose dimension to its inhabitants.

You could have 'knocked them all down with a feather', they swore. Liz Aitken of *all* people. And her set to sit for the bursary. She was just about the *last*! Not that anybody was perfect, of course. But Liz Aitken was . . .

'As liable as the *next* one!' Teen Rait had snapped, in an attempt to keep her own image of perfection intact. God help whoever was the father, they agreed. It was bound to be somebody. *That* was for certain. Though it had happened *once*. Just once. But that was two thousand years ago. And, though they were regular kirk-goers, and believed in every word the psalms uttered, they'd just never quite managed to 'swallow *that* one'. It was for papes. Although Cis Coutts, the simple creature, had tried it on when *she* was pregnant. And syne forgot. And admitted to the doctor that she 'had pink knickers on at the time'. Still, and seriously, though, God help whoever was the father when Rob Aitken got his hands on him. He couldn't get a word out of Liz herself. She wouldn't say a cheep. There was a rumour. Only a rumour, mind you! But then, there always was. They would have died without one. A 'speak'. Oh! A *whisper* just. That it was—*the Gowk*.

'You haven't got Liz to admit it, yet, then?' Kate Riddrie asked.

'No.' Jean Aitken shook her head. 'But she will. The state Liz came home in, that night. Her jumper torn, her legs scratit. And herself, nearly hysterical . . .'

'I can believe *that*! Your Liz would never have had the strength against a brute-beast like the Gowk!'

'Never a one for the lads, Liz. Her head aye buried in some book, just . . .'

'I was saying Liz would never have had the strength! Something will have to be done about the Gowk, *now*! And you've gotten witnesses!'

'Aye some book, just . . .'

'You've gotten witnesses!' Kate Riddrie urged.

'Young Rob. And Peter. They were trying to warn Liz.'

'WELL! THEN! That's *it*!'

'She never crossed the door at night. Except whiles. Down to Chris Forbes for her homework.'

'You've gotten *witnesses*! All it needs now, is to testify before the board!'

'But Liz. Liz is so unwilling. So unwilling to do that! Do you think?

Do you think, maybe . . . ?' Jean Aitken hesitated, unable to put her own apprehension into words. 'Maybe, it's because he is a *Gowk*?'

'That's where you've *got* him!' Kate Riddrie got to her feet, in triumph. 'That's what I'm trying to *tell* you. It's Liz's word against a Gowk's word. And he's got none. At least none that anybody could make any sense out of. Forbye! The whole Howe can testify that the Gowk's forever shambling all over the place. *Exposing* himself!'

'You'll be satisfied *now* then Kate. You've gotten your will. You've gotten rid of Jockie, at last . . .'

'*My* first job. The first fine day. Will be to get that stinking mattress of his outside. And set fire to it.'

'*That* was what you always wanted, Kate . . .'

'It stank the house to high heaven.'

'Wasn't it, Kate?'

'At least we'll get a bit of fresh air into the house, at last . . .'

'Speak! You *bitch*! Or have you lost your tongue! A damned pity you didn't lose it in front of the board!'

'It wasn't *me* that got rid of the . . . *Jocki*e.'

'NO! But you said damn all to prevent it!'

'What could *I* say to prevent it? The board could see for themselves. Liz Aitken's belly was getting big enough!'

'*Jockie* didn't make it so.'

'You've got no *proof* of that.'

'Nor of the *t'other*!' Hugh Riddrie concluded, making for the door. 'All that Jockie ever wanted was for somebody to *speak* to him.'

'*Speak* to him!' Kate Riddrie snorted. 'What on earth can anybody say to a GOWK!'

'I'll tell you what they can say to a Gowk, Kate! I'll *tell* you.'

Hugh Riddrie turned to face her. Searching dumbly for words, that could be put into words. *Knowing* them. Thousands of them. Words that often weren't words at all, but instincts. Transmitted by tone and touch. A language acquired and mastered in a confusion of pain and frustration.

'You can say *anything* to a Gowk, Kate!' The realisation took him by surprise. 'Anything at all. That's the best thing about Gowks. They never tell. And that's the worst thing about them. They cannot tell. But I'll find somebody, Kate. I'll find somebody who *can* tell!'

. . . Liz Aitken O Liz Aitken . . .

'Come on, Liz! Come on, lass,' her mother persuaded. 'Moping around the house like this is doing you no good. No good at all. And it such a fine night. Why don't you take yourself off for a bit walk?'

'Because she's feart!' Young Rob blurted out. Unable to contain his knowledge.

'*FEART*?'

'That's right!' Peter confirmed. '*Feart*!'

. . . Liz Aitken O Liz Aitken . . .

'Feart of what, Liz!' It was her own fear that Jean Aitken probed. Convinced that such a fear had not touched her daughter. Oh, but the young were lucky. One danger at a time. Clear and cut. Over and done with. With little hindsight—and not very much foresight. If only the father had been a normal lad. And not a Gowk. 'Feart of what, Liz?'

'Nothing. Nothing, just.'

. . . Liz Aitken O Liz Aitken . . .

'Well then!' Jean Aitken urged. 'Off into the fresh air with you. Young Rob and Peter will go with you for company.'

'Never *ME*!'

'ME *neither*!' Peter echoed his brother's determination. 'The other bairns will cry after us! "Gowk's bairn! Gowk's bairn!" *That's* what they'll cry.'

'Is that right, Liz?' her Father asked. 'Is that what they cry?'

'Sometimes. It's only the bairns, though.'

. . . Liz Aitken O Liz Aitken . . .

'I wouldn't let that worry you, Liz. Folk have always needed somebody to cry after. And they've got no Gowk, now.'

'If only it had been some other lad . . .' Regret slipped out of Jean Aitken's control. And sounded itself in her voice.

'Some *other* lad!'

Her father's astonishment confirmed Liz's own certainty.

'If it had been some *other* lad, Liz would have been out of here. *Bag* and *baggage*! What happened was no fault of her own. It took half a dozen of us grown men, to hold the Gowk down, till they got him off to the asylum.'

'Come on, Liz. Up you get.' Her mother piloted her towards the door. 'Just you take a turn round the steading. I used to like fine a walk when darkness was coming down,' her mother confided, as they stood on the doorstep. 'I suppose I felt ashamed in daylight. *Not* because I was carrying a bairn, Liz. But just I felt so ungainly. And ugly in myself. Still!' her arm found her daughter's shoulder. 'Every creature's *bonnie* when it's little, Liz.'

A daft thing to say, Jean Aitken thought, as she watched Liz from the door. The wrong words sometimes came out. When you couldn't find the right ones to say.

'Just the length of the steading, Liz!' she called out, reminding her daughter . . .

But the Gowk's father roamed freely enough. On the prowl. Night after night. They said. Neither Gowk to whistle on, nor dog for company. His croft running to wreck and ruin. His oats rotting in the stack. And the threshing mill had gone long since past his road end. His turnips neither howked nor stored for his cattle-beasts. And winter nearly on top of the man. Bad enough when his first wife died, and the Gowk was born. Worse than ever *now* since they'd carted the Gowk off to the asylum.

Come to think of it, they themselves missed the Gowk. You would never believe *that*! But they'd just got used to him, like. Popping up here and there. And everywhere around the Howe. Still. It was an ill wind. And it had fair suited *Katie Riddrie*!

'I'm not so sure that it did!' Meg Tait informed them. 'I'm not so sure at all! Kate Riddrie *herself* was telling me only the other day . . .'

'There's no living with him. No living with him at all. On the prowl all night. And sitting amongst my feet all day. Never taking his eyes off me. And never opening his mouth to me. Just mumbling away yonder to himself. He aye maintained that his first wife was at fault for the bairn being born a Gowk. But I'm beginning to have my doubts. The way he sits mumbling to himself. He'd aye gotten such an *obsession* with that Gowk.'

. . . Liz Aitken O Liz Aitken . . .

LIZ!

So it hadn't been merely in her imagination. Or, maybe it had been created out of her imagination.

LIZ AITKEN!

Strange how prepared she was . . . 'I'm in a hurry, Mr Riddrie.'
'Aye, Liz. You've been in a great hurry this past few weeks!
What is it that you're running from like?
Hold on, Liz! Just hold *on*, there!
You're not *feart* are you, Liz?
No! Of course you're not feart!
You know fine that the *Gowk* canna jump out on you the night.
No. He canna do that. He's far enough away, the night.
You made sure of that!
You *all* made sure of it. The whole bloody jing bang of you!
No! No! Liz! Hold on!
It wasn't Jockie? Was it, now?'

'I told the board . . .'

'I know damned fine what you told the *board*!
You try telling *ME*!
Struck suddenly dumb are you, Liz?
It's some late in the day for that.
It was never Jockie, Liz. *Never* Jockie.
You see Liz, he wouldn't even have kent *where* to *PUT* the bloody thing.
But *I* ken, Liz.

I ken *for* him.'

MURIEL SPARK
1918–

BANG-BANG YOU'RE DEAD

At that time many of the men looked like Rupert Brooke, whose portrait still hung in everyone's imagination. It was that clear-cut, 'typically English' face which is seldom seen on the actual soil of England but proliferates in the African Colonies.

'I must say,' said Sybil's hostess, 'the men look charming.'

These men were all charming, Sybil had decided at the time, until you got to know them. She sat in the dark room watching the eighteen-year-old microfilm unrolling on the screen as if the particular memory had solidified under the effect of some intense heat coming out of the projector. She told herself, I was young, I demanded nothing short of perfection. But then, she thought, that is not quite the case. But it comes to the same thing; to me, the men were not charming for long.

The first reel came to an end. Someone switched on the light. Her host picked the next film out of its tropical packing.

'It must be an interesting experience,' said her hostess, 'seeing yourself after all those years.'

'Hasn't Sybil seen these films before?' said a latecomer.

'No, never—have you, Sybil?'

'No, never.'

'If they had been my films,' said her hostess, 'my curiosity could not have waited eighteen years.'

The Kodachrome reels had lain in their boxes in the dark of Sybil's cabin trunk. Why bother, when one's memory was clear?

'Sybil didn't know anyone who had a projector,' said her hostess, 'until we got ours.'

'It was delightful,' said the latecomer, an elderly lady, 'what I saw of it. Are the others as good?'

Sybil thought for a moment. 'The photography is probably good,' she said. 'There was a cook behind the camera.'

Only the grown-ups mistook one child for another at odd moments. None of her small companions made this mistake. After the school concert Sybil's mother said, 'For a second I thought you were Désirée in the choir. It's strange you are so alike. I'm not a bit like Mrs Coleman and your daddy doesn't resemble *him* in the least.'

Sybil found Désirée unsatisfactory as a playmate. Sybil was precocious, her brain was like a blade. She had discovered that dull children were apt to be spiteful. Désirée would sit innocently cross-legged beside you at a party, watching the conjuror, then suddenly, for no apparent reason, jab at you viciously with her elbow.

By the time Sybil was eight and Désirée nine it was seldom that anyone, even strangers and new teachers, mixed them up. Sybil's nose became more sharp and pronounced while Désirée's seemed to sink into her plump cheeks like a painted-on nose. Only on a few occasions, and only on dark winter afternoons between the last of three o'clock daylight and the coming on of lights all over the school, was Sybil mistaken for Désirée.

Between Sybil's ninth year and her tenth Désirée's family came to live in her square. The residents' children were taken to the gardens of the square after school by mothers and nursemaids, and were bidden to play with each other nicely. Sybil regarded the intrusion of Désirée sulkily, and said she preferred her book. She cheered up, however, when a few weeks later the Dobell boys came to live in the square. The two Dobells had dusky-rose skins and fine dark eyes. It appeared the father was half Indian.

How Sybil adored the Dobells! They were a new type of playmate in her experience, so jumping and agile, and yet so gentle, so unusually courteous. Their dark skins were never dirty, a fact which Sybil obscurely approved. She did not then mind Désirée joining in their games; the Dobell boys were a kind of charm against despair, for they did not understand stupidity and so did not notice Désirée's.

The girl lacked mental stamina, could not keep up an imaginative game for long, was shrill and apt to kick her playmates unaccountably and on the sly; the Dobells reacted to this with a simple resignation. Perhaps the lack of opposition was the reason that Désirée continually shot Sybil dead, contrary to the rules, whenever she felt like it.

Sybil resented with the utmost passion the repeated daily massacre of herself before the time was ripe. It was useless for Jon Dobell to explain, 'Not yet, Désirée. Wait, wait, Désirée. She's not to be shot down yet. She hasn't crossed the bridge yet, and you can't shoot her from there, anyway—there's a big boulder between you and her. You have to creep round it, and Hugh has a shot at you first, and he thinks he's got you, but only your hat. And . . .'

It was no use. Each day before the game started the four sat in conference on the short dry prickly grass. The proceedings were agreed. The game was on. 'Got it all clear, Désirée?' 'Yes,' she said, every day. Désirée shouted and got herself excited, she made foolish sounds even when supposed to be stalking the bandits through the silent forest. A few high screams and then, 'Bang-bang,' she yelled, aiming at Sybil, 'you're dead.' Sybil obediently rolled over, protesting none the less that the game had only begun, while the Dobells sighed, 'Oh, *Désirée*!'

Sybil vowed to herself each night, I will do the same to her. Next time—tomorrow if it isn't raining—I will bang-bang her before she has a chance to hang her panama on the bough as a decoy. I will say bang-bang on her out of turn, and I will do her dead before her time.

But on no succeeding tomorrow did Sybil bring herself to do this. Her pride before the Dobells was more valuable than the success of the game. Instead, with her cleverness, Sybil set herself to avoid Désirée's range for as long as possible. She dodged behind the laurels and threw out a running commentary as if to a mental defective, such as, 'I'm in disguise, all in green, and no one can see me among the trees.' But still Désirée saw her. Désirée's eyes insisted on penetrating solid mountains. 'I'm half a mile away from everyone,' Sybil cried as Désirée's gun swivelled relentlessly upon her.

I shall refuse to be dead, Sybil promised herself. I'll break the rule. If it doesn't count with her why should it count with me? I won't roll over any more when she bangs you're dead to me. Next time, tomorrow if it isn't raining . . .

But Sybil simply did roll over. When Jon and Hugh Dobell called out to her that Désirée's bang-bang did not count she started hopefully to resurrect herself; but 'It does count, it *does*. That's the rule,' Désirée counter-screeched. And Sybil dropped back flat, knowing utterly that this was final.

And so the girl continued to deal premature death to Sybil, losing her head, but never so much that she aimed at one of the boys. For some reason which Sybil did not consider until she was years and years older, it was always herself who had to die.

One day, when Désirée was late in arriving for play, Sybil put it to the boys that Désirée should be left out of the game in future. 'She only spoils it.'

'But,' said Jon, 'you need four people for the game.'

'You need four,' said Hugh.

'No, you can do it with three.' As she spoke she was inventing the game with three. She explained to them what was in her mind's eye. But neither boy could grasp the idea, having got used to Bandits and Riders with two on each side. 'I am the lone Rider, you see,' said Sybil. 'Or,' she wheedled, 'the cherry tree can be a Rider.' She was talking to stone, inoffensive but uncomprehending. All at once she realized, without articulating the idea, that her intelligence was superior to theirs, and she felt lonely.

'Could we play rounders instead?' ventured Jon.

Sybil brought a book every day after that, and sat reading beside her mother, who was glad, on the whole, that Sybil had grown tired of rowdy games.

'They were preparing,' said Sybil, 'to go on a shoot.'

Sybil's host was changing the reel.

'I get quite a new vision of Sybil,' said her hostess, 'seeing her in such a . . . such a *social* environment. Were any of these people intellectuals, Sybil?'

'No, but lots of poets.'

'Oh, *no*. Did they all write poetry?'

'Quite a lot of them,' said Sybil, 'did.'

'Who *were* they all? Who was that blond fellow who was standing by the van with you?'

'He was the manager of the estate. They grew passion-fruit and manufactured the juice.'

'Passion-fruit—how killing. Did *he* write poetry?'

'Oh, yes.'

'And who was the girl, the one I thought was you?'

'Oh, I had known her as a child and we met again in the Colony. The short man was her husband.'

'And were you all off on safari that morning? I simply can't imagine you shooting anything, Sybil, somehow.'

'On this occasion,' said Sybil, 'I didn't go. I just held the gun for effect.'

Everyone laughed.

'Do you still keep up with these people? I've heard that colonials are great letter writers, it keeps them in touch with—'

'No.' And she added, 'Three of them are dead. The girl and her husband, and the fair fellow.'

'Really? What happened to them? Don't tell me *they* were mixed up in shooting affairs.'

'They were mixed up in shooting affairs,' said Sybil.

'Oh, these colonials,' said the elderly woman, 'and their shooting affairs!'

'Number three,' said Sybil's host. 'Ready? Lights out, please.'

'Don't get eaten by lions. I say, Sybil, don't get mixed up in a shooting affair.' The party at the railway station were unaware of the noise they were making for they were inside the noise. As the time of departure drew near Donald's relatives tended to herd themselves apart while Sybil's clustered round the couple.

'Two years—it will be an interesting experience for them.'

'Mind out for the shooting affairs. Don't let Donald have a gun.'

There had been an outbreak of popular headlines about the shooting affairs in the Colony. Much had been blared forth about the effect, on the minds of young settlers, of the climate, the hard drinking, the shortage of white women. The Colony was a place where lovers shot husbands, or shot themselves, where husbands shot natives who spied through bedroom windows. Letters to *The Times* arrived belatedly from respectable colonists, refuting the scandals with sober statistics. The recent incidents, they said, did not represent the habits of the peaceable majority. The

Governor told the press that everything had been highly exaggerated. By the time Sybil and Donald left for the Colony the music-hall comics had already exhausted the entertainment value of colonial shooting affairs.

'Don't make pets of snakes or crocs. Mind out for the lions. Don't forget to write.'

It was almost a surprise to them to find that shooting affairs in the Colony were not entirely a music-hall myth. They occurred in waves. For three months at a time the gun-murders and suicides were reported weekly. The old colonists with their very blue eyes sat beside their whisky bottles and remarked that another young rotter had shot himself. Then the rains would break and the shootings would cease for a long season.

Eighteen months after their marriage Donald was mauled by a lioness and died on the long stretcher journey back to the station. He was one of a party of eight. No one could really say how it happened; it was done in a flash. The natives had lost their wits, and, instead of shooting the beast, had come calling 'Ah-ah-ah,' and pointing to the spot. A few strides, shouldering the grass aside, and Donald's friends got the lioness as she reared from his body.

His friends in the archaeological team to which he belonged urged Sybil to remain in the Colony for the remaining six months and return to England with them. Still undecided, she went on a sight-seeing tour. But before their time was up the archaeologists had been recalled. War had been declared. Civilians were not permitted to leave the continent, and Sybil was caught, like Donald under the lioness.

She wished he had lived to enjoy a life of his own, as she intended to do. It was plain to her that they must have separated had he lived. There had been no disagreement but, thought Sybil, given another two years there would have been disagreements. Donald had shown signs of becoming a bore. By the last, the twenty-seventh, year of his life, his mind had ceased to inquire. Archaeology, that thrilling subject, had become Donald's job, merely. He began to talk as if all archaeological methods and theories had ceased to evolve on the day he obtained his degree; it was now only a matter of applying his knowledge to field-work for a limited period. Archaeological papers came out from England. The usual crank literature on roneo foolscap followed them from one postal address to another. 'Donald, aren't you going to look through them?' Sybil said, as the journals and papers piled up. 'No, really, I don't see it's necessary.' It was not necessary because his future was fixed; two years in the field and then a lectureship. If it were my subject, she thought, these papers would be necessary to me. Even the crackpot ones, rightly read, would be, to me, enlarging.

Sybil lay in bed in the mornings reading the translation of Kierkegaard's *Journals*, newly arrived from England in their first, revelatory month of publication. She felt like a desert which had not realized its own aridity till the rain began to fall upon it. When Donald came home in the late afternoons she had less and less to say to him.

'There has been another shooting affair,' Donald said, 'across the valley. The chap came home unexpectedly and found his wife with another man. He shot them both.'

'In this place, one is never far from the jungle,' Sybil said.

'What are you talking about? We are eight hundred miles from the jungle.'

When he had gone on his first big shoot, eight hundred miles away in the jungle, she had reflected, there is no sign of a living mind in him, it is like a landed fish which has ceased to palpitate. But, she thought, another woman would never notice it. Other women do not wish to be married to a Mind. Yet I do, she thought, and I am a freak and should not have married. In fact I am not the marrying type. Perhaps that is why he does not explore my personality, any more than he reads the journals. It might make him think, and that would be hurtful.

After his death she wished he had lived to enjoy a life of his own, whatever that might have been. She took a job in a private school for girls and cultivated a few friends for diversion until the war should be over. Charming friends need not possess minds.

Their motor launch was rocking up the Zambezi. Sybil was leaning over the rail mouthing something to a startled native in a canoe. Now Sybil was pointing across the river.

'I think I was asking him,' Sybil commented to her friends in the darkness, 'about the hippo. There was a school of hippo some distance away, and we wanted to see them better. But the native said we shouldn't go too near—that's why he's looking so frightened—because the hippo often upset a boat, and then the crocs quickly slither into the water. There, look! We got a long shot of the hippo—those bumps in the water, like submarines, those are the snouts of hippo.'

The film rocked with the boat as it proceeded up the river. The screen went white.

'Something's happened,' said Sybil's hostess.

'Put on the light,' said Sybil's host. He fiddled with the projector and a young man, their lodger from upstairs, went to help him.

'I loved those tiny monkeys on the island,' said her hostess. 'Do hurry, Ted. What's gone wrong?'

'Shut up a minute,' he said.

'Sybil, you know you haven't changed much since you were a girl.'

'Thank you, Ella.' I haven't changed at all so far as I still think charming friends need not possess minds.

'I expect this will revive your memories, Sybil. The details, I mean. One is bound to forget so much.'

'Oh yes,' Sybil said, and she added, 'but I recall quite a lot of details, you know.'

'Do you *really*, Sybil?'

I wish, she thought, they wouldn't cling to my least word.

The young man turned from the projector with several feet of the film-strip looped between his widespread hands. 'Is the fair chap your husband, Mrs Greeves?' he said to Sybil.

'Sybil lost her husband very early on,' her hostess informed him in a low and sacred voice.

'Oh, I *am* sorry.'

Sybil's hostess replenished the drinks of her three guests. Her host turned from the projector, finished his drink, and passed his glass to be refilled, all in one movement. Everything they do seems large and important, thought Sybil, but I must not let it be so. We are only looking at old films.

She overheard a sibilant 'Whish-sh-sh?' from the elderly woman in which she discerned, 'Who is she?'

'Sybil Greeves,' her hostess breathed back, 'a distant cousin of Ted's through marriage.'

'Oh yes?' The low tones were puzzled as if all had not been explained.

'She's quite famous, of course.'

'Oh, I didn't know that.'

'Very few people know it,' said Sybil's hostess with a little arrogance.

'OK,' said Ted, 'lights out.'

'I must say,' said his wife, 'the colours are marvellous.'

All the time she was in the Colony Sybil longed for the inexplicable colourings of her native land. The flamboyants were too rowdy, the birds, the native women with their heads bound in cloth of piercing pink, their blinding black skin and white teeth, the baskets full of bright tough flowers or oranges on their heads, the sight of which everyone else admired ('How I wish I could paint all this!'), distressed Sybil, it bored her.

She rented a house, sharing it with a girl whose husband was fighting in the north. She was twenty-two. To safeguard her privacy absolutely, she had a plywood partition put up in the sitting-room, for it was another ten years before she had learnt those arts of leading a double life and listening to people ambiguously, which enabled her to mix without losing identity, and to listen without boredom.

On the other side of the partition Ariadne Lewis decorously entertained her friends, most of whom were men on leave. On a few occasions Sybil attended these parties, working herself, as in a frenzy of self-discipline, into a state of carnal excitement over the men. She managed to do this only by an effortful sealing-off of all her critical faculties except those which assessed a good male voice and appearance. The hangovers were frightful.

The scarcity of white girls made it easy for any one of them to keep a number of men in perpetual attendance. Ariadne had many boy friends but no love affairs. Sybil had three affairs in the space of two years, to put herself to the test. They started at private dances, in the magnolia-filled gardens that smelt like a scent factory, under the Milky Way which looked like an overcrowded jeweller's window. The affairs ended when she succumbed to one of her attacks of tropical flu, and lay in a twilight of the senses on a bed which had been set on the stone stoep and overhung with a white mosquito net like something bridal. With damp shaky hands she would write a final letter to the man and give it to her half-caste maid to post. He would telephone next morning, and would be put off by the house boy, who was quite intelligent.

For some years she had been thinking she was not much inclined towards sex. After the third affair, this dawned and rose within her as a whole realization, as if in the past, when she had told herself, 'I am not predominantly a sexual being,' or 'I'm rather a frigid freak, I suppose,' these were the sayings of an illiterate, never quite rational and known until now, but after the third affair the notion was so intensely conceived as to be almost new. It appalled her. She lay on the shady stoep, her fever subsiding, and examined her relations with men. She thought, what if I married again? She shivered under the hot sheet. Can it be, she thought, that I have a suppressed tendency towards women? She lay still and let the idea probe round in imagination. She surveyed, with a stony inward eye, all the women she had known, prim little academicians with cream peter-pan collars on their dresses, large dominant women, a number of beauties, conventional nitwits like Ariadne. No, really, she thought; neither men nor women. It is a not caring for sexual relations. It is not merely a lack of pleasure in sex, it is dislike of the excitement. And it is not merely dislike, it is worse, it is boredom.

She felt a lonely emotion near to guilt. The three love affairs took on heroic aspects in her mind. They were an attempt, thought Sybil, to do the normal thing. Perhaps I may try again. Perhaps, if I should meet the right man . . . But at the idea 'right man' she felt a sense of intolerable desolation and could not stop shivering. She raised the mosquito net and reached for the lemon juice, splashing it jerkily into the glass. She sipped. The juice had grown warm and had been made too sweet, but she let it linger on her sore throat and peered through the net at the backs of houses and the yellow veldt beyond them.

Ariadne said one morning, 'I met a girl last night, it was funny. I thought it was you at first and called over to her. But she wasn't really like you close up, it was just an impression. As a matter of fact, she knows you. I've asked her to tea. I forget her name.'

'I don't,' said Sybil.

But when Désirée arrived they greeted each other with exaggerated warmth, wholly felt at the time, as acquaintances do when they meet in another hemisphere. Sybil had last seen Désirée at a dance in Hampstead, and there had merely said, 'Oh, hallo.'

'We were at our first school together,' Désirée explained to Ariadne, still holding Sybil's hand.

Already Sybil wished to withdraw. 'It's strange,' she remarked, 'how, sooner or later, everyone in the Colony meets someone they have known, or their parents knew, at home.'

Désirée and her husband, Barry Weston, were settled in a remote part of the Colony. Sybil had heard of Weston, unaware that Désirée was his wife. He was much talked of as an enterprising planter. Some years ago he had got the idea of manufacturing passion-fruit juice, had planted orchards and set up a factory. The business was now expanding wonderfully. Barry Weston also wrote poetry, a volume of which, entitled *Home Thoughts*, he had published and sold with great success within the confines of the Colony. His first wife had died of blackwater fever. On one of his visits to England he had met and married Désirée, who was twelve years his junior.

'You *must* come and see us,' said Désirée to Sybil; and to Ariadne she explained again, 'We were at our first little private school together.' And she said, 'Oh, Sybil, do you remember Trotsky? Do you remember Minnie Mouse, what a hell of a life we gave her? I shall never forget that day when . . .'

The school where Sybil taught was shortly to break up for holidays; Ariadne was to visit her husband in Cairo at that time. Sybil promised a visit to the Westons. When Désirée, beautifully dressed in linen suiting, had departed, Ariadne said, 'I'm so glad you're going to stay with them. I hated the thought of your being all alone for the next few weeks.'

'Do you know,' Sybil said, 'I don't think I shall go to stay with them after all. I'll make an excuse.'

'Oh, why not? Oh, Sybil, it's such a lovely place, and it will be fun for you. He's a poet, too.' Sybil could sense exasperation, could hear Ariadne telling her friends, 'There's something wrong with Sybil. You never know a person till you live with them. Now Sybil will say one thing one minute, and the next . . . Something wrong with her sex-life, perhaps . . . odd . . .'

At home, thought Sybil, it would not be such a slur. Her final appeal for a permit to travel to England had just been dismissed. The environment mauled her weakness. 'I think I'm going to have a cold,' she said shivering.

'Go straight to bed, dear.' Ariadne called for black Elijah and bade him prepare some lemon juice. But the cold did not materialize.

She returned with flu, however, from her first visit to the Westons. Her 1936 Ford V8 had broken down on the road and she had waited three chilly hours before another car had appeared.

'You must get a decent car,' said the chemist's wife, who came to console her. 'These old crocks simply won't stand up to the roads out here.'

Sybil shivered and held her peace. Nevertheless, she returned to the Westons at mid-term.

Désirée's invitations were pressing, almost desperate. Again and again Sybil went in obedience to them. The Westons were a magnetic field.

There was a routine attached to her arrival. The elegant wicker chair was always set for her in the same position on the stoep. The same cushions, it seemed, were always piled in exactly the same way.

'What will you drink, Sybil? Are you comfy there, Sybil? We're going to give you a wonderful time, Sybil.' She was their little orphan, she supposed. She sat, with very dark glasses, contemplating the couple. 'We've planned—haven't we, Barry?—a surprise for you, Sybil.' 'We've planned—haven't we, Désirée?—a marvellous trip . . . a croc hunt . . . hippo . . .'

Sybil sips her gin and lime. Facing her on the wicker sofa, Désirée and her husband sit side by side. They gaze at Sybil affectionately. 'Take off your smoke glasses, Sybil, the sun's nearly gone.' Sybil takes them off. The couple hold hands. They peck kisses at each other, and presently, outrageously, they are entwined in a long erotic embrace in the course of which Barry once or twice regards Sybil from the corner of his eye. Barry disengages himself and sits with his arm about his wife; she snuggles up to him. Why, thinks Sybil, is this performance being staged? 'Sybil is shocked,' Barry remarks. She sips her drink, and reflects that a public display between man and wife somehow is more shocking than are courting couples in parks and doorways. 'We're very much in love with each other,' Barry explains, squeezing his wife. And Sybil wonders what is wrong with their marriage since obviously something is wrong. The couple kiss again. Am I dreaming this? Sybil asks herself.

Even on her first visit Sybil knew definitely there was something wrong with the marriage. She thought of herself, at first, as an objective observer, and was even amused when she understood they had chosen her to be their sort of Victim of Expiation. On occasions when other guests were present she noted that the love scenes did not take place. Instead, the couple tended to snub Sybil before their friends. 'Poor little Sybil, she lives all alone and is a teacher, and hasn't many friends. We have her here to stay as often as possible.' The people would look uneasily at Sybil, and would smile. 'But you must have *heaps* of friends,' they would say politely. Sybil came to realize she was an object of the Westons' resentment, and that, nevertheless, they found her indispensable.

Ariadne returned from Cairo. 'You always look washed out when you've been staying at the Westons,' she told Sybil eventually. 'I suppose it's due to the late parties and lots of drinks.'

'I suppose so.'

Désirée wrote continually. 'Do come, Barry needs you. He needs your advice about some sonnets.' Sybil tore up these letters quickly, but usually went. Not because her discomfort was necessary to their wellbeing, but because it was somehow necessary to her own. The act of visiting the Westons alleviated her sense of guilt.

I believe, she thought, they must discern my abnormality. How could they have guessed? She was always cautious when they dropped questions about her private life. But one's closest secrets have a subtle way of communicating themselves to the resentful vigilance of opposite types. I do believe, she thought, that heart speaks unto heart, and deep calleth unto deep. But rarely in clear language. There is a misunderstanding here. They imagine their demonstrations of erotic bliss will torment my frigid soul, and so far they are right. But the reason for my pain is not envy. Really, it is boredom.

Her Ford V8 rattled across country. How bored, she thought, I am going to be by their married tableau! How pleased, exultant, they will be! These thoughts consoled her, they were an offering to the gods.

'Are you comfy, Sybil?'

She sipped her gin and lime. 'Yes, thanks.'

His pet name for Désirée was Dearie. 'Kiss me, Dearie,' he said.

'There, Baddy,' his wife said to Barry, snuggling close to him and squinting at Sybil.

'I say, Sybil,' Barry said as he smoothed down his hair, 'you ought to get married again. You're missing such a lot.'

'Yes, Sybil,' said Désirée, 'you should either marry or enter a convent, one or the other.'

'I don't see why,' Sybil said, 'I should fit into a tidy category.'

'Well, you're neither one thing nor another—is she, honey-bunch?'

True enough, thought Sybil, and that is why I'm laid out on the altar of boredom.

'Or get yourself a boy-friend,' said Désirée. 'It would be good for you.'

'You're wasting your best years,' said Barry.

'Are you comfy there, Sybil? . . . We want you to enjoy yourself here. Any time you want to bring a boy-friend, we're broad-minded—aren't we, Baddy?'

'Kiss me, Dearie,' he said.

Désirée took his handkerchief from his pocket and rubbed lipstick from his mouth. He jerked his head away and said to Sybil, 'Pass your glass.'

Désirée looked at her reflection in the glass of the french windows and said, 'Sybil's too intellectual, that's her trouble.' She patted her hair, then looked at Sybil with an old childish enmity.

After dinner Barry would read his poems. Usually, he said, 'I'm not going to be an egotist tonight. I'm not going to read my poems.' And usually Désirée would cry, 'Oh do, Barry, do.' Always, eventually, he did. 'Marvellous,' Désirée would comment, 'wonderful.' By the third night of her visits, the farcical aspect of it all would lose its fascination for Sybil, and boredom would fill her near to bursting point, like gas in a balloon. To relieve the strain, she would sigh deeply from time to time. Barry was too engrossed in his own voice to notice this, but Désirée was watching. At first Sybil worded her comments tactfully. 'I think you should devote more of your time to your verses,' she said. And, since he looked puzzled, added, 'You owe it to poetry if you write it.'

'Nonsense,' said Désirée, 'he often writes a marvellous sonnet before shaving in the morning.'

'Sybil may be right,' said Barry. 'I owe poetry all the time I can give.'

'Are you tired, Sybil?' said Désirée. 'Why are you sighing like that; are you all right?'

Later, Sybil gave up the struggle and wearily said, 'Very good,' or 'Nice rhythm' after each poem. And even the guilt of condoning Désirée's 'marvellous . . . wonderful' was less than the guilt of her isolated mind. She did not know then that the price of allowing false opinions was the gradual loss of one's capacity for forming true ones.

Not every morning, but at least twice during each visit Sybil would wake to hear the row in progress. The nanny, who brought her early tea, made large eyes and tiptoed warily. Sybil would have her bath, splashing a lot to drown the noise of the quarrel. Downstairs, the battle of voices descended, filled every room and corridor. When, on the worst occasions, the sound of shattering glass broke through the storm, Sybil would know that Barry was smashing up Désirée's dressing table; and would wonder how Désirée always managed to replace her crystal bowls since goods of that type were now scarce, and why she bothered to do so. Sybil would always find the two girls of Barry's former marriage standing side by side on the lawn frankly gazing up at the violent bedroom window. The nanny would cart off Désirée's baby for a far-away walk. Sybil would likewise disappear for the morning.

The first time this happened, Désirée told her later, 'I'm afraid you unsettle Barry.'

'What do you mean?' said Sybil.

Désirée dabbed her watery eyes and blew her nose. 'Well, of *course*, it stands to reason, Sybil, you're out to attract Barry. And he's only a man. I know you do it *unconsciously*, but . . .'

'I can't stand this sort of thing. I shall leave right away,' Sybil said.

'No, Sybil, no. Don't make a *thing* of it. Barry needs you. You're the only person in the Colony who can really talk to him about his poetry.'

'Understand,' said Sybil on that first occasion, 'I am not at all interested in your husband. I think he's an all-round third-rater. That is my opinion.'

Désirée looked savage. 'Barry,' she shouted, 'has made a fortune out of passion-fruit juice in eight years. He has sold four thousand copies of *Home Thoughts* on his own initiative.'

It was like a game for three players. According to the rules, she was to be in love, unconsciously, with Barry, and tortured by the contemplation of Désirée's married bliss. She felt too old to join in, just at that moment.

Barry came to her room while she was packing. 'Don't go,' he said. 'We need you. And after all, we are only human. What's a row? These quarrels only happen in the best marriages. And I can't for the life of me think how it started.'

'What a beautiful house. What a magnificent estate,' said Sybil's hostess.

'Yes,' said Sybil, 'it was the grandest in the Colony.'

'Were the owners frightfully grand?'

'Well, they were rich, of course.'

'I can see that. What a beautiful interior. I adore those lovely old oil lamps. I suppose you didn't have electricity?'

'Yes, there was electric light in all the rooms. But my friends preferred the oil-lamp tradition for the dining-room. You see, it was a copy of an old Dutch house.'

'Absolutely charming.'

The reel came to an end. The lights went up and everyone shifted in their chairs.

'What were those large red flowers?' said the elderly lady.

'Flamboyants.'

'Magnificent,' said her hostess. 'Don't you miss the colours, Sybil?'

'No, I don't, actually. There was too much of it for me.'

'You didn't care for the bright colours?' said the young man, leaning forward eagerly.

Sybil smiled at him.

'I liked the bit where those little lizards were playing among the stones. That was an excellent shot,' said her host. He was adjusting the last spool.

'I rather *liked* that handsome blond fellow,' said her hostess, as if the point had been in debate. 'Was he the passion-fruiter?'

'He was the manager,' said Sybil.

'Oh yes, you told me. He was in a shooting affair, did you say?'

'Yes, it was unfortunate.'

'Poor young man. It sounds quite a dangerous place. I suppose the sun and everything . . .'

'It was dangerous for some people. It depended.'

'The blacks look happy enough. Did you have any trouble with them in those days?'

'No,' said Sybil, 'only with the whites.'

Everyone laughed.

'Right,' said her host. 'Lights out, please.'

Sybil soon perceived the real cause of the Westons' quarrels. It differed from their explanations: they were both, they said, so much in love, so jealous of each other's relations with the opposite sex.

'Barry was furious,' said Désirée one day, '—weren't you, Barry?—because I smiled, merely smiled, at Carter.'

'I'll have it out with Carter,' muttered Barry. 'He's always hanging round Désirée.'

David Carter was their manager. Sybil was so foolish as once to say, 'Oh surely David wouldn't—'

'Oh wouldn't he?' said Désirée.

'Oh wouldn't he?' said Barry.

Possibly they did not themselves know the real cause of their quarrels.

These occurred on mornings when Barry had decided to lounge in bed and write poetry. Désirée, anxious that the passion-fruit business should continue to expand, longed for him to be at his office in the factory at eight o'clock each morning, by which time all other enterprising men in the Colony were at work. But Barry spoke more and more of retiring and devoting his time to his poems. When he lay abed, pen in hand, worrying a sonnet, Désirée would sulk and bang doors. The household knew that the row was on. 'Quiet! Don't you see I'm trying to think,' he would shout. '*I suggest*,' she would reply, 'you go to the library if you want to write.' It was evident that her greed and his vanity, facing each other in growling antipathy, were too terrible for either to face. Instead, the names of David Carter and Sybil would fly between them, consoling them, pepping-up and propagating the myth of their mutual attraction.

'Rolling your eyes at Carter in the orchard. Don't think I didn't notice.'

'Carter? That's funny. I can easily keep Carter in his place. But while we're on the subject, what about you with Sybil? You sat up late enough with her last night after I'd gone to bed.'

Sometimes he not only smashed the crystal bowls, he hurled them through the window.

In the exhausted afternoon Barry would explain, 'Désirée was upset—weren't you, Désirée?—because of you, Sybil. It's understandable. We shouldn't stay up late talking after Désirée has gone to bed. You're a little devil in your way, Sybil.'

'Oh well,' said Sybil obligingly, 'that's how it is.'

She became tired of the game. When, in the evenings, Barry's voice boomed forth with sonorous significance as befits a hallowed subject, she no longer thought of herself as an objective observer. She had tired of the game because she was now more than nominally committed to it. She ceased to be bored by the Westons; she began to hate them.

'What I don't understand,' said Barry, 'is why my poems are ignored back in England. I've sold over four thousand of the book out here. Feature articles about me have appeared in all the papers out here; remind me to show you them. But I can't get a single notice in London. When I send a poem to any of the magazines I don't even get a reply.'

'They are engaged in a war,' Sybil said.

'But they still publish poetry. Poetry so-called. Utter rubbish, all of it. You can't understand the stuff.'

'Yours is too good for them,' said Sybil. To a delicate ear her tone might have resembled the stab of a pin stuck into a waxen image.

'That's a fact, between ourselves,' said Barry. 'I shouldn't say it, but that's the answer.'

Barry was over-weight, square and dark. His face had lines, as of anxiety or stomach trouble. David Carter, when he passed, cool and fair through the house, was quite a change.

'England is finished,' said Barry. 'It's degenerate.'

'I wonder,' said Sybil, 'you have the heart to go on writing so cheerily about the English towns and countryside.' Now, now, Sybil, she thought; business is business, and the nostalgic English scene is what the colonists want. This visit must be my last. I shall not come again.

'Ah, that,' Barry was saying, 'was the England I remember. The good old country. But now, I'm afraid, it's decadent. After the war it will be no more than . . .'

Désirée would have the servants into the drawing-room every morning to give them their orders for the day. 'I believe in keeping up home standards,' said Désirée, whose parents were hotel managers. Sybil was not sure where Désirée had got the idea of herding all the domestics into her presence each morning. Perhaps it was some family-prayer assembly in her ancestral memory, or possibly it had been some hotel-staff custom which prompted her to 'have in the servants' and instruct them beyond their capacity. These half-domesticated peasants and erstwhile small-farmers stood, bare-footed and woolly-cropped, in clumsy postures on Désirée's carpet. In pidgin dialect which they largely failed to comprehend, she enunciated the duties of each one. Only Sybil and David Carter knew that the natives' name for Désirée was, translated, 'Bad Hen'. Désirée complained much about their stupidity, but she enjoyed this morning palaver as Barry relished his poetry.

'Carter writes poetry too,' said Barry with a laugh one day.

Désirée shrieked. 'Poetry! Oh, Barry, you can't call that stuff *poetry*.'

'It is frightful,' Barry said, 'but the poor fellow doesn't know it.'

'I should like to see it,' Sybil said.

'You aren't interested in Carter by any chance, Sybil?' said Désirée.

'How do you mean?'

'Personally, I mean.'

'Well, I think he's all right.'

'Be honest, Sybil,' said Barry. Sybil felt extremely irritated. He so often appealed for frankness in others, as if by right; was so dishonest with himself. 'Be honest, Sybil—you're after David Carter.'

'He's handsome,' Sybil said.

'You haven't a chance,' said Barry. 'He's mad keen on Désirée. And anyway, Sybil, you don't want a beginner.'

'You want a mature man in a good position,' said Désirée. 'The life you're living isn't natural for a girl. I've been noticing,' she said, 'you and Carter being matey together out on the farm.'

Towards the end of her stay David Carter produced his verses for Sybil to read. She thought them interesting but unpractised. She told him so, and was disappointed that he did not take this as a reasonable criticism. He was very angry. 'Of course,' she said, 'your poetry is far better than Barry's.' This failed to appease David. After a while, when she was meting him in the town where she lived, she began to praise his poems, persuading herself that he was fairly talented.

She met him whenever he could get away. She sent excuses in answer to Désirée's pressing invitations. For different reasons, both Sybil and David were anxious to keep their meetings secret from the Westons. Sybil did not want the affair mythologized and gossiped about. For David's part, he valued his job in the flourishing passion-fruit concern. He had confided to Sybil his hope, one day, to have the whole business under his control. He might even buy Barry out. 'I know far more about it than he does. He's getting more and more bound up with his poetry, and paying next to no attention to the business. I'm just waiting.' He is, Sybil remarked to herself on hearing this, a true poet all right.

David reported that the quarrels between Désirée and Barry were becoming more violent, that the possibility of Barry's resigning from business to devote his time to poetry was haunting Désirée. 'Why don't you come,' Désirée wrote, 'and talk to Barry about his poetry? Why don't you come and see us now? What have we done? Poor Sybil, all alone in the world, you ought to be married. David Carter follows me all over the place, it's most embarrassing, you know how furious Barry gets. Well, I suppose that's the cost of having a devoted husband.' Perhaps, thought Sybil, she senses that David is my lover.

One day she went down with flu. David turned up unexpectedly and proposed marriage. He clung to her with violent, large hands. She alone, he said, understood his ambitions, his art, himself. Within a year or two they could, together, take over the passion-fruit plantation.

'Sh-sh. Ariadne will hear you.' Ariadne was out, in fact. David looked at her somewhat wildly. 'We must be married,' he said.

Sybil's affair with David Carter was over, from her point of view, almost before it had started. She had engaged in it as an act of virtue done against the grain, and for a brief time it had absolved her from the reproach of her sexlessness.

'I'm waiting for an answer.' By his tone, he seemed to suspect what the answer would be.

'Oh, David, I was just about to write to you. We really must put an end to this. As for marriage, well, I'm not cut out for it at all.'

He stooped over her bed and clung to her. 'You'll catch my flu,' she said. 'I'll think about it,' she said, to get rid of him.

When he had gone she wrote him her letter, sipping lemon juice to ease her throat. She noticed he had brought for her, and left on the floor of the stoep, six bottles of Weston's Passion-fruit Juice. He will soon get over the affair, she thought, he has still got his obsession with the passion-fruit business.

But in response to her letter David forced his way into the house. Sybil was alarmed. None of her previous lovers had persisted in this way.

'It's your duty to marry me.'

'Really, what next?'

'It's your duty to me as a man and a poet.' She did not like his eyes.

'As a poet,' she said, 'I think you're a third-rater.' She felt relieved to hear her own voice uttering the words.

He stiffened up in a comical melodramatic style, looking such a clean-cut settler with his golden hair and tropical suiting.

'David Carter,' wrote Désirée, 'has gone on the bottle. I think he's bats, myself. It's because I keep giving him the brush-off. Isn't it all silly? The estate will go to ruin if Barry doesn't get rid of him. Barry has sent him away on leave for a month, but if he hasn't improved on his return we shall have to make a change. When are you coming? Barry needs to talk to you.'

Sybil went the following week, urged on by her old self-despising; driving her Ford V8 against the current of pleasure, yet compelled to expiate her abnormal nature by contact with the Westons' sexuality, which she knew, none the less, would bore her.

They twisted the knife within an hour of her arrival.

'Haven't you found a man yet?' said Barry.

'You ought to try a love affair,' said Désirée. 'We've been saying—haven't we, Barry?—you ought to, Sybil. It would be good for you. It isn't healthy, the life you lead. That's why you get flu so often. It's psychological.'

'Come out on the lawn,' Barry had said when she first arrived. 'We've got the ciné camera out. Come and be filmed.'

Désirée said, 'Carter came back this morning.'

'Oh, is he here? I thought he was away for a month.'

'So did we. But he turned up this morning.'

'He's moping,' Barry said, 'about Désirée. She snubs him so badly.'

'He's psychological,' said Désirée.

'I love that striped awning,' said Sybil's hostess. 'It puts the finishing touch on the whole scene. How carefree you all look—don't they, Ted?'

'*That* chap looks miserable,' Ted observed. He referred to a shot of David Carter who had just ambled within range of the camera.

Everyone laughed, for David looked exceedingly grim.

'He was caught in an off-moment there,' said Sybil's hostess. 'Oh, there goes Sybil. I thought you looked a little sad just then, Sybil. There's that other girl again, and the lovely dog.'

'Was this a *typical* afternoon in the Colony?' inquired the young man.

'It was and it wasn't,' Sybil said.

Whenever they had the camera out life changed at the Westons. Everyone, including the children, had to look very happy. The house natives were arranged to appear in the background wearing their best whites. Sometimes Barry would have everyone dancing in a ring with the children, and the natives had to clap time.

Or, as on the last occasion, he would stage an effect of gracious living. The head cookboy, who had a good knowledge of photography, was placed at his post.

'Ready,' said Barry to the cook, 'shoot.'

Désirée came out, followed by the dog.

'Look frisky, Barker,' said Barry. The Alsatian looked frisky.

Barry put one arm round Désirée and his other arm through Sybil's that late afternoon, walking them slowly across the camera range. He chatted with amiability and with an actor's lift of the head. He would accentuate his laughter, tossing back his head. A sound track would, however, have reproduced the words, 'Smile, Sybil. Walk slowly. Look as if you're enjoying it. You'll be able to see yourself in later years, having the time of your life.'

Sybil giggled.

Just then David was seen to be securing the little lake boat between the trees. 'He must have come across the lake,' said Barry. 'I wonder if he's been drinking again?'

But David's walk was quite steady. He did not realize he was being photographed as he crossed the long lawn. He stood for a moment staring at Sybil. She said, 'Oh hallo, David.' He turned and walked aimlessly face-on towards the camera.

'Hold it a minute,' Barry called out to the cook.

The boy obeyed at the moment David realized he had been filmed.

'OK,' shouted Barry, when David was out of range. 'Fire ahead.'

It was then Barry said to Sybil, 'Haven't you found a man yet . . . ?' and Désirée said, 'You ought to try a love affair. . . .'

'We've made Sybil unhappy,' said Désirée.

'Oh, I'm quite happy.'

'Well, cheer up in front of the camera,' said Barry.

The sun was setting fast, the camera was folded away, and everyone had gone to change. Sybil came down and sat on the stoep outside the open french windows of the dining-room. Presently, Désirée was indoors behind her, adjusting the oil lamps which one of the houseboys had set too high. Désirée put her head round the glass door and remarked to Sybil, 'That Benjamin's a fool, I shall speak to him in the morning. He simply will not take care with these lamps. One day we'll have a real smoke-out.'

Sybil said, 'Oh, I expect they are all so used to electricity these days. . . .'

'That's the trouble,' said Désirée, and turned back into the room.

Sybil was feeling disturbed by David's presence in the place. She wondered if he would come in to dinner. Thinking of his sullen staring at her on the lawn, she felt he might make a scene. She heard a gasp from the dining-room behind her.

She looked round, but in the same second it was over. A deafening crack from the pistol and Désirée crumpled up. A movement by the inner door and David held the gun to his head. Sybil screamed, and was aware of running footsteps upstairs. The gun exploded again and David's body dropped sideways.

With Barry and the natives she went round to the dining-room. Désirée was dead. David lingered a moment enough to roll his eyes in Sybil's direction as she rose from Désirée's body. He knows, thought Sybil quite lucidly, that he got the wrong woman.

'What I can't understand,' said Barry when he called on Sybil a few weeks later, 'is why he did it.'

'He was mad,' said Sybil.

'Not all that mad,' said Barry. 'And everyone thinks, of course, that there was an affair between them. That's what I can't bear.'

'Quite,' said Sybil. 'But of course he was keen on Désirée. You always said so. Those rows you used to have. . . . You always made out you were jealous of David.'

'Do you know,' he said, 'I wasn't, really. It was a sort of . . . a sort of . . .'

'Play-act,' said Sybil.

'Sort of. You see, there was nothing between them,' he said. 'And honestly, Carter wasn't a bit interested in Désirée. And the question is *why* he did it. I can't bear people to think . . .'

The damage to his pride, Sybil saw, outweighed his grief. The sun was setting and she rose to put on the stoep light.

'Stop!' he said. 'Turn round. My God, you did look like Désirée for a moment.'

'You're nervy,' she said, and switched on the light.

'In some ways you *do* look a little like Désirée,' he said. 'In some lights,' he said reflectively.

I must say something, thought Sybil, to blot this notion from his mind. I must make this occasion unmemorable, distasteful to him.

'At all events,' she said, 'you've still got your poetry.'

'That's the great thing,' he said, 'I've still got that. It means everything to me, a great consolation. I'm selling up the estate and joining up. The kids are going into a convent and I'm going up north. What we need is some good war poetry. There hasn't been any war poetry.'

'You'll make a better soldier,' she said, 'than a poet.'

'What do you say?'

She repeated her words fairly slowly, and with a sense of relief, almost of absolution. The season of falsity had formed a scab, soon to fall away altogether. There is no health, she thought, for me, outside of honesty.

'You've always,' he said, 'thought my poetry was wonderful.'

'I have said so,' she said, 'but it was a sort of play-act. Of course, it's only my opinion, but I think you're a third-rater poet.'

'You're upset, my dear,' he said.

He sent her the four reels of film from Cairo a month before he was killed in action. 'It will be nice in later years,' he wrote, 'for you to recall those good times we used to have.'

'It has been delightful,' said her hostess. 'You haven't changed a bit. Do you *feel* any different?'

'Well yes, I feel rather differently about everything, of course.' One learns to accept oneself.

'A hundred feet of one's past life!' said the young man. 'If they were mine, I'm sure I should be shattered. I should be calling "Lights! Lights!" like Hamlet's uncle.'

Sybil smiled at him. He looked back, suddenly solemn and shrewd.

'How tragic, those people being killed in shooting affairs,' said the elderly woman.

'The last reel was the best,' said her hostess. 'The garden was entrancing. I should like to see that one again; what about you, Ted?'

'Yes, I liked those nature-study shots. I feel I missed a lot of it,' said her husband.

'Hark at him—nature-study shots!'

'Well, those close-ups of tropical plants.'

Everyone wanted the last one again.

'How about you, Sybil?'

Am I a woman, she thought calmly, or an intellectual monster? She was so accustomed to this question within herself that it needed no answer. She said, 'Yes, I should like to see it again. It's an interesting experience.'

ELSPETH DAVIE

1919–

A MAP OF THE WORLD

After each holiday our cousin Robert pays us a visit. Three times a year he comes, and he doesn't leave it till long after he gets back either. Obviously he isn't the kind that just drops in for nothing. He likes to have something to tell and he comes while his experiences are still fresh in his mind. In the Spring he stays in Britain—a different place each time, though it's more often the hills than the sea. He's never been a man for the same place twice, nor the same people either—so as often as not it'll be some out-of-the-way hotel he's found with a bedroom overhanging this waterfall where you can almost catch the leaping fish in your arms. Or maybe he's come across some marvellous cook who'd once owned a big place in the south of France and for reasons best known to himself is now whipping up exotic dishes in some remote valley in the north of Scotland. His winter holiday simply amounts to a few days in London around Christmastime for the theatres, so it's more a description of plays and we enjoy that too. For he talks a bit about these actors and he'll sometimes act out a line or two of some play he's seen—perhaps comparing two different actors he's seen at some time or other in the same part. 'Who, in your opinion, would you say is the best for that particular character?' he'll say to my sister, and she'll mention names and parts better or worse played, lines botched or matchlessly spoken. For she reads reviews and keeps up with everything that's said about the stage as though she were the most regular theatre-goer. Sometimes he listens for a bit, looking rather blank, and then murmurs vaguely: 'Oh yes—reviews . . . You've *read* the play? . . . But, then, the only thing that matters is to be sitting right there up in front with the lights dimming down . . . You've got to really feel this thing between actor and audience before you can know for yourself what the play's all about . . .'

However it's really the autumn visit we enjoy most, for this comes directly after his holiday abroad. And he always brings his map along. It's always a different map, because it's not just that he never goes to the same place twice. Now that he's got more money he never even goes to the same country. We're used to it—it's almost a ritual—the way he spreads out the map on the table after it's cleared, smoothing it out with his hand and sometimes flipping it up impatiently when he feels a crumb underneath or a teaspoon, and at last having it the way he wants it—dead flat—so that he can point out the places to my sister and me. My brother is seldom around at this time. He works all day in an office and in the evening he shuts himself up in his room to study for his exams. But to get back to maps. It should be possible to go out and buy a map of any country in the world, Robert says, but it isn't—such is the narrowness of life here, the parochial island outlook. But he's made up for that. He buys a map of every country he visits and of course he buys it inside that country. This cousin of ours, I should point out, is a widower in his thirties. That's rather unusual, and it's to be hoped he'll marry again before long—at least I hope so. I can imagine him bringing his wife one day and how they'll both smooth the map out with their hands, though she being domesticated by that time would no doubt be more patient about crumbs and teaspoons. Though, again, it might change things for him. She might not like the travelling.

I think the first map we studied was the Scandinavian one which he'd bought along the sea-front in Oslo. It makes a good start—that great, ragged-backed bear pouncing down towards Scotland. It's not so far after all. Nothing is far now, as everyone keeps saying. The world's a small place, and so on and so on. Robert has his little transistor set with him when he comes here, though he's not one that travels everywhere with it. When he's *not* travelling he's afraid of being bored. Sometimes when he's in the middle of a sentence he'll put it on and have to raise his voice a bit over the accompaniment of music, and when this happens Paul comes running downstairs and opens the door a slit.

'What a racket!' he exclaims. 'Keep it low, can't you?'

'What is it?' Janet, my sister, says, looking up startled. 'Is mother all right?'

'Yes, but I'm not. I've got to work. Remember?' He shuts the door again. It's true that working this time of the night is hard after a day in an office. Only the idea of a different and better job keeps him slogging on. But he doesn't smile about it, believe me. Nothing of the gay student about Paul. And by this time Janet has got worried about something.

'Shall I go up?' she says, raising herself from the map, her fingers pressing into the crackly paper around the Arctic Ocean with a sound as though thin ice flakes were breaking up.

'If you like,' I say. 'But she'll be all right.' Janet jumps up and goes quickly off upstairs.

'And how *is* your mother?' says Robert, reluctantly turning the transistor knob round a fraction.

'She's all right, if she doesn't exert herself. Mornings or afternoons in bed. No lifting. No stooping. No excitement.'

'But she's always done that, hasn't she?' he asks. 'Since I've known her. Hasn't she always had these mornings in bed? Or evenings?'

'It used to be just the odd day. But since her last attack . . .'

'And *how* is your father?' he interrupts. The different emphasis makes it sound ironical, and it's meant to be. At once I see this man as he appears in my cousin's mind for that split second in which he allows him to be there. But for that time the older man flounders, blunders, and throws himself about, scandalously selfish, in his own world with his own friends. Defenceless too. How do you defend a spinning, swinging object? If he would walk through this door I could defend him from the ironical smile of my cousin. As it is, I say nothing. Or simply: 'He'll be back late tonight. This is the night he goes round to Foster's place.' And Robert nods gently and goes on nodding for a long time like one of those toy mandarins who, once touched, continue to nod and smile interminably over nothing.

Everything is all right upstairs but my sister comes down looking pale. Is that what anxiety is—looking pale over nothing, or rather over what hasn't happened but might happen tomorrow or the next day, or next month or in ten years' time? This kind of pallor has been spread through our house for as long as I can remember. Even my father at his most turbulent is touched with it, and my brother wears it constantly—a white, constricted patch on either side of his sharp nose. Anxiety can mix with the colours of a house so that no matter what red and yellow cushion covers, striped rugs or checked cloths are brought in to cheer, they begin to look faded in a very short time. In such a setting it's no wonder our cousin's cheeks look ruddy by contrast, as though perpetually sunburnt. And perhaps they are, judging by the short interval between one holiday and the next. The second map he brought a few months later was a full-scale one of Italy. I remember there was a lot of purplish-red about it and when he bent over it you could see all that sunshine and fruit and wine he'd been describing suffusing his whole face. It fairly glowed—almost purple. He was not talking to absolute ignoramuses though. My sister especially has read a lot about this part of the world and this city in particular, and once or twice she interrupted him to ask enthusiastically about some object—maybe a fountain or a tiled patch of floor, or some figure high up on the corner of a building. She looked up from the map and stared hard at these things in mid-air as she spoke, reflecting, it seemed, on her eyeballs the pale, polished marbles and the sparks of light flung from fountains. But he answered none of her questions—certainly not one about the shade of certain walls and pavements. Honey? Or was it more an apricot? And at what time—morning or evening? A shadow of annoyance crossed her face at his silence. But she was stubborn. Later she insisted that he must have seen a particular painting in one of the galleries.

'But it's not there. You're thinking of another one,' he replied.

'I'm absolutely certain . . .'

'I tell you, it was not there!'

'Well, I'm not blaming you. How could you possibly see them all—or remember them?'

'It's definitely not there! Probably not in Florence at all—not even in Italy!'

Perhaps she felt then her exacting manner was a bit out of place, too pernickety in the winey atmosphere of the red and purple map. For she said more mildly: 'Of course, I'd forgotten they're sometimes removed for exhibitions. Naturally there'd be blanks here and there.' It was clear he admitted no such thing. But he said no more. Once in a while he looked puzzled and somewhat taken aback. He had forgotten about books—forgotten the way some people still travel if they can manage to push back a suffocating day and start revving up for the night's reading—a revving up which, if you're to get off the ground at all at that hour, takes nerve, speed, and an absolute jet-engined willpower.

Next came Switzerland. Switzerland was tougher for some reason. It wasn't, anyway, a country we'd ever thought very much about. We were unprepared for Switzerland. And the map was more detailed. In fact, though he's no climber himself, our cousin had got hold of a climber's map and the mountains were shown, coiled tight at the peaks, like dark brown spirals of wire ready to spring. But below this the lines grew less taut, gradually loopier, until at last they relaxed into great, gentle swathes of green valley and blue river. No, my sister was not like herself as she stared at this map, silently steeping herself in it as though in some mysterious way, unlike the other maps, it had an extra dimension, a softness and hardness which she was gradually discovering on her own. Here and there she must have sunk pretty deeply into the snow. And what clear lakes did she slowly let herself down into, recklessly leaving clothes, bags, letters, laundry, medicine bottles, hot water bottles, whisky bottles, typewriters, tin-openers, shovels, dusters, pillows, mothballs, pills, bills, art books, travel books, blankets and Bibles—on the banks?

For the first time Robert drew out a small magnifying glass from his pocket and gave it a polish with his handkerchief before letting us look. It was useful on this map especially for finding the source of rivers and spotting the smallest lakes.

'I read somewhere . . .' Janet began. But he was staring through the glass at the peak of a mountain.

'Well, what?' he said.

'I'm thinking of those fantastic blue cracks,' she said, frowning a bit. 'Have you seen that?'

'Those what?' said her cousin. His hands were flat on the map and he turned his wrist for a second so that he could take a look at his watch.

'Those blue—there's no colour like it anywhere else—those blue cracks in the ice. Did you see that?'

'Cracks? Oh, crevasses you mean?'

'From a distance they're like cracks.'

'Yes, crevasses.'

'Did you see that?'

'I wasn't climbing, you know.'

'So you didn't see this blue?'

'Refraction. No, I was never really high enough.'

'Not even through binoculars?'

'No, I didn't have binoculars. There was plenty else to see though.'

'Oh well, so you missed that!'

'I can't see everything, can I?'

'No, I suppose not,' said Janet, smiling at him.

'I think I've done pretty well.'

'You certainly have,' she said, continuing to smile, and not her pleasantest smile either.

'What exactly is she talking about?' he said, turning to me.

'She's talking about the colour of ice.'

'Yes, I gather that. But I don't see why it should become some sort of issue.'

'It's just this man we've been reading—"an unearthly blue" is how he puts it.'

'Oh, unearthly!' Janet interrupted contemptuously. 'Don't let's get bogged down amongst the adjectives! But I can see it was a blue all right. And anyway let's be thankful it is a blue you *can* find on earth. It seems to give us some chance, doesn't it, of actually seeing it some day?' All this flew over Robert's head like sleet in the wind, but he seemed to feel the cold.

'This was a book of Alpine photos and articles we were looking into not so long ago,' I said quickly in case she should start again.

'So you were following me, were you?' said Robert, brightening up.

'No, not *following* you at all,' said Janet. 'Just reading.'

There was silence for a moment and then he turned away and gave his full attention to me again. 'I imagine some of those travel books can be a bit misleading without the actual colour. Or were they colour photos?'

'Yes. And they were stunning. I'm not so sure about the descriptions.'

'Get it out of your head that we were *following* you!' said Janet.

'Oh, for God's sake!' he exclaimed, turning round angrily.

'Because as you know reading's different,' she went on. 'Reading isn't climbing or driving or walking or flying or sailing or skiing. And it certainly isn't following! Not leading either. Misleading perhaps. But not following!'

Eventually when she'd worried it down to a last wretched rag of a word she calmed down and we all had coffee. She became more than usually polite to her cousin and listened to the rest of what he had to tell like someone sitting still in a darkened room waiting for the bang of the stick announcing the next slide. All the same she's always had the ability to see what's described as though it were there before her—throwing her head back at towers, peaks and skyscrapers, staring plumb down through the carpet into creeks and bottomless lochs. She doesn't waste much. Every scrap is examined and given full value. And some of them are no more than scraps, there's no doubt about that. For all his experience, Robert isn't exactly an Othello when it comes to describing his adventures. You'd say my sister was the better traveller, despite the fact that she hasn't yet set foot in plane or ship.

Later in the evening he described this waterfall. I've no idea when or where he saw it or why, but it was a spectacular sight. Yet it wasn't he who made it spectacular though he gave us its height and its relative volume compared to other waterfalls and its position among trees and rocks. Yes, he gave it its position all right. But Janet gave it something else. She stared with eyes that took in the whole awful glassy sweep of it. But no, she wasn't looking at this waterfall, she was inside it. She was leaning on that green slide, her fingers prised apart, her face transformed with the catastrophic shock as it crashed upon her. And she was still staring through the deluge when there came a knock overhead. Not a spectacular knocking—it never is that—not the thump, thump with the clumping great stick at the side of the bed familiar in accounts of tyrannical invalids. This was a reasonable knocking, quiet, but not apologetic either. The firm, gentle knocking of someone who has taken it for granted for months and years past and who is going to take it for granted for months and years to come that this knock will be heard and answered *whatever happens*. That is to say, whether her daughter is standing inside a waterfall or out of it.

Janet heard the knock but she didn't respond at once. It isn't easy to emerge from a cataract into the thin suburban air. She had to burst from the depths and, like a sea-creature, hurl herself out on to land. But at last she made it. She left the room quickly and we heard her running upstairs. Almost as soon as she'd gone my brother came down with a book in his hand and a thumb in the book to keep the place. He has a horror of getting involved in anything going on upstairs or downstairs, but particularly upstairs, which hasn't anything to do with his work. So he's restless and on bad nights roams from room to room, giving the impression that he may even be forced to sit and work on the stairs until the disturbance is over. This was a bad night. He didn't sit down. He stood, his thumb still in the book, waiting without speaking. After a while, as my sister still did not come down, he put his head back and rolling it round towards the table, exclaimed: 'Ah—another map!' The position of his head, tilted back, forced him to droop his eyes, to look sideways down his long nose.

Our cousin was obviously annoyed by what he felt was patronising in this attitude. But you could see he was determined to be patient with this family. So he merely said: 'I've quite a collection now. Yes, I think I've got just about everything you could ask for in maps. Everything but a relief map and a globe,' he added with a smile.

'Relief map!' exclaimed Paul. He put the back of his free hand against his mouth and gave a burst of laughter.

'I said everything *but*,' answered Robert, this time really annoyed and showing it by the spreading redness on his forehead.

'Relief!' cried Paul again, rolling his head to and fro against the wall. I thought he was going to worry the word, worry it to death as my sister had done. But he let it go. Soon after we heard her coming downstairs and almost before she'd reached the bottom Paul barged up past her and we heard the suppressed bang of his door at the top. How will he get these half-bangs out of his system, semi-slams that have to cut out just before they reach the crash?

Our cousin stayed away for quite a long time after that. Then late the next autumn, long after everyone had forgotten about holidays, he returned. A windy night had been forecast, but he arrived before there was any sign of it. We had a quiet supper, no one talking much, not even Robert, for he likes to save what he has to say till later when we can give it our whole attention. He ate well, the meal was long-drawn-out and it was nearly nine before we were finished. Usually this is the moment when he unfolds his latest map and spreads it out on the cleared table. But this time he unfolded it and laid it out on the floor. It was made of thick, hard paper, still with the new creases ridged across it, and he smoothed it with his big, white hands, crackling it down flat on the carpet. As he did this we sensed the first rustlings of the night begin outside. It was nothing yet—scarcely a movement, or only as though the wind were gently turning the leaves over, one by one, and once in a while lifting some scrap of paper from the street and discreetly laying it down a few inches further on.

It was a map of the world. I've no idea why he should bring the world on this particular night, when it was Spain he had last visited. Maybe he'd lost his Spanish map. Or maybe Spain wasn't enough for him this time of the year. Anyway it was the world all right and rather cheap it looked too, lying directly under the ceiling light like some crude abstract painting in pink, green and yellow, surrounded by a hard sea-blue. Although he still had his finger on Spain he kept his eye on the rest of the world, and after a bit when he got really tired of Spain his finger moved off it and he began to trace devious routes round other countries. It was bound to be a rougher journey than usual, seeing the map was laid out on the floor. We may be proud but there's no time to be house-proud, and every now and then, or let's say every two thousand miles or so, his moving finger would feel some gritty part of the carpet, or maybe as he drew his nail across the Atlantic Ocean it would subside into a ragged patch—a real Sargasso Sea of weedy threads. Once his nail notched on a halfpenny and with an exclamation of impatience he flipped up the map and spun the coin away till it struck the wainscoting on the other side of the room, then smoothed the map down again with a careful hand.

There was now more than a restlessness outside. The wind was really getting up and through the uncurtained window I could see the tops of the trees across the road swing riotously every now and then against the street-lamps, and then as suddenly grow still. My sister was kneeling on the floor the way people kneel to look at maps—I mean those who are really keen—her ankles close together, arms folded round her body so that nothing loose about her clothes should dangle down to disturb the brazen pink and yellow face of the paper. She made herself neat and tight. Even her chin was tucked in, but her eyes travelled the world slowly, absorbedly, as though she were in love with it. Round the coastlines she went, slowing down to follow rivers inland and bending closer for their source, then back again to the coast, holding closely to it through thick and thin, never mind the treacherously splintered inlets or those parts where it was necessary to navigate hundreds of miles of water dotted with quaky islands. She moved at an explorer's pace, steady, dogged, missing nothing, wide-eyed yet cautious—whole-heartedly dedicated to adventure. Long ago she had given up listening to her cousin, who was now explaining how he had nearly lost his camera in a waiting-room of the Paris airport, and by the time he was standing gesticulating on the tarmac beside the waiting plane she was working round the coast of Greenland, steering through gigantic chunks of ice. A lonely route, but she was determined on it. Our cousin, who had stopped for a moment in his story to follow the direction of her eyes, remarked that when he flew to America—and he certainly intended to go there soon, if not at the firm's expense then under his own steam—the route, depending on the winds, might not necessarily be the most direct one but possibly over the southern part of Greenland, or at any rate near enough to see it. 'Though these flights can sometimes be disappointing,' he added. Whatever route Janet was taking she was not disappointed. Nor satisfied. She had a fierce absorption in her travels which was painful to watch. It made me wish that he had brought a globe if he had to discuss the world, something that would spin gently at a touch or could be held between the hands like a child's head. Anything but this uncompromising square of paper sealed to the floor. Yet, after all, it was not fixed and before long our cousin grew tired of speaking to the air and he put his hand on the map to draw it aside and fold it up again.

It was as though he had touched a spring in the person kneeling beside him. With one, abrupt twist of her body she instantly turned, stretched herself out and lay flat on her back, her head on the map and her arms folded across her chest. The change from reverent kneeling to stubborn lying was all accomplished in a couple of seconds. Although I was taken aback I wasn't apprehensive. There was no question of a faint and she certainly wasn't ill. She looked, as she lay there, like one of those figures on a tomb—some overlooked noblewoman stubbornly determined on survival, even if it was to be only a survival in stone.

'What's the matter?' said Robert.

For answer she closed her eyes tight. A thick strand of hair which had come loose was curled about the South American continent, her collar ruffling along its western coast like a great wave.

'What's the matter with her?' he said again impatiently, this time turning to me.

'Nothing's the matter,' I said.

'Isn't she interested in maps?'

'Oh yes, of course.' Because it seemed odd that we should be talking over her as if she were deaf, I moved through the open door into the adjoining kitchen. He followed me uneasily.

'What is it then? Is she annoyed at something? I suppose the map of the world bores her?'

'She's objecting to it.'

'The world?'

'No. The map.'

'It's a funny thing,' he said, looking annoyed himself.

'What is?'

'If she's going to start objecting to the world. She's taken most other countries.'

'That isn't the world. It's a map.'

'Well, what about it? Doesn't she like maps after all?'

'The difference has just struck her. This time it's really pushed her over!'

'What's the matter with this place tonight?' he exclaimed. 'If it wasn't a convenient night you could have said so. The difference? The difference between what?'

'How should I know? Between paper and prairie, maybe.'

'You're out of your wits!'

'Or the difference, if you like, between paper and desert, or paper and blue ice, paper and palm-trees. Between paper and people! That's the worst. Paper and people really is the worst!'

'Oh, if I'd *known* it was to be this sort of night!' he groaned.

'What sort? The wind? I don't mind that wind. I love it.'

'The trouble with this family is it's always lacked proportion. Does there have to be this obsession with every single thing you pick up? And to get worked up over words!'

'Oh we take the words seriously all right. Books! Did you know she's starting on Russian now? She's got the records, the dictionary, the pamphlets and grammars. Everything. Last time it was German. Who knows? It may be Swedish next.'

'All right. That's fine. What else do you want me to say?'

'We're waiting for all those books and maps to change. The dictionary made flesh. What kind of juice can you squeeze from print? Paper's skin-thin. Paper's got no fat at all. It's flesh and blood we need—sun and salt and marble and hot sand!'

He looked frightened, then sympathetic, and at last exasperated. He said nothing but peered through the door into the other room. After a bit he said: 'Of course I know you have a quiet enough life of it here . . .' Janet was still lying there, giving an impression of this quietness—no doubt about that—but it was an alarming quietness all the same, a quietness that dared anyone to disturb it—stubborn and forbidding. It must have been pretty chilly lying there though, for by this time draughts from all sides were stirring the corners of the map. Who would have thought that in the space of an hour or so and from its first, soft dust-siftings, the wind could have reached this uproar? Between two gusts he repeated: '. . . a quiet enough life, but I always imagined that was the way you liked it . . .' Heavens, how the wind began to tear the trees, taking great crackling handfuls of twigs from the upper branches! 'At any rate,' he said, 'I haven't heard either of you two girls . . .' He had to lift his voice to repeat: '. . . never heard either of you complain!' above the sudden, roaring, rending lament outside. I pulled up the window and as I snibbed out the complaining wind I heard the knocking begin upstairs. Or perhaps it hadn't just begun—perhaps it had been going on for some time. We both automatically moved back into the other room and looked expectantly towards Janet. She had risen on to her elbow and was staring intently out of the window. You'd think from the look of her that the summons had come from outside, from such a distance that she had to strain and stare. If she'd been recording sounds from the ends of the earth she couldn't have listened harder. Yet the knocking was very clear now and persistent. She lay down again with her head on the map.

'Why don't *you* go up?' I said to my cousin.

'Me? She seldom sets eyes on me!'

'All the more reason.'

'But she might want something.'

'Then you could give it to her.'

He stood looking towards the ceiling, then said cautiously: 'Well, I suppose I could.' He waited another moment to see if either of us would make a move. But there wasn't a flicker. I may have smiled but it was not a smile which could keep him lingering long. He went upstairs.

'You'll have to get up,' I said to my sister. 'What can you do for yourself lying there? Get up on your feet! From now on things will be different. I promise you. We're really going to move around!' At this her hands moved slightly along the carpet as though she were testing out this promise in her fingertips for a start. And I went on: 'We're going to move around—and I mean travel—travel the world. There must be ways. Money never defeated anybody who really . . . Can't someone come in and take over? If not for both of them, then surely someone else will offer . . .' My voice trailed off at the crucial points. How I despised it for losing all bite on this last word. Yet the idea of anyone offering to put up the man of the house for days, let alone weeks, was about as imaginable as expecting the decent front doors to open up willingly to a battering-ram.

Comfortings and cajolings went out of my head. I abruptly exclaimed: 'He's going to want that map when he comes down!' Her head moved slightly, rolling across God knows what depth of ocean. 'He's coming now! Isn't it time you were on your feet?' But she lay there, sinking deeper into the fabulous underwater world lying, still unmapped, beneath the other. Robert entered the room.

'Yes, that was all right,' he said briskly, not looking at either of us, but staring about the room. He had his coat over his arm. 'She was very glad to see me. I must go up oftener. No, it was nothing. Just the company. I think she just wanted a bit of company—and she was very pleased to see me. I must go oftener.' His glance as it skated swiftly over us made it clear that he would be only too glad another time to go where he was wanted. It had been a bad evening with relations—coming right out of the blue—the kind he'd heard other people tell about but had never yet experienced himself. 'It's high time I was off,' he said, putting on his coat. He drew on both his gloves and walking over to my sister he bent down, took one corner of the map between finger and thumb and very deftly, smoothly, as though practising sleight of hand, he slid it, with his gloved fingers, out from under her head. Her head rolled a bit again but more roughly, as though jarred on some gravelly sea-bed. He did remove one glove to fold the map—smoothing it very carefully and slowly along the correct creases. 'I've always contended and I always will: women can do everything men can do—except fold maps the right way,' he said, shoving it in his pocket with a slight smile. It didn't fit there properly and he had to pull it out. Again he shoved it down and pulled it out, and finally he had to lay it on a chair behind him while he buttoned his coat and rearranged his pockets. Even then he was a long time getting off. At last I accompanied him to the door which he reached without looking back.

'Well, I can't imagine what the next map could be, unless it's a star-map,' I said. I knew there would be a long gap between this visit and the next. He paused with his hand on the knob of the front door. 'Don't think I'm not worried myself,' he said. 'I knew she was upset from the word go. But I wouldn't let her lie there much longer. When people like that flop it takes more to get them on their feet again.'

'People like what?'

'I mean so stiff and straight,' he said.

'But if it *is* a map of the universe,' I said, 'she knows about stars too. She really has read it up. She knows her way about the sky.'

'Look, I daresay she needs a break,' he said. 'I daresay you both need a break.'

'A break? Oh yes—a holiday. I really do mean that about the stars. I'm not joking. She can point out whole star continents.'

He had the door open now and we were standing on the edge of the garden which every now and then swirled like a whirlpool—gravel, leaves, pale flowers on stalks—all circling and then all still again. But there was also a strong draught blowing from inside the house, for Paul, who works with his window open in all weathers, had come out of his room in the last few minutes and was standing at the top of the stairs. My cousin gave a nod and a wave and as he stepped out, the door left my fingers and slammed shut in a sudden gust from inside. An almighty slamming! The kick and slam and crash of pent-up fury could hardly have made it more. I had to wrench it open and shout after him. But the roaring night swept up the apology and tossed it about. I doubt if a word reached him. He was striding off at a terrific pace.

At the top of the stairs Paul stood, laughing angrily, and behind me my sister was slowly raising herself on her elbow from the carpet. Above her on the table lay the folded map which Robert had left. In all the time we've known him he's never before left any part of himself behind. When he goes he removes himself utterly. If he doesn't actually shake the dust from his feet—not a scrap, ash, crumb of him ever remains. But there it was.

'Oh, how he deserved that!' Janet exclaimed. 'Oh, how I hope he heard that crash you sent after him!'

'You're wrong. It was the wind.'

'What was that he called me? Stiff and what—?'

'Straight. You're hardly that tonight.'

'I've had enough! Him and his damned maps!'

'It's there above you then.' She looked up at it and was about to lower herself down again on her elbow as I exclaimed: 'You'd never lie down again!'

'Take it away then. Can't someone go after him? What'll we do with it till he comes back?'

'What do you want me to do with it? Go out and scatter it to the wind? Don't for pity's sake lie down! Do you want me to wrap you up in it—or what?'

She sat up slowly and stared at me, arms round her knees. 'How severe you sound. Are you angry?'

'I'm angry all right.'

'With him?'

'No, with us.'

'With me, you mean of course.'

'No. With us. The lot of us in this box. Book-ridden, bedridden, beer-battered, bogged-down lot!'

'It's a box all right! Oh, if I could breathe in it! Why should he bring his maps to us?'

'Well, it's here now and it'll have to stay. He'll not be back in a hurry.'

'Take it away. Put it on the shelf in the bedroom cupboard—under the blankets. It'll be safe there.'

'And silent. You want it muffled?'

'I can't take another word about it. I'm finished. Oh, he's exhausted me—him and his world! You'd better go up and see if everything's all right. Finished! Say I'll come later if she needs me. I'm stiff too. Better go up then.'

'In a while.' I found nails for the job in the kitchen, came back and shook open the folded map in one jerk. Like a sail it flipped out stiffly, sending paper napkins skiffing along the sideboard and shaking the withered petals out of a jug of chrysanthemums.

'What now?' she cried.

'It's going up,' I said. 'Up here on the wall.'

'But it's the last thing . . . !'

'I want it up. So do you. It's better than a window.'

'Oh, but just look at it! What do you mean—so do I? That pink and green! You're joking. What does it look like with the pictures?'

'Awful. Maybe they'll have to come down.'

'You can't mean to keep it there?'

'I'm keeping it here. I'm going to nail it in.'

'Till he comes back?'

'No, till we get out. I don't much care which country it is. The last nail comes out when we're on our way. Which will it be first? France?'

I started to hammer the first tack in—maybe with more force than was necessary, for I struck it hard several times and at the fifth dunt I heard a door fly open at the top of the stairs. Paul again, shouting this time. I started on the second nail. Maybe it was a tougher piece of wall. It needed more and harder blows before the top edge of the map was fixed. Then I started on the bottom right-hand corner—really swinging it. I felt strong enough to hammer the house down. When the third nail was well and truly sunk I started on the last. I'd got two blows in before I heard the first bang from above—a real, powerful dunt it was—on the floor of her room. I hammered my nail again, and from above came an identical bang as though it were determined to have exactly the same force, the same weight and timing as the one below. Again I pounded and again it was echoed above, stroke for stroke, will for will—steady, strong, competitive. The last nail was far in now, but I went on. It developed into a steady rhythm, like the rhythm maintained by two workmen hammering on a metal post—one stroke below, one dunt above, and a second's silence; one stroke below, another above, and the wait again. I could have gone on all night if necessary, but suddenly I found I was hammering alone. No answering bang came from above. For a few minutes more I went on, to make absolutely sure, but more and more quickly and lightly. For now the hammer weighed next to nothing and nothing divided us from the outer air, as though the pull of gravity inside the house had grown dramatically less. Then I flung the hammer triumphantly down behind me on the carpet. There was dead silence—silence above, silence at the top of the stair. My sister was silent and white, staring at me from the floor.

'All right,' I said. 'You can go up now. That job's done anyway. The map's up.'

She still sat, listening and staring, never taking her eyes off mine. Then slowly, very cautiously she got up, stared from the map to me, from the map to the ceiling, and back to the map again. She really focused on it now. There was life in her eyes.

'And what shall I say?' she said.

'It's all right. Don't worry. The map's up. We've got it up now, and there it stays.'

'But what shall I *say*?' she asked again. 'What'll I say about your hammering?'

'*My* hammering? Weren't there two lots—below and above. Ask her what *hers* was for!'

But again she said: 'Listen, what'll I say about that hammering?'

All the same, she was stiffening up now. The nerve was coming back.

'Tell her about the world,' I said.

'Just like that?'

'Yes, tell her about it. Tell her you mean to look at it one of these days. No, no—as soon as possible. If it's not possible, you'll go anyway. What else would you say?'

'Yes, what else?' she muttered. She wasn't looking anywhere but at the map now, as though, given another chance, she might this time make the journey round the Southern Hemisphere. She might have been regretting even the few minutes longer spent on the carpet, when she could have been on her way. One more look into the centre of Australia to test the dust and space, and a moment to recover herself along the shores of the Pacific Ocean—then she turned to the door and I watched her go upstairs. This time she set each foot down determinedly, unhurriedly. It was the explorer's tread again—the patient, plodding pace of someone who had trained a hope for years, who now had the mind's eye focused so dead straight on the map that, at first hint of opposition, she could reel off—if only to sustain her till she got there—every sea, river, lake; every city, port and mountain village that came to mind. She would make her point. If need be she could pinpoint, bringing forward more and more detailed maps to her aid. She went steadily on past her brother's open door, on and up into her mother's bedroom.

George Mackay Brown

1921–

Andrina

Andrina comes to see me every afternoon in winter, just before it gets dark. She lights my lamp, sets the peat fire in a blaze, sees that there is enough water in my bucket that stands on the wall niche. If I have a cold (which isn't often, I'm a tough old seaman) she fusses a little, puts an extra peat or two on the fire, fills a stone hot-water bottle, puts an old thick jersey about my shoulders.

That good Andrina—as soon as she has gone, after her occasional ministrations to keep pleurisy or pneumonia away—I throw the jersey from my shoulders and mix myself a toddy, whisky and hot water and sugar. The hot water bottle in the bed will be cold long before I climb into it, round about midnight: having read my few chapters of Conrad.

Towards the end of February last year I did get a very bad cold, the worst for years. I woke up, shuddering, one morning, and crawled between fire and cupboard, gasping like a fish out of water, to get a breakfast ready. (Not that I had an appetite.) There was a stone lodged somewhere in my right lung, that blocked my breath.

I forced down a few tasteless mouthfuls, and drank hot ugly tea. There was nothing to do after that but get back to bed with my book. Reading was no pleasure either—my head was a block of pulsing wood.

'Well,' I thought, 'Andrina'll be here in five or six hours' time. She won't be able to do much for me. This cold, or flu, or whatever it is, will run its course. Still, it'll cheer me to see the girl.'

* * * * *

Andrina did not come that afternoon. I expected her with the first cluster of shadows: the slow lift of the latch, the low greeting, the 'tut-tut' of sweet disapproval at some of the things she saw as soon as the lamp was burning. . . . I was, though, in that strange fatalistic mood that sometimes accompanies a fever, when a man doesn't really care what happens. If the house was to go on fire, he might think, 'What's this, flames?' and try to save himself: but it wouldn't horrify or thrill him.

I accepted that afternoon, when the window was blackness at last with a first salting of stars, that for some reason or another Andrina couldn't come. I fell asleep again.

I woke up. A gray light at the window. My throat was dry—there was a fire in my face—my head was more throbbingly wooden than ever. I got up, my feet flashing with cold pain on the stone floor, drank a cup of water, and climbed back into bed. My teeth actually clacked and chattered in my head for five minutes or more—a thing I had only read about before.

I slept again, and woke up just as the winter sun was making brief stained glass of sea and sky. It was, again, Andrina's time. Today there were things she could do for me: get aspirin from the shop, surround my grayness with three or four very hot bottles, mix the strongest toddy in the world. A few words from her would be like a bell-buoy to a sailor lost in a hopeless fog. She did not come.

She did not come again on the third afternoon.

I woke, tremblingly, like a ghost in a hollow stone. It was black night. Wind soughed in the chimney. There was, from time to time, spatters of rain against the window. It was the longest night of my life. I experienced, over again, some of the dull and sordid events of my life; one certain episode was repeated again and again like an ancient gramophone record being put on time after time, and a rusty needle scuttling over worn wax. The shameful images broke and melted at last into sleep. Love had been killed but many ghosts had been awakened.

When I woke up I heard, for the first time in four days, the sound of a voice. It was Stanley the postman speaking to the dog of Bighouse. 'There now, isn't that loud big words to say so early? It's just a letter for Minnie, a drapery catalogue. There's a good boy, go and tell Minnie I have a love letter for her. . . . Is that you, Minnie? I thought old Ben here was going to tear me in pieces then. Yes, Minnie, a fine morning, it is that. . . .'

I have never liked that postman—a servile lickspittle to anyone he thinks is of consequence in the island—but that morning he came past my window like a messenger of light. He opened the door without knocking (I am a person of small consequence). He said, 'Letter from a long distance, skipper.' He put the letter on the chair nearest the door. I was shaping my mouth to say, 'I'm not very well. I wonder. . . .' If words did come out of my mouth, they must have been whispers, a ghost appeal. He looked at the dead fire and the closed window. He said, 'Phew! It's fuggy in here, skipper. You want to get some fresh air. . . .' Then he went, closing the door behind him. (He would not, as I had briefly hoped, be taking word to Andrina, or the doctor down in the village.)

I imagined, until I drowsed again, Captain Scott writing his few last words in the Antarctic tent.

In a day or two, of course, I was as right as rain; a tough old salt like me isn't killed off that easily.

But there was a sense of desolation on me. It was as if I had been betrayed—deliberately kicked when I was down. I came almost to the verge of self-pity. Why had my friend left me in my bad time?

Then good sense asserted itself. 'Torvald, you old fraud,' I said to myself. 'What claim have you got, anyway, on a winsome twenty-year-old? None at all. Look at it this way, man—you've had a whole winter of her kindness and consideration. She brought a lamp into your dark time: ever since the Harvest Home when (like a fool) you had too much whisky and she supported you home and rolled you unconscious into bed. . . . Well, for some reason or another Andrina hasn't been able to come these last few days. I'll find out, today, the reason.'

It was high time for me to get to the village. There was not a crust or scraping of butter or jam in the cupboard. The shop was also the Post Office—I had to draw two weeks' pension. I promised myself a pint or two in the pub, to wash the last of that sickness out of me.

It struck me, as I trudged those two miles, that I knew nothing about Andrina at all. I had never asked, and she had said nothing. What was her father? Had she sisters and brothers? Even the district of the island where she lived had never cropped up in our talks. It was sufficient that she came every evening, soon after sunset, and performed her quiet ministrations, and lingered awhile; and left a peace behind—a sense that everything in the house was pure, as if it had stood with open doors and windows at the heart of a clean summer wind.

Yet the girl had never done, all last winter, asking me questions about myself—all the good and bad and exciting things that had happened to me. Of course I told her this and that. Old men love to make their past vivid and significant, to stand in relation to a few trivial events in as fair and bold a light as possible. To add spice to those bits of autobiography, I let on to have been a reckless wild daring lad—a known and somewhat feared figure in many a port from Hong Kong to Durban to San Francisco. I presented to her a character somewhere between Captain Cook and Captain Hook.

And the girl loved those pieces of mingled fiction and fact; turning the wick of my lamp down a little to make everything more mysterious, stirring the peats into new flowers of flame. . . .

One story I did not tell her completely. It is the episode in my life that hurts me whenever I think of it (which is rarely, for that time is locked up and the key dropped deep in the Atlantic: but it haunted me—as I hinted—during my recent illness).

On her last evening at my fireside I did, I know, let drop a hint or two to Andrina—a few half-ashamed half-boastful fragments. Suddenly, before I had finished—as if she could foresee and suffer the end—she had put a white look and a cold kiss on my cheek, and gone out at the door; as it turned out, for the last time.

Hurt or no, I will mention it here and now. You who look and listen are not Andrina—to you it will seem a tale of crude country manners: a mingling of innocence and heartlessness.

In the island, fifty years ago, a young man and a young woman came together. They had known each other all their lives up to then, of course—they had sat in the school room together—but on one particular day in early summer this boy from one croft and this girl from another distant croft looked at each other with new eyes.

After the midsummer dance in the barn of the big house, they walked together across the hill through the lingering enchantment of twilight—it is never dark then—and came to the rocks and the sand and sea just as the sun was rising. For an hour and more they lingered, tranced creatures indeed, beside those bright sighings and swirlings. Far in the north-east the springs of day were beginning to surge up.

It was a tale soaked in the light of a single brief summer. The boy and the girl lived, it seemed, on each other's heartbeats. Their parents' crofts were miles apart, but they contrived to meet, as if by accident, most days; at the crossroads, in the village shop, on the side of the hill. But really these places were too earthy and open—there were too many windows—their feet drew secretly night after night to the beach with its bird-cries, its cave, its changing waters. There no one disturbed their communings—the shy touches of hand and mouth—the words that were nonsense but that became in his mouth sometimes a sweet mysterious music—'Sigrid'.

The boy—his future, once this idyll of a summer was ended, was to go to the university in Aberdeen and there study to be a man of security and position and some leisure—an estate his crofting ancestors had never known.

No such door was to open for Sigrid—she was bound to the few family acres—the digging of peat—the making of butter and cheese. But for a short time only. Her place would be beside the young man with whom she shared her breath and heartbeats, once he had gained his teacher's certificate. They walked day after day beside shining beckoning waters.

But one evening, at the cave, towards the end of that summer, when the corn was taking a first burnish, she had something urgent to tell him—a tremulous perilous secret thing. And at once the summertime spell was broken. He shook his head. He looked away. He looked at her again as if she were some slut who had insulted him. She put out her hand to him, her mouth trembling. He thrust her away. He turned. He ran up the beach and along the sand-track to the road above; and the ripening fields gathered him soon and hid him from her.

And the girl was left alone at the mouth of the cave, with the burden of a greater more desolate mystery on her.

The young man did not go to any seat of higher learning. That same day he was at the emigration agents in Hamnavoe, asking for an urgent immediate passage to Canada or Australia or South Africa—anywhere.

Thereafter the tale became complicated and more cruel and pathetic still. The girl followed him as best she could to his transatlantic refuge a month or so later; only to discover that the bird had flown. He had signed on a ship bound for furthest ports, as an ordinary seaman: so she was told, and she was more utterly lost than ever.

That rootlessness, for the next half century, was to be his life: making salt circles about the globe, with no secure footage anywhere. To be sure, he studied his navigation manuals, he rose at last to be a ship's officer, and more. The barren years became a burden to him. There is a time, when white hairs come, to turn one's back on long and practised skills and arts, that have long since lost their savours. This the sailor did, and he set his course homeward to his island; hoping that fifty winters might have scabbed over an old wound.

And so it was, or seemed to be. A few remembered him vaguely. The name of a certain vanished woman—who must be elderly, like himself, now—he never mentioned, nor did he ever hear it uttered. Her parents' croft was a ruin, a ruckle of stones on the side of the hill. He climbed up to it one day and looked at it coldly. No sweet ghost lingered at the end of the house, waiting for a twilight summons—'Sigrid. . . .'

I got my pension cashed, and a basket full of provisions, in the village shop. Tina Stewart the postmistress knew everybody and everything; all the shifting subtle web of relationship in the island. I tried devious approaches with her. What was new or strange in the island? Had anyone been taken suddenly ill? Had anybody—a young woman, for example—had to leave the island suddenly, for whatever reason? The hawk eye of Miss Stewart regarded me long and hard. No, said she, she had never known the island quieter. Nobody had come or gone. 'Only yourself, Captain Torvald, has been bedridden, I hear. You better take good care of yourself, you all alone up there. There's still a grayness in your face. . . .' I said I was sorry to take her time up. Somebody had mentioned a name—Andrina—to me, in a certain connection. It was a matter of no importance. Could Miss Stewart, however, tell me which farm or croft this Andrina came from?

Tina looked at me a long while, then shook her head. There was nobody of that name—woman or girl or child—in the island; and there never had been, to her certain knowledge.

I paid for my messages, with trembling fingers, and left.

I felt the need of a drink. At the bar counter stood Isaac Irving the landlord. Two fishermen stood at the far end, next the fire, drinking their pints and playing dominoes.

I said, after the third whisky, 'Look, Isaac, I suppose the whole island knows that Andrina—that girl—has been coming all winter up to my place, to do a bit of cleaning and washing and cooking for me. She hasn't been for a week now and more. Do you know if there's anything the matter with her?' (What I dreaded to hear was that Andrina had suddenly fallen in love; her little rockpools of charity and kindness drowned in that huge incoming flood; and had cloistered herself against the time of her wedding.)

Isaac looked at me as if I was out of my mind. 'A young woman,' said he. 'A young woman up at your house? A home help, is she? I didn't know you had a home help. How many whiskies did you have before you came here, skipper, eh?' And he winked at the two grinning fishermen over by the fire.

I drank down my fourth whisky and prepared to go.

'Sorry, skipper,' Isaac Irving called after me. 'I think you must have imagined that girl, whatever her name is, when the fever was on you. Sometimes that happens. The only women I saw when I had the flu were hags and witches. You're lucky, skipper—a honey like Andrina!'

I was utterly bewildered. Isaac Irving knows the island and its people, if anything, even better than Tina Stewart. And he is a kindly man, not given to making fools of the lost and the delusion-ridden.

* * * * *

Going home, March airs were moving over the island. The sky, almost overnight, was taller and bluer. Daffodils trumpeted, silently, the entry of spring from ditches here and there. A young lamb danced, all four feet in the air at once.

I found, lying on the table, unopened, the letter that had been delivered three mornings ago. There was an Australian postmark. It had been posted in late October.

'I followed your young flight from Selskay half round the world, and at last stopped here in Tasmania, knowing that it was useless for me to go any farther. I have kept a silence too, because I had such regard for you that I did not want you to suffer as I had, in many ways, over the years. We are both old, maybe I am writing this is vain, for you might never have returned to Selskay; or you might be dust or salt. I think, if you are still alive and (it may be) lonely, that what I will write might gladden you, though the end of it is sadness, like so much of life. Of your child—our child—I do not say anything, because you did not wish to acknowledge her. But that child had, in her turn, a daughter, and I think I have seen such sweetness but rarely. I thank you that you, in a sense (though unwillingly), gave that light and goodness to my age. She would have been a lamp in your winter, too, for often I spoke to her about you and that long-gone summer we shared, which was, to me at least, such a wonder. I told her nothing of the end of that time, that you and some others thought to be shameful. I told her only things that came sweetly from my mouth. And she would say, often, 'I wish I knew that grandfather of mine. Gran, do you think he's lonely? I think he would be glad of somebody to make him a pot of tea and see to his fire. Some day I'm going to Scotland and I'm going to knock on his door, wherever he lives, and I'll do things for him. Did you love him very much, gran? He must be a good person, that old sailor, ever to have been loved by you. I *will* see him. I'll hear the old stories from his own mouth. Most of all, of course, the love story—for you, gran, tell me nothing about that. . . .' I am writing this letter, Bill, to tell you that this can never now be. Our granddaughter Andrina died last week, suddenly, in the first stirrings of spring. . . .'

Later, over the fire, I thought of the brightness and burgeoning and dew that visitant had brought across the threshold of my latest winter, night after night; and of how she had always come with the first shadows and the first star; but there, where she was dust, a new time was brightening earth and sea.

IAIN CRICHTON SMITH

1928–

AN AMERICAN SKY

He stood on the deck of the ship looking towards the approaching island. He was a tall man who wore brownish clothes: and beside him were two matching brown cases. As he stood on the deck he could hear Gaelic singing coming from the saloon which wasn't all that crowded but had a few people in it, mostly coming home for a holiday from Glasgow. The large ship moved steadily through the water and when he looked over the side he could see thin spitlike foam travelling alongside. The island presented itself as long and green and bare with villages scattered along the coast. Ahead of him was the westering sun which cast long red rays across the water.

He felt both excited and nervous as if he were returning to a wife or sweetheart whom he had not seen for a long time and was wondering whether she had changed much in the interval, whether she had left him for someone else or whether she had remained obstinately true. It was strange, he thought, that though he was sixty years old he should feel like this. The journey from America had been a nostalgic one, first the plane, then the train, then the ship. It was almost a perfect circle, a return to the womb. A womb with a view, he thought and smiled.

He hadn't spoken to many people on the ship. Most of the time he had been on deck watching the large areas of sea streaming past, now and again passing large islands with mountain peaks, at other times out in the middle of an empty sea where the restless gulls scavenged, turning their yellow gaunt beaks towards the ship.

The harbour was now approaching and people were beginning to come up on deck with their cases. A woman beside him was buttoning up her small son's coat. Already he could see red buses and a knot of people waiting at the pier. It had always been like that, people meeting the ship when it arrived at about eight, some not even welcoming anyone in particular but just standing there watching. He noticed a squat man in fisherman's clothes doing something to a rope. Behind him there was a boat under green canvas.

The ship swung in towards the harbour. Now he could see the people more clearly and behind them the harbour buildings. When he looked over the side he noticed that the water was dirty with bits of wooden boxes floating about in an oily rainbowed scum.

After some manœuvring the gangway was eventually laid. He picked up his cases and walked down it behind a girl in yellow slacks whose transistor was playing in her left hand. Ahead of her was a man in glasses who had a BEA case with, stamped on it, the names of various foreign cities. There were some oldish women in dark clothes among the crowd and also some girls and boys in brightly coloured clothes. A large fat slow man stood to the side of the gangway where it touched the quay, legs spread apart, as if he had something to do with the ship, though he wasn't actually doing anything. Now and again he scratched a red nose.

He reached the shore and felt as if the contact with land was an emotionally charged moment. He didn't quite know how he felt, slightly empty, slightly excited. He walked away from the ship with his two cases and made his way along the main street. It had changed, no doubt about it. There seemed to be a lot of cafés, from one of which he heard the blare of a jukebox. In a bookseller's window he saw *From Russia with Love* side by side with a book about the Highlands called *The Misty Hebrides.* Nevertheless the place appeared smaller, though it was much more modern than he could remember, with large windows of plate glass, a jeweller's with Iona stone, a very fashionable-looking ladies' hairdressers. He also passed a supermarket and another bookseller's. Red lights from one of the cafés streamed into the bay. At the back of the jeweller's shop he saw a church spire rising into the sky. He came to a cinema which advertised Bingo on Tuesdays, Thursdays and Saturdays. Dispirited trailers for a Western filled the panels.

He came to a Chinese restaurant and climbed the steps, carrying his two cases. The place was nearly empty and seemed mostly purplish with, near the ceiling, a frieze showing red dragons. Vague music—he thought it might be Chinese—leaked from the walls. He sat down and, drawing the huge menu towards him, began to read it. In one corner of the large room an unsmiling Chinaman with a moustache was standing by an old-fashioned black telephone and at another table a young Chinese girl was reading what might have been a Chinese newspaper. A little bare-bottomed Chinese boy ran out of the kitchen, was briefly chased back with much giggling, and the silence descended again.

For a moment he thought that the music was Gaelic, and was lost in his dreams. The Chinese girl seemed to turn into Mary who was doing her homework in the small thatched house years and years before. She was asking their father about some arithmetic but he, stroking his beard, was not able to answer. At another table an old couple were solidly munching rice, their heads bowed.

The music swirled about him. The Chinese girl read on. Why was it that these people never laughed? He had noticed that. Also that Chinese restaurants were hushed like churches. A crowd of young people came in laughing and talking, their Highland accents quite distinct though they were speaking English. He felt suddenly afraid and alone and slightly disorientated as if he had come to the wrong place at the wrong time. The telephone rang harshly and the Chinaman answered it in guttural

English. Perhaps he was the only one who could speak English. Perhaps that was his job, just to answer the phone. He had another look at the menu, suddenly put it down and walked out just as a Chinese waitress came across with a notebook and pencil in her hand. He hurried downstairs and walked along the street.

Eventually he found a hotel and stood at the reception desk. A young blonde girl was painting her nails and reading a book. She said to the girl behind her, 'What does "impunity" mean?' The other girl stopped chewing and said, 'Where does it say that?' The first girl looked at him coolly and said, 'Yes, Sir?'

Her voice also was Highland.

'I should like a room,' he said. 'A single room.'

She leafed rapidly through a book and said at last, 'We can give you 101, Sir. Shall I get the porter to carry your bags?'

'It's not necessary.'

'That will be all right then, Sir.'

He waited for a moment and then remembered what he was waiting for. 'Could I please have my key?' he asked.

She looked at him in amazement and said, 'You don't need a key here, Sir. Nobody steals anything. Room 101 is on the first floor. You can't miss it.' He took his cases and walked up the stairs. He heard them discussing a dance as he left.

He opened the door and put the cases down and went to the window. In front of him he could see the ship and the bay with the red lights on it and the fishing boats and the large clock with the greenish face.

As he turned away from the window he saw the Gideon Bible, picked it up, half smiling, and then put it down again. He took off his clothes slowly, feeling very tired, and went to bed. He fell asleep very quickly while in front of his eyes he could see Bingo signs, advertisements for Russian watches, and seagulls flying about with open gluttonous beaks. The last thought he had was that he had forgotten to ask when breakfast was in the morning.

II

The following day at two o'clock in the afternoon he took the bus to the village that he had left so many years before. There were few people on the bus which had a conductress as well as a driver, both dressed in uniform. He thought wryly of the gig in which he had been driven to the town the night he had left; the horse was dead long ago and so was his own father, the driver.

On the seat opposite him there was sitting a large fat tourist who had a camera and field-glasses slung over his shoulder and was wearing dark glasses and a light greyish hat.

The driver was a sturdy young man of about twenty or so. He whistled a good deal of the time and for the rest exchanged badinage with the conductress who, it emerged, wanted to become an air stewardess. She wore a black uniform, was pretty in a thin, sallow way, and had a turned-up nose and black hair.

After a while he offered a cigarette to the driver who took it. 'Fine day, Sir,' he said and then, 'Are you home on holiday?'

'Yes. From America.'

'Lots of tourists here just now. I was in America myself once. I was in the Merchant Navy. Saw a baseball team last night on TV.'

The bus was passing along the sparkling sea and the cemetery which stood on one side of the road behind a grey wall. The marble of the gravestones glittered in the sun. Now and again he could see caravans parked just off the road and on the beach men and children in striped clothing playing with large coloured balls or throwing sticks for dogs to retrieve. Once they passed a large block of what appeared to be council houses, all yellow.

'You'll see many changes,' said the driver. 'Hey, bring us some of that orangeade,' he shouted to the conductress.

'I suppose so.'

But there didn't seem all that many in the wide glittering day. The sea, of course, hadn't changed, the cemetery looked brighter in the sun perhaps, and there were more houses. But people waved at them from the fields, shielding their eyes with their hands. The road certainly was better.

At one point the tourist asked to be allowed out with his camera so that he could take a photograph of a cow which was staring vaguely over a fence.

If the weather was always like this, he thought, there wouldn't be any problem . . . but of course the weather did change . . . The familiar feeling of excitement and apprehension flooded him again.

After a while they stopped at the road end and he got off with his two cases. The driver wished him good luck. He stood staring at the bus as it diminished into the distance and then taking his cases began to walk along the road. He came to the ruins of a thatched house, stopped and went inside. As he did so he disturbed a swarm of birds which flew out of the space all round him and fluttered out towards the sky which he could see quite clearly as there was no roof. The ruined house was full of stones and bits of wood and in the middle of it an old-fashioned iron range which he stroked absently, making his fingers black and dusty. For a moment the picture returned to him of his mother in a white apron cooking at such a stove, in a smell of flour. He turned away and saw carved in the wooden door the words MARY LOVES NORMAN. The hinges creaked in the quiet day.

He walked along till he came to a large white house at which he stopped. He opened the gate and there, waiting about ten yards in front of him, were his brother, his brother's daughter-in-law, and her two children, one a boy of about seventeen and the other a girl of about fifteen. They all seemed to be dressed in their best clothes and stood there as if in a picture. His brother somehow seemed dimmer than he remembered, as if he were being seen in a bad light. An observer would have noticed that though the two brothers looked alike the visitor seemed a more vivid version of the other. The family waited for him as if he were a photographer and he moved forward. As he did so his brother walked quickly towards him, holding out his hand.

'John,' he said. They looked at each other as they shook hands. His niece came forward and introduced herself and the children. They all appeared well dressed and prosperous.

The boy took his cases and they walked towards the house. It was of course a new house, not the thatched one he had left. It had a porch and a small garden and large windows which looked out towards the road.

He suddenly said to this brother, 'Let's stay out here for a while.' They stood together at the fence gazing at the corn which swayed slightly in the breeze. His brother did not seem to know what to say and neither did he. They stood there in silence.

After a while John said, 'Come on, Murdo, let's look at the barn.' They went into it together. John stood for a while inhaling the smell of hay mixed with the smell of manure. He picked up a book which had fallen to the floor and looked inside it. On the fly-leaf was written:

Prize for English
John Macleod

The book itself in an antique and slightly stained greenish cover was called *Robin Hood and His Merry Men*. His brother looked embarrassed and said, 'Malcolm must have taken it off the shelf in the house and left it here.' John didn't say anything. He looked idly at the pictures. Some had been torn and many of the pages were brown with age. His eye was caught by a passage which read, 'Honour is the greatest virtue of all. Without it a man is nothing.' He let the book drop to the floor.

'We used to fight in that hayloft,' he said at last with a smile, 'and I think you used to win,' he added, punching his brother slightly in the chest. His brother smiled with pleasure. 'I'm not sure about that,' he answered.

'How many cows have you got?' said John looking out through the dusty window.

'Only one, I'm afraid,' said his brother. 'Since James died . . .' Of course. James was his son and the husband of the woman he had met. She had looked placid and mild, the kind of wife who would have been suited to him. James had been killed in an accident on a ship: no-one knew very much about it. Perhaps he had been drunk, perhaps not.

He was reluctant for some reason to leave the barn. It seemed to remind him of horses and bridles and bits, and in fact fragments of corroded leather still hung here and there on the walls. He had seen no horses anywhere: there would be no need for them now. Near the door he noticed a washing machine which looked quite new.

His brother said, 'The dinner will be ready, if it's your pleasure.' John looked at him in surprise, the invitation sounded so feudal and respectful. His brother talked as if he were John's servant.

'Thank you.' And again for a moment he heard his mother's voice as she called them in to dinner when they were out playing.

They went into the house, the brother lagging a little behind. John felt uncomfortable as if he were being treated like royalty when he wanted everything to be simple and natural. He knew that they would have cooked the best food whether they could afford it or not. They wouldn't, of course, have allowed him to stay at a hotel in the town during his stay. That would have been an insult. They went in. He found the house much cooler after the heat of the sun.

III

In the course of the meal which was a large one with lots of meat, cabbages and turnip and a pudding, Murdo suddenly said to his grandson:

'And don't you forget that Grandfather John was very good at English. He was the best in the school at English. I remember in those days we used to write on slates and Mr Gordon sent his composition round the classes. John is very clever or he wouldn't have been an editor.'

John said to Malcolm, who seemed quietly unimpressed: 'And what are you going to do yourself when you leave school?'

'You see,' said Murdo, 'Grandfather John will teach you . . .'

'I want to be a pilot,' said Malcolm, 'or something in science, or technical. I'm quite good at science.'

'We do projects most of the time,' said his sister. 'We're doing a project on fishing.'

'Projects!' said her grandfather contemptuously. 'When I was your age I was on a fishing boat.'

'There you are,' said his grandson triumphantly. 'That's what I tell Grandfather Murdo I should do, but I have to stay in school.'

'It was different in our days,' said his grandfather. 'We had to work for our living. You can't get a good job now without education. You have to have education.'

Straight in front of him on the wall, John could see a photograph of his brother dressed in army uniform. That was when he was a corporal in the Militia. He had also served in Egypt and in the First World War.

'They don't do anything these days,' said Murdo. 'Nothing. Every night it's football or dancing. He watches the TV all the time.'

'Did you ever see Elvis Presley?' said the girl who was eating her food very rapidly, and looking at a large red watch on her wrist.

'No, I'm sorry, I didn't,' said John. 'I once saw Lyndon Johnson though.'

She turned back to her plate uninterested.

The children were not at all as he had expected them. He thought they would have been shyer, more rustic, less talkative. In fact they seemed somehow remote and slightly bored and this saddened him. It was as if he were already seeing miniature Americans in the making.

'Take some more meat,' said his brother, piling it on his plate without waiting for an answer.

'All we get at English,' said Malcolm, 'is interpretations and literature. Mostly Shakespeare. I can't do any of it. I find it boring.'

'I see,' said John.

'He needs three Highers to get anywhere, don't you, Malcolm,' said his mother, 'and he doesn't do any work at night. He's always repairing his motor bike or watching TV.'

'When we got the TV first,' said the girl giggling, 'Grandfather Murdo thought . . .'

'Hist,' said her mother fiercely, leaning across the table, 'eat your food.'

Suddenly the girl looked at the clock and said, 'Can I go now, Mother? I've got to catch the bus.'

'What's this?' said her grandfather and at that moment as he raised his head, slightly bristling, John was reminded of their father.

'She wants to go to a dance,' said her mother.

'All the other girls are going,' said the girl in a pleading, slightly hysterical voice.

'Eat your food,' said her grandfather, 'and we'll see.' She ate the remainder of her food rapidly and then said, 'Can I go now?'

'All right,' said her mother, 'but mind you're back early or you'll find the door shut.'

The girl hurriedly rose from the table and went into the living room. She came back after a while with a handbag slung over her shoulder and carrying a transistor.

'Goodbye, Grandfather John,' she said. 'I'll see you tomorrow.' She went out and they could hear her brisk steps crackling on the gravel outside.

When they had finished eating Malcolm stood up and said, 'I promised Hugh I would help him repair his bike.'

'Back here early then,' said his mother again. He stood hesitating at the door for a moment and then went out, without saying anything.

'That's manners for you,' said Murdo. 'Mind you, he's very good with his hands. He repaired the tractor once.'

'I'm sure,' said John.

They ate in silence. When they were finished he and his brother went to sit in the living room which had the sun on it. They sat opposite each other in easy chairs. Murdo took out a pipe and began to light it. John suddenly felt that the room and the house were both very empty. He could hear quite clearly the ticking of the clock which stood on the mantelpiece between two cheap ornaments which looked as if they had been won at a fair.

Above the mantelpiece was a picture of his father, sitting very upright in a tall narrow chair, his long beard trailing in front of him. For some reason he remembered the night his brother, home from the war on leave, had come in late at night, drunk. His father had waited up for him and there had been a quarrel during which his brother had thrown the Bible at his father calling him a German bastard.

The clock ticked on. His brother during a pause in the conversation took up a *Farmers' Weekly* and put on a pair of glasses. In a short while he had fallen asleep behind the paper, his mouth opening like that of a stranded fish. Presumably that was all he read. His weekly letters were short and repetitive and apologetic.

John sat in the chair listening to the ticking of the clock which seemed to grow louder and louder. He felt strange again as if he were in the wrong house. The room itself was so clean and modern with the electric fire and the TV set in the corner. There was no air of history or antiquity about it. In a corner of the room he noticed a guitar which presumably belonged to the grandson. He remembered the nights he and his companions would dance to the music of the melodeon at the end of the road. He also remembered the playing of the bagpipes by his brother.

Nothing seemed right. He felt as if at an angle to the world he had once known. He wondered why he had come back after all those years. Was he after all like those people who believed in the innocence and unchangeability of the heart and vibrated to the music of nostalgia? Did he expect a Garden of Eden where the apple had not been eaten? Should he stay or go back? But then there was little where he had come from. Mary was dead. He was retired from his editorship of the newspaper. What did it all mean? He remembered the night he had left home many years before. What had he been expecting then? What cargo was he bearing with him? And what did his return signify? He didn't know. But he would have to find out. It was necessary to find out. For some reason just before he closed his eyes he saw in the front of him again the cloud of midges he had seen not an hour before, rising and falling above the fence, moving on their unpredictable ways. Then he fell asleep.

IV

The following day which was again fine he left the house and went down to a headland which overlooked the sea. He sat there for a long time on the grass, feeling calm and relaxed. The waves came in and went out, and he was reminded of the Gaelic song *The Eternal Sound of the Sea* which he used to sing when he was young. The water seemed to stretch westward into eternity and he could see nothing on it except the light of the sun. Clamped against the rocks below were the miniature helmets of the mussels and the whelks. He remembered how he used to boil the whelks in a pot and fish the meat out of them with a pin. He realised as he sat there that one of the things he had been missing for years was the sound of the sea. It was part of his consciousness. He should always live near the sea.

On the way back he saw the skull of a sheep, and he looked at it for a long time before he began his visits. Whenever anyone came home he had to visit every house, or people would be offended. And he would have to remember everybody, though many people in those houses were now dead.

He walked slowly along the street, feeling as if he were being watched from behind curtained windows. He saw a woman standing at a gate. She was a stout large woman and she was looking at him curiously. She said, 'It's a fine day.' He said, 'Yes.'

She came towards him and he saw her red beefy face. 'Aren't you John Macleod?' she asked. 'Don't you remember me?'

'Of course I do,' he replied. 'You're Sarah.'

She shouted jovially as if into a high wind, 'You'll have to speak more loudly. I'm a little deaf.' He shouted back, 'Yes, I'm John Macleod,' and it seemed to him as if at that moment he were trying to prove his identity. He shouted louder still, 'And you're Sarah.' His face broke into a large smile.

'Come in, come in,' she shouted. 'Come in and have a cup of milk.'

He followed her into the house and they entered the living room after passing through the scullery which had rows of cups and saucers and plates on top of a huge dresser. In a corner of the room sat a man who was probably her son trapped like a fly inside a net which he was repairing with a bone needle. He was wearing a fisherman's jersey and his hands worked with great speed.

'This is George,' she shouted. 'My son. This is John Macleod,' she said to George. George looked up briefly from his work but said nothing. He was quite old, perhaps fifty or so, and there was an unmarried look about him.

'He's always fishing,' she said, 'always fishing. That's all he does. And he's very quiet. Just like his father. We're going to give John a cup of milk,' she said to her son. She went into the scullery for the milk and though he was alone with George the latter didn't speak. He simply went on repairing his net. This room too was cool and there was no fire. The chairs looked old and cracked and there was an old brown radio in a corner. After a while she came back and gave him the milk. 'Drink it up,' she instructed him as if she were talking to a boy. It was very cold. He couldn't remember when he had last drunk such fine milk.

'You were twenty-four when you went away,' she said, 'and I had just married. Jock is dead. George is very like him.' She shouted all this at the top of her voice and he himself didn't reply as he didn't want to shout.

'And how's that brother of yours?' she shouted remorselessly. 'He's a cheat, that one. Two years ago I sold him a cow. He said that there was something wrong with her and he got her cheap. But there was nothing wrong with her. He's a devil,' she said approvingly. 'But he was the same when he was young. After the penny. Always asking if he could run messages. You weren't like that. You were more like a scholar. You'd be reading books sitting on the peat banks. I remember you very well. You had fair hair, very fair hair. Your father said that you looked like an angel. But your brother was the cunning one. He knew a thing or two. And how are you?'

'I'm fine,' he shouted back.

'I hope you've come to stay,' she shouted again. He didn't answer.

'You would be sorry to hear about your mother,' she shouted again. 'We were all fond of her. She was a good woman.' By 'good' she meant that she attended church regularly. 'That brother of yours is a devil. I wonder if your mother liked him.' George looked at her quickly and then away again.

He himself shouted, 'Why do you ask that?' She pretended not to hear him and he had to shout the words again.

'It was nothing,' she said. 'I suppose you have a big job in America.'

He was wondering what she had meant and felt uneasy, but he knew that he wouldn't get anything more out of her.

'They've all changed here,' she shouted. 'Everything's changed. The girls go about showing their bottoms, not like in my day. The boys are off to the dances every night. George here should get married but I wouldn't let him marry one of these trollops. And you can't visit your neighbours any more. You have to wait for an invitation. Imagine that. In the old days the door would be always open. But not any more. Drink up your milk.'

He drank it obediently as if he were a child.

'Jock died, you know. A stroke it was. It lasted for three years. But he never complained. You remember Jock.'

He didn't remember him very well. Was he the one who used to play football or the one who played tricks on the villagers? He couldn't summon up a picture of him at all. What had she meant by his mother and his brother? He had a strange feeling as if he were walking inside an illusion, as if things had happened here that he hadn't known of, though he should have. But who would tell him? They would all keep their secrets. He even had the feeling that this large apparently frank woman was in fact treacherous and secretive and that behind her huge façade there was lurking a venomous thin woman whose head nodded up and down like a snake's.

She laughed again. 'That brother of yours is a businessman. He is the one who should have gone to America. He would have got round them all. There are no flies on him. Did you not think of coming home when your mother died?'

'I was . . . I couldn't at the time,' he shouted.

George, entrapped in his corner, the net around his feet, plied his bone needle.

'It'll be good to come home again,' she shouted. 'Many of them come back. Donny Macdonald came back seven years ago and they hadn't heard from him for twenty years. He used to drink but he goes to church regularly now. He's a man of God. He's much quieter than he used to be. He used to sing a lot when he was young and they made him the precentor. He's got a beautiful voice but not as good as it was. Nobody knew he was coming home till he walked into the house one night off the bus. Can you imagine that? At first he couldn't find it because they had built a new house. But someone showed it to him.'

He got up and laid the cup on the table.

'Is Mr Gordon still alive?' he shouted. Mr Gordon was his old English teacher.

'Speak up, I can't hear you,' she said, her large bulging face thrust towards him like a crab.

'Mr Gordon?' he shouted. 'Is he still alive?'

'Mr Gordon,' she said. 'Yes, he's alive. He's about ninety now. He lives over there.' She took him over to the window and pointed out a house to him. 'Oh, there's the lad,' she said. 'He's always sitting on the wall. He's there every day. His sister died, you know. She was a bit wrong in the head.'

He said goodbye and she followed him to the door. He walked out the gate and made his way to where she had pointed. The day seemed heavy and sleepy and he felt slightly drugged as if he were moving through water. In the distance a man was hammering a post into the ground. The cornfields swayed slightly in the breeze and he could see flashes of red among them. He remembered the days when he would go with a bucket to the well, and smelt again the familiar smell of flowers and grass. He expected at any moment to see the ghosts of the dead stopping him by the roadway, interrogating him and asking him, 'When did you come home? When are you going away?' The whole visit, he realised now, was an implicit interrogation. What it was really about was: What had he done with his life? That was the question that people, without realising it, were putting to him, simply because he had chosen to return. It was also the question that he himself wanted answered.

Ahead of him stretched the moors and in the far distance he could see the Standing Stones which could look so eerie in the rain and which had perhaps been used in the sacrifice of children in Druid times. Someone had to be knifed to make the sun appear, he thought wryly. Before there could be light there must be blood.

He made his way to see Mr Gordon.

V

Gordon recognised him immediately: it was almost as if he had been waiting for him. He came forward from behind a table on which were piled some books and a chessboard on which some pieces were standing, as if he had been playing a game.

'John,' he said, 'John Macleod.'

John noticed that standing beside the chair was a small glass in which there were the remains of whisky.

'Sit down, sit down,' said Gordon as if he hadn't had company for a long time. He was still spry, grey-haired of course, but thin in the body. He was wearing an old sports jacket and a shirt open at the neck. There was a slightly unshaven look about him.

'I play chess against myself,' he said. 'I don't know which of us wins.' His laugh was a short bark. John remembered himself running to school while Gordon stood outside the gate with a whistle in his hand looking at his watch impatiently.

'I suppose coming from America,' he said, 'you'll know about Fischer. He's about to do the impossible, beat the Russian World Champion at chess. It's like the Russians beating the Americans at baseball—or us at shinty,' he added with the same self-delighting barking laugh. 'He is of course a genius and geniuses make their own rules. How are you?'

'Very well. And how are you?' He nearly said 'Sir' but stopped himself in time.

'Oh, not too bad. Time passes slowly. Have you ever thought about time?' Beside his chair was a pile of books scattered indiscriminately. 'I belong to dozens of book clubs. This is a book on Time. Very interesting. From the point of view of physics, psychiatry and so on.' He pointed to a huge tome which looked both formidable and new. 'Did you know, for instance, that time passes slowly for some people and rapidly for others? It's a matter of personality, and the time of year you're born. Or that temperature can affect your idea of time? Very interesting.' He gave the impression of a man who devoured knowledge in a sterile way.

John looked out of the window. Certainly time seemed to pass slowly here. Everything seemed to be done in slow motion as if people were walking through water, divers with lead weights attached to them.

'Are you thinking of staying?' said Gordon, pouring out a glass of whisky for his guest.

'I don't know that yet.'

'I suppose you could buy a house somewhere. And settle down. Perhaps do some fishing. I don't do any myself. I read and play chess. But I suppose you could fish and do some crofting. Though I don't remember that you were particularly interested in either of these.'

'I was just thinking,' said John, 'of what you used to tell us when we were in your English class. You always told us to observe. Observation, you used to say, is the secret of good writing. Do you remember the time you took us out to the tree and told us to smell and touch it and study it and write a poem about it? It was a cherry tree, I recall. We wrote the poem in the open air.'

'I was in advance of my time,' said Gordon. 'That's what they all do now. They call it Creative Writing. But of course they can't spell nowadays.'

'And you always told us that exactitude was important. Be observant and exact, you said, above all be true to yourselves.'

'Drink your whisky,' said Gordon. 'Yes, I remember it all. I've kept some of your essays. You were gifted. In all the years I taught I only met two pupils who were really gifted. How does one know talent when one sees it? I don't know. Anyway, I recognised your talent. It was natural, like being a tiger.'

'Yes, you kept telling us about exactitude and observation. You used to send us out of the room and change objects in the room while we were out. You made Sherlock Holmeses out of us.'

'Why do you speak about that now? It was all so long ago.'

'I have a reason.'

'What is your reason?' said Gordon sharply.

'Oh, something that happened to me. Some years ago.'

'And what was that? Or don't you want to talk about it?'

'I don't see why not. Not that it's very complimentary to me.'

'I have reached the age now,' said Gordon, 'when I am not concerned with honour, only with people.'

'I see,' said John, 'but suppose you can't separate them. Well, I'll tell you anyway.' He walked over to the window, standing with his back to the room and looking out at the empty road. It was as if he didn't want to be facing Gordon.

'I was an editor for some time as you know,' he said. 'Your training stood me in good stead. It was not a big paper but it was a reasonable paper. It had influence in the largish town in which I stayed. It wasn't Washington, it wasn't New York, but it was a largish town. I made friends in this town. One was a lecturer in a university. At least that is what we would call it here. As a matter of fact, he wasn't a lecturer in English. He was a lecturer in History. It was at the time of the McCarthy trials when nobody was safe, nobody. Another of my friends went off his head at that time. He believed that everyone was persecuting him and opening his mail. He believed that planes were pursuing him. In any case this friend of mine, his name was Mason, told me that files had been dug up on him referring to the time when he was a student and had belonged to a Left Wing university club. Now there were complaints that he was indoctrinating his students with Communism and, of course, being a History lecturer, he was in a precarious position. I told him that I would defend him in my paper, that I would write a hard-hitting editorial. I told him that I would stand up for principles, humane principles.' He stretched out his hand for the whisky and decided against drinking it. 'I left him on the doorstep at eleven o'clock on a Monday night. He was very disturbed because of course he was innocent, he wasn't a Communist and anyway he had great integrity as a teacher and lectured on Communism only theoretically as one ideology among others. But the McCarthy people of course were animals. You have no conception. Not here. Of the fog of lies. Of the quagmire. No conception.' He paused. A cow outside had bent its head to the grass and was eating.

'Anyway this was what happened. I walked home because I needed the exercise. The street was deserted. There were lampposts shining and it was raining. A thin drizzle. I could hear the echo of my feet on the road. This was the kind of thing you taught us, to remember and listen and observe, to be aware of our surroundings sensuously. By then it had become a habit with me.

'As I was walking along two youths came towards me out of the shadow, from under the trees. I thought they were coming home from the cinema or from a dance. They wore leather jackets and were walking towards me along the sidewalk. They stayed on the sidewalk and I made as if to go round them since they were coming straight for me without deviating. One of them said, "Daddy." I stopped. I thought he was going to ask me for a light. He said, "Your wallet, daddy." I looked at him in amazement. I looked at the two of them. I couldn't understand what was going on. And something happened to me. I could feel everything very intensely, you see. At that moment I could have written a poem, everything was so clear. They were laughing, you see, and they were very casual. They walked like those cowboys you see on the films, physically at ease in their world. And their eyes sparkled. Their eyes sparkled with pure evil. I knew that if I protested they would beat me up. I knew that there was no appeal. None at all. One of them had a belt, and a buckle on it sparkled in the light. My eyes were at the level of the buckle. I took out my wallet and gave them the money. I had fifty dollars. I observed everything as you had trained us to do. Their boots which were shining except for the drizzle: their neckties: their leather jackets. Their legs which were narrow in the narrow trousers. And their faces which were looking slightly upwards and shining. Clear and fine almost, but almost innocent though evil. A rare sort of energy. Pure and bright. They took the wallet, counted the money and gave me back the wallet. They then walked on. The whole incident took perhaps three minutes.

'I went into the house and locked the door. The walls seemed very fragile all of a sudden. My wife had gone to bed and I stood downstairs thinking, now and again removing a book from the shelves and replacing it. I felt the house as thin as the shell of an egg: I could hear, I thought, as far away as San Francisco. There was a tap dripping and I turned it off. And I didn't write the editorial, I didn't write anything. Two weeks after that my friend killed himself, with pills and whisky.'

The whisky which Gordon had given him was still untouched.

'Observation and exactitude,' he said, 'and elegance of language.' There was a long silence. Gordon picked up a chess piece and weighed it in his hand.

'Yes,' he said, 'and that's why you came home.'

'Perhaps. I don't know why I came home. One day I was walking along a street and I smelt the smell of fish coming from a fish shop. And it reminded me of home. So I came home. My wife, of course, is dead.'

'Many years ago,' said Gordon, still holding the chess piece in his hand, 'I was asked to give a talk to an educational society in the town. In those days I used to write poetry though of course I never told anyone.

I was working on a particular poem at the time: it was very difficult and I couldn't get it to come out right. Well, I gave this talk. It was, if I may say so myself, a brilliant talk for in those days I was full of ideas. It was also very witty. People came and congratulated me afterwards as people do. I arrived home at one o'clock in the morning. When I got home I took out the poem and tried to do some work on it. But I was restless and excited and I couldn't get into the right mood. I sat and stared at the clock and I knew quite clearly that I would never write again. Odd, isn't it?'

'What are you trying to say?'

'Say? Nothing. Nothing at all. I don't think you'd better stay here. I don't think this place is a refuge. People may say so but it's not true. After a while the green wears away and you are left with the black. In any case I don't think you'd better settle here: that would be my advice. However, it's not my business. I have no business now.'

'Why did you stay here?' said John slowly.

'I don't know. Laziness, I suppose. I remember when I was in Glasgow University many years ago we used to take the train home at six in the morning after the holiday started. At first we were all very quiet, naturally, since we were half-asleep, most of us. But then as the carriages warmed and the sun came up and we came in sight of the hills and the lochs we began to sing Gaelic songs. Odd, and Glasgow isn't that far away. What does it all mean, John? What are you looking at?'

'The broken fences.'

'Yes, of course. There's a man here and he's been building his own house for ten years. He carries stone after stone to the house and then he forgets and sits down and talks to people. Time is different here, no doubt about it.'

'I had noticed.'

'If you're looking for help from me, John, I can't give you any. In the winter time I sit and look out the window. You can see the sea from here and it can look very stormy. The rain pours down the window and you can make out the waves hitting the islands out there. What advice could I give you? I have tried to do my best as far as my work was concerned. But you say it isn't enough.'

'Perhaps it wasn't your fault.'

John made his way to the door.

'Where are you going?'

'I shall have to call on other people as well. They all expect one to do that, don't they?'

'Yes, they still feel like that. That hasn't changed.'

'I'll be seeing you then,' said John as he left.

'Yes, yes, of course.'

He walked towards the sea cliffs to a house which he had visited many times when he was a boy, where he had been given many tumblers of milk, where later in the evening he would sit with others talking into the night.

The sea was large and sparkling in front of him like a shield. No, he said automatically to himself, it isn't like a shield, otherwise how could the cormorants dive in and out of it? What was it like then? It was like the sea, nothing else. It was like the sea in one of its moods, in one of its sunny gentle moods. As he walked pictures flashed in front of his eyes. He saw a small boy running; then a policeman's arm raised, the baton falling in a vicious arc, the neon light flashing from his shield. The boy stopped in midflight, the picture frozen.

VI

He knocked at the door of the house and a woman of about forty, thin and with straggly greying hair, came to the door.

She looked at him enquiringly.

'John, John Macleod,' he said. 'I came to see your mother.' Her face lighted up with recognition and she said, 'Come in, come in.' And then inexplicably, 'I thought you were from the BBC.'

'The BBC?'

'Yes, they're always sending people to take recordings of my mother singing and telling stories, though she's very old now.'

He followed her into a bedroom where an old white-faced white-haired woman was lying, her head against white pillows. She stretched out her prominently veined hand across the blankets and said, 'John, I heard Anne talking to you. There's nothing wrong with my hearing.'

They were left alone and he sat down beside the bed. There was a small table with medicine bottles and pills on it.

'It's true,' she said, 'the BBC are always sending people to hear me sing songs before I die.'

'And how are you?'

'Fine, fine.'

'Good, that's good.' Her keen wise eyes studied his face carefully. The room had bright white wallpaper and the windows faced the sea.

'I don't sleep so well now,' said the old woman. 'I waken at five every morning and I can hear the birds twittering just outside the window.'

'You look quite well,' he said.

'Of course I'm not well. Everybody says that to me. But after all I'm ninety years old. I can't expect to live forever. And you're over sixty but I can still see you as a boy.' She prattled on but he felt that all the time she was studying him without being obvious.

'Have you seen the BBC people? They all have long hair and they wear red ties. But they're nice and considerate. Of course everybody wears long hair now, even my daughter's son. Would you like to hear my recording? My grandson took it down on a tape.'

'I would,' he said.

She tapped on the head of the bed as loudly as she could and her daughter came in.

'Where's Hugh?' she asked.

'He's outside.'

'Tell him to bring in the machine. John wants to hear my recording.' She turned to John and said, 'Hugh is very good with his hands, you know. All the young people nowadays know all about electricity and cars.'

After a while a tall quiet long-haired boy came in with a tape recorder. He plugged it into a socket beside the bed, his motions cool and competent and unflurried. He had the same neutral quizzical look that John had noticed in his brother's two grandchildren. They don't want to be deceived again, he thought. This generation is not interested in words, only in actions. Observation, exactitude, elegance. The universe of the poem or the story is not theirs, their universe is electronic. And when he thought of the phrase 'the music of the spheres' he seemed to see a shining bicycle moving through the heavens, or the wheels of some inexplicable machine.

Hugh switched on the tape recorder and John listened.

'Tonight,' the announcer began, 'we are going to hear the voice of a lady of ninety years old. She will be telling us about her life on this far Hebridean island untouched by pollution and comparatively unchanged when it is compared with our own hectic cities. This lady has never in all her life left the island on which she grew up. She has never seen a train. She has never seen a city. She has been brought up in a completely pastoral society. But we may well ask, what will happen to this society? Will it be squeezed out of existence? How can it survive the pollution of our time, and here I am speaking not simply of physical but of moral pollution? What was it like to live on this island for so many years? I shall try to elicit some answers to that question in the course of this programme. But first I should like you to hear this lady singing a Gaelic traditional song. I may interpolate at this point that many Gaelic songs have apparently been anglicised musically, thus losing their traditional flavour. But Mrs Macdonald will sing this song in the way in which she was taught to, the way in which she picked it up from previous singers.'

There followed a rendering of *Thig Tri Nithean Gun Iarraidh* ('Three things will come without seeking . . .'). John listened to the frail voice: it seemed strange to hear it, ghostly and yet powerful in its own belief, real and yet unreal at the same time.

When the singing was over the interviewer questioned her:

I: And now, Mrs Macdonald, could you please tell me how old you are?
Mrs M: I am ninety years old.
I: You will have seen a lot of changes on this island, in this village even.
Mrs M: O yes, lots of changes. I don't know much about the island. I know more about the village.
I: You mean that you hardly ever left the village itself?
Mrs M: I don't know much about the rest of the island.
I: What are your memories then of your youth in the village?
Mrs M: Oh, people were closer together. People used to help each other at the peat gathering. They would go out with a cart and they would

put the peats on the cart. And they would make tea and sing. It was very happy times especially if it was a good day.

I: Do they not do that any more? I mean, coal and electricity . . .

Mrs M: No, they don't do that so much, no. Nowadays. And there was more fishing then too. People would come to the door and give you a fish if they had caught one.

I: You mean herring?

Mrs M: No, things like cod. Not herring. They would catch them in boats or off the rocks. Not herring. The herring were caught by the drifters. And the mackerel. We used to eat herring and potatoes every day. Except Sunday of course.

I: And what did you eat on Sunday?

Mrs M: We would always have meat on Sunday. That was always the fashion. Meat on Sundays. And soup.

I: I see. And tell me, when did you leave school, Mrs Macdonald?

Mrs M: I left school when I was fourteen years old. I was in Secondary Two.

I: It was a small village school, I take it.

Mrs M: Oh, yes, it was small. Perhaps about fifty pupils. Perhaps about fifty. We used to write on slates in those days and the children would bring in a peat for a fire in the winter. Every child would bring in a peat. And we had people called pupil-teachers.

I: Pupil-teachers? What were pupil-teachers?

Mrs M: They were young people who helped the teacher. Pupils. They were pupils themselves.

I: Then what happened?

Mrs M: I looked after my father and mother. We had a croft too. And then I got married.

I: What did your husband do?

Mrs M: He was a crofter. In those days we used to go to a dance at the end of the road. But the young people go to the town now. In those days we had a dance at the end of the road.

I: Did you not know him before, your husband I mean?

Mrs M: Yes but that was where I met him, at the dance.

I: What did they use for the dance?

Mrs M: What do you mean?

I: What music did they use?

Mrs M: Oh, you mean the instrument. It was a melodeon.

I: Can you remember the tunes, any of the tunes, any of the songs?

Mrs M: Oh yes, I can remember *A Ribhinn Oig bheil cuimhn' agad*?

I: Could you tell our listeners what that means, Mrs Macdonald?

Mrs M: It's a love song. That's what it is, a sailors' song. A love song.

I: I see. And do you think you could sing it?

And she proceeded to sing it in that frail voice. John listened to the evocation of nights on ships, moonlight, masts, exile, and he was strangely moved as if he were hearing a voice speaking to him from the past.

'I think that will be enough,' she said to the boy. He switched off the tape recorder without saying anything, put it in its case and took it away, closing the door behind him.

John said, 'You make it all sound very romantic.'

'Well, it was true about the peats.'

'But don't you remember the fights people used to have about land and things like that?'

'Yes but I remember the money they collected when Shodan was drowned.'

'But what about the tricks they used to play on old Maggie?'

'That was just young boys. And they had nothing else to do. That was the reason for it.'

There was a silence. A large blue fly buzzed in the window. John followed it with his eyes. It was restless, never settling, humming loudly with an angry sound. For a moment he nearly got up in order to kill it, he was so irritated by the booming sound and its restlessness.

'Would you like to tell me about my mother?' said John.

'What about your mother?'

'Sarah said something when I was speaking to her.'

'What did she say?'

'I felt there was something wrong, the way she talked. It was about my brother.'

'Well you know your brother was fond of the land. What did you want to know?'

'What happened. That was all.'

'Your mother went a bit odd at the end. It's quite common with old people. Perhaps that's what she was talking about. My own brother wouldn't let the doctor into the house. He thought he was poisoning him.'

'You say odd. How odd?'

'She accused your brother of wanting to put her out of the house. But I wouldn't pay any attention to that. Old people get like that.'

'I see.'

'You know your brother.'

'Yes. He is fond of land. He always was. He's fond of property.'

'Most people are,' she said. 'And what did you think of my singing?'

'You sang well. It's funny how one can tell a real Gaelic singer. It's not even the way they pronounce their words. It's something else.'

'You haven't forgotten your Gaelic.'

'No. We had societies. We had a Gaelic society. People who had been on holiday used to come and talk to us and show us slides.' The successful and the failed. From the lone sheiling of the misty island. Smoking their cigars but unable to go back and live there. Since after all they had made their homes in America. Leading their half lives, like mine. Watching cowboys on TV, the cheapness and the vulgarity of it, the largeness, the spaciousness, the crowdedness. They never really belonged to the city, these Highlanders. Not really. The skyscrapers were too tall, they were surrounded by the works of man, not the works of God. In the beginning was the neon lighting . . . And the fake religions, the cheap multitudinous sprouting so-called faiths. And they cried, some of them, at these meetings, in their large jackets of fine light cloth, behind their rimless glasses.

He got up to go.

'It's the blood, I suppose,' she said.

'Pardon.'

'That makes you able to tell. The blood. You could have seen it on my pillow three months ago.'

'I'm sorry.'

'Oh, don't be sorry. One grows used to lying here. The blood is always there. It won't allow people to change.'

'No, I suppose not.'

He said goodbye awkwardly and went outside. As he stood at the door for a moment, he heard music coming from the side of the house. It sounded American. He went over and looked. The boy was sitting against the side of the house patiently strumming his guitar, his head bent over it. He sang the words in a consciously American way, drawling them affectedly. John moved quietly away. The sun was still on the water where some ducks flew low. He thought of the headland where he was standing as if it were Marathon. There they had combed each other's long hair, the effeminate courageous ones about to die.

As he walked back he couldn't get out of his mind an article about Billy Graham he had read in an American magazine not long before. It was all about the crewcut saint, the electric blue eyed boy perched in his mountain eyrie. The Victorian respect shown by the interviewers had been, even for him with a long knowledge of American papers, nauseating. Would you like these remarks off the record, and so on. And then that bit about his personal appearances at such shows as *Laugh-in* where the conversation somehow got round to Jesus Christ every time! In Africa a corps of black policemen, appointed to control the crowd, had abandoned their posts and come forward to make a stand for Jesus!

Mad crude America, Victorian and twentieth century at the one time. Manic country of the random and the destined. What would his father or his mother have thought of Billy Graham? The fundamentalist with the stereophonic backing. For the first time since he came home he laughed out loud.

VII

It was evening when he got back to his brother's house and the light was beginning to thicken. As he turned in at the gate his brother, who must have seen him coming, walked towards it and then stopped: he was carrying a hammer in his right hand as if he had been working with posts. They stood looking at each other in the half-light.

'Have you seen everybody then,' said his brother. 'Have you visited them all?' In the dusk and carrying the hammer he looked somehow more authoritative, more solid than his brother.

'Most of them. Sarah was telling me about the cow.'

'Oh, that. There was something wrong with the cow. But it's all right now. She talks too much,' he added contemptuously.

'And also,' said John carefully, 'I heard something about our mother.'

'What about her? By God, if that bitch Sarah has been spreading scandal I'll . . .' His hands tightened on the hammer and his whole body seemed to bulge out and bristle like a fighting cock. For a moment John had a vision of a policeman with a baton in his hand. John glimpsed the power and energy that had made his brother the dominant person in the village.

After a while he said, 'I didn't want to worry you.'

'About what?' said John coldly.

'About our mother. She went a bit queer at the end. She hated Susan, you see. She would say that she was no good at the housework and that she couldn't do any of the outside work. She accused her of smoking and drinking. She even said she was trying to poison her.'

'And?'

'She used to say to people that I was trying to put her out of the house. Which of course was nonsense. She said that I had plotted to get the croft, and you should have it. She liked you better, you see.'

'Why didn't you tell me any of this?'

'I didn't want to worry you. Anyway I'm not good at writing. I can dash off a few lines but I'm not used to the pen.' For that moment again he looked slightly helpless and awkward as if he were talking about a gift that he half envied, half despised.

John remembered the letters he would get—'Just a scribble to let you know that we are well and here's hoping you are the same . . . I hope you are in the pink as this leaves me.' Clichés cut out of a half world of crumbling stone. Certainly this crisis would be beyond his ability to state in writing.

'She was always very strong for the church. She would read bits of the Bible to annoy Susan, the bits about Ruth and so on. You know where it says, "Whither thou goest I will go . . ." She would read a lot. Do you know it?'

'I know it.' John said, 'I couldn't come back at the time.'

'I know that. I didn't expect you to come back.'

As he stood there John had the same feeling he had had with Sarah, only stronger, that he didn't know anything about people at all, that his brother, like Sarah, was wearing a mask, that by choosing to remain where he was his brother had been the stronger of the two, that the one who had gone to America and immersed himself in his time was really the weaker of the two, the less self-sufficient. He had never thought about this before, he had felt his return as a regression to a more primitive place, a more pastoral, less exciting position, lower on the scale of a huge complex ladder. Now he wasn't so sure. Perhaps those who went away were the weaker ones, the ones who were unable to suffer the slowness of time, its inexorable yet ceremonious passing. He was shaken as by a vision: but perhaps the visions of artists and writers were merely ideas which people like his brother saw and dismissed as of no importance.

'Are you coming in?' said his brother, looking at him strangely.

'Not yet. I won't be long.'

His brother went into the house and John remained at the gate. He looked around him at the darkening evening. For a moment he expected to see his mother coming towards him out of the twilight holding a pail of warm milk in her hand. The hills in the distance were darkening. The place was quiet and heavy.

As he stood there he heard someone whistling and when he turned round saw that it was Malcolm.

'Did you repair the bike?' said John.

'Yes, it wasn't anything. It'll be all right now. We finished that last night.'

'And where were you today, then?'

'Down at the shore.'

'I see.'

They stood awkwardly in each other's presence. Suddenly John said, 'Why are you so interested in science and maths?'

'It's what I can do best,' said Malcolm in surprise.

'You don't read Gaelic, do you?'

'Oh, that's finished,' said Malcolm matter-of-factly.

John was wondering whether the reason Malcolm was so interested in maths and science was that he might have decided, perhaps unconsciously, that his own culture, old and deeply rotted and weakening, was inhibiting and that for that reason he preferred the apparent cleanness and economy of equations without ideology.

'Do you want to go to America?' he asked.

'I should like to travel,' said Malcolm carelessly. 'Perhaps America. But it might be Europe somewhere.'

John was about to say something about violence when it suddenly occurred to him that this village which he had left also had its violence, its buried hatreds, its bruises which festered for years and decades.

'I want to leave because it's so boring here,' said Malcolm. 'It's so boring I could scream sometimes.'

'It can seem like that,' said John. 'I shall be leaving tomorrow but you don't need to tell them that just now.'

He hadn't realised that he was going to say what he did till he had actually said it.

Malcolm tried to be conventionally regretful but John sensed a relief just the same.

They hadn't really said anything to each other.

After a while Malcolm went into the house, and he himself stood in the darkening light thinking. He knew that he would never see the place again after that night and the following morning. He summoned it up in all its images, observing, being exact. There was the house itself with its porch and the flowers in front of it. There was the road winding palely away from him past the other houses of the village. There was the thatched roofless house not far away from him. There were the fields and the fences and the barn. All these things he would take away with him, his childhood, his pain, into the shifting world of neon, the flashing broken signals of the city.

One cannot run away, he thought to himself as he walked towards the house. Or if one runs away one cannot be happy anywhere any more. If one left in the first place one could never go back. Or if one came back one also brought a virus, an infection of time and place. One always brings back a judgment to one's home.

He stood there for a long time before going into the house. He leaned over the fence looking out towards the fields. He could imagine his father coming towards him, in long beard and wearing wellingtons, solid, purposeful, fixed. And hadn't his father been an observer too, an observer of the seasons and the sea?

As he stood thinking he saw the cloud of midges again. They were rising and falling in the slight breeze. They formed a cloud but inside the cloud each insect was going on its own way or drifting with the breeze. Each alive and perhaps with its own weight, its own inheritance. Apparently free yet fixed, apparently spontaneous yet destined.

His eyes followed their frail yet beautiful movements. He smiled wryly as he felt them nipping him. He'd have to get into the house. He would have to find out when the bus left in the morning. That would be the first stage of the journey: after that he could find out about boats and trains and planes.

ALASDAIR GRAY
1934–

PROMETHEUS

It was unkind of the Jews to give the job of building the world to one man for it made him very lonely. Earlier people saw the creator as a woman giving birth, which is sore, but not sore on the head, and fulfils body and soul until the empty feeling starts. But these wandering shepherds were so used to featureless plains under a vast sky (not even the sea is vaster than the sky) that they thrust a naked man into formless void and left him there forever with nothing to remember, not even the sweetness of a mother's breast.

Roman Catholics and the English parliamentary poet Milton evade the horror of this by placing the void below a mansion where God lives in luxury among angelic flunkeys. Satan, his sinister head waiter, provokes a palace rebellion resulting in a serious staff-shortage; so God, without leaving his throne, gives orders which create a breeding and testing ground (the earth) for a new race of servants (mankind). This notion is very reassuring to people with power and to those weaklings and parasites who admire them. Most citizens with a religion really do believe that heaven is a large private property, and that without a boss to command them they would be nobody. I reject this bourgeois image of God. If God is the first cause of things then he started in a vacuum with no support and no ideas except those arising from his passions. Some commentators present the void as a sort of watery egg on which God broods like a hen until it hatches. Oh yes, why not? This sweet notion is easily reconciled with the splitting of that grand primordial atom which scientists have made so popular. But I am better than a scientist. The Jewish Genesis intoxicates me by attributing all creation to a mind like mine, so to understand God I need only imagine myself in his situation.

First, then, black void, pure and unflawed by sensations. No heat, no cold, no pressure, no extent. What is there to do? Be. Being is all that can be done. But gradually a sensation does occur, the sensation of duration. We perceive that we have been for a long time, that we will be forever in this darkness unless we do something. The more we endure of our dark self the less we can bear it. We move from boredom to unease and then to panic-horror of an eternity like this. We are in Hell. So the cry 'Let there be light' is not an order but a desperate prayer to our own unknown powers. It is also a scream rejecting everything we know by committing us to an unimaginable opposite. And there is *light*. And oh, what appalling vertigo we feel when eternity becomes infinity also and we find ourselves floating beside above beneath that dazzling blank bright breadth, height, depth with no content but ourselves. The light is too much for us, we turn to darkness again. And the evening and the morning are the first day.

Genesis says God saw the light, that it was good, but I cannot imagine him standing happily upon that boundless floor of light before he has peopled it with creatures. His first creature is water, a body compatible with his agitation and as formless as his thinking at this stage. Its sparkling movement reflects and refracts the light into every possible tone and tint, there is a rainbow in each drop of it. With this water he makes the sliding architecture of the sea and the steady, starry flood of the firmament. Unscientific? Good. I would have it so. I will skip most of the other stages. By the sixth day God is almost wholly incarnate. We taste earth and dew through a million roots, our leaves and blossoms sense and scent the air, we graze on our herbs and strike beaks into our squealing flesh while our unutterable doubt of the whole enterprise sneaks searchingly through sunlit grass in the body of the serpent. Our largest intellectual powers are almost (but not quite) realized in Adam, who kneels to study, in a puzzled way, his reflection in a quiet stream. The reflection causes a stiffening in his ureter which has to do with the attached seedballs, but the stiffening is not sufficient to impregnate the image in the water or the moist gravel under it. What other body do we need? Eve, of course, our last and most intricate creature. So Adam knew Eve, his wife, and she conceived and bare Cain. And Cain knew *his* wife, who conceived and bare Enoch, who builded a city. And after more generations of knowing and conceiving, a Seventh-Day Adventist, Joseph Pollard, cleaved to his far more liberal wife Marie, who conceived and bare myself, the poet. If your education is adequate you already know I have been paraphrasing the start of my *Sacred Sociology*, printed privately at Dijon in 1934.

My infancy resembled that of God, my ancestor. I only dimly recall the dark time before I screamed into light, but I was in that dark, like all of us, and I screamed, and there was light. I may have found the light emptier than most. My mother once told me, in an amused voice, that as a baby I screamed continually until one day they sent for a doctor. He examined me minutely then said, 'Madam, what you have here is a screaming baby.' Clearly she had never wondered what I was screaming *for*. Herself, probably. But soon my vocal chaos acquired the rhythm and colours of articulate speech and I named and commanded a child's small universe. My command was not absolute. In my tenth year Marie Pollard eloped to Algiers with one of her husband's business acquaintances. I sympathized with at least half the feelings which compelled her. Illness had made Joseph Pollard hard to live with. His fits of blinding rage destroyed a great deal of furniture and did not always spare the human body. But I am grateful to him. Paul Cézanne once said, 'My father was the real genius. He left me a million francs.' Father Pollard was not such a genius as Father Cézanne, but in my eighteenth year he freed me from himself and the curse of earning my bread by succumbing to cancer of the spleen. The consequent income did not permit me to marry, or support a housekeeper, or to frequent respectable brothels; but I silenced the desperate hunger in my young heart by studying it, and the world containing it, and by learning to read all the great sacred books in their original tongues. And I depressed my professors at the Sorbonne by finally submitting no thesis. A poet need not truck with bureaucrats.

I am shy, fastidious and arrogant. I am unattractive, but do not need friends. I am a close reasoner, and love language. My poetic vision is deep, but lacks breadth. It is the drama of their infancy which makes men poets, but the writers of the greatest divine and human comedies are men of the world, they discover and represent that drama in commonplace streets, bedrooms and battlefields. I can only represent Gods, and lonely intelligences, and multitudes viewed from a very great distance. I will never be popular. To pay the printers of the *Sacred Sociology* and *Child's Dictionary of Abstractions** I went shabby and hungry for many days and these books made no great stir. An early act of folly cured me of seeking fame in the reviews. I sent Gide the *Sacred Sociology* with a letter indicating that his protestant education had made him capable of appreciating it. He returned the copy with a seven-word comment: 'Literature cannot be founded on Larousse Encyclopedias.' His rage, when his wife burned all his letters to her, still amuses me extremely. Bravo, Madame Gide! You hoarded these scribblings as long as you believed he had no other way of making love, but thrust them in the stove when you discovered he enjoyed that passion, physically, elsewhere. You refused to be a postbox through which the great man despatched himself to posterity, bravissimo. I am the opposite of Gide. I now address the public in order to be read by one woman I can reach in no other way. Love drives me to this. Gide was driven by vanity.

I am as old as my century. In the late sixties the respectable working men who frequent the café where I dine began to be ousted by students and other members of the lower intellectual classes. This led to an increase of prices and one day I told the manageress that I could no longer afford to patronize her establishment. A shade of unease came to her face and was instantly quelled. After a moment she indicated that, to a customer of long standing, a reduction of five per cent was permissible. She was not being friendly. She had been friendly twenty years earlier, but I then made certain detailed proposals which she construed as insults. She is one of those strict atheists who determine themselves far more completely than a priest determines a good catholic. Over the years her

* M. Pollard clearly wishes to consign to oblivion his translation of Carlyle's *French Revolution* into heroic Alexandrines, published privately at Dijon in 1927.

splendid body had come to depend on the corset for its shape but I still found the sight of it entertaining; she knew this and cordially detested me. I told her that a fifteen per cent reduction might ensure my continued custom and, after quelling a distinct flicker of wrath, she agreed. I left the café proud to be a Frenchman. The change in clientele was due to myself. Though unpopular I had clearly become famous, and where else in the world would intellectual eminence receive such tactful regard? I remembered also that my mail had recently become abundant, though I only open envelopes from publishers and from the bank which manages my estate. I decided to give myself a holiday. I usually study in a small useful library containing no publications after 1765. Today, in a spirit of sheer caprice, I visited the Bibliothèque Nationale and investigated the history of my reputation.

My books had suffered from an absence of agreement upon how to regard them. In the thirties, the only period when I associated with a political movement, my support of the National Front led the surrealists and left wing generally to regard the *Sacred Sociology* as a satire against religion in the fashion of Anatole France; but Claudel called it a grand heresy revealing the truth through the agony of estrangement, Celine praised it as hilarious anti-semitic comedy, and Saint-Exupéry noticed that it did not seek to deface or replace the scriptures, but to be bound in with them. In the forties the existentialists had just begun to bracket me with Kierkegaard when I printed *A Child's Plainchant Dictionary of Abstractions*. This was thought an inept satire against dictionaries and final proof that I was not a serious thinker. Twelve years later a disciple of Levi-Strauss discovered that, though printed as prose, each definition in my dictionary was a pattern of assonance, dissonance, half-rhyme and alliteration invoking the emotions upon which words like *truth, greed, government, distaste* and *freedom* depend for their meanings. My definition of *digestion*, for example, if spoken aloud, soothes stomachs suffering from indigestion. This realization brought me the reverence of the structuralists who now used my dictionary as a text in three universities. I was often quoted in controversies surounding the American linguist, Chomsky. It seemed that among my unopened mail lay an invitation to join the French Academy and the offer of a Nobel prize for literature. There was widespread speculation about my current work. My first two books were of different kinds and I still pursued the habits of study which had produced them. It was noted that five years earlier I had begun subscribing to a journal devoted to classical Greek researches. All things considered, there was a chance that, before the century ended, my name might be attached to a Métro terminus.

I left the Bibliothèque Nationale knowing a new epoch was begun. I had become magnetic. In the café, when I raised a finger to order a Pernod, the manageress brought it, then turned her haunches on me in a manner less suggestive of the slamming of a door. The glances of the other customers kept flickering toward my meagre person in a way which showed it reassured them. I was filled with social warmth which I did not need to express, dismissing importunate journalists and research students with a blank stare or aloof monosyllable. This procedure greatly entertained the respectable working men. I came to notice a nearby face, a well-exercised face of the sort I like. The fine lines between the brows and the corners of the mouth and eyes showed it was accustomed to smiling and scowling and was often near to tears; the main expression was eager and desperate. My own face is too big for my body and bland to the point of dullness. My only lines are some horizontal ones on the brow which show I am sometimes surprised, but not often and not much. I arranged my features to indicate that if I was approached I would not be repellent, and after hesitating a moment she left her table and sat down facing me, causing a vast disturbance in the surrounding field of attention. We were silent until it settled.

She was in the mid-thirties with long, rather dry, straw-coloured hair topped by a defiant red beret. Her other clothes (baggy sweater, trousers, clogs) had been chosen to muffle rather than display her not very tall figure, which was nonetheless good. My personality is modelled upon August Dupin. I eventually said, 'There is a book in your handbag?' She opened the bag and laid a thin pamphlet on the table between us. I could not read the author's name, but it was not by me, so must be her own, or by her lover, or perhaps child. I said, 'Poems?'

She nodded. I said, 'Why do you approach someone so famous for taciturnity as myself? Have you been rejected by the more accessible celebrities?'

She said, 'In approaching you I have not been guided by reason. You are an almost complete reactionary, I ought to despise you. But as a young girl I read your dictionary, it gave me an ineradicable respect for the meanings, the colours, the whole *sense* of our language. If your talent is not dead from disuse, if you are not wholly the dotard you impersonate, you may help me, perhaps. I so much want to be a good poet.'

I am not accustomed to challenges, my usual habits prevent them. Her words released into my veins an utterly intoxicating flood of adrenalin. I gazed on her with awe and gratitude. She had turned her face sideways and tears slid down the curve of her cheek. I lifted the little book and said playfully, 'If I dislike this you will think me a dead man; if I love it there is hope for me?'

She said coldly, 'I am not a fool.'

Her proud chin was at an angle as defiant as the beret, her small nose was as tip-tilted as a sparrow's beak. The bridge may have been broken. I, who have never touched a woman tenderly in my life, longed to lift her in my arms and cradle her gently and protectingly. So much new experience, arriving so suddenly, could overstrain an old heart. I pocketed the book, climbed down to the floor and said, 'Madam, my habits are invariable, you may find me here whenever you please. I will have read your book in two days.'

For the second time in four years I allowed myself a holiday. On the sunlit pavement I was gripped by a walking frenzy. She had given me a precious part of herself (I stroked the book in my pocket) and defied me, and asked for help, and wept. I loved her, of course, and did not regret that this state, for me, would be painful and perhaps impossible. Romeo and Juliet, Anthony and Cleopatra found love painful and almost impossible. Phaedra and Medea found it quite impossible, but nobody doubts they were enhanced by it. I was enhanced by it. I am a wholly suburban Parisian who distrusts, as much as any provincial, that collection of stale imperial pie-crusts calling itself *Paris*. We require it to awe the foreigner who would otherwise menace our language and culture, but I refuse to be awed. So I was amused to find myself standing on one of the curving benches of the Pont-Neuf, my arms on the parapet, staring downstream at one of our ugliest buildings with a strong sentiment of admiration and delight. Since I was physically unable to seduce her I must persuade her to seduce me. This required me to remain aloof, while feeding her with increasingly useful parts of my mind, cunningly sauced with flattery to induce addiction. Everything depended on her poems. If they were entirely bad I would lose interest and botch the whole business. I sat down and read.

I was lucky. The person of the book was intelligent, tough, and on the way to being a good writer. She was a feminist, vivisecting her mauled sexual organs to display the damage and making the surgery icily comic by indicating, in a quiet lip-licking way, 'It's even more fun when I slice up him.' Some poems showed her embracing and embraced by a lover and unable to tell him something essential, either because of his indifference or from her fear that big truths are too destructive to be shared by a society of two. It was entertaining to see a woman in these Byronic postures but she was potentially greater than Byron. Her best work showed respect for human pain at a profounder level than sexual combat. It was spoiled by too many ideas. A good poem is a tautology. It expands one word by adding a number which clarify it, thus making a new word which has never before been spoken. The seed-word is always so ordinary that hardly anyone perceives it. Classical odes grow from *and* or *because*, romantic lyrics from *but* and *if*. Immature verses expand a personal pronoun ad nauseam, the greatest works bring glory to a common verb. Good poems, therefore, are always close to banality, above which, however, they tower like precipices. My woman avoided banality (which has, indeed, swallowed hordes of us) by turbulent conjunctions. Her book was filled with centaurs because she had not fully grasped the complexity of actual people, actual horses. Her instinct to approach me had been sound. I could teach her a great deal.

I bought a writing pad from a stationer in the Place Dauphin. It was Sunday so I entered the Louvre and fought my way through the polyglot mobs to the Maria de Medici salon, where I always feel at home. The canvases adorning this temple to female government bubble with enough good-humoured breasts to suckle a universe. My favourite painting, which always gives me a wicked thrill, shows the Italian banker's fat daughter handing over the tiller of state to her son, a boy with the clothes, rigid stance and far too solemn face of a very small adult. The ship of state has a mast with Athene beside it pointing the way, her curves compressed by armour which recalls the corsets of my manageress. The motive power of the vessel is provided by lusty women representing Prudence, Fortitude, etc., who toil at heavy oars with pained, indignant expressions which suggest that work comes to them as a horrible surprise. Unluckily there is no sofa near this painting. I settled before the canvases which show Maria's coronation inaugurating a new golden age and there, in the severe language of literary criticism, wrote the first love-letter of my life. I was inspired. I filled twenty-four pages with minute writing before closing time, then walked home, corrected them in red ink, typed them, recorrected, retyped, then sealed them in a large manilla envelope of the sort used for preserving legal documents. My heart palpitated as I inscribed her name upon it.

I had said I would read her poems within two days. For a week I extended my lunch hour beyond the normal and on the eighth day her shadow fell across the print of the book I was reading. Without raising my eyes I placed the letter between us, saying, 'You may wish to digest this in private.'

I heard the envelope torn, then looked up. She was giving the letter an attention which excluded myself and everything else. She read slowly, and some passages more than once. An hour elapsed. She slid the pages into her handbag, gave me a full, sincere smile and said, 'Thanks. You have misunderstood my work almost completely, but your warped picture of it conveys insights which I will one day find useful. Thanks.'

She tapped the tabletop with her fingers, perhaps preparing to leave. I grew afraid. I said, 'Will you now explain why, in our first encounter, you called me reactionary?'

'Your work explains that.'

'The *Sacred Sociology*?'

'Yes, for the most part.'

I sighed and said, 'Madam, it is not my custom to justify myself or criticize others. Both practices indicate insufficiency. But to you I surrender. Your poems suggest you love freedom, and want a just communism to release human souls from the bank-vaults of the West, the labour camps of the East.'

'And from the hospitals and asylums!' she cried ardently. 'And from the armies, churches and bad marriages!'

'Good. You desire a world-wide anarchic commonwealth where government may be safely left to a committee of retired housekeepers chosen by lot, like a jury.'

She smiled and nodded. I said, 'Madam, I wish that also. For centuries men have been misled by words like God, fate, nature, necessity, world, time, civilization and history: words which hide from us our cause and condition. The bourgeois say that because of these things our state can change very little, except for the worse. But these words are nothing but names for *men. We* are our God, fate, nature, necessity, world, time, civilization and history. Common men achieved these limbs, this brain, the emotions and the skills and the languages which share them. Man is the maker of every blessing we enjoy, including sunlight, for the sun would be a meaner thing without our eyes to reflect it. The fact that man is infinitely valuable—that man is essentially God—underlies every sacred code. And when I say *man is God* I refer least of all to God the landlord, God the director, God the ruler with the power to kill everyone for the good of the rest. To hell with these overpaid demiurges! My gratitude is to God the migrant labourer, the collectivized peasant, the slave of Rio Tinto Zinc and the American Fruit Company. *He* is the heavenly host whose body is broken day after day to nourish smart people like you and me. The *Sacred Sociology* tried to make news of this ancient truth. Did it fail?'

'Yes!' she said. 'It failed. The good wine of truth cannot be poured out of filthy old bottles. I will quote something. *Our largest intellectual powers are almost realized in Adam who kneels to study, in a puzzled way, his reflection in a quiet stream. It causes a stiffening in his ureter which has to do with the attached seedballs, but the stiffening is not sufficient to impregnate the image in the water or the moist gravel under it. What other body does he need? Eve, of course, our last and most intricate creation.* Ha! You have done nothing but reaffirm the old lie that a big man made the world, then created a small man to take charge of it, then begot a woman on him to mass-produce replicas of himself. What could be more perverse? You have been deceived, Mr Pollard. There is a great garden in your brain which is in total darkness. You have been taken in by the status quo of men and women and what sex is about, much as people were taken in by the Empire and the Church. You don't know *how* you have been oppressed because you have a penis.'

I was silent for a while then said, 'Correct. You have put your finger exactly upon my weakness, which is sexual. Speak of work which does not refer to the sexes. Speak of my dictionary.'

She said, 'It liberated me. My education was thoroughly religious in the worst sense but a cousin lent me Beauvoir's memoirs and your dictionary and they liberated me. At university you were my special study. Do you know how the professors use you? Not to free, but to bind. You are understood to support their systems. The students study commentaries on your book, not the book itself. I defended your assertion of the radical, sensual monosyllable. I was not allowed to complete the course.'

I nodded sadly and said, 'Yes. I support common sense with uncommon intelligence so the bourgeois have appropriated me, as they appropriate all splendid things. But my book is called A *Child's Plainchant Dictionary of Abstractions*. I wanted it set to music and sung in primary schools throughout France. Impossible, for I have no friends in high places. But if children sang my definitions with the voices of thrushes, larks and little owls they would get them by heart, they would easily detect the fools and rascals who use words to bind and blind us. The revolution we require would be many days nearer.'

'A dictator!' she cried scornfully. 'You! A dwarf! Would dictate language to the children of a nation!'

I laughed aloud. It is a rare relief when an interlocutor refers to my stature. She blushed at her audacity, then laughed also and said, 'Intellectually you are a giant, of course, but you do not live like one. You live like the English authors who all believe the highest civic virtue is passivity under laws which money-owners can manipulate to their own advantage. The second Charlemagne has made our country a near dictatorship. In Algeria, Hungary, Vietnam and Ireland governments are employing torturers to reinforce racial, social and sexual oppression. The intelligent young hate all this and are looking for allies. And you, whose words would be eagerly studied by every intelligence in Europe, say nothing.'

I said, 'I have no wish to become the everyday conscience of my tribe. Our Sartre can do that for us.'

I was sublimely happy. She saw me as a position to be captured. I longed for captivity; and if I was mistaken, and she only saw me as a barricade to be crossed, might she not, in crossing, be physically astride me for a few moments? My reference to Sartre was making her regard me with complete disdain. I raised an imploring hand and said, 'Pardon me! I have no talent for immediate events. My art is solving injustice through historical metaphor and even there I may be defeated.'

'Explain that.'

I looked directly into her eyes. I had expected sharp blue ones, but they were mild golden-brown and went well with the straw-coloured hair. I said, 'You asked for my help to become a better poet. I need yours to finish my last and greatest work. I lack the knowledge to complete it myself.'

She whispered, 'What work?'

'*Prometheus Unbound.*'

I hoped this conversation would be the first of a series lasting the rest of my life. Her curt, impetuous words, together with a haunted look, as if she must shortly run away, had led me to speak of Prometheus at least a year before I intended, but it was now too late to speak of less important things. I asked her to be patient if I told her a story she perhaps knew already. She glanced at her wristwatch then nodded.

The early Greeks (I said) believed the earth was a woman who, heated by his lightningstrokes, fertilized by his rain, undulated beneath her first offspring, the sky. She gave birth to herbs, trees, beasts and titans. The titans can be named but never clearly defined. There is Atlas the maker of space, and Cronos whom Aristotle identifies with time. There is also Prometheus, whose name means *foresight* and *torch*. He was a craftsman, and moulded men from the dust of his mother's body.

The multiplying children of earth could not leave her. She tired of her husband's lust, needing room for her family, room to think. She persuaded Cronos to castrate his dad with a stone sickle. The sky recoiled from her and *time* became master of the universe. When people came to live in cities they looked back on the reign of Cronos as a golden age, for in those days we were mainly shepherds and food-gatherers and shared the goods of the earth equally, without much warfare. But we had cyclops too, great men who worked in metal. Cronos feared those and locked them in hell, a place as far below the earth as the sky is above her. And when Cronos mated with his sister Rhea he became a cruel husband. He knew how dangerous a man's children can be and swallowed his own as soon as they were born. The earth disliked that. She advised Rhea to give her man a stone when the next child came. Time, who has no organs of taste, swallowed this stone thinking it was yet another son. The boy's mother called him Zeus and had him privately educated. When he was old enough to fight his father for the government of the universe he tricked the old man into drinking emetic wine and vomiting up the other children he had swallowed. These were the gods, and Zeus became their leader. The gods were more cunning than the titans, but less strong, and only Prometheus saw that cunning would replace strength as master of the universe. He tried to reconcile the two sides. When this proved impossible he joined the rebels.

The war which followed lasted ten years. Prometheus advised Zeus to release the cyclops from hell and when this was done they equipped the gods with helmet, trident and thunderbolt. Zeus won, of course, being supported by his brothers, by the earthmother, by the cyclops, by Prometheus and by men. What followed? The new boss of the universe confirmed his power by threatening mankind with death. Prometheus saved us by giving us hope (which allows us to despise death) and fire (which the gods wanted to keep to themselves). So Zeus punished Prometheus by crucifying him on a granite cliff. But Prometheus is Immortal. He writhes there to the present day.

'Madam,' I asked my woman, 'do these matters seem savage and remote from you? This oppressed mother always plotting with a son or daughter against a husband or father, yet breeding nothing but a new generation of oppressors? This new administration crushing a clumsy old one with the help of the skilled workers, common people and a radical intellectual, and then taking control with the old threats of prison and bloody punishment?'

She nodded seriously and said, 'It is savage, but not remote.'

I said, 'Exactly. Our political theatres keep changing but the management always presents the tragedy of Prometheus or *foresight abused*. The ancient titans are the natural elements which shape and govern us when we live in small tribes. Foresight helps us build cities which give protection from the revolving seasons and erratic crops. Unluckily these states are also formed through warfare. They are managed by winners who enrich themselves at the expense of the rest and pretend their advantages are as natural as the seasons, their mismanagement as inevitable as bad weather. In these states the fate of Prometheus warns clear-sighted people not to help the commoners against their bosses. But wherever we notice that poverty is not natural, but created by some of us unfairly distributing what the rest have made, *democracy* is conceived. The iron wedges nailing Prometheus to his rock begin to loosen. This is why the poem which presents Prometheus as a hero was written for the world's first and greatest democratic state. I mean Athens, of course.'

'Ancient Athens,' said my woman firmly, 'oppressed women, kept slaves, and fought unjust wars for gain.'

'Yes!' I cried. 'And in that it was like every other state in the history of mankind. But what made Athens different was the unusual freedom enjoyed by most men in it. When these men compared themselves with the inhabiters of the great surrounding empires (military Persia, priest-ridden Egypt, Carthage with its huge navy and stock-exchange) they were astonished by their freedom.'

She said, 'Define freedom.'

I said, 'It is the experience of active people who live by work they do best, are at ease with their neighbours, and responsible for their government.'

She said, 'You have just admitted that your free, active Athenians oppressed their neighbours and more than half their own people.'

I said, 'Yes, and to that extent they were not free, and knew it. Their popular drama, the first plays which the common memory of mankind has seen fit to preserve, shows that warfare and slavery—especially sexual slavery—are horrible things, and at last destroy the winners and the empires who use them.'

'Which means,' cried my woman, looking more like a tragic heroine with every utterance, 'that the Athenians were like our educated bourgeois of Western Europe and North America, who draw unearned income from the poor of their own and other countries, yet feel superior to the equivalent class in Russia, because we applaud writers who tell us we are corrupt.'

After a silence I said, 'Correct, madam. But do not be offended if I draw a little comfort from just one Athenian achievement: the tragic poem *Prometheus Bound* which was written by Aeschylus and is the world's second oldest play. It shows Prometheus, creative foresight, being crucified and buried by the cunning lords of this world after they have seized power. But Prometheus prophesies that one day he will be released, and tyranny cast down, and men will see their future clear. Aeschylus wrote a sequel, *Prometheus Unbound*, describing that event. It was lost, and I can tell you why.

'The democracy of Athens, great as it was, flawed as it was, tried to become an empire, was defeated, and finally failed. All the great states which followed it were oligarchies. Some, like Florence and Holland, claimed to be republics, but all were oligarchies in which poets and dramatists were so attached to the prosperous classes that they came to despise, yes really despise, the commoners. They saw them as incurably inferior, deserving a tear and a charitable breadcrust in bad times, but potentially dangerous and at best merely comic, like the grave-diggers in *Hamlet*. No doubt the rulers of states thought *Prometheus Unbound* was seditious, but it must also have annoyed educated people by showing how slavish their best hopes had become. They could no longer imagine a good state where intelligence served everyone equally. In twenty-three centuries of human endurance and pain only one hero, Jesus of Nazareth, declared that a common man was the maker of all earthly good, and that by loving and sharing with him we would build the classless kingdom of heaven. And, madam,' I told her, 'you know what the churches have made of *that* message. How cunning the winners are! How horrible!'

And I, who had not wept since I was a baby, shed passionate tears. I felt her grip my hand across the tabletop and though I had never before felt such pure grief for abused humanity, I have never felt such happiness and peace. It was a while before I could continue.

'But one day France, madam, yes our own France declared that democracy must return; that liberty, equality, fraternity are indivisible; that what the Athenians started, we will achieve. We have not achieved it yet, but the world will never know peace until we have done so. The main task of poetry today is to show the modern state the way to liberty and peace by remaking the lost verse-drama, *Prometheus Unbound*. I have completed half of it.'

She stared at me. I hoped she was fascinated. I described my play.

It starts with the supreme God (spelt with a capital G to separate him from lesser gods) standing on a mountaintop after the defeat of the titans. The sky behind him is a deep dark blue, his face and physique are as Michelangelo painted him, only younger—he looks about thirty. Round his feet flows a milky cloud and under the cloud, on a curving ridge, stands the committee of Olympus: Juno, Mars, Venus etcetera. These are the chorus. On two hills lower down sit Pan and Bacchus among the small agents of fertility and harvest: nymphs, fauns, satyrs and bacchantes with fiddles, drums, bagpipes and flutes. This orchestra makes music for the scene-changes. A dark vertical cleft divides the two hills. At the base of it is spread out a great tribe of common people who may as well be played by the audience. Their task is to enhance the play with their attention and applause until, at the end, the release of Prometheus releases them too. But at the start God's gravely jubilant voice addresses the universe while the sky grows light behind him.

He speaks like any politician who has just come to power after a struggle. Together (he tells us) we have destroyed chaos and oppression. Prosperity and peace are dawning under new rules which will make everybody happy. Even the wildest districts are now well-governed. My brother Neptune commands the sea, storms and earthquake. My brother Pluto rules the dead. Let us praise the cyclops! The powers of reason would have been defeated without the weapons they made. They have been sent back to hell, but to an improved, useful hell managed by my son Vulcan. He is employing them to make the thunderbolts I need to coerce law-breakers. For alas, law-breakers exist, hot-heads who protest because my new state is not equally good to everyone. It is true that, just now, some must have very little so that, eventually, everyone has more; but those who rage at this are prolonging sufferings which I can only cure with the help of time . . . God is interrupted here by a voice from the ridge below him. Minerva-Athene, his minister of education, or else Cupid his popular clown, points out that the recent war was fought to destroy old time yet now God says he needs time to let him do good. Yes! (cries God) for time is no longer your tyrant, he is my slave. Time will eventually show how kind I am, how good my laws are, how well I have made everything.

Throughout this speech God's nature is clearly changing. From sounding like the spokesman of a renewed people he has used the language of a law-maker, dictator, and finally creator. At his last words the cloud under him divides and floats left and right uncovering the shining black face of the earth-mother. It is calm and unlined, with slanting eyes under arched brows like a Buddha, and flat negro lips like the Sphinx. The white cloud is her hair, the ridge where the gods stand is her collarbone, the orchestra sits on her breasts, the audience in her lap. God, erect on top of her head, with one foot slightly advanced and arms firmly folded, looks slightly ridiculous but perfectly at home. When she parts her lips a soft voice fills the air with melodious grumbling. She knows that God's claim to be a creator is false but she is endlessly tolerant and merely complains instead of shaking him off. Her grammar is difficult. She is twisting a huge statement into a question and does not divide what she knows into separate sentences and tenses.

EARTH
Who was before I am dark
without limbs, dancing, spinning
space without heat who
was before I am alight
without body, blazing, dividing
continents on rocking mud who
was before I am breathing

without eyes, floating, rooted
bloody with outcry who
was before I am a singing
ground, wormy-dark, alight
aloud with leaves, eyes and
gardeners, the last plants I grew who
uplifted you who
was before I am?

GOD
Who is before you now!
I grasp all ground, mother.
The gardeners you grew were common men, a brood
too silly and shapeless to be any good
outside my state, which has made them new.
They cannot remember being born by you.
They are my image now. I am who
they all want to obey, or if not obey, be.

EARTH
Not Prometheus.

GOD
Yes, Prometheus! Punishment is changing him
into a cracked mirror of me.

There is a sudden terrible cry of pain. Two great birds with dripping beaks fly out of the cleft between the earth's breasts. Light enters it and shows the crucified Prometheus, a strong man of middle age with a bleeding wound in his side. Though smaller than his mother he is a giant to the God who stands high above him and declares that this is the end of the titan who made men, and made them hard to govern, by giving them hope of better life. The great mother, with a touch of passion, tells God that though he is supreme he is also very new, and his state will perish one day, like all states, and only Prometheus knows how. God does not deny this. He says he has a lot of work to do and will reconsider the case of Prometheus when he has more time. He turns and goes down behind the earth's head. The cloud closes over her. Prometheus, twisting his face up, asks the gods on the ridge to tell him the present state of mankind. They sing a chorus describing the passage of over two thousand years. Men combine into rich empires by many submitting to a few. They discover the world is vaster than they thought, and add new realms to tyranny. Liberators are born who create new religions and states, and the rulers of the world take these over and continue to tighten their grip. At last human cunning grasps, not just the world but the moon and the adjacent planets, yet half mankind dies young from bad feeding, and young courage and talent are still warped and killed by warfare. The controllers of the world fear the people under them as much as each other, and are prepared to defend their position by destroying mankind and the earth which bore them. This is the final state to which we have been brought by cunning without foresight. Prometheus cries out, 'This cannot last!' From the middle of her cloud this cry is repeated by the great mother, then by the chorus and orchestra, and then (the cloud clearing) by God himself, who stands on the height with his arms flung sideways in a gesture which resembles the crucified Prometheus. God is also now a middle-aged man. He walks down from the height, sits on the edge of the cleft and tries to engage Prometheus in friendly conversation. He is sorry he punished Prometheus so harshly and promises not to set vultures on him again. When he came to power he had to be harsh, to keep control. People needed strong government, in those days, to drag them out of the idiocy of rural life. But the whole world now belongs to the city states. He is sure Prometheus knows that neither of them is completely good or completely bad, and have a lot to give each other. If they cooperate they can save mankind. He asks Prometheus for the secret of the force which will destroy him. Prometheus asks to be released first. God is sorry, but he cannot release Prometheus. If he did Prometheus would seize power.

GOD
I am not the stark power who chained you here.
I am softened by what you endured, while my laws
have made you a hard reflection of the tyrant I once was.
It cannot be right to enthrone
a killing revenge the world should have outgrown,
or if right, then right will make greater wrong.

PROMETHEUS
It is right to give back what you stole—liberty.
You see me as I am. You cannot see
who I will become when I am free.
Why do you think I will kill?

GOD
Your every glance threatens me terrible ill.

PROMETHEUS
I am in pain! My illness, the illness you dread,
is yours, is you!

GOD
Then endure my terrible nature!
I must endure it too.

(*God has lost his temper. Prometheus laughs bitterly.*)

PROMETHEUS
At last you unmask, old man
and show what you are again:
the ruler of a kingdom kept by pain.

All history has added nothing to you
but a mad wish to be pitied for what you do.

I paused. My woman said, 'What happens then?' I said, 'I cannot imagine.'

She started laughing. I said, 'To end happily my play needs a new character, someone we have already seen, without much interest, in the chorus, or even audience. The action until now is between a man, a big woman, and another man. To strike a balance the fourth character must be a woman. She is a new wisdom who will unite our imprisoned intelligence with the productive earth, reducing government from a form of mastery to a form of service. She is sensuous, for both governments and rebels keep asking us to crush our senses in order to gain an ultimate victory which never arrives. But she is not disorderly, not a beatnik, not careless. She is living proof that when our senses are freed from fear our main desire is to make the world a good home for everyone. I cannot conceive such a heroine. Can you conceive her? Could we conceive her together?'

My woman looked thoughtful then said slowly, 'I am qualified to assist you. I have been a daughter and a mother, a victim and a tyrant. I saw my father torment his wife into her grave. I have driven a man to suicide, or very nearly. I know how love heats and warps us, but I feel there is still hope for me, and the world.'

I said, 'That indicates a kind of balance.'

'I have climbed mountains in Scotland and Germany. I have swum underground rivers in the Auvergne.'

I said, 'That also indicates balance, but a balance of extremes. The tension you feel must be nearly unbearable. We must connect the extremes where you squander so much energy with the centre where my knowledge lies chained and stagnant.'

Her mouth and eyes opened wide, she raised her chin and gazed upward like the Pythoness on the tripod when Apollo enters her. For nearly a minute she became pure priestess. Then her gaze shrank, descended and focussed on the table where my great, droll, attentive head rested sideways on my folded arms. A look of incredulity came upon her face. I had never before seemed to her so improbably grotesque. She pretended to glance at her wristwatch, saying, 'Excuse me, I must go.'

'May I write to you, madam? A literary collaboration is perhaps best prosecuted by letter.'

'Yes.'

'Your address?'

'I don't know—I am moving elsewhere, I don't know where yet. I have many arrangements to make. Leave your letters with the management here. I will find a way to collect them.'

I said, 'Good,' and achieved a smile. She arose, came to my side and hesitated. I signalled by a small headshake that condescension would be unwelcome. She turned and hurried out. I sat perfectly still, attending to the beaks of the vultures tearing at my liver. They had never felt so sharp. The manageress came over and asked if I felt well? I grinned at her and nodded repeatedly until she went away.

After that I waited. I could do almost nothing else. Study was impossible, sleep difficult. I addressed to her a parcel of worknotes for *Prometheus Unbound* and it lay on the zinc beside the till, but I was always sitting nearby for I wanted not to leave the only place where I might see her again. I waited a day, a week, three weeks. I was dozing over my book and glass one afternoon when I grew conscious of her talking to the manageress. She seemed to have been doing it for some time. They frowned, nodded, glanced towards me, shrugged and smiled. I was very confused and prayed God that when she sat facing me I would be calm and firm. She patted the manageress's arm and walked straight out through the door. I screamed her name, scrambled down from the chair, charged into the crowded street and ran screaming to the right, banging against knees, treading on feet and sometimes trodden on. Not seeing her I turned and ran to the left. As I passed the café door I was seized and lifted, yes, lifted up by one who held her face to mine so that our noses touched, and whispered, 'Mister Pollard, this conduct does you no good. I have a letter.'

I became very icy and hissed, 'Put me down, madam.'

I should have asked to be taken home. I could suddenly hardly walk. I got to my table and opened the letter, noticing that my parcel lay uncollected beside the till.

My dear friend,

I no longer wish to be a poet. It requires an obsessional balancing of tiny phrases and meanings, an immersion in language which seems to me a kind of cowardice. As a man and poet I can respect you but only because you are also a dwarf. For people of ordinary health and height, with a clear view of the world and a wish to do well, it is a waste of time making signboards pointing to the good and bad things in life. If we do not personally struggle towards good and fight the bad, people will merely praise or denounce our signs and go on living as usual. I must make my own life the book where people read what I believe. I decided this years ago when I became a socialist, but I still grasped, like a cuddly toy, my wish to be a poet. That wish came from the dwarfish part of me, the frightened lonely child who hoped that a DECLARATION *would bring the love of mother earth, the respect of daddy god, the admiration of the million sisters and brothers who normally do not care if I live or die. Your critical letter had an effect you did not intend. It showed me that my declarations are futile. It has taken a while for the message to sink in. I am grateful to you, but also very bitter. I cannot be completely logical.*

My sweet, you are the cleverest, most deluded man I ever met. Rewriting PROMETHEUS UNBOUND *is like rewriting* GENESIS, *it can be done but who needs it? It is just another effort to put good wine in a filthy old bottle. I was touched when you poured over me your adolescent enthusiasm for ancient Athens but I also wanted to laugh or vomit. I am educated. I have been to Greece. I have stood on the Acropolis facing the Erichtheon and can tell you that Greece represents:*

men	*against*	*women*
war	"	*peace*
business	"	*play*
intellect	"	*emotions*
authority	"	*anarchy*
hierarchy	"	*equality*
discipline	"	*sensuality*
property-inheritance	"	*sexuality*
patriarchy	"	*everything*

Yet you see civilization as an unfinished story the Athenians started and which a few well-chosen words will help to a satisfactory finish! You are wrong. The best state in the world was that primitive matriarchy which the Athenians were foremost in dismantling. Men were happy and peaceful when women ruled them, but so naturally wicked that they turned our greatest strength (motherhood) into weakness by taking advantage of it and enslaving us. Men have made hell of the world ever since and are now prepared to destroy all life in it rather than admit they are wrong. Masculine foresight cannot help our civilization because it is travelling backward. Even our enemies realize this. In the last fifty years they have driven us to the brink of the dark age. The rational Greek foundation of things has been unbuilt, unlearned. And you did not notice! My poor dwarf, you are the last nineteenth-century romantic liberal. That is why a corrupt government wishes to make you a national institution.

Which brings me, beloved, to what you really *want from me: cunt. In your eyes it probably looks like an entrance to the human race. Believe me, you are human enough without. No good was ever done by those who thought sexual pleasure a goal in life. I speak from experience. I divorced a perfectly nice husband who could only give me that stultifying happiness, that delicious security which leads to nothing but more of itself. But if you require that delight you can have it by merely relaxing. As a national institution—a blend of tribal totempole and pampered baby—you are ringed by admirers you have so far had the sense and courage to ignore. Weaken, enjoy your fame and get all the breasts you want: except mine. When I first spoke to you I accused you of impersonating a dead man. That was jealousy speaking. I admired you then and I regret I unhinged you so easily. I did not want to do that. I love you, but in a way you cannot perceive and I cannot enjoy. So I also hate you.*

I am a monster. The cutting words I write cut my heart too. I am under unusual strain. I am about to do something difficult and big which, if discovered, will end my freedom forever. My friends will think me insane, an unstable element, a traitor if they learn I have told you this. But you love me and

deserve to know what I am leaving you for, and I do trust you, my teacher, my liberator.
Adieu.

Is printing the above letter for the world to read a betrayal of her trust? Is a secret police computer, as a result of this story, stamping the card of every female, blonde, brown-eyed, snub-nosed poet with a number which means *suspect political crime investigate*? No. This story is a poem, a wordgame. I am not a highly literate French dwarf, my lost woman is not a revolutionary writer manqué, my details are fictions, only my meaning is true and I must make that meaning clear by playing the wordgame to the bitter end.

Having read the letter I sat holding it, feeling paralysed, staring at the words until they seemed dark stains on a white surface like THIS one, like THIS one. I was broken. She had made me unable to bear loneliness. And though we had only met twice I had shown the world that women could approach me. I sat at the table, drinking, I suppose, and in the evening a girl sat opposite and asked what I thought of de Gaulle's latest speech? I asked her to inform me of it. Later we were joined by another girl and a young man, students, all of them. It seemed we were on the brink of revolution. I ordered wine.

'Tomorrow we will not protest, we will occupy!' said the young man.

'You must come with us, Mister Pollard!' cried the girls, who were very excited. I agreed and laughed and bought more wine, then grew enraged and changed my position. I quoted Marx to support de Gaulle and Lenin to condemn the students. The uselessness of discourse became so evident that at last I merely howled like a dog and grew unconscious. And awoke with a bad headache, in darkness, beside a great soft cleft cliff: the bum of my manageress. I had been conveyed into her bed. I was almost glad.

In the morning she said, 'Mister Pollard, you know I have been a widow for seven months.'

I said nothing. She said, 'Some years ago you made to me certain detailed proposals which, as a respectable, newly-married, very young woman I could not entertain. What you suggested then is now perfectly possible. Of course, we must marry.'

Lucie, you have made me need you, or if not you, someone. Lucie, if you do not return I must fall forever into her abyss. Lucie, she makes me completely happy, but only in the dark. Oh Lucie Lucie

Lucie save me from her. The one word this poem
exists to clarify is *lonely*. I am Prometheus.
I am lonely.

William McIlvanney

1936–

Performance

Fast Frankie White didn't go into a bar. He entered. He felt his name precede him like a fanfare he had to live up to. As with a lot of small criminals, he had no house of his own, no money in the bank, no deposit account of social status to draw on. He had no fixed place in the scheme of things that could feed back a clear sense of himself, be a mirror. His only collateral was his reputation, a whiff of mild scandal that clung round him like eau-de-Cologne.

Being an actor, he needed applause. His took the form of mutterings of 'Fast Frankie White' most places he went because he chose to go places where they would mutter it. Without that reminder of who he was, he might forget his lines. His favourite lines were cryptic throwaways that reverberated in the minds of the gullible with vaguely dark potential.

'Been doin' a wee job,' he said.

'Checkin' out a couple of things,' he said.

'A good thing I don't pay income tax.'

'This round's on the Bank of Scotland.'

'He's got his own style, Frankie,' some people said. But that was a less than astutely critical observation. It was really a lot of other people's styles observed from the back row of the pictures, a kind of West of Scotland American. Once, when he was twenty, he had seen a Robert Taylor film about New York where some people were wearing white suits. He had snapped his fingers and said, 'I'm for there!' A few weeks later, by a never-explained financial alchemy, he was. A few weeks later, he was back but he liked to talk about New York. 'There's half-a-million people in the Bronx,' he would say. 'And most of them's bandits.' It wasn't Fodor's Guide to the USA but it sounded impressive, said quickly. And Frankie said everything quickly.

'The Akimbo Arms' was one of the pubs where he liked to make his entrance. He was originally from Thornbank, a village near Graithnock, but he lived in Glasgow now, people said. Frankie didn't say where he lived. He would simply appear in a Graithnock bar, dressed, it seemed, in items auctioned off from the wardrobe department of some bankrupt Hollywood studio and produce a wad of notes.

This time he was wearing a light blue suit, pink shirt, white tie and grey shoes. He looked as if he had stepped out of a detergent advert.

'Where's ma sunglasses?' somebody said.

But Frankie was already flicking a casual hand in acknowledgement of people who didn't know who he was. He walked round to the end of the bar where he could have his back against the wall, presumably in case the G-men burst in on him, and he prepared to give his performance.

It was a poor house. Matinées usually were. Mick Haggerty was standing along the bar from him, in earnest debate with an unsuspecting stranger, who probably hadn't realised Mick's obsession until it was too late.

'Give me,' Mick was saying, 'the four men that've played for Scotland an' their names've only got three letters in them.'

Frankie hoped for the stranger's sake that he didn't get the answer right. Doing well in one of Mick's casual football quizzes was a doubtful honour, earning you the right to face more and more obscure questions the relevance of which to football wasn't easy to see. 'Tell me,' Frankie had once said to Mick by way of parody. 'In what Scotland–England game did it rain for four-and-a-half minutes at half-time? And how wet was the rain?'

Over in the usual corner Gus McPhater was sitting with two cronies. Frankie hated the big words Gus used. That left Big Harry behind the bar, besides three others Frankie didn't know. Big Harry had finally noticed him and was approaching with the speed of a mirage.

'Frankie,' Big Harry said.

'Harry. I'll have a drop of the wine of the country.'

'Whit?'

'A whisky, Harry. Grouse. And what you're havin' yourself?'

Big Harry turned down the corners of his mouth even further. He looked at Frankie as if dismayed at his insensitivity.

'Me?' Big Harry said. 'Ye kiddin'? Wi' ma stomach? Ye want a death on yer conscience? Still.' His face assumed a look of martyred generosity. 'Tell ye what. Ah'll take the price of it an' have it when Ah finish. Probably no' get a wink of sleep the night. But ye've got to get some pleasure.'

Frankie remembered Harry's nickname—Harry Kari. He wasn't sure whether the nickname was because that was what everybody felt like trying after a conversation with Harry or because that was what people thought Harry should do. No wonder Gus McPhater was quoted as saying, 'Harry does for conversation what lumbago does for dancin'.' Harry was the kind of barman who told you *his* problems.

'Religion?' Gus McPhater was saying. He was always saying something. 'Don't waste ma time. The opium of the masses. It's done damage worse than a gross of atomic bombs. Chains for the brainbox, that's religion. Ministers? Press agents for the rulin' classes. Ye'll no' catch me in a church. If Ah could, Ah'd cancel ma christenin' retrospectively. Take yer stained-glass windaes. Whit's a windae for? To see through. Right? So what do they do? They cover it in pictures. So that when ye look at the light. The light, mind ye. That's how ye see, ye know. Light refractin' on yer pupils. When ye look for the light, it gets translated intae what they want ye tae see. How's that for slavery? An' whit d'ye see? A lot of holy mumbo-jumbo. People Ah don't know from Adam. What've a bunch of first-century Jewish fanatics got to do wi' me? Ah'll tell ye what. Know when Ah'll go intae a church? When it's man's house. When the stained-glass windaes are full of holy scenes of rivetters in bunnets and women goin' the messages wi' two weans hangin' on to their arse an' auld folk huddled in at one bar of an electric fire after fifty years o' slavin' their guts out for a society that doesny care if they live or die. Those would be windaes worth lookin' at. That's what art should be. Holy pictures of the people. Or a mosaic even. How about that? See when they made that daft town centre. The new precinct. The instant slum. See instead o' that fountain. Why not a big mosaic? Showin' the lives of the people here an' now. How about that? The Graithnock mosaic. Why no?'

Frankie had no desire to join in. He contented himself with a mime of his superior status. Gus McPhater depressed him. People listened to him as if the noises he made with his mouth meant something. He was a balloon. A lot of stories were told about him. He was supposed to have travelled all round the world. He was supposed to be writing a novel or short stories or something. Frankie didn't believe any of them.

Gus seemed to Frankie an appropriate patron saint for Graithnock. He was like the town itself—over the hill and sitting in dark pubs inventing the past. Frankie could remember this place when the industry was still going strong. There had been some vigour about the place then. They were all losers now—phoneys, like Gus McPhater.

Frankie couldn't believe this place. The only kind of spirit in it was bottled. He felt like an orchid in a cabbage-patch. Where was the old style, the old working-class gallousness? Since the Tory government had come to power, it had really done a job on them, slaughtering all the major industry. They believed they were as useless as the government had told them they were. These men were the cast-offs of capitalism. They were pathetic.

Well, he was different. If the system was trying to screw him, he would screw it. He had his own heroes and they weren't kings of industry. He thought of McQueen. He wondered how long it would be before McQueen got back out. McQueen, there was a man. He was more free in the nick than most men were outside it.

That was what you had to do: defy your circumstances. You were what you declared yourself to be. Frankie looked round the bar and made a decision. He would buy a drink for someone. He pulled his wad of money from his pocket. In the flourish of the gesture he became a successful criminal.

He decided on Gus McPhater's group. His distaste for them somehow made the gesture grander. He felt like Robin Hood giving the poor a share of his spoils. Besides, Gus was a great talker. Buying him a drink was as good as a photograph in the paper. They would know he had been. He threw a fiver on the counter.

'Harry,' he said loudly. 'Give Gus an' the boys whatever they want.'

He noticed a boy who was drinking alone watching him interestedly. It was all the encouragement Frankie needed. He made an elaborate occasion of getting the drinks and taking them over to Gus's table. He dismissed their thanks with a wave. He took his change and put some of the silver into the bottle where they collected for the old folks. The whole thing became a mini-epic, a Cecil B. De Mille production called 'The Drink'.

'Okay,' Frankie said, saluting the room. 'Don't do anythin' Ah wouldn't do. If ye can think of anythin' Ah wouldn't do.'

As he went out, he heard the boy asking, 'Who *is* that?'

Stepping into the street, he felt the gulped whisky sting his stomach. It was a twinge that matched the bad feeling the pub had given him. Hopeless, he thought. But maybe he was wrong. He remembered the admiration on the boy's face as he had asked who Frankie was. Frankie lightened his step and started to whistle.

A good actor never entirely knows the impact he is having. Perhaps in the thinnest house, unnoticed beyond the glare of the actor's preoccupation, a deep insight is being experienced or young ambitions being formed for life.

He would try 'The Cock and Hen'. There might be some real people in there. He side-stepped into a shop doorway and checked his wad of money. He had three fivers left and he repositioned them carefully to make sure they were concealing the packing of toilet paper inside that made them look like a hundred. He would try 'The Cock and Hen'.

ERIC MCCORMACK

1940–

THE ONE-LEGGED MEN

You're on their ground now. Don't be fooled by their looks. Watch out for the following: abnormal frequency of stumbling over kerbs and door-steps, studied balance in walking, reliance upon arm-strength in getting up out of chairs, stiffness in lower limbs, hesitancy in picking up dropped things. A sure sign is one sharp crease and one dull in a pair of trousers. You can often go by shoes: in a suspect pair, each will seem well used, but one of them not by any natural foot. Remember this: whenever you note great animation and expressiveness in conversation by the use of hands and upper body, you're probably on to something.

From 'A Description of Muirton', James R. Ross, Ayrshire Today *(Glasgow, 1950)*

Muirton, East Ayrshire, nestles in a treeless valley beyond which the Lead Hills stutter towards the Border with England. It is one of those villages that cluster around a coal-mine. The mine itself looks, to the visitor, like a suppuration from the bare, but otherwise healthy flesh of the timeless moorlands. Yet, in the usual paradoxical way, the huge slag-heaps of these ugly mines that abound in this part of Ayrshire may also be likened to breasts that nourish a breed of men and women capable of great physical tenacity and intellectual vigour. Like the scrawny heather that grows all around, these people have their seasons of beauty.

The boys of Muirton don't like school. The boys of Muirton don't like poetry. The boys of Muirton do like: playing soccer; trout-fishing; pitch-and-toss. So far as talking goes, they like: talking about girls; talking about getting a job in the mine when they're fifteen. The six-times-a-day hoot of the mine's siren is like the Pied Piper's tune to them. It calls, coaxes, cajoles them into an exclusive abyss.

From Annals of the Modern Ayrshire Parish, *Rev. James O'Connell (Dumfries, 1979)*

The day that brought Muirton fame for its excess of one-legged inhabitants occurred during my first year as minister there. It was a day not blessed in its beginnings so far as weather was concerned. A murky morning carried the threat of rain—the Lead Hills, so appropriately named, in my view, were hidden in mist—even though summer had ostensibly begun. June in Muirton, I was to learn, is only slightly warmer than January. The day-shift workers, many of whom were my parishioners, set out along the shrub-lined roadway to the mine at about six o'clock. There were fathers, sons, uncles, nephews, cousins, in truth 'by the dozens'. On their way, they passed the homeward-bound night-shift workers, all black-faced, even the young boys—God bless them—bound for dinner and bed, their world topsy-turvy. No one, of course, had any premonition of the trial some of them were to undergo that very day in fulfilling the mysterious will of their Maker.

The men and boys of Muirton. They stand in clusters at the gates of the mine elevator. The gloomy tower creaks eerily above them. A giant Ferris wheel. They await the daily, commonplace, infernal descent.

From Coalminer to Honourable Member, *Tom Kennedy, MP (Glasgow, 1946)*

The first few times you went down in the cage it scared the wits out of you. Your stomach never seemed to get back to its usual place. Some of the new boys used to wet their trousers and everybody would make wisecracks about the poor lads for many weeks afterwards. But we all got used to the cage eventually, and most of the time it never crossed your mind how far down you were going—1,000 or 10,000 feet, it didn't make any difference. As you went deeper, the air used to get warmer. Even in the middle of winter, the sweat would be lashing off you.

What I miss now—I wish we had some of it here in the House of Commons—is the great spirit we used to have in those days.

This one jokes, this one laughs, this one yawns, this one spits, this one whistles, this one day-dreams, this one scratches his neck, this one turns his head and looks up to the shroud of hills. The final squad entering the cage.

From The Ayr Daily News, *July 4*

MUIRTON HORROR UPDATE

No one knows yet for sure how the disaster happened that shook this little Ayrshire town yesterday morning. One thing is certain, hardly a family has been spared.

Experts from the National Coal Board are examining at this very instant the wrecked elevator cage at the bottom of the shaft to determine why the braking mechanism failed. They expect to have a report ready for next month's Inquiry. However, they reconstruct the tragedy as follows:

At approximately 6:40 a.m., forty men and boys, the final day-shift squad, entered the cage for their descent to the coal face. Some time in the next few minutes as the cage descended to the 4,000 feet level, they would hear the alarm bell and see the flashing red light, warning them that the cage was out of control.

At this point they would take the normal precautionary measures. Inspector David McCann, 54, of the Miners' Safety Board, describes these: 'The miners are drilled once a week in safety procedures. In the case of a cage going out of control they all know what to do. They have to reach up and grasp firmly one of the leather hand-straps attached to the roof of the cage. Then they lift one leg off the floor.

'In this way, with any luck, depending on how far the cage has to fall, they might save one leg when they hit the bottom even though the cage comes down quite hard.'

This afternoon I met two of the rescue team who were present when the smashed cage was opened. They were, Orderly John McCallum, 43, and Orderly Tom McLeary, 29. McLeary had no comment to make, but McCallum gave me this brief reaction:

'It's the worst thing I've seen in years at this job. It's hard to believe anybody survived it. The cage was all twisted so we had a very hard time with the acetylene torches getting inside it. We could hear noises so we knew some of them were still alive. When we got the front off, it wasn't easy to know what to do, it was such a sight.'

It is now established that there were thirteen survivors out of the forty, so it seems that the safety measures worked in part. At the point of impact, the leather handstraps absorbed some of the shock, but then snapped—that was to be expected. The rest of the body-weight was then transferred completely to the supporting leg. That leg was smashed immediately to pulp.

The Jury is Out. The Jury is In. Boards of Inquiry, Commissions of Inquiry, Industrial Tribunals, Insurance Adjustments, Compensation Committees, Medical Examinations, Burdens of Proof, Amendments to Safety Subsections, Revised Regulations, Compulsory Hard Hats.

From 'Your Letters', The Edinburgh Times, *November 6*
Sir,
I would like to add some pertinent data to your recent correspondence on the issue of the physiological appearance of any part of the human anatomy upon its being smashed to a pulp.

As one of the physicians who examined the victims of the recent Muirton disaster, I feel qualified to comment with some authority upon the effects on the lower limbs of what is now called 'the pulping phenomenon'.

I would like, first, to dispel the speculative notion that pulped limbs after a massive fall look, as one of your correspondents hypothesised, 'like minced beef à la Sweeney Todd'. Nonsense. Aside from the inordinate amount of blood on the cage floor at Muirton, the signs were much less obvious.

Let us consider for example the pulped limbs of the thirteen survivors. It appears that at the moment of impact, the femur, patella, tibia, and fibula disintegrated, though, in one case, the femur was pushed up into the lower intestinal tract causing severe, but not fatal injury.

But, after the disintegration of the bones, the limbs appear to have sprung back into their accustomed positions with skin and muscle seemingly intact. At first glance, therefore, it was impossible to assess the damage. It was when the Orderlies tried to move the bodies they discovered that they were attempting to lift limbs that were completely gelid. Immediate amputation was, of course, absolutely necessary.

I might add that, in the cases of the dead miners, one or both legs of each had been similarly traumatised. Why death resulted instantly in their cases, even though their upper bodies remained quite functional, is worthy, surely, of some research.

I trust these remarks will be of help in settling the issue,

Sincerely,
J. Blair, MD

Tell me, do you know? can you say? would you mind? could you try? got the message? get the picture? walking on four feet, and two feet, and one foot, and three feet, the fearful, fearful beast.

Bits and Pieces, Odds and Ends

1. A survivor, one year later: 'I'm going to carry on living in Muirton. At least when I'm here, people don't look at me as if I'm a freak.'

2. A Muirton children's skipping song:

Skip me one,
Skip me two,
Tell me what does your daddy do?
He lies in his bed,
And eats fried egg,
Because he's only got one leg.

3. Stephen Neil, Headmaster of Muirton Primary School: 'I know the families of most of the survivors. I get the impression that some of the men and boys have never recovered: they're under psychiatric care for constant nightmares and depression. And impotence.'

4. A survivor, one year later: 'Did you ever hear tell that a drowning man is supposed to see his whole life passing before him? I often say to the wife that's what happened to me as we went down. I even saw my grandmother, and she's been dead for twenty years now.'

5. A survivor, one year later: 'I never thought I'd have to do it. Even when we practised the leg-drill, we all used to laugh, because we thought, "if we ever have to rely on this we'll know we're dead men." Big Jock McCutcheon used to say, "Do they think we're all daft? Standing on one leg like a chicken is no way to die."'

6. A survivor, one year later: 'I couldn't make up my mind which leg to give up.'

7. A member of the rescue team: 'I think they all hoped they were dead.'

SHENA MACKAY

1944–

VIOLETS AND STRAWBERRIES IN THE SNOW

As he lay reflecting on the procession of sad souls who had occupied this bed before him, the door burst open with an accusing crack.

'You know smoking is forbidden in the dormitories!'

'I'm terribly sorry, nurse. I must have misread the notice. I thought it said, "Patients are requested to smoke at all times, and whenever possible to set fire to the bedclothes."'

In the leaking conservatory which adjoined the lounge, puddles marooned the pots of dead Busy Lizzies and the brown fronds of withered Tradescantias, and threatened with flood the big empty doll's house that stood incongruously, and desolate, with dead leaves blown against its open door; a too-easy metaphor for lost childhoods and broken homes and lives. At seven o'clock in the morning in the lounge itself, the new day's cigarette smoke refreshed the smell of last night's butts, whose burnt-out heads clustered in the tall aluminium ashtrays. A cup, uncontrollable by a shaking hand, clattered in a saucer. The Christmas-tree lights were winking red and green and yellow and blue, and on the television, creatures from another sphere were sampling mince pies and sipping sherry in an animated consumer guide to the delights of the worst day of the year. There was port and wine and whiskey and whisky too, and Douglas Macdougal sat among the casualties of alcohol and watched what once would have been his breakfast vanish down the throats of those to whom nature, or something, had granted a mandate or dispensation, those who were paid in money and fame as well as in the satisfaction, which had brought a virtuous glow to their cheeks, that they were imbibing in the national interest. There were no saucer-like erosions under their eyes, no pouchy sacs of unshed tears; and in subways and doorways, on station forecourts and in phoneboxes, in suburban kitchens a thousand bottles clinked in counterpoint. Cheers.

'You were as high as a kite last night when they brought you in.' The man seated on his left pushed a pack of cigarettes towards Douglas.

'Well I've been brought down now. Somebody cut my string. Or the wind dropped.'

A coloured kite crashed to earth; a grotesquely broken bird among the ashtrays and dirty cups, trailing clouds of ignominy.

Although a poster in the hall showed a little girl, her face all bleared with tears and snot, the victim of a parent's drunkenness, it became apparent that not everybody was here for the same reason. A woman with wild, dilated bright eyes glided back and forward across the room, as if on castors, with a strange stateliness, passing and repassing the television screen, and from time to time stopping to ask someone for a cigarette, from which she took one elegant puff before stubbing it out in the ashtray and continuing her somnambulistic progress. No one refused her a cigarette; Douglas had noticed already a kindness towards one another among the patients, and no one objected when she blocked the television screen where peaches bloomed in brandy and white grapes were frosted to alabaster. No one was watching it. The inmates sat, bloated and desiccated, rotten fruit dumped on vinyl chairs, viewing private videos; reruns of the ruins they had made of their lives, soap operas of pain and shame, of the acts which had brought them to be sitting between these walls bedecked with institutional gaiety, or fastforwarding to scenes of Christmases at home without them; waiting for breakfast time, waiting for the shuffling queue for medication.

To his right a woman was crying, comforted by a young male member of staff.

'Just because a person hears voices in her head, it doesn't give anyone the right to stop them being with their kids on Christmas Day.'

'But I'm sure they'll come up to see you, Mary.'

'But I won't be there when they open their presents . . .'

'But they'll come to see you, I'm sure . . .'

Her voice rose to a wail, 'But they don't like coming here!'

The tissue was a dripping ball in her hand. He patted her fist.

'They'll be coming to see you, Mary—it's Christmas.'

Douglas felt like screaming, 'She knows it's fucking Christmas, that's the point, you creep!' but who was he to say anything?

'You don't understand,' she said, and male, childless, half her age, an adolescent spot still nestling in the fair down on his chin, how could he have understood?

Douglas was shaking. He didn't want any breakfast. Although the routine had been explained by two people he hadn't been able to take it in. He was afraid to go into the kitchen where the smell of dishcloths mingled with the steam from huge aluminium kettles simmering on an old gas cooker. He hovered in the doorway for a minute, taking in the plastic tub of cutlery on the draining board, the smeary plastic box of margarine, the cups and plates, inevitably pale green, that belonged to nobody. *Timor mortis conturbat me*. He had been saying that in the ambulance, but mercifully could remember little else. A faint sickly smell clung to his shirt. He had refused to let them undress him, clinging to a spurious shred of vomity dignity that was all he felt he had left, aware of his bloated stomach, and had slept in his clothes. Up and down staircases, down windowless corridors whose perspectives tapered to madness, past toothless old men who mimed at him asking for cigarettes, repassing the women with heavy-duty vacuum cleaners, past the closed but festive occupational therapy unit with a plate of cold and clayey mince pies on its windowsill, past the locked library, he ranged on his aching legs, until at last he found a bathroom. As he washed himself, and the front of his shirt and his stubbly face, avoiding the mirror, the words of a song doubled him up with pain, 'Oh Mandy, will you kiss me and stop me from shaking . . .' He used to sing 'Oh Mandy, will you kiss me and stop me from shaving . . .' when his daughter ran into the bathroom and he picked her up and swung her round and dabbed a blob of foam on her nose. If he had had a razor now he would have drawn it across his aching throat, across the intolerable ache of remembered happiness.

Downstairs again he was given some coloured capsules in a transparent cup, and then it seemed that his time was his own. It was apparent that, a long time ago, a severely disturbed patient had started to paint the walls with shit and the management had been so pleased with the result that they had asked him to finish the job, and then had been reluctant to break up the expanse of ochreous gloss with the distraction of a lopsided still life painted in occupational therapy and framed in dusty plywood or even one of the sunny postcards which are pinned to most hospital walls, exhorting the reader to smile. The television, with the sound turned down, was showing open-heart surgery; the naked dark red organ fluttered, pulsated and throbbed in its harness of membranes. Douglas turned to the man sitting next to him.

'What do we do now?'

'We could hang ourselves in the tinsel.'

Like several of the residents, he was wearing a grey tracksuit, the colour of the rain, the colour of despair. He held out his hand.

'I'm Peter.'

'Douglas.'

'I was going to walk down to the garage to get some cigarettes, if you feel like a walk.'

Douglas shook his head. His pockets were empty. If he had had any money at the start of this débâcle, he had none now.

'Do you want anything from the shop, then?'

'Just get me a couple of bottles of vodka, a carton of tomato juice, and a hundred Marlboro.'

'No Worcester?'

'Hold the Worcester. A couple of lemons, maybe, and some black pepper.'

'You're on, mate.'

Peter was taking orders for chocolate and cigarettes from the others, then he set out into the sheet of rain. Douglas was summoned to see the doctor, a severe lady in a sari: afterwards he remembered nothing of the interview.

Back in the lounge a ghostly boy watched him with terrified eyes, and gibbered in fear when Douglas attempted a smile; whatever the reason for his being here was, it had been something that life had done to him, and not he to himself; some gross despoliation of innocence had brought him to this state. Douglas watched a nurse crouch beside him for half an hour or more, coaxing him to take one sip of milk from a straw held against his clamped bloodless lips; the milk ran down his white chin and she wiped it with a tissue. This sight of one human being caring for another moved him in part of his mind, but he felt so estranged from them, as if he had been watching on television a herd of elephants circling a sick companion. He might have wept then; he might have wept when Peter returned battered by the rain and dropped a packet of cigarettes in his lap; he might have wept when he held Mary's hand while she cried for her imprisonment and for her children, thinking also of his own, but he couldn't cry.

'Perhaps all my tears were alcohol,' he thought. He picked up a magazine. *Don't let Christmas Drive you Crackers*, he read. *Countdown to Christmas*. He thought about his own countdown to Christmas, which had started in good time some two weeks ago, in the early and savage freeze which had now been washed away by the grey rain.

* * * * *

'Slip slidin' away, slip slidin' away . . .' the Paul Simon song was running through his head as he skidded and slid down the icy drives of the big houses where he delivered free newspapers. His route was Nob Hill and the houses were large and set back from the road. 'This is no job for a man' he thought, but it was the only job that this man could find. He had seen women out leafleting, using shopping trolleys to carry their loads, and he had considered getting one himself as the strap of his heavy PVC bag bit into his shoulder, but that would have been the final admission of failure; and the suspicion that nobody wanted the newspapers anyway crystallized his embarrassment to despair.

What struck him most about the houses was the feeling that no life took place behind those windows; standing in front of some of them, he could see right through; it was like looking at an empty film set where no dramas were played out. Beyond the double-glazed and mullioned windows his eye was drawn over the deep immaculate lawn of carpet, the polished frozen lake of dining table with its wintry branched silver foliage of candelabra, past the clumps of Dralon velvet furniture and the chilly porcelain flowers and birds, through the locked french windows to the plumes of pampas grass, the stark prickly sticks of pruned roses in beds of earth like discarded Christmas cake with broken lumps of icing, the bird table thatched and floored with snow and the brown rushes keening round the invisible pond. Latterly, ghostly hands had installed, by night,

Christmas trees festooned with electric stars that sparkled as coldly and remotely as the Northern Lights. Douglas conceived the idea that the inhabitants of these houses were as cold and metallic as the heavy cutlery on their tables, as hollow as the waiting glasses.

It was late one morning, on a day that would never pass beyond a twilight of reflected snowlight, that he got his first glimpse of life beyond the glass; there had been tyre marks and sledge marks in the silent drives before, but never a sight of one of the inhabitants in this loop of time. Her hair was metallic, falling like foil, heavy on the thin shoulders of her cashmere sweater; he knew that it was cashmere, just as he knew that the ornate knives and forks that she set on the white tablecloth were pewter. Pewter flatware. He had found the designation for these scrolled and fluted implements in an American magazine filched from one of the cornucopias, or dustbins, concealed at the tradesmen's entrance to one of these houses. He stood and watched her as she folded napkins and cajoled hothouse flowers into an acceptable centrepiece. What was her life, he wondered, that so early in the day she had the time, or perhaps the desperation, to set the table so far in advance of dinner. She looked up, startled like a bird, or like one whose path has been powdered with snow from the feathery skirt of a bird, and Douglas retreated. He retrieved a real newspaper, not one of the local handouts which he delivered, from next door's bin and stood in the wide empty road, glancing at the headlines, with a torn paper garland, consigned to the wind, leaking its dyes into the snow at his feet.

Glasgow—World's Cancer Capital, he read. Nicotine and alcohol had given to his native city this distinction.

'Christ. Thank God I left Glasgow when I did.'

He poured the last drop down his throat and threw the little bottle into the snow, taking a deep drag on the untipped cigarette, which was the only sort which gave him any satisfaction now. He coughed, a heavy painful cough, like squashed mistletoe berries in his lungs, and returned to the room he had rented since he had left his wife, and children. The next time that he saw the woman she was unloading some small boys in peaked prep school caps from a Volvo Estate. She was wearing a hard tweed hat with a narrow brim, a quilted waistcoat and tight riding breeches, like a second skin, so that at a distance it looked as if she wasn't wearing any trousers above her glossy boots. Douglas was tormented by her. He looked for her everywhere, seeing her metallic hair reflected in shop windows, in the unlikely mirrors of pubs which she would never patronize. He stood in the front garden staring at her sideboard which had grown rich with crystallized fruits, dates and figs, a pyramid of nuts and satsumas, some still wrapped in blue and silver paper, Karlsbad plums in a painted box, a bowl of Christmas roses; his feet were crunched painfully in his freezing, wet shoes, his shoulders clenched against the wind; he wanted to crack open the sugary shell of one of those crystallized fruits and taste the syrupy dewdrop at its heart. Once he met her, turning from locking the garage, and handed her the paper. He couldn't speak; his heart was sending electric jolts of pain through his chest and down his arms. He stretched his stubbly muzzle, stippled with black, into what should have been a smile, but which became a leer. She snatched the paper and hurried to the house. If a man hates his room, his possessions, his clothes, his face, his body, whom can he expect not to turn from him in hatred and fear? There was nothing to be done, except to wrap himself in an overcoat of alcohol.

Whisky warmed the snow, melted the crystals of ice in his heart; he skidded, slip slidin' away, home to the dance of the sugar-plum fairy tingling on a glassy glockenspiel of icicles, to find the woman who organized the delivery round of the free newspapers on his doorstep. She was demanding his bag. He perceived that she was wearing acid yellow moonboots of wet acrylic fur. She blocked the door like a Yeti.

'There have been complaints,' she was saying. 'We do have a system of spot-checks, you know, and it transpires that half the houses on your round simply haven't been getting their copies. We rely on advertising, and it simply isn't good enough if the papers aren't getting through to potential customers, not to mention the betrayal of trust on your part. There has also been a more serious allegation, of harassment, but I don't want to go into that now. I had my doubts about taking you on in the first place. I blame myself, I shouldn't have fallen for your sob story . . . so if you'll just give me that bag, please . . . and calling me an abominable snowman is hardly going to make me change my mind . . .'

She hoisted the bag, heavy with undelivered papers, effortlessly on to her shoulders and stomped on furry feet out of his life.

So Pewter Flatware had betrayed him. He turned back from the door and went out to re-proof his overcoat of alcohol, to muffle himself against that knowledge, and the interview with the Yeti on the doorstep, and its implications.

A year ago, when he had had a short stay in hospital for some minor surgery, his voice had been the most vehement in the ward expressing a desire to get out of the place. He remembered standing in his dressing gown at the window of the day room, staring across the asphalt specked with frost, at the smoke from the incinerators and the row of dustbins, and saying, 'What a dump.' The truth was that he had loved it. When he had been told that he could go home, they had to pull the curtains round his bed, but the flowery drapes had not been able to conceal the shameful secret that he sobbed into his pillow. The best part had been at night, after the last hot drinks and medication had been dispensed from the trolley, and the nurse came to adjust the metal headboard and arrange the pillows and make him comfortable for the night. Tucked up by this routine professional tenderness into a memory of hitherto forgotten peace and acceptance, he felt himself grow childishly drowsy, and turned his face into the white pillow and slept. He restrained the impulse to put his thumb in his mouth.

Now he was lying in a bank of snow under the copper beech hedge of the woman with tinfoil hair, a lost dissolute baby, guzzling a bottle. The kind white pillow was soft and pure and accepting; he turned his face into it, into the nurse's white bosom, and slept, deaf to the siren that brought the Silver and Pewter people to their leaded windows at last, and blind to the blue lights spinning over the snow. He was now on the other side of sleep, on a clifftop, wrestling with a huge red demon which towered out of the sea, unconquerable and entirely evil. He woke in the ambulance, gibbering of the fear of death, and was taken to the interrogation room of the mental hospital, in whose lounge he sat now, reading a magazine article on how to prevent Christmas from driving you crackers.

At some stage in the interminable morning, one of the nurses brought into the lounge her own set of Trivial Pursuit, and divided the hungover, the tearful, the deranged, the silent and the illiterate into two teams, but the game never really got off the ground. The red demon of his dream came into Douglas's mind, and at once he realized that it had been the Demon Drink; a diabolical manifestation, a crude and hideous personification of the liquid to which he had lost every battle. But the demon assumed other disguises by day; liquefying into seductive and opalescent and tawny amber temptresses who whispered of happiness, that it would be all right this time, they promised; they would make everything all right and each time that he succumbed he couldn't have enough of them, and their promises were broken like glass, and at night as Douglas lay neither asleep nor awake, the demon took his true shape and led him to glimpses of Hell, or at least to the most grotesque excesses of the human mind. He had not dared to go to sleep in his dormitory bed; all night thin ribbons of excelsior had glittered round the doorframe and the barred windows; it sparkled pink and phosphorescent and crackled in nosegays on the snores of the sleeping men, and danced in haloes of false fire above their restless heads.

'I hate going to bed,' he heard Peter say to a man called Bob, 'it's like stepping into an open grave.' And then, 'I'm so terrified of drinking myself to death that I have to drink to stop myself from thinking about it.' Bob was a big gentle man with broken teeth, and his bare forearms were garlanded with tattooed hearts and flowers. Peter asked him what had happened to his teeth, and about a scar on his hand.

'They sent me up to D ward and the nurses broke my teeth. They broke two of my ribs as well.'

He said it quite without rancour: this is what happens when you are sent up to D ward. Unable to bear the implications of Bob's statement, Douglas concentrated on a somewhat haphazard game of Give Us a Clue that was in progress across the room. Charades had been proposed, and abandoned in favour of this idiosyncratic version of the television game, and Douglas was invited to join in. As he rose from his chair, he saw that Peter was crying, and he saw Bob reach out his scarred and flowery hand and place it on Peter's knee and say gently through his broken teeth, 'I wish I could help you with your troubles, Peter.'

In his shamed and demoralized state Douglas felt that he had come as near as he ever would to a saint, or even to Jesus Christ. The sight of the big broken man giving a benediction on the other's self-inflicted wounds moved him so that he sat silent and clueless in the game, unable to weep for anyone else, or for his own worthlessness. Then an old man stood up, his trousers hoisted high over his stomach to his sagging breasts. He extended his arm, closed thumb and forefinger together, and undulated his arm.

'What's that then?' he demanded.

'Snake,' said Douglas.

'Yep. Your go.'

Douglas sat; the embodiment of the cliché: he didn't know whether to laugh or to cry.

'For When the One Great Scorer comes
To write against your name,
He marks—not that you won or lost—
But how you played the game.'

The debauched Scottish pedant swayed to his feet, grinning uncertainly through stained teeth, and played the game.

In the afternoon his daughters came to visit. He would have done anything to prevent them, if he had known of their intention. He wanted to hide, but they came in, smelling of fresh air and rain, with unseasonal daffodils and chocolates, like children, he thought, in a fairytale, sent by their cruel stepmother up the mountainside to find violets and strawberries in the snow. He took them to the games room which was empty. Here, too, the ashtrays overflowed; those deprived of drink had dedicated themselves to smoking themselves to death instead. The girls had been to his room, and had brought him clean clothes in a carrier bag, and cigarettes. He was so proud of them, and they, who had so much cause to be ashamed of him, made him feel nothing but loved and missed. They laughed and joked, and played a desultory game of table tennis on the dusty table with peeling bats, and mucked about on the exercise bicycle and rowing machine which no one used, and picked out tunes on the scratched and stained and tinselled piano. There was an open book of carols on its music stand: that will be the worst, he thought, when we gather round on Christmas Day to emit whatever sounds come from breaking hearts. Two of the girls lit cigarettes, which made him feel better about the ash-strewn floor, and Mandy, who did not smoke, let no flicker of disapproval cross her face; all in all they acted as if visiting their father in a loony bin was the most normal and pleasurable activity that three young girls could indulge in on a Saturday afternoon. It was only when the youngest said that she was starving, and he said that there were satsumas, which another patient had given him, in his locker, and she made a face and replied 'Satsumas are horrible this year' that they all looked at each other in acknowledgement that her words summed up the whole rotten mess that he had made of Christmas. The fathers have eaten sour grapes, and the children's teeth are set on edge. Douglas broke the silence that afflicted them by saying, 'Good title for a story, eh?' A reminder that in another life he had been a writer. Someone was waiting for the girls in a car, and as he led them to the front door, he hurried them past a little side room where Bob was hunched in a chair, his great head in his hands, his body rocking in grief. Douglas heard the laughter of staff, a world away, behind the door that separates the drunks from the sober. In his carrier bag he found a razor, electric. Now I can shave myself to death, he thought, as opposed to cutting my throat. There were also some envelopes and stamps, a writing pad and pen. There were no letters that he wished to write, but he took the paper and pen, and wrote 'Satsumas Are Horrible This Year', as if by writing it down he could neutralize the pain; turn the disgrace to art. It would not be very good, he knew, but at least it would come from that pulpy, sodden satsuma that was all that remained of his heart.

Later, he went into the kitchen to make a cup of tea for himself and for several of the others: like ten-pence pieces for the phone, and cigarettes, coffee was at a premium here. He was hungry, not having eaten for days, and thought of making a piece of toast, but he did not know if he was allowed to take any bread, and the grill pan bore the greasy impressions of someone else's sausages. He realized then what all prisoners, evil or innocent, learn; that what seems such a little thing, and which he had forfeited, the act of making yourself a piece of toast under your own lopsided grill, is in fact one of life's greatest privileges. He stood in the alien kitchen that smelled of industrial detergent and fat and old washing-up cloths, seeing in memory his children smiling and waving at the door, their resolute backs as they walked to the car concealing their wounds under their coats, forgiving and brave, and carrying his own weak and dissolute genes in their young and beautiful bodies. Violets and strawberries in the snow.

JAMES KELMAN

1946–

HOME FOR A COUPLE OF DAYS

Three raps at the door. His eyes opened and blinked as they met the sun rays streaming in through the slight gap between the curtains. 'Mister Brown?' called somebody—a girl's voice.

'Just a minute.' He squinted at his wristwatch. 9 o'clock. He walked to the door and opened it, poked his head out from behind it.

'That's your breakfast.' She held out the tray as if for approval. A boiled egg and a plate of toast, a wee pot of tea.

'Thanks, that's fine, thanks.' He took it and shut the door, poured a cup of tea immediately and carried it into the bathroom. He was hot and sweaty and needed a shower. He stared at himself in the mirror. He was quite looking forward to the day. Hearing the girl's accent made it all even more so. After the shower he started on the grub, ate all the toast but left the egg. He finished the pot of tea then shaved. As he prepared to leave he checked his wallet. He would have to get to a bank at some point.

The Green Park was a small hotel on the west side of Sauchiehall Street. Eddie had moved in late last night and taken a bed and breakfast. Beyond that he was not sure, how long he would be staying. Everything depended.

He was strolling in the direction of Partick, glancing now and then at the back pages of the *Daily Record*, quite enjoying the novelty of Scottish football again. He stopped himself from smiling, lighted a cigarette. It was a sunny morning in early May and maybe it was that alone made him feel so optimistic about the future. The sound of a machine, noisy—but seeming to come from far away. It was just from the bowling greens across the street, a loud lawn-mower or something.

He continued round the winding bend, down past the hospital and up Church Street, cutting in through Chancellor Street and along the lane. The padlock hung ajar on the bolt of the door of the local pub he used to frequent. Farther on the old primary school across the other side of the street. He could not remember any names of teachers or pupils at this moment. A funny feeling. It was as if he had lost his memory for one split second. He had stopped walking. He lighted another cigarette. When he returned the lighter and cigarette packet to the side pockets of his jacket he noticed a movement in the net curtains of the ground floor window nearby where he was standing. It was Mrs McLachlan. Who else. He smiled and waved but the face disappeared.

His mother stayed up the next close. He kept walking. He would see her a bit later on. He would have to get her something too, a present, she was due it.

Along Dumbarton Road he entered the first cafe and he ordered a roll and sausage and asked for a cup of tea right away. The elderly woman behind the counter did not look twice at him. Why should she? She once caught him thieving a bar of Turkish Delight, that's why. He read the *Daily Record* to the front cover, still quite enjoying it all, everything, even the advertisements with the Glasgow addresses, it was good reading them as well.

At midday he was back up the lane and along to the old local. He got a pint of heavy, sat in a corner sipping at it. The place had really changed. It was drastic—new curtains!

There were not many customers about but Eddie recognized one, a middle-aged man of average build who was wearing a pair of glasses. He leaned on the bar with his arms folded, chatting to the bartender. Neilie Johnston. When Eddie finished his beer he walked with the empty glass to the counter. 'Heavy,' he said and he pointed at Neilie's drink. The bartender nodded and poured him a whisky. Neilie looked at it and then at Eddie.

'Eddie!'

'How's it going Neilie?'

'Aw no bad son no bad.' Neilie chuckled. The two of them shook hands. 'Where've you been?'

'London.'

'Aw London; aw aye. Well well.'

'Just got back last night.'

'Good . . .' Neilie glanced at Eddie's suit. 'Prospering son eh?'

'Doing alright.'

'That's the game.'

'What about yourself? still marking the board?'

'Marking the board! Naw. Christ son I've been away from that for a while!' Neilie pursed his lips before lifting the whisky and drinking a fairly large mouthful. He sniffed and nodded. 'With Sweeney being out the game and the rest of it.'

'Aye.'

'You knew about that son?'

'Mm.'

'Aye well the licence got lost because of it. And they'll no get it back either neither they will. They're fucking finished—caput! Him and his brother.'

Both of them were silent for a time. The bartender had walked farther along and was now looking at a morning paper. Neilie nudged the glasses up his nose a bit and he said, 'You and him got on okay as well son, you and Sweeney, eh?'

Eddie shrugged. 'Aye, I suppose.' He glanced at the other men ranged about the pub interior, brought his cigarettes and lighter out. When they were both smoking he called the bartender: 'Two halfs!'

'You on holiday like?' said Neilie.

'Couple of days just, a wee break . . .' He paused to pay for the two whiskies.

Neilie emptied the fresh one into the tumbler he already had. 'Ta son,' he said, 'it's appreciated.'

'You skint?'

'Aye, how d'you guess! Giro in two days.'

'Nothing doing then?'

'Eh well . . .' Neilie sniffed. 'I'm waiting the word on something, a wee bit of business. Nothing startling right enough.' He pursed his lips and shrugged, swallowed some whisky.

'I hope you're lucky.'

'Aye, ta.'

'Cheers.' Eddie drank his own whisky in a gulp and chased it down with a mouthful of heavy beer. 'Aw Christ,' he said, glancing at the empty tumbler.

'You should never rush whisky son!' Neilie chuckled, peering along at the bartender.

'I'm out the habit.'

'Wish to fuck I could say the same!'

Eddie took a long drag on the cigarette and he kept the smoke in his lungs for a while. Then he drank more beer. Neilie was watching him, smiling in quite a friendly way. Eddie said, 'Any of the old team come in these days?'

'Eh . . .'

'Fisher I mean, or Stevie Price? Any of them? Billy Dempster?'

'Fisher drinks in T.C.'s.'

'Does he? Changed days.'

'Och there's a lot changed son, a lot.'

'Stevie's married right enough eh!'

'Is that right?'

'He's got two wee lassies.'

'Well well.'

'He's staying over in the south side.'

'Aw.'

A couple of minutes later and Eddie was swallowing the last of his beer and returning his cigarettes and lighter to the side pockets. 'Okay Neilie, nice seeing you.'

Neilie looked as if he was going to say something but changed his mind.

'I'm taking a walk,' said Eddie.

'Fair enough son.'

'I'll look in later.' Eddie patted him on the side of the shoulder, nodded at the bartender. He glanced at the other customers as he walked to the exit but saw nobody he knew.

It was good getting back out into the fresh air. The place was depressing and Neilie hadnt helped matters. A rumour used to go about that he kept his wife on the game. Eddie could believe it.

There was a traffic jam down at Partick Cross. The rear end of a big articulated lorry was sticking out into the main road and its front seemed to be stuck between two parked cars near to The Springwell Tavern. The lines of motors stretched along the different routes at the junction. Eddie stood at the Byres Road corner amongst a fair crowd of spectators. Two policemen arrived and donned the special sleeves they had for such emergencies and started directing operations. Eddie continued across the road.

In T.C.'s two games of dominoes were in progress plus there was music and a much cheerier atmosphere. It was better and fitted in more with the way Eddie remembered things. And there was Fisher at the other end of the bar in company with another guy. Eddie called to him: 'Hey Tam!'

'Eddie!' Fisher was delighted. He waved his right fist in the air and when Eddie reached the other end he shook hands with him in a really vigorous way. 'Ya bastard,' he said, 'it's great to see ye!' And then he grinned and murmured, 'When did you get out!'

'Out—what d'you mean?'

Fisher laughed.

'I'm being serious,' said Eddie.

'Just that I heard you were having a holiday on the Isle of Wight.'

'That's garbage.'

'If you say so.'

'Aye, fuck, I say so.' Eddie smiled.

'Well, I mean, when Sweeney copped it . . . Then hearing about you . . . Made me think it was gen.'

'Ah well, there you are!'

'That's good,' said Fisher and he nodded, then jerked his thumb at the other guy, 'This is Mick . . .'

After the introductions Eddie got a round of drinks up and the three of them went to a table at the wall, the only one available. An elderly man was sitting at it already; he had a grumpy wizened face. He moved a few inches to allow the trio more space.

There was a short silence. And Eddie said, 'Well Tam, how's Eileen?'

'Dont know. We split.'

'Aw. Christ.'

'Ah,' Fisher said, 'she started . . . well, she started seeing this other guy, if you want to know the truth.'

'Honest?' Eddie frowned.

Fisher shook his head. 'A funny lassie Eileen I mean you never really fucking knew her man I mean.' He shook his head again. 'You didnt know where you were with her, that was the fucking trouble!'

After a moment Eddie nodded. He lifted his pint and drank from it, waiting for Fisher to continue but instead of continuing Fisher turned and looked towards the bar, exhaled a cloud of smoke. The other guy, Mick, raised his eyebrows at Eddie who shrugged. Then Fisher faced to the front again and said, 'I was surprised to hear that about Sweeney but, warehouses, I didnt think it was his scene.'

Eddie made no answer.

'Eh . . . ?'

'Mm.'

'Best of gear right enough,' Fisher added, still gazing at Eddie.

Eddie dragged on his cigarette. Then he said, 'You probably heard he screwed the place well he never, he just handled the stuff.'

'Aw.'

'It was for screwing the place they done him for, but . . .' Eddie sniffed, drank from his pint.

'Aye, good.' Fisher grinned. 'So how you doing yourself then Eddie?'

'No bad.'

'Better than no bad with that!' He gestured at Eddie's clothes. He reached to draw his thumb and forefinger along the lapel of the jacket. 'Hand stitched,' he said, 'you didnt get that from John Collier's. Eh Mick?'

Mick smiled.

Eddie opened the jacket, indicated the inner pocket. 'Look, no labels.'

'What does that mean?'

'It means it was fucking dear.'

'You're a bastard,' said Fisher.

Eddie grinned. 'Yous for another? A wee yin?'

'Eh . . . Aye.' Fisher said, 'I'll have a doctor.'

'What?'

'A doctor.' Fisher winked at Mick. 'He doesnt know what a doctor is!'

'What is it?' asked Eddie.

'A doctor, a doctor snoddy, a voddy.'

'Aw aye. What about yourself?' Eddie asked Mick.

'I'll have one as well Eddie, thanks.'

Although it was busy at the bar he was served quite quickly. It was good seeing as many working behind the counter as this. One of the things he didnt like about England was the way sometimes you could wait ages to get served in their pubs—especially if they heard your accent.

He checked the time of the clock on the gantry with his wristwatch. He would have to remember about the bank otherwise it could cause problems. Plus he was wanting to get a wee present for his mother, he needed a couple of quid for that as well.

When he returned to the table Fisher said, 'I was telling Mick about some of your exploits.'

'Exploits.' Eddie laughed briefly, putting the drinks on the table top and sitting down.

'It's cause the 2,000 Guineas is coming up. It's reminding me about something!'

'Aw aye.' Eddie said to Mick. 'The problem with this cunt Fisher is that he's loyal to horses.'

'Loyal to fucking horses!' Fisher laughed loudly.

'Ah well if you're thinking about what I think you're thinking about!'

'It was all Sweeney's fault!'

'That's right, blame a guy that cant talk up for himself!'

'So it was but!'

Eddie smiled. 'And Dempster, dont forget Dempster!'

'That's right,' said Fisher, turning to Mick, 'Dempster was into it as well'

Mick shook his head. Fisher was laughing again, quite loudly.

'It wasnt as funny as all that,' said Eddie.

'You dont think so! Every other cunt does!'

'Dont believe a word of it,' Eddie told Mick.

'And do you still punt?' Mick asked him.

'Now and again.'

'Now and again!' Fisher laughed.

Eddie smiled.

'There's four races on the telly this afternoon,' said Mick.

'Aye,' said Fisher, 'we were thinking of getting a couple of cans and that. You interested?'

'Eh, naw, I'm no sure yet, what I'm doing.'

Fisher nodded.

'It's just eh . . .'

'Dont worry about it,' said Fisher, and he drank a mouthful of the vodka.

'How's Stevie?'

'Alright—as far as I know, I dont see him much; he hardly comes out. Once or twice at the weekends, that's about it.'

'Aye.'

'What about yourself, you no married yet?'

'Eh . . .' Eddie made a gesture with his right hand. 'Kind of yes and no.'

Fisher jerked his thumb at Mick. 'He's married—got one on the way.'

'Have you? Good, that's good.' Eddie raised his tumbler of whisky and saluted him. 'All the best.'

'Thanks.'

'I cant imagine having a kid,' said Eddie, and to Fisher he said: 'Can you?'

'What! I cant even keep myself going never mind a snapper!'

Mick laughed and brought out a 10-pack of cigarettes. Eddie pushed it away when offered. 'It's my crash,' he said.

'Naw,' said Mick, 'you bought the bevy.'

'I know but . . .' He opened his own packet and handed each of them a cigarette and he said to Fisher: 'You skint?'

Fisher paused and squinted at him, 'What do you think?'

'I think you're skint.'

'I'm skint.'

'It's a fucking dump of a city this, every cunt's skint.'

Fisher jerked his thumb at Mick. 'No him, he's no skint, a fucking millionaire, eh!'

Mick chuckled, 'That'll be fucking right.'

Eddie flicked his lighter and they took a light from him. Fisher said, 'Nice . . .'

Eddie nodded, slipping it back into his pocket.

'What you up for by the way?'

'Och, a couple of things.'

'No going to tell us?'

'Nothing to tell.'

Fisher winked at Mick: 'Dont believe a word of it.'

'It's gen,' said Eddie, 'just the maw and that. Plus I was wanting to see a few of the old faces. A wee while since I've been away, three year.'

'Aye and no even a postcard!'

'You never sent me one!'

'Aye but I dont know where the fuck you get to man I mean I fucking thought you were inside!'

'Tch!'

'He's supposed to be my best mate as well Mick, what d'you make of it!'

Mick smiled.

Not too long afterwards Eddie had swallowed the last of his whisky and then the heavy beer. 'That's me,' he said, 'better hit the road. Aw right Tam! Mick, nice meeting you.' Eddie shook hands with the two of them again.

Fisher said, 'No bothering about the racing on the telly then . . .'

'Nah, better no—I've got a couple of things to do. The maw as well Tam, I've got to see her.'

'Aye how's she keeping? I dont see her about much.'

'Aw she's fine, keeping fine.'

'That's good. Tell her I was asking for her.'

'Will do . . .' Eddie edged his way out. The elderly man shifted on his chair, made a movement towards the drink he had lying by his hand. Eddie nodded at Mick and said to Fisher, 'I'll probably look in later on.'

A couple of faces at the bar seemed familiar but not sufficiently so and he continued on to the exit, strolling, hands in his trouser pockets, the cigarette in the corner of his mouth. Outside on the pavement he glanced from right to left, then the pub door banged behind him. It was Fisher. Eddie looked at him. 'Naw eh . . .' Fisher sniffed. 'I was just wondering and that, how you're fixed, just a couple of quid.'

Eddie sighed, shook his head, 'Sorry Tam but I'm being honest, I've got to hit the bank straight away; I'm totally skint.'

'Aw. Okay. No problem.'

'I mean if I had it . . . I'm no kidding ye, it's just I'm skint.'

'Naw dont worry about it Eddie.'

'Aye but Christ!' Eddie held his hands raised, palms upwards. 'Sorry I mean.' He hesitated a moment then said, 'Wait a minute . . .' He dug out a big handful of loose change from his trouser pockets and arranged it into a neat sort of column on his left hand, and presented it to Fisher. 'Any good?'

Fisher gazed at the money.

'Take it,' said Eddie, giving it into his right hand.

'Ta Eddie. Mick's been keeping me going in there.'

'When's the giro due?'

'Two more days.'

'Garbage eh.' He paused, nodded again and patted Fisher on the side of the shoulder. 'Right you are then Tam, eh! I'll see ye!'

'Aye.'

'I'll take a look in later on.'

'Aye do that Eddie. You've actually just caught me at a bad time.'

'I know the feeling,' said Eddie and he winked and gave a quick wave. He walked on across the street without looking behind. Farther along he stepped sideways onto the path up by the Art Galleries.

There were a lot of children rushing about, plus women pushing prams. And the bowling greens were busy. Not just pensioners playing either, even young boys were out. Eddie still had the *Record* rolled in his pocket and he sat down on a bench for a few minutes, glancing back through the pages again, examining what was on at all the cinemas, theatres, seeing the pub entertainment and restaurants advertised.

No wind. Hardly even a breeze. The sun seemed to be beating right down on his head alone. Or else it was the alcohol; he was beginning to feel the effects. If he stayed on the bench he would end up falling asleep. The hotel. He got up, paused to light a cigarette. Along Sauchiehall Street there was a good curry smell coming for somewhere. He was starving. He turned into the entrance to The Green Park, walking up the wee flight of stairs and in to the lobby, the reception lounge. Somebody was hoovering carpets. He pressed the buzzer button, pressed it again when there came a break in the noise.

The girl who had brought him breakfast. 'Mrs Grady's out the now,' she told him.

'Aw.'

'What was it you were wanting?'

'Eh well it was just I was wondering if there's a bank near?'

'A bank. Yes, if you go along to Charing Cross. They're all around there.'

'Oh aye. Right.' Eddie smiled. 'It's funny how you forget wee details like that.'

'Mmhh.'

'Things have really changed as well. The people . . .' He grinned, shaking his head.

She frowned. 'Do you mean Glasgow people?'

'Aye but really I mean I'm talking about people I know, friends and that, people I knew before.'

'Aw, I see.'

Eddie yawned. He dragged on his cigarette. 'Another thing I was wanting to ask her, if it's okay to go into the room, during the day.'

'She prefers you not to, unless you're on full board.'

'Okay.'

'You can go into the lounge though.'

He nodded.

'I dont know whether she knew you were staying tonight . . .'

'I am.'

'I'll tell her.'

'Eh . . .' Eddie had been about to walk off; he said, 'Does she do evening meals as well like?'

'She does.' The girl smiled.

'What's up?'

'I dont advise it at the moment,' she said quietly, 'the real cook's off sick just now and she's doing it all herself.'

'Aw aye. Thanks for the warning!' Eddie dragged on the cigarette again. 'I smelled a curry there somewhere . . .'

'Yeh, there's places all around.'

'Great.'

'Dont go to the first one, the one further along's far better—supposed to be one of the best in Glasgow.'

'Is that right. That's great. Would you fancy coming at all?'

'Pardon?'

'It would be nice if you came, as well, if you came with me.' Eddie shrugged. 'It'd be good.'

'Thanks, but I'm working.'

'Well, I would wait.'

'No, I dont think so.'

'It's up to you,' he shrugged, 'I'd like you to but.'

'Thanks.'

Eddie nodded. He looked towards the glass-panelled door of the lounge, he patted his inside jacket pocket in an absent-minded way. And the girl said, 'You know if it was a cheque you could cash it here. Mrs Grady would do it for you.'

'That's good.' He pointed at the lounge door. 'Is that the lounge? Do you think it'd be alright if I maybe had a doze?'

'A doze?'

'I'm really tired. I was travelling a while and hardly got any sleep last night. If I could just stretch out a bit . . .'

He looked about for an ashtray, there was one on the small half-moon table closeby where he was standing; he stubbed the cigarette out, and yawned suddenly.

'Look,' said the girl, 'I'm sure if you went up the stair and lay down for an hour or so; I dont think she would mind.'

'You sure?'

'It'll be okay.'

'You sure but I mean . . .'

'Yeh.'

'I dont want to cause you any bother.'

'It's alright.'

'Thanks a lot.'

'Your bag's still there in your room as well you know.'

'Aye.'

'Will I give you a call? about 5?'

'Aye, fine. 6 would be even better!'

'I'm sorry, it'll have to be 5—she'll be back in the kitchen after that.'

'I was only kidding.'

'If it could be later I'd do it.'

'Naw, honest, I was only kidding.'

The girl nodded.

After a moment he walked to the foot of the narrow, carpeted staircase.

'You'll be wanting a cheque cashed then?'

'Aye, probably.'

'I'll mention it to her.'

Up in the room he unzipped his bag but did not take anything out, he sat down on the edge of the bed instead. Then he got up, gave a loud sigh and took off his jacket, draping it over the back of the bedside chair. He closed the curtains, lay stretched out on top of the bedspread. He breathed in and out deeply, gazing at the ceiling. He felt amazingly tired, how tired he was. He had never been much of an afternoon drinker and today was just proving the point. He raised himself up to unknot his shoelaces, lay back again, kicking the shoes off and letting them drop off onto the floor. He shut his eyes. He was not quite sure what he was going to do. Maybe he would just leave tomorrow. He would if he felt like it. Maybe even tonight! if he felt like it. Less than a minute later he was sleeping.

Alan Spence

1947–

Its Colours They are Fine

Billy pulled on the trousers of his best (blue) suit, hoisting the braces over his shoulders, and declared that without a doubt God must be a Protestant. It was no ponderous theology that made him say it, but simple observation that the sun was shining. And a God who made the sun shine on the day of the Orange Walk must surely be a Protestant, in sympathy at least.

From the front room Lottie mumbled responses he couldn't quite make out, but which he recognized as agreement. Over the 23 years they'd been married, she had come to accept his picture of God as a kind of Cosmic Grand Master of the Lodge. It seemed probable enough.

Billy opened the window and leaned out.

The smell of late breakfasts frying; music from a radio; shouted conversations; traffic noises from the main road. A celebration of unaccustomed freedom. Saturday had a life and a character all of its own.

Sunlight shafted across the tenement roofs opposite, cleaved the street in two. A difference of greys. The other side in its usual gloom, this side warmed, its shabbiness exposed. Sun on stone.

Directly below, between a lamp-post and the wall, a huddle of small boys jostled in this improvised goalmouth while another, from across the road, took endless glorious corner kicks, heedless of traffic and passers-by.

One of the most noticeable things about a Saturday was the number of men to be seen in the street, waiting for the pubs to open, going to queue for a haircut, or simply content to wander about, enjoying the day. For them, as for Billy, a Saturday was something to be savoured. He would willingly work any amount of overtime—late nights, Sundays, holidays—but not Saturdays. A Saturday was his. It was inviolable. And this particular Saturday was more than that, it was sacred. In Glasgow the Walk was always held on the Saturday nearest the 12th of July, the anniversary of the Battle of the Boyne. It was only in Ulster that they observed the actual date, no matter what day of the week that might be.

Billy closed the window and went through to the front room, which was both living-room and kitchen.

Lottie was laying out his regalia in readiness for the Walk—the sash, cuffs, white gloves and baton. She had laid them out flat on a sheet of brown paper and was wrapping them into a parcel.

'Whit's this?' he asked.

'Ah'm wrappin up yer things. Ye kin pit them oan when ye get tae Lorne School.'

'Not'n yer life! D'ye think ah'm frightened tae show ma colours?'

'That's jist whit's wrang wi ye. Yer never DONE showin yer colours! Look whit happened last year. Nearly in a fight before ye goat tae the coarner!'

'Look, wumman, this is a Protestant country. A Protestant queen shall reign.' He rapped on the table. 'That's whit it says. An if a Marshal in the Ludge canny walk the streets in is ain regalia, ah'll fuckin chuck it. Ah mean wu've goat tae show these people! Ah mean whit wid HE say?'

He gestured towards the picture of King William III which hung on the wall—sword pointing forward, his white stallion bearing him across Boyne Water. In a million rooms like this he was hung in just that pose, doomed to be forever crossing the Boyne. This particular ikon had been bought one drunken afternoon at the Barrows and borne home reverently and miraculously intact through the teatime crowds. Its frame was a single sheet of glass, bound around with royal blue tape. Fastened on to one corner was a Rangers rosette which bore a card declaring NO SURRENDER.

'An you're askin me tae kerry this wrapped up lik a fish supper!'

'Ach well,' she said, shoving the parcel across the table. 'Please yersel. But don't blame me if ye get yer daft heid stoved in.'

Billy grinned at the picture on the wall. Underneath it, on the mantelpiece, was the remains of what had been a remarkable piece of sculpture. One night in the licensed grocer's, Billy had stolen a white plastic horse about ten inches high, part of an advertising display for whisky. On to its back had been fitted a Plasticine model of King William, modelled by Peter, a young draughtsman who was in Billy's Lodge.

But one night Billy had come home drunk and knocked it over, squashing the figure and breaking one of the front legs from the horse. So there it sat. A lumpy Billy on a three-legged horse.

He picked up the splintered leg and was wondering if it could be glued back in place when there was a knock at the door.

'That'll be wee Robert,' he said, putting the leg back on the mantelpiece.

'Ah'll get it,' said Lottie.

Robert came in. He was actually about average height but he just looked small beside Billy. He and Billy had been friends since they were young men. They were both welders, and as well as working together, they belonged to the same Lodge. Robert was not wearing his sash. Under his arm he carried a brown paper parcel which looked remarkably like a fish supper.

'Is that yer sash?' asked Billy.

'Aye. Ach the wife thought it wid be safer like, y'know.'

'Well ah'm glad some'dy's goat some sense!' said Lottie.

'Ach!'

Billy buttoned his jacket and put on his sash, gloves and cuffs.

'Great day orra same!' said Robert. He was used to being caught between them like this and he knew it would pass.

'It is that,' said Billy. He picked up his baton.

'Right!' he said.

'Ye better take this,' said Lottie, handing him his plastic raincoat.

'O ye of little faith, eh!' He laughed, a little self-conscious at setting his tongue to a quote, but he took the coat nevertheless.

'Noo mind an watch yersels!'

Lottie watched from the window as they walked along the street and out of sight. At least this year they'd got that far without any trouble.

As they rounded the corner, in step, Billy turned to Robert.

'Ah'm tellin ye Robert,' he said. 'God's a Protestant!'

Emerging into the sunlight from the subway at Cessnock, they could hear some of the bands warming up. Stuttered rolls and paradiddles on the side drums, deep throb of the bass, pipes droning, snatches of tunes on the flutes.

'Dis yer heart good tae hear it, eh!' Billy slapped Robert on the back.

Robert carefully unwrapped his sash and put it on, then defiantly screwed up the paper into a ball and threw it into the gutter.

'At's the stuff!'

Billy caught the strains of 'The Bright Orange and Blue' and started to whistle it as he marched along.

'Ther's the bright orange an blue for ye right enough,' said Robert, gesturing towards the assembly of the faithful.

So much colour, on uniforms, sashes and banners. The bright orange and blue, the purple and the red, the silver and the gold, and even (God forgive them!) the green.

The marchers were already forming into ranks. It must be later than they'd thought. They hurried up to where their Lodge was assembled and took their positions, Billy at the side, Robert up behind the front rank, carrying one of the cords which trailed from the poles of their banner. Purple and orange silk, King William III, Loyal and True. Derry, Aughrim, Inneskillin, Boyne. These were the four battles fought by William in Ireland, their magic names an incantation, used now as rallying cries in the everlasting battle against popery.

They were near the front of the procession and their Lodge was one of the first to move off, a flute band from Belfast just in front of them.

Preparatory drumroll. 'The Green Grassy Slopes.' Sun glinting on the polished metal parts of instruments and the numerals on sashes and cuffs.

To Billy's right marched Peter, long and thin with a wispy half-grown beard. Billy caught his eye once and looked away quickly. He was still feeling guilty about ruining the Plasticine model that Peter had so carefully made. A little further on, Peter called over to him. 'The band's gaun ther dinger, eh!'

'Aye they ur that. Thull gie it laldy passin the chapel!'

It was as if they were trying to jericho down the chapel walls by sheer volume of sound, with the bass drummer trying to burst his skins. (He was supposed to be paid a bonus if he did, though Billy had never seen it happen.) And the drum major, a tight-trousered shaman in a royal blue jumper, would leap and birl and throw his stick in the air, the rest of the band strutting or swaggering or shuffling behind. The flute band shuffle. Like the name of a dance. It was a definite mode of walking the bandsmen seemed to inherit—shoulders hunched, body swaying from the hips, feet scuffling in short, aggressive steps.

Billy's own walk was a combination of John Wayne and numberless lumbering cinema-screen heavies. He'd always been Big Billy, even as a child. Marching in the Walk was like being part of a liberating army. Triumph. Drums throbbing. Stirring inside. He remembered newsreel films of the Allies marching into Paris. At that time he'd been working in the shipyards and his was a reserved occupation, 'vital to the war effort' which meant he couldn't join up. But he'd marched in imagination through scores of Hollywood films. From the sands of Iwo Jima to the beachheads of Normandy. But now it was real, and instead of 'The Shores of Tripoli', it was 'The Sash My Father Wore'.

They were passing through Govan now, tenements looming on either side, people waving from windows, children following the parade, shoving their way through the crowds along the pavement.

The only scuffle that Billy saw was when a young man started shouting about civil rights in Ireland, calling the marchers fascists. A small sharp-faced woman started hitting him with a union jack. Two policemen shoved their way through and led the man away for his own safety as the woman's friends managed to bustle her, still shouting and brandishing her flag, back into the crowd.

'Hate tae see bother lik that,' said Peter.

'Ach aye,' said Billy. 'Jist gets everyb'dy a bad name.'

Billy had seen some terrible battles in the past. It would usually start with somebody shouting or throwing something at the marchers. Once somebody had lobbed a bottle from a third-storey window as the Juvenile Walk was passing, and a mob had charged up the stairs, smashed down the door and all but murdered every occupant of the house. Another common cause of trouble was people trying to cross the road during the parade. The only time Billy had ever used his baton was when this had happened as they passed the war memorial in Govan Road, with banners lowered and only a single drumtap sounding. A tall man in overalls had tried to shove his way through, breaking the ranks. Billy had tried to stop him, but he'd broken clear and Billy had clubbed him on the back of the neck, knocking him to the ground. Another Marshal had helped him to pick the man up and bundle him back on to the pavement.

But this year for Billy there was nothing to mar the showing of the colours and he could simply enjoy the whole brash spectacle of it. And out in front the stickman led the dance, to exorcise with flute and drum the demon antichrist bogeyman pope.

They turned at last into Govan Road and the whole procession pulsed and throbbed and flaunted its way along past the shipyards. Down at the river, near the old Elder cinema, buses were waiting to take them to the rally, this year being held in Gourock. Billy and Robert found seats together on the top deck of their bus and Peter sat opposite, across the passage. As the bus moved off there was a roar from downstairs.

'Lik a fuckin Sunday-school trip!' said Robert, and he laughed and waved his hanky out the window.

They were in a field somewhere in Gourock and it was raining. Billy had his raincoat draped over his head. He was eating a pie and listening to the speeches from the platform, specially erected in the middle of the field. The front of the platform was draped with a union jack and like the banners it drooped and sagged in the rain.

Robert nudged him. 'Wher's yer proddy god noo!'

Proddy god. Proddy dog.

(A moment from his childhood—on his way home from school—crossing wasteground—there were four of them, all about his own age—the taunt and the challenge—'A Billy or a Dan or an auld tin can'—They were Catholics, so the only safe response would be 'A Dan'—To take refuge in being 'An auld tin can' would mean being let off with a minor kicking. Billy stood, unmoving, as they closed round him. One of them started chanting—

'Auld King Billy
had a pimple on is wully
an it nip nip nipped so sore
E took it tae the pictures
an e gave it dolly mixtures
an it nip nip nipped no more.'

Jeering, pushing him. 'A Billy or a Dan . . .' Billy stopped him with a crushing kick to the shin—heavy parish boots—two of the others jumped on him and they fell, struggling, to the ground. They had him down and they would probably have kicked him senseless but about half a dozen of Billy's friends appeared round a corner, on their way to play football. They ran over, yelling, and the Dans, outnumbered, ran off—and as they ran, their shouts drifted back to Billy where he lay—'Proddy dog! Proddy dog!' fading on the air.)

On the platform were a number of high-ranking Lodge officials. One of them, wearing a dog-collar, was denouncing what he called the increasing support for church unity and stronger links with Rome.

'The role of the Order,' he went on, 'must increasingly be to take a firm stand against this pandering to the popery, and to render the strongest possible protest against moves towards unity.'

Billy was starting to feel cold because of the damp and he wished the rally was over.

'Wish e'd hurry up,' said Robert, rain trickling down his neck.

Billy shuffled. His legs were getting stiff.

The speaker pledged allegiance. Loyal address to the crown.

Applause. At last. The national anthem.

Billy straightened up. The blacksuited backs of the men in front. Long live our noble. Crumb of piecrust under his false teeth. Rain pattering on his coat. Huddled. Proddy god. Happy and glorious. Wet grass underfoot, its colour made bright by the rain.

Billy poured the dregs of his fourth half into his fourth pint. The discomforts of the interminable return journey and the soggy dripping march back to Lorne School were already forgotten as Billy, Robert and Peter sat drying off in the pub. Soon the day would form part of their collective mythology, to be stored, recounted, glorified.

Theirs was one of four rickety tables arranged along the wall facing the bar. Above them, rain was still streaking the frosted glass of the window but they no longer cared as the night grew loud and bright around them.

'Didye see that wee lassie?' said Robert. 'Cannae uv been merr than six year auld, an ther she wis marchin alang in the rain singin "Follow Follow". Knew aw the words as well. Magic so it wis.'

'Bringin them up in the Faith,' said Billy.

Over at the bar an old man was telling the same joke for the fifth time.

'So ther's wee Wullie runnin up the wing, aff the baw like, y'know. So ah shouts oot tae um "Heh Wullie, make a space, make a space!" An he turns roon an says "If ah make a space Lawrence'll build fuckin hooses oan it!"'

A drummer from one of the accordion bands was standing next to him at the bar, still wearing his peaked cap. Addressing the bar in general, he said, 'Aye, if Lawrence wid stoap tryin tae run Rangers like is bloody builders we might start gettin somewherr!'

Robert hadn't been listening. He was still thinking about children and the Faith. He turned to Peter.

'Is that wee burd ae yours no a pape?'

'Ach she disnae bother,' said Peter, and added quickly, 'Anywey, she's gonnae turn when we get merried.'

'Ah should think so tae,' said Billy. 'See thae cathlicks wi weans. Fuckin terrible so it is. Tell'n ye, see at that confirmation, the priest gies them a belt in the mouth! Nae kiddin! A wee tiny wean gettin punched in the mouth! It's no right.'

'Soldiers of Christ for fucksake,' said Peter.

'D'ye know whit ah think?' said Robert.

'You tell us,' said Peter.

'Ah think it's because thur families ur that big they don't bother wae them. D'ye know whit ah mean? Ah mean it stauns tae reason. It's lik money. The likes a some'dy that hisnae goat much is gonnae look efter whit e's goat. Well! Ther yar then! It's the same wi families. Folk that's only goat wan or two weans ur gonnae take kerr ae them. But thae cathlicks wi eight or nine weans, or merr, they're no gonnae gie a bugger, ur they?'

'They eat babies an aw!' said Peter, mocking.

'You kin laugh,' said Billy. 'But ah'm tellin ye, that's how they huv such big families in the furst place. It disnae happen here mindye, but see in some a thae poor countries wher ye've goat famine an that, they widnae think twice aboot eatin a baby or two. Usetae happen aw the time in the aulden days. Likes a the middle ages, y'know.'

Peter didn't argue. It might well be true and anyway it didn't really matter.

'Whit d'ye make a that cunt this mornin then?' asked Robert. 'Cryin us aw fascists!'

'Ach ah've goat a cousin lik that,' said Peter. 'Wan a they students y'know. Wurraw fascists except him like. E wis layin intae me the other day aboot the Ludge, sayin it was "neo-fascist" and "para-military" an shite lik that. E says the Juveniles is lik the Hitler Youth an Ian Paisley's another fuckin Hitler. Ah don't know wher they get hauf thur ideas fae, neither ah dae. E used tae be a nice wee fulla tae. But is heid's away since e went tae that Uni. Tell'n ye, if is heid gets any bigger, e'll need a fuckin onion bag fur it!'

'Sounds lik mah nephew,' said Billy. 'Tryin tae tell me that King Billy an the pope wur oan the same side at the Boyne! Talks a lotta shite so e dis. Ah jist cannae understaun them ataw. Ah mean, if wurraw supposed tae be fascists, whit wis the war supposed tae be aboot?'

(Those old newsreels again. Nuremberg rally. Speeches. Drums. The liberation of Paris. VE Day.)

'Ach fuck them all!' said Peter 'Smah round, intit?' He made his way to the bar. 'Three haufs'n three pints a heavy, Jim!' He made two journeys, one for the whisky, one for the beer, and when he sat down again Robert was telling a joke.

'Huv ye heard that wan aboot the proddy that wis dyin? Well e's lyin ther oan is death bed an e turns roon an asks fur a priest. Well, is family thoaght e wis gawn aff is heid, cause e'd always been a right bluenose. But they thoaght they better humour im like, in case e kicked it. So anywey the priest comes.'

'Impossible!' said Peter.

Billy shooshed, but Robert was carrying on anyway.

'An the fulla says tae the priest, "Ah want tae turn father". So the priest's as happy as a fuckin lord, an e goes through the ceremony right ther, an converts um intae a cathlick. Then e gies um the last rites, y'know, an efter it e says tae the fulla, "Well my son, ah'm glad ye've seen the light, but tell me, what finally decided ye?" An e lifts imself up, aw shaky an that, an wi is dyin breath e turns tae the priest an says, "This'll be another durty cathlick oot the road!"'

'Very good!' said Billy, laughing. 'Very good!'

'Ah heard a cathlick tellin it,' said Peter, 'only the wey he tellt it, it wis a cathlick that wis dyin an e sent fur a minister!'

'Typical!' said Robert.

'Ah'll away fur a pee,' said Peter.

While he was gone, Billy asked Robert if he'd seen any of Peter's cartoons.

'Ah huv not,' said Robert.

'Great, so they ur. E's a bitty an artist like. Anywey, e goat this headline oot the paper—y'know how the pope's no been well—an this headline says something aboot im gettin up, an Peter's done this drawin a the pope humpin this big blonde. Ye wanty see it!'

Peter came back.

'Ah've jist been tellin Robert aboot that drawin a yours, wi the pope.'

'Ah think ah've goat it wae me,' said Peter. He rummaged through his wallet and brought out a piece of paper. Gummed on to the top was the headline POPE GETS UP FOR FIRST TIME, and underneath Peter had drawn the pope with an utterly improbable woman.

He passed it round the table.

'Terrific!' said Robert. 'Fuckin terrific!'

Peter put it back in his wallet.

The talk of Peter's artistic talent reminded Billy, yet again, of the ruined Plasticine model. He quickly slopped down some more beer.

He set down his glass and held it in place as the room swayed away from him then rocked back to rest. He was looking at the glass and it was suddenly so clearly there, so sharply in focus. All the light of the room seemed gathered in it. Its colour glowed. The gold of the beer. Light catching the glass and the glistening wet mesh of froth round the rim. He was aware of his glass and his thick red hand clutching it. The one still point in the room. And in that moment he knew, and he laughed and said, 'Ah'm pished!'

The group over at the bar were singing 'Follow Follow' and Billy shouted 'Hullaw!' and the three of them joined in.

'For there's not a team
Like the Glasgow Rangers
No not one
No not one.
An there's not a hair
On a baldy-heided nun.
No not one!
There never shall be one!'

The barman made the regulation noises of protest, fully aware that they would have no effect.

'C'mon now gents, a wee bit order therr!'

'Away an fuck ya hun!'

The accordion band drummer produced his sticks and somebody shouted 'Give us The Sash!' The barman gave up even trying as the drummer battered out the rhythm on the bartop, and the drunken voices rose, joyful, and on past closing time they sang.

> 'Sure it's old but it is beautiful
> And its colours they are fine
> It was worn at Derry, Aughrim,
> Inneskillin and the Boyne,
> My father wore it as a youth
> In the bygone days of yore
> And it's on the twelfth I love to wear
> The Sash My Father Wore.'

Billy was vaguely aware of Robert going for a carry-out and Peter staggering out into the street. He swayed back and aimed for the door, lurching through a corridor of light and noise, getting faster as he went, thinking he would fall at every step. Out of the chaos odd snatches of song and conversation passed somewhere near.

'C'mon now sir, clear the bar . . .'

'An ah didnae even know the cunt . . .'

'RIGHT gents!'

'So ah shouts oot tae um Make a space Make a space . . .'

'Ah'm gonnae honk . . .'

No not one.

Into the street and the sudden rush of cold air. Yellow haloes round the streetlamps. The road was wet but the rain had stopped. He leaned back against the cold wall and screwing up his eyes to focus, he looked up at the sky and the stars. All he knew about astronomy was what he had learned from an article in the *Mail* or the *Post* called 'All about the Heavens', or 'The Universe in a Nutshell', or something like that. Millions of stars like the sun making up the galaxy and millions of galaxies making up the universe and maybe millions of universes.

Robert came out of the pub clutching his carry-out under his arm. Peter emerged from a closemouth where he'd just been sick.

'Yawright son?'

'Aye Robert, ah'll be awright noo ah've goat it up.'

Robert handed him a quarter bottle of whisky.

'Huv a wee snifter.'

'Hanks Robert.' He sipped some and shuddered, screwing up his face. Then he shook hands lingeringly with each of them, telling them they were the greatest.

'Ah fancy some chips!' said Robert.

'Me tae,' said Billy.

'Ah've took a helluva notion masel,' said Peter, and the three of them swayed off towards the chip shop, passing the bottle between them as they went.

At the next corner, Billy stopped to drain the last drops, head tilted back to catch the dregs. Tenements looming. The night sky. Dark. Galaxies. He looked at Robert and asked him, earnestly, if he'd ever smelt fall.

'Smelt whit?'

Recovering from a coughing spasm, he tried again, this time enunciating his words very carefully.

'Robert.'

'Aye.'

'Whit ah meant tae say wis, have ye ever felt small?'

Robert looked thoughtful for a moment, before replying, emphatically, 'Naw!' Grabbing Billy's lapel, he continued, 'An you're the biggest cunt ah know, so ah don't see whit YOU'RE worried aboot!'

'An neither dae ah!' said Billy, laughing. 'It's fuckin hilarious!'

And in all the stupid universe there was not a man like himself, not a city like Glasgow, not a team like the Rangers, not a hair on a nun, not a time like the present, not a care in the world.

Telling it, he shouted.

'God Bless King Billy!'

'EE-ZAY!'

And he hurled his bottle, arching, into the air, into the terrible darkness of it all.

MARGARET ELPHINSTONE
1948–

AN APPLE FROM A TREE

You ask for the whole story, which seems a large return for the loan of a small towel, but since you helped me out when I needed it, I'll tell you what I think happened, which is, after all, as near to a whole story as anyone can get.

As with many significant changes, the story began in the Botanic Gardens. I have always enjoyed walking there, not only to commune with nature, such as it is, but also to watch people in the throes of conversations and encounters that are obviously about to change their lives. However, I always thought of myself as a mere observer. I thought I was as likely to become part of the drama as I was to start sprouting leaves in spring, or to receive a proposal under the lilac. Not only was I not expecting anything, I didn't want anything much either.

One day in early September, the gardens were almost deserted. It was a damp Monday evening and the place seemed separate and enclosed. After the weekend, the city had gone back to whatever it thought mattered, and a thick fog lay over it, obliterating the distinctive skyline. The muffled sound of traffic could have come from anywhere, and might as easily have been made by live creatures roaring in a wilderness of their own devising. Meanwhile in this oasis of calm, I wandered under the trees. The first leaves were beginning to come down. I wove to and fro across the grass, kicking as many leaves as I could, but they were still sparse. At least no one had attempted to rake them up. A balsam poplar grows just beside the gate, and when I reached it I scooped up a handful of leaves and held them to my face. Damp and yellow, they smelt exotic and flower-like. I wandered on through the mist.

Twice I circled the lawn at the top, which is itself encircled by an ancient hedge of yew. Then I slipped in through a gap, and found myself under the great beech trees at the end. Perhaps their tops were above the mist. I couldn't tell, but there seemed to be a wind up there that spoke of cleaner air, sighing as it did among the branches, sending the first brown leaves scudding into the sky. I looked up and saw the mist clearing, the sky turning blue, like seeing the sea from an aeroplane when the cloud breaks apart. Hearing the sea up there made me nostalgic. It was autumn, and I had a whole winter to face in the city. The Firth hardly counted. It was brown and smelly, and when the tide went down it left a thick black scum, like the tideline from a diabolic bath.

I went to and fro under the beeches, as though by waiting long enough I could make something happen. That was foolish, of course, but the wind up there seemed so real and so near, I almost expected the waves of it to come lapping at my feet. I don't know what I was thinking of but, as the leaves came down, I jumped and tried to catch them. Seven years' good luck, we used to say. I don't know who taught me that. You don't catch luck falling off a tree; at least, I never supposed so.

Something came down hard that was not a leaf, falling straight towards me. I reacted instinctively, as though playing a celestial form of cricket. I cupped my hands together and caught it, letting my hands go with it so it didn't hurt me. It was round and solid, like nothing from a beech tree. I unfolded my hands slowly, and looked.

It was an apple.

Up I looked again, but there was nothing, only the beech trees tossing in a wind I could not feel. The last mist wreathed away like the tail of an unidentified creature whisking out of sight. Shadows that I had not known existed slid out of vision, and the leaves, both green and gold, were washed in sudden brightness.

I turned the apple over. It was not very big, greenish on one side, red on the other, hard and firm. There was a lack of emphasis about it I recognised as being old-fashioned. On a supermarket shelf it would have looked small and misshapen, and would probably have lain untouched until past its sell-by date. That gave me reason to suppose it might taste good. I smelled it cautiously. It smelled more like an apple than any apple I'd come across for years. It was more like the apple-scented bottles in the Body Shop. I'd forgotten that apple was astringent, sweet certainly, but almost as sharp as pine at the same time. I sniffed it again. Then, in spite of the fact that I'd no idea where it had been, I bit into it.

It seems tragic that a taste more subtle than anything we have come to know on earth should be wasted, but I hardly noticed it, for the very good reason that I was caught up in something that resembled, more closely than anything else I can think of, a Victorian steam roundabout. I once rode such a roundabout. I rode a great grey dapple horse with staring eyes and flaring nostrils; we tore up and down in strides like no horse ever made, faster and faster, with a wild music clinging to us that never dropped behind for all our speed.

This was not quite like that. There was no horse and no music. At least, I think there was not but, now I try to remember more carefully, I cannot say for sure. We did go round and round, in a spiral of leaves all caught up and rolled up, spun into a cone shape that whizzed over the lawn like a bobbin gone mad, until it fell apart in the long tall grass under the apple tree, flinging me down on knobbly ground that turned out to be wet soil strewn with windfalls. A painful landing, but I was too surprised to think about my bruises. I was still clutching my apple. I stood up, shakily, and realised that I was not alone.

The woman who stood watching me was quite naked. That was the first thing that I noticed about her. If that makes you think that I pay too much attention to trivia, you should read the regulations governing the conduct of visitors to the Botanics, and you will see at once why this should seem so startling. Also, it was September, and none too warm. She stood in thigh-high grass. Her dark hair fell right down her back, and was woven through with stems of bryony. She was brown, so evenly tanned that she might never have worn clothes at all, and quite slender, though not skinny. She seemed unmoved by my sudden appearance and stood regarding me without moving. It was a disconcerting gaze. Her eyes were so black they looked almost hollow.

'Hello,' I said.

'Hello,' she replied, so that it sounded like my own voice coming back at me. I thought she was mocking me and stiffened, but her gaze was not unfriendly. If she spoke English, I thought, it seemed all the odder that she should be naked. In case you think there is prejudice in that, let me remind you that it could mean only that she was able to read the regulations as well as I could. Assuming she could read, of course. I looked at her askance.

'Where have you come from?' she asked me next.

That flummoxed me. Surely she was taking the words from my mouth? 'Me?' I said. 'I was here all the time. Where have you come from?'

'I was here all the time.'

I don't like being mimicked. She even had the same accent, but she seemed to be using it quite innocently. I held out the apple.

'I caught it,' I said. 'Does it belong to you?'

'It came off the tree.' She pointed, and I looked.

The beech trees were gone, and the lawn, and the yew hedge. But it was still autumn and the leaves were still turning red and gold. This time they were apple leaves, for before me rose the hugest apple tree I ever saw, crowning the hill like a wreath of berries round an old man's head. The grass was long and unkempt, beginning to die back, but over it the apple-laden tree bent its branches low, so I could have reached up and picked and picked, though never got a tithe of all that lay beyond my reach. I never saw such a tree, nor so many apples, red and green and gold. I looked away, and saw a line of craggy hills across the skyline, softened by the ranks of trees.

'Where is this? What is that tree?'

She looked at me, apparently puzzled. 'This is the world,' she said, 'and that is an apple tree.'

'Thank you,' I said. I can be sardonic when I choose, but sarcasm seemed to wash over her like a summer breeze.

'That's all right,' she said, and then she asked a question of her own. 'What is wrong with you?'

'Me?' I was indignant. 'Nothing. Why? What should be wrong?'

'You're all wrapped about. Even your feet.'

I looked down at my feet. I was wearing trainers, somewhat damp and down at heel but perfectly respectable.

'My feet are fine, and so is the rest of me. I just happen to be wearing clothes. I'm surprised you managed to get here without, yourself.'

There was no doubt about it then. She was laughing at me. I would have walked away with dignity, but I had the disadvantage of not being sure any more where anything was. In fact, I didn't know where I was myself except that I hadn't left the Botanics and was, presumably, still in them.

'Are you upset? Don't you like the apple?'

I had almost forgotten the apple. I looked at it a mite suspiciously, not being quite sure how far it was responsible. I had an idea and held it out to her. 'Eat it yourself.'

'I was going to, but then you caught it.'

'Did you throw it?'

'No, he did.'

'He?' I looked round nervously. 'Who?'

'He,' she repeated with emphasis. 'Didn't you hear? I thought you said you'd been here all the time?'

'If I was, I wasn't listening.' I looked at the apple as though it might bite me, rather than vice versa. 'Why?' I asked, and failed to keep misgiving entirely out of my voice. 'Perhaps you'd better tell me what happened.'

'Nothing,' she said. 'After all that. I picked it, and I offered it to him. And he said, "I'm not hungry", and threw it. Then he went, and you came. That was all.'

'Where's he gone?'

'I neither know nor care. He showed a lamentable want of curiosity. I don't think he's ever really interested in anything but sex. And food, of course. I'm not sure whether he never had any imagination, or whether nameless fears subsequently drove it out of his head. What sex are you, incidentally?'

'Me?' For the third time I was indignant. 'I'm a woman, same as you. Can't you tell?'

'I thought so. It was the wrappings that confused me. What happened after you bit it?'

'I was here.'

'Damn.'

'I'm sorry?'

'I had hoped,' she explained patiently, 'that it might get one to somewhere else. Evidently not. Never mind. Shall we go now?'

'Go? Where?'

'Well, we can't stay here.'

She was right. At any moment there would be voices crying out 'Closing ti-ime' from one end of the gardens to the other. Such was the deception of the Botanics. Refuge seemed to be offered, a way out through the very heart of the city, then, just as the shadows lengthened and the truth seemed near at hand, policemen came shouting along the paths, expelling everybody, then closing the great iron gates behind them. I was never sure whether it was worse to leave before the expulsion began, or wait as long as possible, and be harried to the gates while the voices circled round me like sheepdogs driving an unwilling flock.

'Where do you live?' I asked her.

'In the world. Are you coming?'

If we passed through the gates, I never noticed. Once we left the apple tree behind, alone on the summit of the hill, we were down among the rest of the trees. The smell of balsam was still there, permeating the evening like a promise of something new. But the trees seemed to crowd more thickly than usual, pressing together and lining our route, like people watching a procession. We hardly made a procession, she and I, and there was no other human being to be seen anywhere. The path wound this way and that until quite soon I had lost all bearings. Golden leaves were strewn underfoot. They rustled as we walked. Where there were gaps in the trees, the long evening light broke through, bright on the fading leaves above, then speckled further down, where the leaves interrupted its path.

Brambles grew beside the way, thick with berries. She picked and ate them as she passed, absent-mindedly, as though browsing were automatic and her mind far away. But it slowed her down, and I was glad of that. The trailing stems caught my jacket, and I had to keep stopping to disentangle them. It was warm and humid among so many trees. I felt the sweat gather under my shirt, and trickle down my back. It was quiet in the wood, but not silent. The squirrels were everywhere, up and down the trees and running across the path in front of us, stopping when they saw us, and regarding us with bright eyes from about a yard distant, perched on their hind legs, jaws chattering. I was pleased to see them; they helped me to grasp that I was in the Botanics, which is the only place I know where squirrels are that tame and numerous. But when we saw the deer, I began to wonder again. I know there is more wildlife in city parks than ever meets the eyes, but a herd of roe deer? The little group we saw bounded away through the trees as we approached, disappearing among the birch stems in a pattern of brown and silver.

Presently we crossed a burn. Luckily there hadn't been much rain lately; even so, we had to wade. A curtain of willow hung down over the water, stems trailing even in the diminished flow. She waited while I took my shoes and socks off, and rolled up my trousers. My clothes seemed to amuse her. To distract her from that, and because our brief halt renewed the chance for conversation, I tried to pin her down a little more.

'What is your name?' I asked her.

'Nosila.'

I thought she might ask mine, but she didn't.

After that there was little opportunity for talk. We were ascending a steep hill, weaving our way through thick trunks and twining stems. The canopy was so dense here that the sun was quite blocked out. I would have said that it was getting darker, but it could not be less than an hour before sunset, for no one had called 'Closing ti-ime'. I wondered if I would wake up enough to hear them. I had had strange dreams in the Botanics before now, I told myself, and the police could surely be trusted not to ignore a sleeping body?

The hill hardly seemed to slow her down at all. I stopped even trying to think but stumbled after her, tripping over roots, and pulling myself uphill by thin stems. The forest badly needed clearing. No one had thinned out the young saplings, or cleared away the masses of fallen trunks and dead wood. Sometimes we climbed over these obstacles, sometimes we forced our way round, ducking under horizontal branches that were now pushing up new shoots from their sides, straight towards the sky. Even the sky shocked me; it was not pale, but a deep velvety blue, with one star showing. Even in the heart of the Botanics the dark could not look like that, even if one were allowed to stay and see. There had been times when I had circumnavigated the gardens at night, two sides of houses and two sides of endless iron railing, hoping for a glimmering of the night sky that might look down on the silent trees within. Shut out in an unremitting orange glow like tortured fire, I could only imagine it. The stars were lost to me. I could have clung to those railings and shaken them, hammered at the gates of the garden like a demented being, but I was afraid of the police, and of the danger of walking alone at night, so I walked quickly and never stopped.

Now there was no walking quickly, although the dusk filtered in soft and fast, swallowing up the tree tops, while the sky began to prick out stars, like the lighting of old-fashioned lamps. I could see better now. We were at the top of the hill and the slope below us stretched down to open ground. When I looked again I saw it was not ground but water. It was black in the half-light, uncannily calm, and on the far side it was lost in shadow where a great precipice towered over it. I found it weird, as though the place were full of hidden things I did not wish to think about.

'Where are we?'

'By the loch.'

It occurred to me that she could not help being completely literal. Although she spoke as fluently as I did, perhaps her grasp of language did not go so far. If she sought for nuances in anything, it was clearly not in words.

'Where are we going?' I asked patiently.

'There is shelter. It will soon be night.'

'I would rather go home.'

She looked at me with concern. A slight breeze blew up, and I heard the water lap on the shore below, invisible in the thickening air. I identified a difference then, which had been present from the moment when

I bit the apple. The trees smelled so sharp they overwhelmed me. The air was full of scents: leaves, water, soil, a tinge of salt on the breeze. Just as when I tasted the apple, I felt I had never known such richness. I sniffed again; and the breeze was damp and resinous, heavy with tree smells.

She answered just as I was about to speak again. The rhythm of her conversation was quite different from mine. She left long pauses for thought between each statement as though the fear of interruption was quite unknown to her.

'Are you tired?' she asked. 'Are you hungry? You still have the apple.'

I remembered, and took it out of my pocket. I was not sure what that apple was responsible for, and I distrusted it. 'I'm not sure about it,' I said. 'You have it.'

She took it. 'You want to give it to me?'

She seemed to attach undue importance to my purpose. I only wanted to be rid of the thing, and it hardly seemed such a momentous gift. After all, she could have picked as many as she liked.

'Yes,' I said lightly. 'Why not?'

Her eyes widened and she looked at it almost fearfully, as though I had challenged her to something. I was about to explain that nothing was further from my thoughts, and she need feel herself under no imaginary obligations, when she bit into it.

There was a roar like an approaching train from somewhere deep below us, and the ground seemed to quiver and to shake. I heard rushing water, like a great waterfall bursting its confines and engulfing the whole forest. We were flung together, and I felt her flesh under my hands as I clung to her, whirling downwards as the sky vanished from over us, until we landed hard on concrete, crashing in a heap amidst a tumble of wet leaves.

'Christ,' I muttered, clutching my bruised head in my hands. There was an orange glow which hurt my eyes, and the roaring was still in my ears, with a stench like the pits of a furnace. I felt a clutching at my arm, which I gradually realised was becoming painful. I looked up.

We were sitting on the pavement at the foot of Waverley Bridge, and there were taxis turning in and out of the station. We must have just missed the iron railings that enclosed Princes Street Gardens behind us. It was dark in there, and silent, all locked away for the night. I found myself staring at a fat white airport bus, that ground towards us like a giant caterpillar.

Nosila screamed in my ear.

'Ow!' I tried to calm her, or myself, I wasn't sure which. 'Now hold on. At least we know where we are.' I turned and looked at her, and the appalling truth dawned on me: she was still stark naked.

I looked up and down the street in horror. There was no one on the pavement except an old wino huddled against the railings, and a shuffling woman weighed down with carrier bags. We were still unseen.

'Here,' I said, struggling out of my jacket. 'No, no, that won't do, it's too short. Oh Christ!' I thought desperately. 'You'll have to take my jumper. You can hold it down at the ends.' I tore it off.

'What are you doing?'

'Taking my clothes off,' I snapped. 'Here, put this on.'

She stared at it as though it might shrivel her up. 'Why? What is it?'

'It's a jumper.' I realised, even in my desperate embarrassment, that more explanation was going to be necessary. 'You have to wear it. Put it on. Like I did. You can't walk around here stark naked.'

She looked round, her eyes dilating with fear. 'What is naked?'

'I'll tell you when we get home. Now look, this is Edinburgh. Anywhere else, they'd probably take no notice. Put it on, for God's sake. Trust me.'

Well, I got it on to her, and explained the necessity for holding the ends down so that they reached to her thighs. She was puzzled, but docile. I put my jacket on again.

'Come on, quick.'

'Where are we going?'

'Home. It still looks most odd, though at least you're decent. But hurry.'

She didn't seem to have any idea what the trouble was. In Princes Street the shops were shut, and the crowds had gone, but the evening groups of kids with nowhere to go were just starting to assemble. We had to pass a gang of boys with motor bikes as we scurried through the shadow of the Scott monument, and sure enough they stopped and stared, and muttered as we passed. I'd rather they'd shouted something lewd, it would have meant they didn't intend to do anything else, but they just looked after us in silence. Nosila's bare feet left marks on the pavement, still being wet from the leaves. I didn't look at her. I didn't want my worst fears confirmed. But I remembered the wreaths of bryony, and shuddered.

In Hanover Street there were people passing to and fro, and cafés open, as well as long queues at the bus stop. I hesitated, wondering if a bus would be easier, but then, I thought, there was no way I could allow Nosila to sit down, and it would be light inside. I grabbed her elbow and hurried her on.

That's when I thought of you. The idea of walking all the way home appalled me but your house was quite close. True, I didn't know you very well then but I thought you'd be willing to help, and that was the most important thing. As we hurried downhill I tried to think how to explain it to you. You hadn't struck me as being particularly quick on the uptake, when we met. Perhaps the less said, the better.

Nosila broke into the tail of my thoughts. We were just passing Queen Street Gardens when she pulled away from me. The next thing I knew, she was trying to climb the railings, abandoning every vestige of decency in the process. I grabbed her by the tail of her jersey.

'What are you doing, for God's sake? Not that way!'

I had to prise her hands away from the railings. I couldn't have done it if she had gone on resisting, but she seemed to give in suddenly and turned to face me. She looked wild-eyed and desperate, like a sheep that has been separated from the rest of the flock being brought in by dogs. I would have preferred her to cry. It would have made her seem less alien.

'What are you trying to do?'

'I want to go back where it's dark!'

'You can't. It's locked up. And it would be dangerous. Come on, I'll take you home.'

'What is home?'

'Safe,' I told her, and hurried her on before she got the chance to argue. We reached your street, I pressed your buzzer and thank God you were in. I thought for a moment, then decided to leave her at the foot of the stairs. I just couldn't see myself explaining everything.

'I'll be right back. Stand against the wall and don't move. No one can get in without a key. If they do, just stand back against the wall and don't say anything.'

She stood at the bottom watching me. I tore up the five flights of stone steps as fast as I could, and found your door open at the top.

It would have been easier to explain if I hadn't been so out of breath, but I can't say you were helpful. The way you said, 'Trousers?' as if I'd asked you for a time bomb, I could have hit you. That's why I gave up. You seemed to have a bee in your bonnet about lending a simple pair of trousers, as if I were about to go off and do something diabolical or humiliating with them. For Christ's sake, I thought, he thinks it's Rag Week or something. Anyway, that's why I ended up just asking for a towel, and that was hard enough. I never met anyone as suspicious as you seemed to be that night. And when you did produce a towel, it wasn't exactly generous, was it?

Anyway, it served to get us home. I tied it round her like a kilt, and tucked it in. She looked as though she'd been interrupted in the middle of a sauna, but no one accosted us. I realised when I got home that I was completely shattered, and so was she. I fried up herring and tatties for supper, and that made the world seem slightly more tenable again.

About halfway through the evening the phone rang. It was Kate. 'Alison?' she said. 'I was just phoning to see if you wanted a lift to-morrow.'

'A lift to what?'

She was talking about that party at the gallery. I'd forgotten all about it. It was the last thing I felt like dealing with, but I can't afford to miss any possible opportunities, so I said I'd go.

I offered Nosila a bath, and that intrigued her. She seemed to think it was funny. I found her attitude slightly irritating, but at least amusement brought the colour back to her cheeks and restored her equanimity. After that I made a cup of tea and we sat down to discuss the situation. At least, that's what I'd planned to do. She didn't seem to know what a discussion was. She was more interested in my houseplants. I watched her wandering from one to another, apparently whispering sweet nothings into their leaves. It seemed like a caricature of myself going round with the Baby Bio, but I'd never behave like that if I were a guest in someone's house.

'Nosila, we have to talk about things.'

She ran her fingers along the mantelpiece. She might have been testing for dust, but I understood by now she was only intrigued by the feel of paintwork. 'You talk a lot,' she remarked. 'Does it make you happy?'

'I don't. In fact I'm remarkably anti-social. But we have a problem.'

She looked round, but evidently saw nothing that might be thus defined.

'Do you think one of us is dreaming?' I asked her next.

She shook her head helplessly.

'How do you think you got here?'

'Like you,' she said. Then she sighed. 'Perhaps he was right to be content.'

'I don't know what you mean. But think about this: if I bite the apple, we get where you were, and if you do, we get to where I was. Am. I mean we are.'

'Yes,' she said.

'But we can't either of us get back to where we were. Not both at once.'

She seemed to be listening carefully. 'Entwined,' she said.

'I beg your pardon?'

'Hold out your hand.'

I held out my right hand. She came round beside me and laid her own against it, palm to palm. Our hands were the same size, small but square and firm. Hers were more roughened than mine but there seemed to be no other difference. Moreover, her hand was quite real, flesh and blood. I could feel the warmth of it against mine. 'That's touching,' she said.

'I see it is.'

'You want to let go?'

'I'm torn.'

'Yes,' she said. 'We can't have two places at once.'

'Then we have to find a way of separating.'

'No.'

'What do you mean, no?'

'It's all one.'

I gave up then, and suggested that we went to bed.

The morning was sunny and hopeful. She was still there. I found her drifting naked round the kitchen, tasting the food from the jars.

'You can't eat raw flour. We'll have breakfast.'

I was in a practical mood and, as the major problem seemed no nearer a solution, I applied myself to the more immediate task of acclimatising her to the world as it was. I took her shopping, and explained to her about traffic lights, and crossing roads. She kept bumping into people, and I taught her to say 'Sorry' every time. I had cause to regret that, as it left very little opportunity for saying anything else. She just didn't seem to see people coming. There were other problems. She stopped and wept over the greengroceries displayed outside the shop on the corner and, when we passed a row of gardens, she kept trying to climb the low stone walls. By the end of the morning I was ready to try anything to be rid of her, even if I had to bite that wretched apple all over again.

In spite of it all, I decided to take her with me to the party that night. I didn't dare leave her alone in my flat and I didn't want to miss the party. I was too desperate for work to miss such an opportunity: it was the opening of a new exhibition. I knew the people vaguely and the more they saw my face the more likely they were to employ me.

I don't know what I'd have done if you hadn't been there. You hadn't distinguished yourself over the towel episode, but that evening you saved me from social disaster. I still don't know whether that was just a ploy but I suppose I had some idea what you were hoping for. Anyway, in my eyes you redeemed yourself.

I was trying to talk to the person I most needed to impress, but I could hear your conversation out of the corner of my ear. It made my own somewhat disjointed.

'A friend of Alison's,' you were saying to her, 'I thought you must be a relative.'

'Yes,' she said.

'Are you involved with art too?' you asked her.

'What is art?'

'Help,' I heard you say, 'I don't know that I'm in the mood for intelligent conversation. Are you a student?'

'No,' she said. 'I'm frightened already. I don't want more.'

'. . . And of course,' the important person was saying to me, 'you'll know the work of so and so and such and such.'

'Well,' I said cautiously, 'naturally the names are familiar . . . Wasn't that . . . ? Oh, yes, when would it be . . . ?'

'1985.'

'Of course,' I said, 'a breakthrough.'

'You don't think that was mostly hype?'

'It's amazing what marketing can do,' I said.

I had to concentrate for a bit. When I heard you and Nosila again, you were talking about the sex life of plants. She looked relaxed and happy, almost as I had seen her first. I took a deep breath, and turned back to my own conversation. 'No,' I said, 'I haven't actually read it, but of course I saw the reviews . . .'

There was a hush, as the tape of soft music wound to an end, and for a moment conversation died with it.

Nosila's voice came loud and clear, 'Why do they sway together so, like trees in a gale? Why do they express anger and sex at the same time? What are they trying to do to one another? Do you know?'

There was a silence.

'We must all have asked those questions,' you replied easily. 'A highly ambiguous work of art. Everyone thought so.'

Nosila looked puzzled but I could have kissed you. A month later I did, of course, but by that time all was well.

'She must have seen the preview,' murmured my companion. 'It's booked out now, more or less until the millennium, I believe.' She chortled at her own joke. 'Luckily I was sent tickets.'

'What is?' I said, and realised that I sounded like Nosila.

'The performance to which that girl referred. Surely you've come across it? It was reviewed everywhere.'

'Oh yes,' I said, subdued.

I had no idea what chivalrous impulse had prompted you but when I managed to get a word with you, just before I left, I said, 'Thank you for taking care of my friend. She really isn't used to this.'

'I admire her for it,' you said, and left me feeling confused.

I was even more confused when you phoned me three weeks later to say you'd been meditating in a tent in the Cairngorms ever since, and you wanted me to come and live on an uninhabited island with you to grow potatoes and study the ancient philosophers. However, I was tempted. I'm glad we came to a more reasonable compromise than that, but that's not part of this story.

I woke next morning from a deep sleep and saw that the day was drear and grey. I pressed my face into the pillow, trying to recapture the dark. There seemed to be no greater gift than oblivion. Evading me, it hovered at the corners of my consciousness, tempting me onwards into a desire for something too close to nothingness for me to pursue with any courage. Reluctantly I sat up and faced the day.

It was raining. That upset all my plans. But then again, why not? At least the gardens would be deserted. I got up and surveyed Nosila. She was sleeping flat on her back, the covers flung away, her hair spread over the pillow like a black halo. To my jaundiced eyes, she looked like something out of Aubrey Beardsley. Irritated, I nudged her with one slippered foot. 'It's morning,' I said. 'I'm going to make a cup of tea.'

She was fully awake at once, jumping up eagerly to look through the window at the rain. It evidently moved her: she began to chant and then to dance around the room, thumping on the floor in a manner calculated to get the man downstairs balanced on a chair, thumping back on the ceiling with the end of a broom handle.

'Stop! You can't do that here.'

'But it's raining.'

'You can't do that even if it's Noah's flood. Here, borrow my dressing-gown.'

She didn't want to get dressed that day. She kept murmuring that it was raining and what was the point? She seemed to be in a fever to get outside and, sure enough, once we were on the pavement she went leaping down the street, banging her fists against her chest and ululating at the sodden sky. The curtains of the flat below mine were seen to twitch and to remain poised half held, like a blocked-in question mark.

'Nosila, stop. You can't do that here.'

Once we were on the path by the water, I let her go. She went dancing and leaping under the trees, her bare feet light as fallen leaves. I never did manage to make her wear anything on her feet, although otherwise I persuaded her to dress fairly normally. It was a relief when we finally turned in at the gates of the garden. I cast a doubtful glance at the policeman in his box as we passed, but he was reading the *Express* and didn't see us. Not that we were contravening anything, but Nosila certainly looked as if she might.

I felt I needed time to think, out of the rain, so I took her into the plant houses. Now, when I think of it, I am more satisfied about that than about any of the rest. I would like to think that she gained something from her experiences, and nothing else that I could provide seemed to make her any wiser or happier. But for her that place was enchanted. She touched everything, while I kept a wary eye out for patrols. She ran her hands up and down the stems of the palms, and poked her fingers gently inside the orchids. When she dabbled her hand in the pool, the carp came up and nibbled her fingers, and she laughed. The ferns astounded her. She stood in their house for a long time, as if she were listening. I was so near I could hear her breathing.

'But they are so old,' she said at last. 'Now there at last is a story worth the telling.'

'What story?' But she was away, her hands moving up the vines.

By the time I got her out of there, the clouds had cleared a little. We walked up to the crown of the hill, and stood beneath the beech trees. The sound of water dripping from trees was cold and mournful. Dew clung to the grass, so that we left green footprints over a lawn that had been white. Nosila took the apple from her pocket. It was wrinkled now, and the two bite marks had gone quite brown.

'You try first,' I said to her.

She bit, and screwed up her face. 'It's sour now.'

Nothing else happened.

'I was afraid of that,' I said, and held out my hand. The apple looked thoroughly unappetising. I was aware of a pit of fear yawning inside me, and I think my hand was shaking.

'There's no other way,' she said quietly. I sensed her desperation. I could refuse, I supposed, but then I'd feel responsible for her ever after. I hate responsibility, which left me no choice. I bit.

Since it was wet this time, it felt like being run through the rinse programme of an automatic washing-machine. We were caught up in a blur of spinning water, gyrating wildly. Sky, grass and trees melted away into rapids, all turning water-coloured. We were too entangled to keep our balance and when we fell we stayed entwined, like a four-legged monster gasping in its lair of thick grass, for the grass was long again.

When we sat up we could see nothing but feathered seedheads bending before the breeze above our heads. The broken stems gave us a softer landing than before. Once we had reclaimed our respective limbs, we stood up cautiously.

The rain had gone. Sun dappled the grass, patterned by the shadows of the leaves. The air was heady with the smell of apples. All round us the forest was turning golden under the morning sun. The sky was pale blue, with a thumbnail moon rising high over the ridge to the south. This time I studied the horizon more closely. There were precipices over there, surrounding a craggy bluff topped by trees and, further east, a higher hill whose shape was more than familiar. The expanse of forest in between set it far beyond my reach. A line of grey cliff, surmounted by autumn gold, all still untrodden. I touched Nosila's shoulder. 'Over there.' I pointed. 'You've been up there?'

She nodded, but her eyes searched the nearer woods restlessly and she hardly looked where I was pointing.

'What do you see from there?'

She glanced again. 'From there? The sea, of course. Beaches.'

'What colour is the sea?'

She laughed at that, as though it were a foolish question. 'All colours,' she said. 'Like the sky, like the forest, like the loch. All those. Changing.'

'I see.'

She cried out then, throwing back her head and making a sound that must have echoed almost to the crags, a long, yodelling call that threw back an echo, then left the woods more silent than they had been before.

I stood trembling, not understanding. There was a movement in the trees, shadows passing, and the flick of branches pushed back. Then a man appeared, standing in the clear space that surrounded the apple tree, naked as the day he was born and brown as a naturist back from the Riviera. Nosila left my side and flung herself upon him.

I turned my back, and tried to steady my pulse by reciting all the regulations governing behaviour in the Botanics, from beginning to end. I had just reached the part about no sketching, painting or picnicking, for the third time, when various animal-like sounds behind me heralded the end of my ordeal. I counted to three hundred and turned round.

Nosila was on her feet again, watching me anxiously. 'Are you ill?' she asked me.

'No,' I said, 'I don't know what I want.'

'No?'

'I want both,' I said, suddenly understanding, 'but I don't know how.'

'Let me think.' She turned back to the man and began to talk to him in a low voice. I hardly listened. The apple was still in my hand. After the three bites we had already taken there was not much left. My time was running out. Was this my world, or hers? I only knew that I wanted to keep it, more than I had wanted anything in my life. I couldn't stay; I thought of you, without intending to. I wanted what Nosila had, but not here. I couldn't inhabit her whole world. There was no place for me in it. I glanced up at the apple tree.

There was my answer, staring me in the face.

'Nosila!'

She looked at me, still troubled. 'What is it?'

'The apples,' I said. 'I'm going back now, but I want my apple.'

She was puzzled for a moment, then she laughed, suddenly understanding. She pulled a branch down until it was level with my face. It was heavy with apples, two or three on every spur. I selected one, and twisted it off.

The man touched her on the shoulder, and muttered something. I wished he would stop hovering around, and leave us in peace.

'Why not?' said Nosila out loud. 'She's free to do what she likes. Who's got the right to stop her?'

He shrugged.

My apple was smooth and sweet-smelling, red and gold and green. I put it carefully in my pocket.

She watched. 'That's the first good reason I've seen for them.'

'For what?'

'Clothes.' She waved her hands in an effort to explain. 'A way of carrying apples. That's all.'

'Yes,' I said, and hugged her.

'Are you ready?'

I nodded.

'Then give me the first apple.'

The man said something incoherent, and she shook her head at him. 'No, it's all right. We'll only be a few minutes.'

He frowned.

'I promise,' she said, and took the remains of our original apple from me.

I was elated, excited by my own cleverness. I could have both. I had an apple of my own, and so all the worlds I wanted were now open to me. I would go home and Nosila would be here, also at home. Separated, but entwined. I took her hand. She bit the apple.

The whirlwind that followed was nothing to the power I felt inside. I was drunk with it, my head still spinning like a top long after the rest of me had fallen with a thud upon the neat mown turf. I hardly noticed that Nosila had lost all her clothes, and here we were back in civilisation. It didn't matter. The familiar Botanic Gardens were all around me, and imprinted on my mind was the image of the wilderness they hid.

Nosila held my hand and we walked quietly down the hill. People stared at us, as well they might. We crossed a shrub bed. Luckily no uniformed official was in sight. There were railings in front of me. Eight-foot-high black railings and the road below. I felt the curve of the apple in my pocket and was reassured. I could see the Christian Science Church opposite, and the line of country to the south all blurred by rain.

'If I hold on tight,' I said, 'whatever happens, when it starts, you go. But I'll hold on. I think it'll work, if you hold the apple.'

Her black eyes looked hollow again, as they had the first time I saw her. I would miss her physical presence, in spite of what I knew. I realised I was crying for the loss of it, although I knew it was illusory. She shed no tears. Perhaps she had none. I looked through the bars to the world outside.

She followed my gaze. 'You belong there,' she remarked. 'Hold tight.' She held the apple to my lips.

I bit, and spat.

The world flung away from me. The bars bucketed like a boat in a storm. The apple was wrenched away, and something else, splitting away from my side like my heart being torn out, but I went on holding. I held so hard the iron bit into me, and I heard someone scream. It was I, and someone other, spinning away from me out of the world. Through the bars I saw a forest turning, leaves torn before a driving gale, and the sky circling below. Then slowly the grey road subsiding, the squat church, and a skyline of spires and defences. I sank to my knees, soaked, and exhausted.

When someone touched my shoulder I swung round, not knowing what to expect.

It was a policeman. Looking down, I saw his black polished shoes sinking into the newly dug soil. For a moment I was disorientated, then I felt the apple in my pocket, firm and real.

'You all right, hen?' he asked me.

I stood up slowly. 'Yes,' I said, without attempting to explain, and smiled at him, secure in my own knowledge.

Ronald Frame

1953–

Merlewood

When a person dies, his portraits change.
His eyes see differently and his mouth smiles different smiles.
Anna Akhmatova

Every picture tells a story. Not a 'picture' in this case, but a photograph. Preserved in the darkness of a trunk, against the odds: faded and worn from the mint black-and-white first impression, but without the damning sepia rinse.

* * * * *

We are eight.

It is a summer's day, in the mid-1960s. We are arranged on the lawn in front of Merlewood. The bay is visible in the left-hand portion of the photograph, and the single rock jutting up two or three hundred yards out, and the thin line of horizon between the sea and the vast West Coast sky.

Gregor stands in the middle of the group, rather formal for the occasion, but that is his way; he's in his late teens, and his face is already set into adult seriousness. On his left is Kirsty, sitting on the travelling rug that's been spread on the scorched grass; her left arm rests on my mother's leg. My mother occupies a low deck-chair, and is her Kelvinside self, seeming to enjoy and welcome the moment, royally receiving this person who has come to photograph us. (I can catch a trace of the lily-of-the-valley toilet water that made a haze round her movements and actions, which she sprayed on herself for a very practical purpose—to keep the Tighnabruaich midges away.) Immediately in front of Gregor, Angus lies on the rug facing forwards, supported on his elbows and with his chin cupped in his hands. Beside him, to his right, Sandy sits cross-legged on the burnt grass, twisting the bamboo pole of a shrimping net and screwing his eyes up against the sun. Behind Sandy, I stand in vague imitation of Gregor, who's on my left, but I can only reach his shoulder for height. (I still have to learn to dress like him: I'm in a t-shirt and shorts, while Gregor wears long white flannels and a Clydella checked shirt open at the neck.) On my right is Rhona, lying in a striped deck-chair and patiently holding a book open in her lap as she waits for the photographer to be done.

That's seven of us.

The eighth person is my father, on the edge of the photograph, on my mother's left. He's standing, one hand on the back of the deck-chair, his other arm hanging loosely by his side. The hand that appears to be doing nothing is, in fact—so I discovered when I looked very closely at the photograph—clenched, the fingers are curled up into the palm. He smiles, as my mother and Kirsty and I all smile, but with him it is a quiet mysterious gesture. He knows or understands something we do not, and while he is one of us he is also a man apart. The rest of us look at the camera: his eyes are focused on some point out of the picture, well out, on the sea side.

* * * * *

Cloud, sea, sand, rock.

And Merlewood.

Merlewood had been my father's father's house. It was built by one of the Clydeside ironmasters to accommodate his Victorian family—wife, children, governess, nanny, cook, chauffeur, coy servant girls brought up from the city. There were eight bedrooms and an attic of rooms, and two turrets, and a scaled-down portico à la Balmoral, and steeped corbie-stanes on the gables, and a flagpole to fly the Union Jack or the Saltire. The building had weathered very slowly, bald grey sandstone against the blue Atlantic sea-skies, the bracken hill behind, the rhododendron bushes studded with purple, pink and white heads.

We spent every summer there, and an occasional Easter or September weekend. A ritual developed.

My parents drove the hundred or so miles from Glasgow.

The day before, the rest of us would sail up from the Broomielaw on the steamer.

We would stand watching at the rails for a first glimpse of Merlewood. On overcast days the chimneys would be smoking, in sunshine the windows shone like mirrors.

Merlewood was waiting: and it was as if—I would think—I only had to see it again to become the 'other' version of myself.

As the distance to the shore grew less and less I had the sensation of sailing into last summer, and the one before, and the one before that.

We nudged against the pier; the holding ropes were thrown, and the gangways were hoisted up to the gates.

Our feet clattered over the soft, salty, echoing wood.

Trunks and cases were loaded into a Humber taxi, not the pony trap of my father's time.

Gregor and Rhona and I would start walking, as we always walked, while Kirsty and Sandy and Angus bundled themselves into the Humber.

Tighnabruaich grass was springy with moss. The sun dazzled off the sea. Wavelets slapped about the rock in the bay, standing in line with the house.

The red road was always redder than I remembered. Bees as big as halfpennies stuck to garden walls. The Scoulers had this year's new model of Daimler. Rhona walked as our mother walked, with a woman's rolling hips. The steamer's hooter blew. The taxi was disappearing out of sight along the coast track, spreading a balloon of white dust behind it.

The real summer had begun.

* * * * *

My father's quiet smile.

He's not smiling as my mother and Rhona and I do, declaring ourselves to the camera.

He smiles so enigmatically because what *he* knows none of the rest of us know. He shares his knowledge with whoever watches from the future and recognises the ironies of the scene: the midsummer sloth, the carelessly splayed limbs, the props (deck-chairs, sun umbrella, shrimping net, tin spade, picnic hamper), the careless expressions and open smiles.

Even holding the photograph to the light, at first I could see no more than is there: there seemed to be no new angle of discovery, no possible plane beyond the flat one presented. Some aspect of the picture was unsettling, though, and it was only after several days of living with the image that I suspected what it might be.

* * * * *

There used to be photographs of my mother in whites, playing tennis on a court in Zanzibar. Servants stood in the background and menials watered flower beds while she lunged for low, devious shots and jumped up on supple ankles to make fiercesome returns.

She began to forget the names of the other schoolgirls and young women she'd played against in the heat, in advantaged circumstances, although once she had been able to remember and could reel them off like a litany, Dutch, Australian, Swiss, Portuguese, Venezuelan.

Her uncle was second-in-charge of His Majesty's interests in the Government Protectorate of Zanzibar. A few months after the War ended, escaping British greyness and austerity she had been shipped out to the sun and plenty of the island. The plan was that she would live with her uncle and aunt and cousins for the next three years, in their colonnaded, fan-ventilated official residence, on the pretext that her education would be broadened by 'the experience of Africa'. There were other reasons, however, less noble-sounding and rather more pressing: later I guessed at them, and the likelihood (the necessity, surely) that the removal had been effected at her uncle's expense.

By then the history of my immediate antecedents was clearer to me. After his Navy days my mother's father had embarked on his married life in the Argentine; with family funds he'd set himself up as a cattle farmer, but after seven or eight years he realised that he had no natural aptitude for the business. He took his family back to Britain and repeated the earlier mistake by buying another farm—sheep grazing and arable land this time—in Perthshire. Without a farmer's sixth sense to rely on, he fared no better than before. Each new season was worse than the one that preceded it. He was a proud man and he blamed his many failures however he could. He was also a susceptible man—or merely given to self-destructiveness in the Scottish way—and he was guilt-racked for the sake of his wife and children. He retreated into the only two enthusiasms the years had left him: painting watercolours 'en plein air' (which excused him from the house for a few hours at a stretch) and, his favourite indoors pursuit, helping himself to generous measures (ever a liberal, in stolen pleasures as in politics) from the decanters on the dining-room sideboard.

Some of the watercolours survived and hung on Merlewood's walls. I grew up with them, hardly noticing them: delicate, understated views of rolling Perthshire, the hills and glens, Lednock, Devon, Farg, and Strath Bran. The perspectives were always very careful, and exact. By a cruel irony, my mother's father's most dependable source of inspiration and consolation was the amber spirit ('peat water' was his euphemism) by which my own father made his living and *his* father before him, and which had afforded Merlewood in the 1920s.

The affairs of the farm were constantly in a crisis state. The bursars of the children's schools (discreetly distanced in neighbouring counties) began to sound rather less gentlemanly in their letters of reminder. Repairs in the house had to be overlooked: social appearances slipped. The watercolour painting stopped when my grandfather couldn't trust his hand to hold the brush steady. All that was left to him then was the other private solace.

The parental roles were reversed after that. His wife became the provider, living on her wits, and he became the dependent. My mother, the youngest daughter, went out to East Africa and lived grandly during these three impressionable, formative years of her middle teens. Employment was found for her two brothers via 'connections', and—while my mother darted about the tennis courts of Zanzibar in cosmopolitan company—on the home front her sisters were directed towards sensible, low-risk and utterly predictable matches of another kind. Wedding bells rang in clean, scrubbed, scrupulous Dunkeld, and this or that field of grazing was sold to settle accounts on the day. On the cedar lawn the same marquee was erected three times—and a fourth time for my mother, when it was her turn—to house the celebrations. As my grandfather struggled to hold his composure till the guests would leave, his wife could remind herself that at least the token rituals of the order were being respected.

My mother's husband came from the West Coast, unlike the others. In Zanzibar she'd been introduced to a London cousin of a Glasgow family who worked in a shipping office. An invitation from the London cousin, who was visiting his relations in Glasgow, reached her a number of months after her return to depressed, grey, wanting Britain, to the far, slumbering backwaters of Perthshire. The family were called Hamilton and lived elegantly in Kelvinside (she had forgotten elegance since Zanzibar), in a stone villa with a high terrace that had views over the flatness of Glasgow, across the miles of tenement streets and smoky works to the cranes and jibs of the shipyards and the giant hulks of foundries. (For reasons best known to herself, my mother never forgot those, the spectacular views.) The Hamiltons were whisky brokers: the father (my grandfather, the purchaser of Merlewood) and a bachelor uncle were joint overseers of the firm. The one son Alasdair had come back into the family fold after his time at Cambridge; the supposition had always been that he would take his place in the business, and when my mother first met him he'd been working there, assiduously, for three years. The difference between stolid, redoubtable Alasdair Hamilton and her defeated father in Perthshire must have reassured her. She neglected the Londoner's attention in enjoying his cousin's grateful, rather sober company. 'I think I changed you,' she always claimed, and he wouldn't fail to acknowledge his debt to her for that.

In the next couple of years he came out of his shell—it could only have been my mother's doing—and he added a personable public exterior to the native temperament relied on by his father and uncle, which was diligent, industrious and (by common regard in whisky broking circles) enterprising. Maybe those were his best years: with the confidence to take financial gambles when his father and uncle had final liability, forcing an entrée with my mother on to the social scene and cutting a dash in all the right places. Much later my father would make rather complicated jokes about his youthful endeavours in those polite, fashionable circles, telling us he'd felt he was a suitor required to prove his worth, like the humble man bid to perform certain labours before he can claim his bride, who turned out in his case to have been a princess in disguise. (My father never mastered the art of telling funny stories: usually the funniness was lost, and came out sounding like the seriousness he spoke for the rest of the time.)

It was a very welcome pairing for all parties, and the announcement of the engagement was greeted with manifest approval. It may have been that Alasdair Hamilton and his fiancée were the two least able at first to believe what was happening, happening *to them*: my mother, the youngest of the four sisters, had clearly outdone the others, and perhaps my father thought he didn't fully deserve this exotic Zanzibar-bred prize for his efforts. Otherwise their feelings for each other did prove to be honest and true, and they made a notable and noticeable—if not quite complementary—partnership: my father was disciplined and hard-working, concerned at all times for the greater good of the firm, sometimes (after his father and uncle retired) over-involving himself and relying too much on the risk factor—while my mother with her superior ways and graces simply preferred the surfaces of things, and was happier attempting to influence and win others through her considerable gifts of charm, with maybe a little of that smooth-talking diplomacy she'd learned from her uncle for good measure. Somehow, as a pair, they had enough give-and-take with each other to be able to survive the mishaps and misadventures that occur in any marriage: neither had forgotten their pleasure and relief at finding the other, and that gratitude developed into a resolve to see and trust to the best in each other and to accept the 'worse', if it came, as only a temporary obverse of the positive—the riches and health of their first years together.

* * * * *

Looking again at the photograph of the eight of us, I'm prey to another suspicion: that even then Merlewood was more than a house and a garden, already our history there was passing from fact to the half-imagined.

Merlewood was becoming a myth before I was out of my adolescence, and its inhabitants—the differences time made to us seeming more marked with each successive year—were also possessed by the myth. The journey to Tighnabruaich was made every summer in search of other selves, *possibilities* of selves we'd outgrown: in our middle teens we were beginning to settle into later versions of ourselves, but somehow the illusion persisted that we could undo that confirming process of time.

In the photograph we seem too close to the images of ourselves we *should* be: the props might have been decided by a professional photographer over the telephone, the composition surveyed in advance, the gestures advised on. At first sight careless and unaffected, the group strikes me more and more as playing to the camera and humouring the expectations behind it.

But what expectations, and whose?

* * * * *

The garden was all things to us: Arizona, the Spanish Main, Swallowdale, Outer Space, the Amazon, Treasure Island.

For Gregor's twenty-first birthday party we hung it with paper lanterns, and we forgot its familiarity. The candles glowed in the trees and among the rhododendrons, and our shadows stretched ahead of us for yards, climbing hedges and walls, scaling the sides of the house. That evening we were all granted strange powers of magic and dissembling, and we hardly as much as thought about it.

Someone took photographs, but the results had nothing to do with my perceptions of the evening. We only looked dazed by the flashbulbs; the males of the company seemed rather foolish in their too generously cut dinner jackets (Gregor wore a kilt), and the females uneasy in long dresses that were at least twenty years behind the fashion. Gregor was in fine form, because Annabel Lavery's arm was linked through his: in their evening attire the pair were the very picture of middle-aged respectability.

For years now their photographs have appeared in the social pages of *Scottish Tatler* and *Scottish Field*, and in *The Scotsman* at legal and Unionist functions, and their expressions and arm-in-arm poses, like the kilt and the cut of dress, have never varied from the original.

That night they could afford to look contented and just a little bland: at last Lavery Senior had the heir he required for the practice. When they were photographed dancing, it was marionettes I was reminded of: the gestures, like the smiles, were extravagant, or merely cautious, as if they were both afraid that the wires they worked on might become entangled. There were strings attached to the arrangement, of course, but they'd known that from the outset, from the moment Mr Lavery had welcomed Gregor to the first 'At Home' in Edinburgh. I could say they had no one to blame but themselves, but I know that, like the victims sacrificed in societies we consider to have been more primitive than our own, their lives were not really theirs to hold.

In the lantern-light, maybe they could have believed to the contrary.

It was Merlewood's finest evening, its supreme deceit. The Laverys had come, and the MacNaughtons, and the Todd Hutchinsons. We were eminently desirable company, and Gregor was forty years old, not twenty-one: you only had to look at him to see him in his wig and gown, or mixing drinks for a colleague in the Heriot Row office, or being lunched at a prospective client's expense in Edinburgh's own Café Royal, in the Oyster Room.

In the end it was Gregor who attended to all the legal complications when the brokers' business was compulsorily wound up, and maybe my father knew that too, seeing more in the situation as it clarified itself that evening than the rest of us could: even the ablest of our number, Gregor—ductor exercitus—wasn't to be spared.

* * * * *

But it's that earlier photograph which intrigues and mystifies me, which draws me back: ourselves on the lawn at Merlewood, one perfect summer's day in what appears to be our innocence.

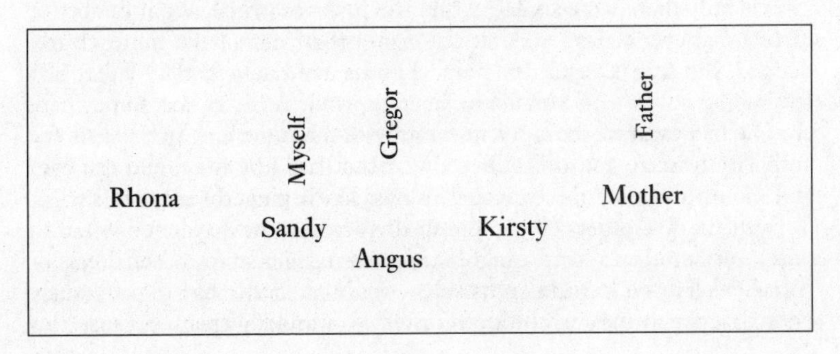

I see that these faces, intercepted at a split second one August afternoon in 1966 (the month and year have been scribbled on the back), are outside the tense in which they are living.

What it has taken me so long to perceive is this: that—at some level—we must have been conscious of the fleetingness of life and its conditions even as we were living through them. It's as if, by some most curious process of misdirection, the psychic instinct of generations past in the Western Isles has passed to us, we've picked up an echo of it in that garden: we see forward to ourselves looking back, with the equivocal aid of the Celts' third eye. Carrying the consciousness of the moment's loss and distance, we are beyond recapture.

The faces confront the lens, and that tunnel refracts the image, concentrates it into the future when these things will have passed: when the shrimping nets are lost or broken, the woollens thin and mothy, the hair of the oldest children starting to show grey, the eyes of all of us acquiring the habit of hesitation to match our sensibly non-committal manners.

The faces invite the imprints of age, they are empty vessels for the future. They are icons of the change to come.

* * * * *

The photographs of Gregor's twenty-first birthday party were mislaid, or destroyed; or maybe Gregor discreetly removed them later, either for my mother's peace of mind or his own.

All the photographs disappeared, that is, except one, which survived—for a dozen years or so, until I lost it by my own carelessness. It showed Gregor and Annabel framed by both sets of parents. What principally interested me about it was my father's face, for two reasons: because we had very few mementoes of how he looked, and (although I didn't realise it then) because the face in the photograph was doing more than presenting a public front, much more, it was betraying his certainty of the end that was coming to him.

On the evening of Gregor's coming-of-age party, something momentous happened, and we none of us paid it any attention. What it was was this: to all intents and purposes *my father wasn't with us, he was absent.*

Certainly there were six faces, and the photographer had got the better of the shadows at last, so that the features of each were quite clearly defined. But that is beside the point. I understand now that my father had left behind only the impression of his face, while really he was somewhere else: he had escaped the insolent tyranny of the camera, which boasts the authority to seize and reproduce. My father had got away, and the eyes and mouth did only the expected things, like a gingerbread man's.

With the five others in the group there was some accidental detail to note—my mother's hand caught straightening the strap of her dress, or Annabel's left foot resting on its side—while my father had too obviously been dragooned into exhibiting himself: he was on display, because this was 'An Occasion', he was affecting ease when no one else could quite manage it. His arms hung straight, his hands were positioned in line with the seams of his trousers, his shirt cuffs showed the respectable distance beneath the jacket's, his pumps were in perfect alignment. If I could examine the image again, I would see that he had disciplined his face to the point where it would reveal only what each moment required and would hazard nothing more, no possibility of a random emotion.

Even the buttoned-up Scots can let slip how they feel, but there was no danger of that with my father. Tradition—chastising and self-denying—had perfected itself in him to the 'n'-th-degree: he was its apogee. Most people might no longer subscribe to religious tenets of the 'elect' and the 'damned', but he had worked his own secular reinterpretation of them. If the damned can say to themselves, 'well, what the hell?' and do as they like—which is only what they must, after all—he preferred to put his trust in the possibility of suffering nobly: the stoic ideal, the victim-renascent as hero and martyr.

* * * * *

In the photograph where we are eight on that August afternoon, past and future cross, momentarily, in the time that was our 'present'.

The house and garden had seen their best days, and the decline was now continual, from season to season: but in the photograph Merlewood wins a reprieve, and someone who knew no better might put a more favourable reading on events already too far gone to be halted.

What seems especially untrue about the photograph is the group pose: how else, except for a camera, could eight people have been distracted from their doings to concentrate their attention so fixedly on one point? That is not how we were, never: it was impossible, except at mealtimes, to hold us, and in the dining-room we would talk over each other and listen or not listen and our simultaneity wasn't really that at all.

In the end, of course, the past caught up with us in another sense and we were shocked into the kind of stillness I see in the photograph. But by then we were seven not eight. Without our being aware, Merlewood was like a colony suddenly liberated, returned to its natural condition—wilderness.

The question only occurred to me after looking at the photograph many times, who was it who took that 'likeness' of us, finger on the shutter button or watching our reflection, squared off, upside down inside the box of the camera? Was it someone hired for the purpose?—or one of my father's visitors who came up from Glasgow to the house? Perhaps one of our neighbours wanted a record?—but there is no evidence among the props of our sharing the day with anyone else, and our visitors (especially my father's) invariably came in couples or posses.

Who was he, or she? The clues are lost. Memory has failed on that one point.

Without the memory, the history of that episode is incomplete. Not knowing, not even remembering the moments of preparing for the photograph and then having it taken, it's hard to believe that it can have happened. Yet it must have done, because here is the evidence before me.

The person is unknown: he or she was only the agent or medium, with a well-versed amateur's skill in organising a group shot. The work that resulted is a study of ghosts, of a kind: children who are already middle-aged—a woman who learned how to keep the facts at a distance just as she defeated the midges with her lily-of-the-valley toilet water, smelling of funeral flowers—and a man who's swimming through these days with one hand held behind his back, his eyes focused on some point out of the picture, well out, past the rock, where the bay becomes sea.

* * * * *

We failed to see when the driveway greened and the lawn sprouted lichen, or else it failed to matter to us.

It was a common predicament. The other houses we visited on the Cowal were also past their best.

We were caught in history's downward spiral. The steamer was taken off the Tighnabruaich route. In the late 'sixties and early 'seventies people openly mentioned money in their conversations, with hostility usually, which was not how it had been in the beginning. Cars (the only mode of transport) were required to last longer, and things that wore out weren't necessarily replaced. In the house the curtains faded in the sun and as the loose-covers frayed with too many washings the Sanderson rose garden turned threadbare; the carpets became thinner under our feet, cracked panes in the windows were sellotaped at their weakest point, paint flaked from the sills. We carried on as if none of it mattered, as if a holiday home should have scuffs and bruises anyway: but Merlewood wasn't as it had been, not in the days when *it* seemed to have expectations of *us*. We had inherited from my father's father what he'd supposed was only our natural prerogative in life, but we found that in fact what we had taken on was the onus of living up to the standards of eighty years before, the Golden Jubilee epoch when the house was built.

It was hopeless.

* * * * *

A man in a shiny suit appeared one afternoon, driving his shiny new Vauxhall Velox up to the front door. He inspected the house and garden and covered several of the pages on his clipboard with notes, and departed. We never saw him again.

Other men came whom we *did* recognise—my father's colleagues from the offices in Glasgow—but they seemed ill-at-ease in their weekend tweeds and safari jackets, eating an afternoon tea of buttered bread with potted paste or date spread in the garden before heading back, the creases of business still on their brows. My mother presided, smiling as if the guests she was obliged to entertain were here for the view of the Sound and Goat Fell and for politeness' sake, not because they knew the worst was about to happen. All the time they were there with us, she didn't let up and played out the scene for all she was worth, with those same skills of dissimulation that always made her the winner at games of charades.

Other visitors came who didn't stay for tea—men in city suits, with briefcases—and she had no option on those days but to leave them and my father to their business in the house and occupy herself outside with a pair of secateurs. As she dead-headed, the group in the sitting-room would watch disinterestedly from the windows. They thought in their wisdom they were instructing on the fate of Merlewood and yet it wasn't really so: they didn't know it, but unseen events were according to a darker design.

* * * * *

We 'children'—Kirsty was eleven, Sandy and Angus were already in their teens, I was seventeen, Rhona was nineteen, and Gregor was officially one of the company of adults—had gone to Kames, with a family of neighbours called the McMinns. My mother saw us off from the front door while my father was elsewhere, making a business call to Glasgow. All his life consisted of nowadays was business calls, at peak rates in the middle of the day.

I thought afterwards I must have had a premonition on the journey back: it seemed to me that the sky was very dark for four o'clock on an afternoon in early September, and peculiarly lit, with a greenish tinge to it. But dark or not, the two black police cars in front of the house were only too visible as we turned in at the gates.

My mother was upstairs, sedated, in the care of a policewoman and the locum doctor. A police sergeant took Gregor into the morning room that served as my father's office while the rest of us stood outside in the hall staring at each other.

Kirsty started to cry. Gregor reappeared, shell-shocked. He told us in a dry, patched voice that our father had been 'carried off to sea'.

We none of us knew at first what he meant: we weren't expecting a metaphor. It almost sounded like a joke.

Kirsty must have understood before any of us. Suddenly she ran off, down the hall, for the open front door. We gazed after her. Sandy set off in pursuit, then Angus.

Gregor explained to Rhona and me: he'd gone for a swim and hadn't returned.

'Gone where?' Rhona asked.

Gregor pointed to the strand of sand we always used, where the garden ended.

'Where is he, then?'

My mother had seen it: the current, she'd said, it had taken him out of his depth. I kept shaking my head, I didn't understand how: we'd swum there for years and he'd warned us so often not to 'pass beyond our limitations'. There weren't proper '*currents*' anyway: the word sounded as bookish as Gregor's first ones to us, that the sea had 'carried him off'.

For years afterwards, until there wasn't the need to discuss it in detail, my mother stated quite determinedly and categorically that it was 'the current'. She had *seen*, she would remind us. Her version never varied, it became an oft-repeated tale, although not told to Kirsty and the twins, who suffered a long spell of nightmares.

In the photograph the bay is placid and still, or relatively so. I see no 'current' straining beneath the glassy surface. The sea looks not unlike a length of some shiny cloth, lightly rumpled on a counter top. It is there waiting for us, and in the next few minutes—so the photograph suggests—some of us will have responded to its pull and be breasting the brine. The shrimping nets and Kirsty's tin spade and the constant horizon behind us are the clues. Whereas, so my mother's account of events would have had us believe, the sea had been a treacherous presence all along: a false friend, only reckoning on its moment, baring that stump of rock like a warning tooth.

The body was washed up on Ardlamont Point. Death was by drowning, so medical opinion would have it, not through the failure of any particular organ; the police termed the cause 'misadventure'. His blood was found on the rock in the bay, splattered on the side we looked out on to from the windows of the house. Somehow it failed to signify anything that my father had been a strong swimmer who'd taught us all and instructed us to respect the sea. My mother started shaping the story she would tell, to us, to our friends, then later to strangers, the people she met in shop restaurants, or in the dining-rooms of the hotels where she decided to live after Merlewood and the house in Cleveden Gardens were both finally sold. The death wasn't allowed to be suicide because that made the story less sympathetic: but as the notion took hold of Gregor, then Rhona, and then me, it seemed an equally plausible reading of events and no less tragic. For wholly practical reasons, it had the edge: even after the company went into liquidation, the carefully regulated funds from the life insurance policy permitted my mother to keep on the most unlikely of homes for us, Merlewood. But a willing death would have been too melodramatic a tale for her to recount to strangers, and it ventured on to foreign territory—the realm of the psychological. It would have betrayed all those family summers and turned my father from a hapless victim of the sea to someone she couldn't have contemplated having shared twenty-five years of her life with—a man alone, a knowing pragmatist who had also, with crystal clarity, recognised his own ineffectualness and failure.

* * * * *

So, of course, it couldn't be.

And while we still had the house—until my mother tired of Tighnabruaich and took it into her head that Malvern and the Mendips and Sidmouth were preferable—returning to Merlewood was supposed to reconcile us to the fate that had befallen the family.

'One thing we'll never have again,' my mother would say with relief, 'is men in safari jackets in *this* house.' Her judgments on people had recently become sharper and generally less charitable and I made the discovery from her example that no one's character is ever fixed and finally formed. At least my father had escaped any more dispassionate verdict on *his* character, or at any rate if her opinions about him changed they were never voiced aloud to us. How people *looked* seemed to count even more with her than it had, which struck me as backward thinking: but gradually I appreciated that how 'we', the family, had all appeared, meeting and engaging with the world, had been of the essence all along.

Studying whatever photographs I could find, I realised that they had always composed themselves around *her*, and it was she who had determined both the shape of the groups and our mood. We would take our cue from her, even my father, and in this one last surviving tableau of the family—ranged as we would be for another four or five years against the slowly declining grandeur of Merlewood—we could show whoever would chance to glance at us that it was *we* who were the fortunate ones, living our charmed lives.

* * * * *

Guilt afflicted Gregor later that, if he'd gone into the company as our father had been obliged to after his, the situation might have been otherwise: they could have chosen a moment to sell, even undersell, rather than have had the money slowly and systematically and humiliatingly bled from them.

I remembered mornings or afternoons I'd spent on the premises, being shown this or that process of the business: my father's colleagues smiling, my remarks given attention and answered with a serious nod or a smile. Mr Letham explained the broking business to me on three or four occasions—the ledgers were produced and a nicotine-stained index finger ran down the columns of figures. I was shown round a distillery at Killearn. A friend of my father's took me into the Glasgow Stock Exchange and then bought us both lunch at the Malmaison. Undeclared expectations seemed to accompany me on my visits to the offices, like a shadow falling behind me which I could never catch. But I had no taste for the whisky I was permitted in very small measures, and I didn't care for my father's over-heated room with its outlook of rooftops and chimney pots, although I pretended to be at home there.

I wonder now if I didn't pretend too hard, and only gave him false grounds for hope.

But those hopes failed in the end to count for much, and that guilt at what seemed to be my own fecklessness and duplicity has taken me years to get the better of. There was nothing to be done, though—nothing *I* could have done—and I knew it then, in my bones, in my blood.

After my father's death I used to puzzle why we had persevered with that charade, making a unity of eight for the photographs, when—at that time of my life, smiles and metaphors abounding—the strings and links between us hadn't been strong enough for the work of lifelines.

* * * * *

Merlewood became someone else's (a hairdresser with a chain of salons bought it: my mother specifically asked us not to tell her who the new owner was, and what 'sort of person'), then it was re-sold in inflation days and the house divided and the property put on the market as time-sharing accommodation.

The steamers only ran to Tighnabruaich on nostalgia cruises. I went on one once, with Rhona. It lashed with rain that day, but as we sailed up the Kyles of Bute we doggedly stood at the rails for a first sight of Blair's Ferry, then Kames, then Tighnabruaich. We mistook another house for Merlewood, and then we saw it. The garden was missing the evergreens and most of the rhododendron bushes; it looked naked, and functional and very modern. Someone in a yellow kagoul and sou'wester was walking a dog. The rock in the bay hadn't been claimed yet, as we'd always told ourselves we were going to do: we would row out and christen it with one of our father's bottles and an Arthur Ransome name, and fly the Saltire. The thought of the blood-smears must have been in both our minds.

The upstairs tea-room had gone from the pier, the room was empty. The slippery boards under our feet were very worn and in the gaps the sea sternly churned beneath us. We bought a postcard and sent it to our mother, now resident in a private hotel in Budleigh Salterton. We searched for a post box and eventually found one, in a mossy wall.

The red road with its pot-holes and puddles of rain water was redder than I remembered. It was raining too hard for us to spot any droning bees. Under a golf umbrella we walked along as far as the boatyard, which seemed—oddly, in these doldrum days—to be doing good business.

We imagined how the postcard would be received; my mother's favourite time of the day was the topping-up hour before dinner, and we saw her handing it to favoured fellow-guests over a snifter. In Zanzibar she'd shown her svelte international friends her family's postcards from Scotland and (so she would tell us) they had envied her. Life goes round in a circle, I said to Rhona, and she agreed with me, in a slightly strained voice she'd had all day.

We looked in at Merlewood from the gates. The creeper on the south wall of the house had been taken down—by the builders or on the hairdresser's instructions—and double-glazing fitted to the windows.

We only had an hour between landing and setting sail again. Rhona said she thought we'd done something 'useful' with our day, but I couldn't think quite what.

I felt we'd travelled to a place which had any number of realities, as many as there had been people to live there and days to be lived. But more than that, though, reality blurs—and imagination does a magician's trick on events that may have happened or not happened.

'The past is like a prism,' I said. 'All those reflecting surfaces.'

Rhona looked backwards over the rails, to what was being left behind in the rain.

'But it's *not* empty, a prism, is it?' she said, thinking matters out *her* way. 'It's solid. A thing, it exists.'

I couldn't think how I should reply, if there really was an answer—or a question.

* * * * *

In the photograph it seems that no question could be grave enough to bother or perplex us. The mysteries belong to our future selves.

Meanwhile the sun is too high in the sky to trouble us with shadows.

After noon the shadows will lengthen and obtrude, but we put a family face on what we can do nothing about. The third eye *sees*, but it can't turn time and tide.

Although there is nothing to be done, will-power and light-magic confer a modest substitute for grace. History was raging all about us and at last we were caught in the vortex: but the day blazes out splendent and undimmed, and, with a smile she learned in Zanzibar in the lucky peace after the War, my mother confidently claims the victory—for my father, for Merlewood—against all comers.

DILYS ROSE
1954–

STREET OF THE THREE TERRACES

There is a street in the city, a curving, high-standing terrace with some of the best views in the city. It is divided in three by name and outlook.

Section 1: north-facing. At a respectable distance—beyond a leafy embankment, fenced, and the wide straight London Road—grimy tenements sweep downhill to the newly renovated port where, in conjunction with cheap but not always cheerful pubs, warehouses and cut-price carpet shops, a clutch of smart restaurants have recently opened. Tourists and city people go to these new places, to eat, drink and soak up the local atmosphere. Looking out over the docks, where flat-hulled swans glide across oil slicks, they sip chilled white wine or real ale, and sample locally caught seafood, amongst darkly gleaming gantries, brass fittings and *objets d'art*. Outside is the atmosphere, bands of off-duty seamen and lorry-drivers, looking for bars, brawls, women. There is little interaction between the spectacle and the spectators.

Section 2: it is the back view, to the south, which catches the eye. A huge, shared garden backs on to the The Hill, a prominent mound, famous for its observatory, monument to Nelson, pillared folly and its popularity as a gay rendezvous. In spite of its centrality, the hill is a lonely isolated place where the wind wolf-whistles through the tough grass and musses the carefully-groomed hair of solitary young men.

Section 3: turning away from that high spot in favour of another, facing east and eventually the sea, volcanic rock rises out of parkland. Some say the rock resembles a sleeping giant. From the terrace the view is best in winter, glimpsed through bare trees. In summer, foliage crowds out the view and the street becomes a fragrant retreat. The city's noise, dirt, deprivation is far enough beyond the scrubbed doorsteps to be out of earshot, sight, mind.

To live on the street costs a good deal of money. The four-storey houses are grand but—as many of their owners would take pains to tell you—monsters to keep up. Hotels, offices for professional bodies, art galleries (private, fine art only) and consulates have mostly replaced private dwellings and much of the remainder has been divided up into flats. American CIA men are reputed to jog by day in the shared back garden, in reality a park of several acres where nanny-tended children fight over non-violent toys. The terrace is discreet, aloof, undisturbed. Because it has little value as a thoroughfare, many people are unaware of its existence.

On a warm night in early summer, in the midst of chestnut blossom, rhododendrons and rogue bluebells, while an aria sung by Pavarotti drifted on to the terrace through an open window, a young man was beaten to death.

* * * * *

Words used by the press and media to describe the murder: brutal, vicious, depraved, horrific, fiendish, sadistic, merciless, mindless, motiveless. Words used by the police and the accused: poof-bashing.

The accused: Five teenage boys.

The weapons found in their possession: one machete (blunt), a length of studded rubber tubing, an iron bar, two craft knives. The Dr Martens boots which several boys wore were fitted with steel toecaps.

HOMOSEXUAL HAUNT MURDER: ACCUSED BLAME EACH OTHER: The victim, a stranger to the city, suffered multiple injuries before he died, including what appeared to be an attempted castration. It was said that the assault continued even after the victim had been beaten unconscious and left hanging upside down with his trouser leg caught on a railing. When it was suggested by one of the accused that an ambulance be called, one of the boys, who cannot be named for legal reasons, refused and went on to 'finish him off'. All five boys deny the murder. All five have also lodged a special defence of incrimination, naming the others accused.

Across the London Road and a few blocks down the hill from the terrace, on the fourth floor of a blackened tenement which has been awaiting renovation under an urban restoration project for ten years, live the family of one of the accused, Billy McBain: his mother, Bet, father Joe, sister Noreen.

—I'm no saying he was an angel or nothing, and Joe and me've been through our share of trouble with the boy. But nothing no other lad his age wouldn't get up to, no nothing over the score, nothing anywhere near this. It can't be right. He tells me—and I have to believe him—I just have to, see, he's my son and I'm his ma and blood's thicker than water, right? Says he was there all right, wi his pals. But no, they weren't pals, Billy's no so daft as to have pals who'd do that kind of thing—and the story, the story they're giving out, the nasty bits, that's no the way it was. Billy says it's no and he should know. He was there. He was. There. When it happened. He stayed, even if he did nothing he stayed and watched, didn't he, looked on while they hacked up that poor bloke. Poor fucking poof, Joe says. And, ay but ye cannae help thinking, why did he take that road home? Must have been up on the hill. Maybe he tried something on with the lads, maybe made some kind of suggestion. Young lads are touchy about their manhood, maybe he said something, the victim, said the wrong thing and things went wrong from there on, maybe he was asking for trouble. But he died. The poor sod died.

Bet's monologue continues, after the reporters and the police and the social workers have gone, after Joe has given in to a Mogodon slumber, in her head, aloud, all through the sleepless nights. My son, my own bloody son. Up for murder. Near killed me bringing the boy into the world and now he's up for murder. Life over before his sixteenth birthday. A life sentence. One way or another. And if they nail him, how long will he get, how long is life? Bet talks but doesn't get many replies. Joe won't speak about the boy. It's as if he's no longer their son but hers alone, her bairn, her big sullen bairn, her aching burden. Bet cleans her house incessantly, as if she might be able to scrub away the trouble which has wormed its way into every aspect of her life. She can't eat without throwing up. Her stomach has clenched into a burning ball of acid. She can't get a word of sense out of Joe. He's off his work. When he's not making expeditions to the jail, he sits in front of the telly, popping pills. Noreen, who has just started her first year at the big school wakes nightly with the terrors, and wets the bed.

Within a radius of half a mile, four other families are disintegrating in a similar way. Jobs are lost through absenteeism and illness. Friends, not knowing how to be, or unwilling to be drawn into the bloody business, stay away. Sisters and brothers are tormented daily in playgrounds and school corridors. Because of the particular nature of the defence each family is isolated, set against each other. Each new development, reported in the papers or passed on by word of mouth or hearsay, every word—no matter how slight or how unlikely—prompts a chain of adverse reactions in every home.

Down the coast some twenty miles, the once patiently tended garden of a beach cottage is showing signs of neglect. Weeds have choked the flower-beds. Tidal debris has blown up from the sand and been left to settle in the long grass. Inside the cottage a woman sits at her kitchen table, littered with the remains of several meals. Amongst the stale crusts, sour milk, the smears of egg-yolk and jam around which wasps and bluebottles feast without interruption, a square of plastic tablecloth has been cleared. The woman is staring at the square of blue plastic. She has been staring at it all day and the day before that, and seeing on it an ever-changing image of her dead son.

As the wheels of justice roll on laboriously, as men in red silk gowns shuffle papers and pass asides, as they squint through pince-nez at a sheaf of conflicting statements, at forensic reports, evidence and possible alibis, as the fate of five boys is considered in acute and frequently tedious detail, as the jurors inch towards a verdict, the trees turn golden on the street of the three terraces. The dog-walking path and short cut to the

London Road remain cordoned off for the time being but the cordons are unnecessary. Nobody other than official visitors—escorted at all times by at least two senior police officers—chooses to push open the squeaking iron gate and walk down three moss-spattered steps to the path which leads to the scene of the crime. The incident is still too fresh in the imagination of even the most sanguine of strollers. Security on the street has been tightened, in spite of the high insurance premium. Locks on gates, extra bolts on doors, electronic alarms. Guard dogs lope and snap behind crackling autumn hedges but still, of an evening, arias by Pavarotti drift through an open window and float above the trees.

Janice Galloway

1956–

Blood

He put his knee up on her chest getting ready to pull, tilting the pliers. Sorry, he said. Sorry. She couldn't see his face. The pores on the backs of his fingers sprouted hairs, single black wires curling onto the bleached skin of the wrist, the veins showing through. She saw an artery move under the surface as he slackened the grip momentarily, catching his breath; his cheeks a kind of mauve colour, twisting at something inside her mouth. The bones in his hand were bruising her lip. And that sound of the gum tugging back from what he was doing, the jaw creaking. Her jaw. If you closed your eyes it made you feel dizzy, imagining it, and this through the four jags of anaesthetic, that needle as big as a power drill. Better to keep her eyes open, trying to focus past the blur of knuckles to the cracked ceiling. She was trying to see a pattern, make the lines into something she could recognise, when her mouth started to do something she hadn't given it permission for. A kind of suction. There was a moment of nothing while he steadied his hand, as if she had only imagined the give. She heard herself swallow and stop breathing. Then her spine lifting, arching from the seat, the gum parting with a sound like uprooting potatoes, a coolness in her mouth and he was holding something up in the metal clamp; great bloody lump of it, white trying to surface through the red. He was pleased.

There you go eh? Never seen one like that before. The root of the problem ha ha.

All his fillings showed when he laughed, holding the thing out, wanting her to look. Blood made a pool under her tongue, lapping at the back of her throat and she had to keep the head back instead. Her lips were too numb to trust: it would have run down the front of her blazer.

Rinse, he said. Cough and spit.

When she sat up he was holding the tooth out on a tissue, roots like a yellow clawhammer at the end, one point wrapping the other.

See the twist? Unusual to see something like that. Little twist in the roots.

Like a deformed parsnip. And there was a bit of flesh, a piece of gum or something nipped off between the crossed tips of bone.

Little rascal, he said.

Her mouth was filling up, she turned to the metal basin before he started singing again. *She's leaving now cos I just heard the slamming of the door* then humming. He didn't really know the words. She spat dark red and thick into the basin. When she resurfaced, he was looking at her and wiping his hands on something like a dishtowel.

Expect it'll bleed for a while, bound to be messy after that bother. Just take your time getting up. Take your time there. No rush.

She had slid to the edge of the chair, dunting the hooks and probes with having to hold on. The metal noise made her teeth sore. Her stomach felt terrible and she had to sit still, waiting to see straight.

Fine in a minute, he said. Wee walk in the fresh air. Wee walk back to school.

He finished wiping his hands and grinned, holding something out. A hard thing inside tissue. The tooth.

You made it, you have it haha. There you go. How's the jaw?

She nodded, and pointed to her mouth. This almost audible sound of a tank filling, a rising tide over the edges of the tongue.

Bleed for a while like I say. Don't worry though. Redheads always bleed worse than other folk. Haha. Sandra'll get you something: stop you making a mess of yourself.

Sandra was already away. He turned to rearrange the instruments she had knocked out of their neat arrangement on the green cloth.

Redheads, see. *Don't take your love to town.*

Maybe it was a joke. She tried to smile back till the blood started silting again. He walked over to the window as Sandra came back with a white pad in her hand. The pad had gauze over the top, very thick with a blue stripe down one side. Loops. A sanitary towel. The dentist was still turned away, looking out of the window and wiping his specs and talking. It took a minute to realise he was talking to her. It should stop in about an hour or so he was saying. Maybe three at the outside. Sandra pushed the pad out for her to take. If not by six o'clock let him know and they could give her a shot of something ok? Looking out the whole time. She tried to listen, tucking the loops at the ends of the towel in where they wouldn't be obvious, blushing when she put it up to her mouth. It was impossible to tell if they were being serious or not. The dentist turned back, grinning at the spectacles he was holding between his hands.

Sandra given you a wee special there. Least said haha. Redheads eh? *Oh Rooooobeee*, not looking, wiping the same lens over and over with a cloth.

The fresh air was good. Two deep lungfuls before she wrapped her scarf round the white pad at her mouth and walked. The best way from the surgery was past the flats with bay windows and gardens. Some had trees, crocuses and bits of cane. Better than up by the building site, full of those shouting men. One of them always shouted things, whistled loud enough to make the whole street turn and look. Bad enough at the best of times. Today would have been awful. This way was longer but prettier and there was nothing to stop her taking her time. She had permission. No need to worry about getting there for some particular ring of some particular bell. Permission made all the difference. The smell of bacon rolls at the café fetched her nose: coffee and chocolate. They spoiled when they reached her mouth, heaped up with sanitary towel and the blood still coming. Her tongue wormed towards the soft place, the dip where the tooth had been, then back between tongue root and the backs of her teeth. Thick fluid. A man was crossing the road, a greyhound on a thin lead, a woman with a pram coming past the phone box. Besides, girls didn't spit in the street. School wasn't that far though, not if she walked fast. She clutched the tooth tight in her pocket and walked, head down. The pram was there before she expected it; sudden metal spokes too near her shoes before she looked up, eyes and nose over the white rim of gauze. The woman not even noticing but keeping on, ploughing up the road while she waited at the kerb with her eyes on the gutter, trying hard not to swallow. Six streets and a park to go. Six streets.

The school had no gate, just a gap in the wall with pillars on either side that led into the playground. The blacked-out window was the staff room; the others showed occasional heads, some white faces watching. The Music block was nearest. Quarter to twelve. It would be possible to wait in the practice rooms till the dinner bell in fifteen minutes and not shift till the afternoon. She was in no mood, though, not even for that. Not even for the music. It wouldn't be possible to play well. But there was no point in going home either because everything would have to be explained in triplicate when the mother got in and she never believed you anyway. It was all impossible. The pad round her mouth was slimy already, the wet going cold further at the far sides. She could go over and ask Mrs McNiven for another towel and just go anyway, have a lie-down or something but that meant going over to the other block, all the way across the playground again and the faces looking out knowing where you were going because it was the only time senior girls went there. And this thing round her mouth. Her stomach felt terrible too. She suddenly wanted to be in the music rooms, soothing herself with music. Something peaceful. Going there made her feel better just because of where it was. Not like at home. You could just go and play to your heart's content. That would be nice now, right now this minute, going up there and playing something: the Mozart she'd been working on, something fresh and clean. Turning, letting the glass door close, she felt her throat thicken, closing over with film. And that fullness that said the blood was still coming. A sigh into the towel stung her eyes. The girls toilets were on the next landing.

Yellow. The light, the sheen off the mirrors. It was always horrible coming here. She could usually manage to get through the days without having to, waiting till she got home and drinking nothing. Most of the girls did the same, even just to avoid the felt-tip drawings on the girls' door—mostly things like split melons only they weren't. All that pretending you couldn't see them on the way in and what went with them, GIRLS ARE A BUNCH OF CUNTS still visible under the diagonal scores of the cleaners' Vim. Impossible to argue against so you made out it wasn't there, swanning past the word CUNTS though it radiated like a black sun all the way from the other end of the corridor. Terrible. And inside, the yellow lights always on, nearly all the mirrors with cracks or warps. Her own face reflected yellow over the nearside row of sinks. She clamped her mouth tight and reached for the towel loops. Its peeling away made her mouth suddenly cold. In her hand, the pad had creased up the centre, ridged where it had settled between her lips and smeared with crimson on the one side. Not as bad as she had thought, but the idea of putting it back wasn't good. She wrapped it in three paper towels instead and stuffed it to the bottom of the wire bin under the rest, bits of paper and God knows what, then leaned over the sinks, rubbing at the numbness in her jaw, rinsing out. Big, red drips when she tried to open her mouth. And something else. She watched the slow trail of red on the white enamel, concentrating. Something slithered in her stomach, a slow dullness that made it difficult to straighten up again. Then a twinge in her back, a recognisable contraction. That's what the sweating was, then, the churning in her gut. It wasn't just not feeling well with the swallowing and imagining things. Christ. It wasn't supposed to be due for a week yet. She'd have to use that horrible toilet paper and it would get sore and slip about all day. Better that than asking Mrs McNiven for two towels, though, anything was better than asking Mrs McNiven. The cold tap spat water along the length of one blazer arm. She was turning it the wrong way. For a frightening moment, she couldn't think how to turn it off then managed, breathing out, tilting forward. It would be good to get out of here, get to something fresh and clean, Mozart and the white room upstairs. She would patch something together and just pretend she wasn't bleeding so much, wash her hands and be fit for things. The white keys. She pressed her forehead against the cool concrete of the facing wall, swallowing. The taste of blood like copper in her mouth, lips pressed tight.

The smallest practice room was free. The best one: the rosewood piano and the soundproofing made it feel warm. There was no one in either of the other two except the student who taught cello. She didn't know his name, just what he did. He never spoke. Just sat in there all the time waiting for pupils, playing or looking out of the window. Anything to avoid catching sight of people. Mr Gregg said he was afraid of the girls and who could blame him haha. She'd never understood the joke too well but it seemed to be right enough. He sometimes didn't even answer the door if you knocked or made out he couldn't see you when he went by you on the stairs. It was possible to count yourself alone, then, if he was the only one here. It was possible to relax. She sat on the piano stool, hunched over her stomach, rocking. C major triad. This piano had a nice tone, brittle and light. The other two made a fatter, fuzzier noise altogether. This one was leaner, right for the Mozart anyway. Descending chromatic scale with the right hand. The left moved in the blazer pocket, ready to surface, tipping something soft. Crushed tissue, something hard in the middle. The tooth. She had almost forgotten about the tooth. Her back straightened to bring it out, unfold the bits of tissue to hold it up to the light. It had a ridge about a third of the way down, where the glaze of enamel stopped. Below it, the roots were huge, matte like suede. The twist was huge, still bloody where they crossed. Whatever it was they had pulled out with them, the piece of skin, had disappeared. Hard to accept her body had grown this thing. Ivory. She smiled and laid it aside on the wood slat at the top of the keyboard, like a misplaced piece of inlay. It didn't match. The keys were whiter.

Just past the hour already. In four minutes the bell would go and the noise would start: people coming in to stake claims on the rooms, staring in through the glass panels on the door. Arpeggios bounced from next door. The student would be warming up for somebody's lesson, waiting. She turned back to the keys, sighing. Her mouth was filling up again, her head thumping. Fingers looking yellow when she stretched them out, reaching for chords. Her stomach contracted. But if she could just concentrate, forget her body and let the notes come, it wouldn't matter. You could get past things that way, pretend they weren't there. She leaned towards the keyboard, trying to be something else: a piece of music. Mozart, the recent practice. Feeling for the clear, clean lines. Listening. She ignored the pain in her stomach, the scratch of paper towels at her thighs, and watched the keys, the pressure of her fingers that buried or released them. And watching, listening to Mozart, she let the music get louder, and the door opened, the abrupt tearing sound of the doorseals seizing her stomach like a fist. The student was suddenly there and smiling to cover the knot on his forehead where the fear showed, smiling fit to bust, saying, Don't stop, it's lovely; Haydn isn't it? and she opened her mouth not able to stop, opened her mouth to say Mozart. It's Mozart—before she remembered.

Welling up behind the lower teeth, across her lips as she tilted forwards to keep it off her clothes. Spilling over the white keys and dripping onto the clean tile floor. She saw his face change, the glance flick to the claw roots in the tissue before he shut the door hard, not knowing what else to do. And the bell rang, the steady howl of it as the outer doors gave, footfalls in the corridor gathering like an avalanche. They would be here before she could do anything, sitting dumb on the piano stool, not able to move, not able to breathe, and this blood streaking over the keys, silting the action. The howl of the bell. This unstoppable redness seeping through the fingers at her open mouth.

DUNCAN MCLEAN
1964–

DOUBLED UP WITH PAIN

John came back from the kitchen with two cans of beer and sat down along the settee from the boy. The first time I met your mother, he said, I was doubled up with pain at the foot of Leith Walk.

The boy looked up from the football comic he was reading.

I hope you're not going to embarrass the laddie, said Freddy, leaning out of his chair to take the can John was offering him.

Not at all, not at all. I never did embarrassing things in these days.

Not like now, eh!

John sighed, tapping the bottom of his can. Give us a break will you Freddy? I'm telling the boy his family history here, let me get it out, eh?

Aye John, go on there, sorry, on you go. He peeled the ringpull off and lifted the beer to his lips.

John looked at his can for a second, then put it down on the floor at the end of the settee. Well, he said. He looked at the boy. The boy shut his comic and put it on the floor. This was outside the Central Bar at the foot of the Walk. I'd been struck by indigestion, see, like a spear through my guts. I was getting it all the time in these days, the grub I was eating and that. And if I'd had a drink the night before as well, ken, that would bring it on in the middle of the next morning, without fail. Anyway, this one morning it had come over me bad, really biting into my belly, like the acid was going to burn right through to my backbone, and I had to stop in the street, stop in my tracks, and I was holding my guts with one hand, leaning with the other on this post-box there. I was staring down at the pavement just, doubled up with pain.

Freddy laughed. And there you saw your bride-to-be, lying on the ground in front of you!

John glared across at him, then quickly grinned along at the boy and kept on talking, looking down at the rug in front of the fire. I was in a wee world of my own there, he said. Just concentrating on this ball of pain, keeping it down in my belly, stopping it from spinning up my throat, ken, waiting for it to pass, and I wasn't hearing the traffic at all, or aware of folk passing me by . . . and it was a hell of a busy place in these days, with the station still on the go and that. But then something did break through the haze and into here. John tapped the side of his forehead with a finger, glancing from the boy to Freddy and back again. Voices speaking in a deliberately loud kind of way, right behind me. So I straightened myself and looked up, and there were two lassies standing there, a big stack of letters and packages each which they started to stick into the post-box as soon as I moved out of their road—not even looking at me any more. And then I heard one of them say, the one that wasn't your mother . . .

So the other one was Agnes was it?

Aye Freddy, that's the point, that was Agnes standing there, sticking letters in the thingmy, and there's me bent over like a half-closed knife beside her!

A touch of the old gutrot, eh? Freddy screwed up his face and grimaced across at the boy. The boy looked away and back to his dad.

The one that wasn't your mother, said John. Well, I heard her saying, And at this time of the morning too! It's a damned disgrace!

The boy smiled at the pan-loaf voice his dad had put on, then listened again to the story.

I saw the both of them looking over their shoulders at the Central, ken, and I thinks to myself, Aye aye, they reckon I've been on the bevvy, enjoying myself, and really I'm chopped in half with the bloody indigestion! And I don't know what came over me to bother, but I just reckoned I'd set them right. I mean nine times out of ten I wouldn't've given a toss, but this particular time for some reason I just stood straight up there, thumbs down the seams of the breeks, looked them in the eye and said, Ladies, excuse me for butting in here, but I fear you're misconstruing me . . .

Misconstruing! said Freddy.

That's what *they* said, said John. And I said, Aye, you've been construing the fact that I'm hytering about here as proof that I'm half-cut, when the fact is I'm sober as a judge and sick with a stomach complaint! Well, the two of them made kind of oh-ing noises, still sticking mail in the slot. So don't be so quick to condemn, I went on. I could've been dying with an exploded ulcer and you'd be standing there tut-tutting!

The boy laughed and Freddy did too, then said, Aye John, you had a sharp tongue on you that day right enough.

John grinned. Well I had had my mid-morning pint ten minutes before, you see, I reckon that's where all my smart comments were coming from. Hih! Anyway. Agnes, your mother, she looks at me, kind of biting her lip, and begins to say, Sorry . . . But the other one grabs her by the elbow saying, Come on now, don't pay him any heed, let's get back over the road. And she starts to pull her off towards the crossing. I look after them, ken, and I'm just about to spit some acidy spit after the back of them when Agnes looks round and sees me watching her rear. So quick as a flash I whip off my cap and nod to her, a big smile across my chops. And sure enough she pulls away from her mate and walks right back towards me.

Here comes the juicy bit, said Freddy.

The boy sighed and leaned forward, elbows on his knees, listening to his dad.

She walks right up to me, a nice-looking lassie she was, she walks up and says, I'm sorry about my friend, her man takes a bucket, she's touchy about it. I felt I had to say something, but I couldn't think what to say, so I just kind of shrugged and said, Ach, and waved my cap at her. Then I stopped waving my cap and put it back on my head and said, Ach, again. She kind of cocked her head on one side and looked me up and down and said, You were telling the truth were you? I mean you weren't in there drinking? Not at this hour? I shrugged again then shook my head—cause I hadn't been in the Central at all, I'd had my pint in the Spey up the road—and then I decided that I really better say something. Anything! So I said, No, not at all, I was in the Job Centre there, looking around the vacancies and that . . . John paused, frowned, then said to the boy, Except I didn't say Job Centre, it wasn't called that in these days. I was in the Labour Exchange, that's what I said to her. John cleared his throat. She looked me up and down again, her head on one side like a bloody bird, and she said to me, So you're out of a job, eh? I nodded, tried to look sorry about it. I was sorry about it! Hmm, she says, Come with us then. So I came with them. Or at least I went with Agnes: the other one walked off ahead with her nose in the air when she saw me coming. Hih. We crossed the road and went down Kirk Street. We're in a shipper's office down here, said Agnes. They've been looking for a man for the stores for weeks now, what do you think? Ho ho, I said, Tell them to look no more!

There was a silence. John looked along the settee at the boy, who smiled briefly, then frowned. John looked across at Freddy.

So that's how you met her, said Freddy. I never knew that one.

John reached down and picked up his can. He tapped the bottom of it. And that's how I got that job as well. I was there for more than a year, it was a fine job. He opened the can and played about with the ringpull for a few seconds, slipping it on and off the fingers of his left hand. Then we got married and got the house out in Porty and the both of us found work out here. Then after a few more years you came along, then after another few more . . . well, that's now. John took the ringpull off his pinkie and flicked it away towards the bin by the fireplace. After a second he lifted the can and took a long drink.

The boy sat back into the depths of the settee, frowning. Dad, he said.

What son?

Where's Mum gone?

John took another drink, then lowered the can and held it in both hands in his lap. She's gone, he said. It doesn't matter where.

Freddy half stood up, then sat down again on the arm of his chair. Here, he said to the boy. How about you taking me out the back for a bit of penalty practice?

Dad . . .

A silence.

Dad!

What?

When's she coming back?

John gazed at the boy for a moment, then shrugged. Ach . . . He looked at the top of his beer can, then bent over and put it on the floor between his feet. He stayed bent over for quite a long time before straightening up.

A. L. Kennedy

1965–

Night Geometry and the Garscadden Trains

One question.

Why do so many trains stop at Garscadden? I don't mean stop. I mean finish. I mean terminate. Why do so many trains terminate at Garscadden?

Every morning I stand at my station, which isn't Garscadden, and I see them: one, two, three, even four in a row, all of them terminating at Garscadden. They stop and no one gets off, no one gets on; their carriages are empty, and then they pull away again. They leave. To go to Garscadden. To terminate there.

I have never understood this. In the years I have waited on the westbound side of my station, the number of trains to Garscadden has gradually increased; this increase being commensurate with my lack of understanding. The trees across the track put out leaves and drop leaves; the seasons and the trains to Garscadden pass and I do not understand.

It's stupid.

So many things are stupid, though. Like the fact that the death of my mother's dog seemed to upset me more than the death of my mother. And I loved my mother more than I loved her dog. The stupidity of someone being killed by the train that might normally take them home, things like that. There seems to be so much lack of foresight, so much carelessness in the world. And people can die of carelessness. They lack perspective.

I do, too. I know it. I am the most important thing in my life. I am central to whatever I do and those whom I love and care for are more vital to my existence than statesmen, or snooker players, or Oscar nominees, but the television news and the headlines were the same as they always are when my mother died and theirs were the names and faces that I saw. Nations didn't hold their breath and the only lines in the paper for her were the ones I had inserted.

Inserted. Horrible word. Like putting her in a paper grave.

To return to the Garscadden trains, they are not important in themselves; they are only important in the ways they have affected me. Lack of perspective again, you see? Naturally, they make me late for work, but there's altogether more to them than that. It was a Garscadden train that almost killed my husband.

Of course you don't know my husband, Duncan, and I always find him difficult to describe. I carry his picture with me sometimes; more to jog my memory than through any kind of sentiment. I do love him. I do love him, even now. I love him in such a way that it seems, before I met him, I was waiting to love him. But I remember what I remember and that isn't his face.

Esau was an hairy man. I remember my mother saying that. It always sounded more important than just saying he was born with lots of hair. I only mention Esau now, because Duncan wasn't hairy at all.

He had almost no eyebrows, downy underarm hair and a disturbingly naked chest. We used to go walking together as newly weds, mainly on moorland and low hills where he'd been as a scout. The summers were usually brief, unsettled, the way you'd expect, but the heat across the moors could be remarkable. It seems to be a quality of moors. The earth is warm and sweaty under the wiry grass, the heather bones are brilliant white and the sun swings, blinding, overhead. You walk in a cloud of wavering air and tiny, black insects.

On such days—hot days—Duncan would never wear a T-shirt. Not anything approaching it. He would put on a shirt, normally pale blue or white, roll the sleeves up high on his arms and wear the whole thing loose and open like a jacket, revealing a thin, vulnerable chest. Sensible boots, socks, faded khaki shorts and the shirt flapping: he would look like those embarrassing forties photographs of working class men at the beach or in desert armies. He had a poverty stricken chest, pale with little boy's skin.

There was hair on his head, undoubtedly, honey brown and cut short enough to subdue the natural curl, but his face was naked. I remember him washing and brushing his teeth, but I don't believe he ever shaved. There was no need.

Duncan, you might also notice, is in the past tense—not because he's dead, because he's over. I call him my husband because I've never had another one and everything I tell you will only show you how he was. Today I am a different person and he will be, too. Whatever I describe will be part of our past. I used to want to own his past. I used to want to look after him retrospectively. This was during the time when our affair had turned into marriage but still had something to do with love. In fact, there was a lot of love about. I mean that.

My clearest memory of him comes from about that time. I don't see it, because I never looked at it. I only remember a feeling, safe and complete, of lying with him, eyes closed, and whispering that I wanted to own his past; that I wanted to own him, too.

It was strange. However we flopped together, however haphazardly we decided to come to rest, the fit would always be the same.

His right arm, cradling my neck.

My head on his shoulder.

My right arm across his chest.

My left arm, tucked away between us with my hand resting quietly on his thigh. Not intending to cause disturbance, merely resting, proprietary.

In these pauses, we would doze together before sleeping and dreaming apart and we would whisper. We always whispered, very low and very soft, as if we were afraid of disturbing each other.

'I love you.'
'Uh hu.'
'I do love you.'
'I know that. I feel that. I love you, too.'
'I want to look after you.'
'You can't.'
'Why not.'
'Because I'm looking after you.'
'That's alright, then.'
'I love you.'
'Uh hu.'
'I do love you.'

And, finally, we would be quiet and sleepy and begin to breathe in unison. I've noticed since, if you're very close to anything for long enough, you'll start to breathe in unison. Even my mother's dog, when he slept with his head on my lap, would eventually breathe in time with me. There was more to it than that with Duncan, of course.

I sometimes imagined our hearts beat together, too. It's silly, I know, but we felt close then. Closer than touch.

This positioning, our little bit of night geometry, this came to be important in a way I didn't like because it changed. I didn't like it then, as much as I now don't like to remember the two of us together and almost asleep, because, by fair means or foul, you can't replace that. Intensity is easy, it's the simple nearness that you'll miss.

The change happened one evening on a Sunday. We had cocoa in bed. I made it in our little milk pan and I whisked it with our little whisk, to make it creamy, and we drank it sitting up against the pillows and ate all butter biscuits, making sure we didn't drop any crumbs. There is nothing worse than a bed full of crumbs. And we put away the cocoa mugs and we turned out the lights and that was fine. Very nice.

But when we slowed to a stop, when we terminated, the geometry had changed. I didn't really think about it because it was so nicely changed.

My right arm around his neck.

His head against my shoulder.

His one arm tucked between us very neat, and the other, just resting, doing nothing much, just being there.

It all felt very pleasant. The good weight of him, snuggled down there, the smell of his hair when I kissed the top of his head. I did that. I told him I could never do enough, or be enough, or give enough back and I kissed the top of his head. I told him I belonged to him. I think he was asleep.

I told him anyway and he was my wee man, then, and I couldn't sleep for wanting to look after him.

The following morning, I waited on the westbound platform and the smell of him was still on me, even having washed. All that day when I moved in my clothes, combed my hair, his smell would come round me as if he'd just walked through the room.

It was good, that. Not unheard of in itself, hardly uncommon, in fact. It wasn't unknown for me to leave my bed and dress without washing in order to keep what I could of the night before, but you'll understand that, this time, I was remembering something special. I thought, unique.

Now I realise that you can never be sure that anything is unique. You can never be sure you know enough to judge. I mean, when Pizzaro conquered the Incas, they thought he was a god—his men, too—when really Spain was full of Spaniards just like him. Eventually you see you were mistaken, but look what you've had to lose in order to learn.

I thought that the way I met Duncan was unique.

Wrong.

Not in the place: a bar. Not in the time: round about eight. Not in the circumstances: two friends of friends, talking at a wee, metal table when the rest of the conversation dipped. It was a bit of a boring evening to tell the truth.

We all left on the bell for last orders and there was the usual confusion about coats—who was sitting on whose jacket, who'd lost gloves. Duncan and I were a little delayed, quite possibly not by chance.

'I'm going to call you tomorrow. Ten o'clock. What's your number?'
'What?'
'What number could I get you on, tomorrow at ten o'clock?'
'In the morning?'
'Yes, in the morning.'
'Well, I would be at work, then.'
'I know that, what's the number?'

I gave him the office number and he went away. I don't even think he said goodbye.

At a quarter to eleven, the following day, he called McSwiggin and Jones and was put through to me. I had some idea that he might be in need of advice. McSwiggin and Jones accepted payment from various concerns with money to call in the debts of various individuals without it. Debt, as Mr McSwiggin often said, could be very democratic—Mrs Gallacher with two small boys, no husband and her loan from the Social Fund turned down was in debt. And so was Peru. Perhaps, I thought, Duncan was in debt.

'What do you mean, in debt?'

'I mean, who do you owe money? I can't help you if it's on our books. I mean I can, but not really, you know.'

'No, I don't know. I owe my brother a fiver, since you ask.'

'Mm hm.'

'And that's it. I don't have any debts, just a bit of an overdraft which doesn't count. I want to see you tonight. I could bring my bank statement with me if you'd like.'

'Look, I'm sorry, but you're wasting my time, aren't you?'

'I'm sorry if you feel that way. I thought we got on well together.'

'Ring me at home tomorrow evening. This is ridiculous and I'm at work.'

He called at the end of the week and we went out for a coffee on the Sunday afternoon. Before I had time to ask he told me that he and Claire, his partner from the pub, were only friends. They'd been at school together which is why they'd seemed so close the other evening.

When Duncan and I were married, quite a while later, Claire was at the little party afterwards. She smiled quietly when she saw me, danced with Duncan once and then left. I had to ask who she was because she looked so familiar, but I couldn't remember her name.

So, Duncan and I were married and we were unique. Although men and women often marry as an expression of various feelings and beliefs and although they often go to bed together before, during and after marriage, the thing with Duncan and me was unrepeatable, remarkable and entirely unique. So I thought.

No one had ever married us before and we had never married each other. It was tactfully assumed that the going to bed had happened with other partners in other times, but they had never managed to reach the same conclusion. We were one flesh, one collection of jokes and habits and one smell. Even now, I know, the smell of my sweat and the taste of my mouth are not the same as they were before I met him. He will always be that much a part of me, whether I like it or not.

Even when two different friends in two different ladies' toilets in two different bars told me that Claire and Duncan had been sleeping and staying awake together for months before I met him, I didn't mind. I didn't mind if they had continued to see each other after we met. I was flattered he had taken the trouble to lie. It didn't matter because he had left her for me and we had made each other unique.

Finally, of course, I realised the most original things about us were our fingerprints. Nothing of what we did was ever new. I repeated the roles that Duncan chose to give me in his head—wicked wife, wounded wife, the one he would always come back to, the one he had to leave and I never even noticed. I always felt like me. For years, I never knew that when he rested with his head on my shoulder, all wee and snuggled up, it was helping him to ease his guilt. Once or twice a year, it was his body's way of saying he'd been naughty, but he was going to be a good boy from now on.

And I was a good wife. I even answered the telephone with a suitably unexpected voice, to give his latest girlfriend her little shiver before she hung up. Like a good wife should.

All the time I thought I was just being married when, really, Duncan was turning me into Claire and the ones before and after Claire.

I lived with the only person I've met who can snore when he's wide awake, who soaked his feet until they looked like a dead man's, then rubbed them to make them peel. I've washed hundreds of towels, scaley with peelings from his feet. I've cooked him nice puddings, nursed him through the 'flu, stopped him trimming his fringe with the kitchen scissors and have generally been a good wife. Never knowing how Duncan saw me inside his head. It seems I was either a victim, an obstacle or a safety net. I wasn't me. He took away me.

But it wasn't his fault, not really. It was the E numbers in his yoghurt, or his role models when he was young. It was a compulsion. Duncan, the wee pet lamb, would chase after anything silly enough to show him a half inch of leg. From joggers to lady bicyclists to the sad-looking Scottish Nationalist who sold papers round the pubs in his kilt. Duncan couldn't help it. It wasn't his fault.

I sound like an idiot, not seeing how things were for so long. I felt like an idiot, too. Nothing makes you feel more stupid than finding out you were wrong when you thought you were loved. The first morning after I discovered, it wasn't good to wake up. Over by the wall in the bedroom there was a wardrobe with a mirror in the door. I swung my legs out of bed and just sat. There I was; reflected; unrecognisable. I looked for a long while until I could tell it was me: pale and slack, round-shouldered and dank-haired, varicose veins, gently mapping their way. You would have to really love me to like that and Duncan, of course, no longer loved me at all. I could have felt sorry for him, if I hadn't felt so sorry for myself.

I considered the night before and letting his head rest on my shoulder, knowing what I finally knew. It was as if I wasn't touching him, only pressing against his skin through a coating of other women. I'd felt his breath on my collar bone and found it difficult not to retch.

It had taken about a month to fit all the pieces together in my head. Nothing silly like lipstick on collars, or peroxide blonde hairs along his lapels: it was all quite subtle stuff. He would suddenly become more crumpled, as if he had started sleeping in his shirts, while his trousers developed concertina creases and needed washing much more regularly. The angle of the passenger seat in the car would often change and, opening the door in the morning, there would be that musty smell. And yet, for all the must and wrinkles, the fluff all over his jackets, as if they'd been thrown on the floor, Duncan would be taking pains with his appearance. When he walked out of the flat he'd never looked better and when he returned he'd never looked worse. Life seemed to be treating him very roughly, which perhaps explained his sudden interest in personal hygiene, the increasingly frequent washing and the purchase of bright, new Saint Michael's underwear. It's all very obvious now, but it wasn't then. Even though it had been repeating itself for years.

Duncan's infidelity didn't have all the implications it might have today. I didn't take a blood test, although I've watched for signs of anything since. Still, you can imagine the situation in the first few weeks with both of us constantly washing away the feel of his current mistress. We went through a lot of soap.

I suppose that I should have left him, or at least made it clear that I'd found him out. I should have made sure that we both knew that I knew what he knew. Or whatever it was you were meant to make sure of. I didn't know. To tell the truth, it didn't really seem important. It was to do with him and things to do with him didn't seem important any more. I couldn't see why he should know what was going on inside my head when, through all the episodes of crumpled shirts and then uncrumpled shirts and even the time when he tried for a moustache, I had never had any idea of what Duncan was thinking.

I stayed and, for a long time, things were very calm. We finished with all of our washing, started to sleep at night and I managed to get the drier to chew up six pairs of rainbow coloured knickers. Duncan went back to being just a little scruffy and always coming home for tea.

It wasn't going to last, I knew. It would maybe be a matter of months before the whole performance started up again and I wasn't sure how I would react to that. In the meantime I sorted out my past. I still worked at McSwiggin and Jones, but only for three days out of seven and instead of spending the rest of my week on housework or other silly things I started to sit on the bed a lot and stare at that mirror door. I bought some books on meditation and, at night, when I felt Duncan sleeping, I used to breathe the way they told me to—independently. It wasn't easy, crumpling up a marriage and throwing it away, looking for achievements I'd made that weren't to do with being a wife, but I don't think I did too badly. For a while I was a bit depressed, but only a bit.

My future, and this surprised me, was much harder to redefine. All the hopes you collect: another good holiday abroad, a proper fitted kitchen, children, a child. Your future creates an atmosphere around you and mine was surprisingly beautiful. Duncan and I, retired, would grow closer and closer, more and more serene, there would be grandchildren, picnics, gardening and fine, white hair. There would be trust and understanding, dignity in sickness and not dying alone. We would leave good things behind us when we were gone. I can't imagine where it all came from, I only know that it was hard to give away.

Then, one Monday morning, there was an incident involving my husband and a Garscadden train.

I went down, as usual, to stand on the westbound platform, this time in a hard, grey wind, the black twigs and branches over the line, oily and dismal with the damp. I waited in the little, orange shelter, read the walls and watched the Garscadden trains. There were three, and a Not In Service and, for the first time in my life, I gave up the wait. I turned around, walked away from the shelter and went home. I wished it would rain. I wanted to feel rain on my face.

The hall still smelt of the toast for breakfast. I took off my coat and went into the bedroom, needing to look in the mirror again, and there they were, in bed with the fire on, nice and cosy: Duncan and a very young lady I had never met before. They seemed to be taking the morning off. Duncan ducked his head beneath the bedclothes, as if I wouldn't know it was him, and she stared at the shape he made in the covers and then she stared at me.

I don't believe I said a single word. There wasn't a word I could say. I don't remember going to the kitchen, but I do remember being there, because I reached into one of the drawers beside the sink and I took out a knife. To be precise, my mother's old carving knife. I was going to run back to the bedroom and do what you would do with a carving knife, maybe to one of them, maybe to both, or perhaps just cut off his prick. That thought occurred.

That thought and several others and you shouldn't pause for thought on these occasions. I did and that was it. In the end I tried to stab the knife into the worksurface, so that he would see it there, sticking up, and know that he'd had a near miss. The point slid across the formica and my hand went down on the blade, so that all of the fingers began to bleed. When Duncan came in, there was blood everywhere and my hand was under the tap and I'm sure he believed I'd tried to kill myself. The idea seemed to disturb him, so I left it at that.

He drove me to and from the hospital and stayed that night in the flat, but, when he was sure I felt stable again, he went away and we began the slow division of our memories and ornaments. It was all done amicably, with restraint, but we haven't kept in touch.

And that, I suppose, is the story of how my husband was almost killed by one Garscadden train too many. It is also the story of how I learned that half of some things is less than nothing at all and that, contrary to popular belief, people, many people, almost all the people, live their lives in the best way they can with generally good intentions and still leave absolutely nothing behind.

There is only one thing I want more than proof that I existed and that's some proof, while I'm here, that I exist. Not being an Olympic skier, or a chat show host, I won't get my wish. There are too many people alive today for us to notice every single one.

But the silent majority and I do have one memorial, at least. The Disaster. We have small lives, easily lost in foreign droughts, or famines; the occasional incendiary incident, or a wall of pale faces, crushed against grillwork, one Saturday afternoon in Spring. This is not enough.

GLOSSARY

Some Scots words appear in different spellings. Others have more than one meaning—which one applies should be clear from the context. All glosses within the text or in footnotes are by the individual authors.

'a: have
a-wat (*also, awat*): truly, indeed
ablich: dwarf
ae: one (only)
agley: off the straight
ahin: behind
aiblins: perhaps, possibly
aifterhin: afterwards
aik: oak
aits: oats
anent: about, concerning
aneth: beneath, under
aneuch: enough
argle-barglet: argued the toss, disputed
ate-meats: parasites
Auld Licht (*New Light*): dissenting branch of Presbyterian Church and its beliefs
ava: at all
ayont: beyond
bail(l)ie: town magistrate
bannock: flat oatmeal cake
bardy: bold
bawbee: halfpenny
bed-stock: front bar of wood in a bed
beel't thoom: festered thumb
beetle: flat piece of wood used in laundering
begoud: began
ben: in, inside
bestial: livestock
betherel: beadle and sexton
beylin: boiling
bield: shelter, haven
billies: comrades, young men
bir: force
birkie: young man
birks: birches
birn: burden
blate: shy, timid
bogle: apparition; a devil
bothy: cottage shared by farm labourers
bowit (*booet*): hand-lantern
brawlie kent I . . .: I was sure of it

brawly: in fine fashion
breet: brute
brittle: financially embarrassed
brunstane spunk: brimstone (i.e. sulphur) match
brunt: burned, burnt
but and ben: two-roomed cottage
by-ordinair: unusual
byous: very
ca' twa pair: drive two pairs of plough horses
caidgy: cheerful
callants: lads
caller: cool, fresh
canny: prudent; fortunate; clever; etc.
cantrips: spells, witches' tricks
carlin: (old) woman
cauld kail: cold (i.e. yesterday's) cabbage
causey-talk: street gossip
chafts: cheeks
change-house: tavern
chape: cheap
chappin': striking (as of a clock)
chaum'er: chamber
chiel (*childe*): man; young man
chirk: make a harsh, non-verbal noise
clachan: village
claes: clothes
clam(n)jamphry: rabble, crowd, riff-raff
clavers: chatter
cleading: clothes
clecking: hatching
cloking: sitting on eggs
closs (*close*): farmyard; courtyard; passageway; tenement entrance
clour: blow
clouts: nappies; cloths, rags, etc.
coggly: unsteady
coling: putting hay in cocks
collieshangie: turbulence
connach: waste
coorse: coarse, rough
corbie craw: raven; carrion crow
corp: body, corpse
corse candle: 'corpse candle', 'dead light', phosphorescence supposed to appear in graveyards
cottars, cotters: inhabitants of small houses, i.e. ordinary people
cowe the gowan: that beats everything
crabbit: short tempered
craft: croft
craig: throat, neck
crap: crept
craw nancy: scarecrow

creagh: Highland raid
crook: hook over fire
crusie: small, open lamp
cummers: contemptuous term for a gathering of women
cushie doos: wood-pigeons
cuttie: young mare
cutty: short-stemmed clay pipe; worthless woman
daddit: beaten
daffin': romping
darger: casual labourer
daured: dared
daurna: dare not
dawtie: pet
deave: deafen
deece: couch used as a bed
deems: young women
devil a: never a
dicht: wipe
dirl: vibrate
disjaskit: dejected
div: do
dominie: schoolmaster
donsie: unlucky; badly behaved
douce: kind; respectable; modest
dowie: mournful
drumly: muddy; gloomy, troubled
dule: grief
dung: beaten
Dunniè-wassel: gentleman
dwam: daydream; mindlessness
dwine: waste away
eenoo: shortly, soon
ein: even
eir: ever
eldritch: uncanny; frightful
erls: payment (in part)
ettle: effort
even me: accuse me of
even-down: downright, confirmed
eydency: industry
fa: who
fader: father
fae: from
fan: when
fat: what
fat wye: how
feck: abundance
feein': hiring, or engaging, for wages
Feint a thing: The devil . . . (i.e. nothing at all)
fell: exceedingly

fenn't: managed to subsist
file: while
fir: pine-torch
fleg (*fley*): fright (frighten)
flittin: removal
flytin: scolding
forbye: besides
forfoughten: worn out
forjeskit: tired
fornent: opposite, facing
frae: from
freats: superstitious imaginings
freely: very
fudder: whether
fuff: cat's hiss
fup: whip
gaed their dinger: expressed themselves passionately
gang: go
gangrel: tramp, vagrant
gangrel bairn: spoiled child; or, child able to walk
gar: make, cause
gate: way, manner
gate-farrin: comely
gaye, gey: very; rather
geet: child
girn: grin; snarl; whimper
glaikit: stupid
glampit: groped
gleg: nimble, clever, bright
glisk: glimpse
Glunamie: Highlander
gousty: gusty
gowan: daisy
gowd: gold
gowk: simpleton; cuckoo
gowpin: handful
gradawa: medical practitioner with a degree
greet (*greit*): cry, weep
grogram: coarse fabric, or garment made of it
grossets: gooseberries
grue: shiver; feeling of horror
grumph[*y*]: grunt; pig
gudeman: husband
guidet: troubled
gweed: good
gyaun oot: going out
habbered: stammered
haffets: locks of hair
hags: moorland
hairst: harvest

haiveless: slovenly; useless
han'lings: handlings—transactions; cases
hanches: haunches
haud'n doun: afflicted; oppressed
haudin: a holding (i.e. of land)
haverel: half-wit
heckle: steel-toothed comb
het: hot
hinna: haven't
hinnie: endear
hire-house: service, i.e. the house or farm to which a servant was engaged
hirsle: move awkwardly
hobble-show: commotion
hooly: gently, softly
horn en: superior part of a house
horse-couper: horse-dealer
howdie: midwife
howff: haunt; resort
howp: hope
huckstry: work of a middleman; bargaining, haggling
huggars: coarse, footless stockings
Humdudgeons: unnecessary fuss, bulks
huz: us
I'se: I shall
ilka: each, every
ill-getted ways: bad habits
ill-pay't: very sorry
immas: gloomy
jalouse: suspect, guess, suppose
jaud: jade
jeyl: jail
keek: peep
keelie: working-class townee, usu. male
keep's a': expression of surprise
kelpie: water-sprite
ken: know
kent-speckle (*kenspeckle*): easily recognized, conspicuous
kirsen't: christened
kist: chest, trunk
kittle: tricky, difficult
kye: cattle
laigh: low
laive (*lave*): remainder
lether: ladder
leuket owre a door: looking or being outside
limmer: girl or woman (playful or contemptuous term); loose woman
linn: waterfall
lintie: linnet
lippen: put trust in
looted: stooped

lowe: blaze
lown: calm
lug: ear
lum: chimney
mark: dark
meikle: much; great
mid-hoose: middle room in a three-roomed house
mim: demure
min: mind, i.e. remember
minaway: minuet-like
minnie: mother
mistrysted: frightened; crossed the path of
mith: might
molloching: time-wasting
moniment: spectacle (monument)
morn's mornin': tomorrow morning
mornin(g): dram before breakfast
Muhme: probably *muime* (Gaelic) foster-mother; nurse; stepmother
neb: nose
neen: none
neep: turnip, swede
neep-reet: growing turnips
neist: next
neives: fists
nieve: first
no cannie: unlucky
nor: than
nowt(e): cattle; black cattle
oe: grandchild
oor: hour; our
orra work: odd jobs
orra-man: farm labourer who does odd jobs
outsrapolous: obstreperous
owre: over
oxter: armpit
pan-loaf: affected Glasgow accent
paumerin: walk idly
pawkie: shrewd
peer: poor
peerie: spinning-top
peesweeps: lapwing
peremptors ('*at her peremptors*'): at her wits' end
perjinct: neat
pick: pitch
pick-thank: ingrate; mischief-maker
pitten: put
pleiter: plight, complaint
ploiterie: wet, muddy
precunnance: understanding, condition
preen: pin

press: wardrobe-like piece of furniture, box bed, etc.
priggit: pleaded
protty: good-looking
puckle (*pickle*): small amount
quags: boglands
quean, quine: young woman
rabiator: violent bully
ramplor (*rampler*): boisterous young man
ranting: roistering
rape-thackit: thatch secured by ropes
rattling: wild; lively
red(*d*): tidy up, set in order
richt: right; = thorough, true
rive: smash; split
roch: rough
saft southert ye . . .: won you round easily
sair: sore; sorely
sang: an oath = 'blood!'
saugh: willow
scad: reflection
scaulin: scolding
sclate: slate
screek: cockcrow
scunner: sickening disgust
sec (*seck*): sex
selt: sold
set: beset
seyt: sieved, strained
shalt (*shelt; sheltie*): pony
shavelin-gabbit: with a twisted mouth
sholtie: pony
sibbest: closest (relation, friends)
siccan: such a, such an
siccar: secure, sure
siller: silver, money
sinsyne: ago, since then
skaithless: harmless, unhurt
skelloch: yell, scream
skeugh: twisted, skew
skimmer: flicker
skri(*e*)*ghed*: screamed
skyeow-fittet: skew-footed
slaugh: hollow between hills
sleekit: sly
slockened: slaked, quenched
smatterie: family of young children
smeddum: spirit, spunk, initiative, gumption
snot: nasal dirt = despised person
soom: swim, float
sore: grief, sorrow

sorra: sorrow
soss: mess, muddle
sough: breeze, wind; rumour
souple: cunning
sowens: a kind of porridge, made with fermented oat husks and meal
spaewife: fortune-teller
speak: talk, gossip
spean't: weaned
speir: ask, say, tell
splore: revels; antics
sprack: lively
spunkie: small fire; will o' the wisp
stammy-gastered [*stammagust*, = *gast*]: unpleasantly surprised
steading: farmhouse and out-buildings
steekit: shut
steen: stone
steer: stir, disturbance
sterns: stars
sticket, stickit: unsuccessful; stunted—i.e. 'stickit minister' = minister without a parish
stirk: steer; fool
stock: term of sympathy (or contempt)—creature, bloke, chap, etc.
straemash: violent disturbance
straighting-board: board on which dead are laid out
streen (*i.e. 'the streen'*): last night
sumph: surly, sullen person
swack: active, fit
swatch: sample
sweir: lazy, slow
syne: ago, since, then
taiglet: delayed
tamboured: embroidered (on a frame)
tawpie: slovenly, foolish person
tee: too
thereoot: outside
thirled: dependent on, bound to
tholed: put up with, tolerated, suffered
thowless: useless, lacking in vigour
thrang: busy
thrapple: windpipe
thrawn: twisted; misshapen
threep: urge, assert, argue
threepit: asserted
through-gaun: active, busy
till: to
tink: tinker, gipsy
tir't: tired out
town: public excitement
tramp-pick: special spade used on hard soils
treacle ale: light beer (made with treacle)

trig: trim, nimble
trimmil: tremble
tsil': child
tyauve: struggle
tyke: dog
tympathy: tumour
unco: strange, extraordinary
unstreakit: dead body not yet prepared for burial
upsitten: proud, aloof
vacance: holidays
vir: energy
wakerife: sleepless
wally-wallying: lamentation
wanchancy: unlucky; dangerous; foreboding
ware: spend
warsle: struggle
waur: worse
weans: children
wechtfu': as much as a 'wecht' will hold: wecht = container for winnowing or carrying grain
weel-a-wat: for sure, certainly
weel-faured: good-looking
westling: western; or, west-moving
wham(m)led: upset, tumbled
whaten: what kind of
wheen: a small quantity; a few
whiles: sometimes
wice: wise
widdifus: scamps
windlestraw: withered grass
winnockie: little window
witters: throats
wud: mad
wull: stray
wyv'n: woven
yam-yammered: complained loudly
yea-threepit: asserted again and again
yellow yite: yellow-hammer
yett: gate

ACKNOWLEDGEMENTS

The editor and publisher are grateful for permission to include the following copyright material.

Iain Crichton Smith: 'An American Sky' from *Listen to the Voice: Selected Stories* (Canongate Press Ltd., 1993). Reprinted by permission of the publishers.

Elspeth Davie: 'A Map of the World' from *The Spark and Other Stories*, published by Calder & Boyars, London, © Elspeth Davie 1968. Reprinted by permission of The Calder Educational Trust, London.

Margaret Elphinstone: 'An Apple from a Tree' from *An Apple from a Tree*, first published by The Women's Press Ltd., 1991, 34 Great Sutton Street, London EC1V 0DX, is used by permission of The Women's Press Ltd.

Ronald Frame: 'Merlewood' from *A Long Weekend With Marcel Proust* (Bodley Head, 1986), © Ronald Frame 1986. Reprinted by permission of Curtis Brown Ltd. on behalf of Ronald Frame.

George Friel: 'I'm Leaving You' from *A Friend of Humanity* (Polygon, 1992). Reprinted by permission of the publishers.

Edward Gaitens: 'Dance of the Apprentices' from *Growing Up* (Canongate Press Ltd., 1990). Reprinted by permission of the publishers.

Janice Galloway: 'Blood' (Secker & Warburg Ltd.). Reprinted by permission of Reed Consumer Books Ltd.

Alasdair Gray: 'Prometheus' from *Unlikely Stories, Mostly* (Canongate Press Ltd., 1983). Reprinted by permission of the publishers.

Neil Gunn: 'The Tax Gatherer' from *The White Hour* (Chambers Harrap Publishers Ltd.). Reprinted by permission of the publishers.

Violet Jacob: 'The Debatable Land' from *Tales Of My Own Country* (John Murray Publishers Ltd., 1922). Reprinted by permission of the publishers.

Robin Jenkins: 'A Far Cry from Bowmore' from *A Far Cry From Bowmore & Other Stories* (Victor Gollancz Ltd., 1973). Reprinted by permission of the publishers.

James Kelman: 'Home For A Couple Of Days' from *Greyhound For Breakfast* (Secker & Warburg, 1987). Reprinted by permission of Reed Consumer Books Ltd.

A. L. Kennedy: *Night Geometry and the Garscadden Trains* (Polygon, 1990). Reprinted by permission of the publishers.

Jessie Kesson: 'The Gowk' from *Where The Apple Ripens* (Chatto & Windus, 1985). Reprinted by permission of Random House UK Ltd.

Eric Linklater: 'Country-born' from *The Stories of Eric Linklater* (Macmillan Publishers Ltd., 1968). Reprinted by permission of the Peters Fraser & Dunlop Group Ltd.

Shena Mackay: 'Violets and Strawberries in the Snow' from *Dreams Of Dead Women's Handbags* (William Heinemann, 1987), © Shena Mackay 1987. Reprinted by permission of Reed Consumer Books Ltd. and Rogers Coleridge and White Ltd.

George Mackay Brown: 'Andrina' from *Andrina and Other Stories* (Chatto & Windus, 1983). Reprinted by permission of Random House UK Ltd.

Eric McCormack: 'The One Legged Man' from *Inspecting The Vaults* (Viking, 1987), © Eric McCormack 1987. Reprinted by permission of Penguin books Canada Ltd.

William McIlvanney: 'Performance' from *Walking Wounded* (Hodder & Stoughton, 1989), © William McIlvanney 1989. Reproduced by permission of the publishers and Curtis Brown, London Ltd.

Duncan McLean: 'Doubled Up With Pain' from *Bucket Of Tongues* (Secker & Warburg, 1992). Reprinted by permission of Reed Consumer Books Ltd.

Robert McLellan: 'The Donegals' from *Linmell Stories* (Canongate Press Ltd.). Reprinted by permission of the publishers.

Norah and William Montgomerie: 'The Wee Bannock' from *The Well At The World's End* (Canongate Press, Ltd., 1986).

Neil Paterson: 'The Life and Death of George Wilson' from *And Delilah* (Hodder & Stoughton, 1951). Reprinted by permission of the author.

Dilys Rose: 'Street of the Three Terraces' from *Red Tides* (Secker & Warburg, 1993). Reprinted by permission of Reed Consumer Books Ltd.

Muriel Spark: 'Bang-bang You're Dead' from *Voices At Play* (Macmillan Publishers Ltd., 1961). Reprinted by permission of David Higham Associates Ltd.

Alan Spence: 'Its Colours They Are Fine' (William Collins & Co., 1977), © Alan Spence 1977. Reprinted by permission of Sheil Land Associates Ltd.

Fred Urquhart: 'The Last Sister' (Methuen, 1950), © Fred Urquhart 1950. Reprinted by permission of the author.

Efforts have been made to trace and contact copyright holders prior to printing. OUP apologizes for any errors or omissions in this list and if notified they will be pleased to make corrections at the earliest opportunity.

Index of Authors